General Chemistry

A Mastery-Oriented Curriculum

Third Edition

Camp Hill, Pennsylvania
2021

General Chemistry

A Mastery-Oriented Curriculum

Third Edition

John D. Mays

NOVARE
Classical Academic Press

Camp Hill, Pennsylvania
2021

General Chemistry
© Classical Academic Press®, 2021
Edition 3.1

ISBN: 978-0-9972845-1-5

All rights reserved. Except as noted below, this publication may not be reproduced, stored in a retrieval system, or transmitted, in any form or by any means, without the prior written permission of Classical Academic Press.

All images attributed to others under any of the Wikimedia Commons licenses, such as CC-BY-SA-3.0 and others, may be freely reproduced and distributed under the terms of those licenses.

Scripture quotations taken from The Holy Bible, English Standard Version, copyright ©2001 by Crossway Bibles, a publishing ministry of Good News Publishers. Used by permission. All rights reserved.

Classical Academic Press
515 S. 32nd Street
Camp Hill, PA 17011
www.ClassicalAcademicPress.com/Novare/

Cover design: Nada Orlic, http://nadaorlic.com

VP.06.22

Acknowledgements

I wish to express my deep appreciation to those who reviewed this text during its development: Chris Mack, PhD; Christina Swan, PhD; and Christian J. Corley. Many thanks also to our excellent copy editor Gerald Tilma. Each of these friends is an expert at what he or she does, and their contributions have been invaluable.

Reviewers

Chris Mack	Adjunct Faculty, University of Texas at Austin
	PhD, University of Texas at Austin, Chemical Engineering
	MS, Electrical Engineering, University of Maryland
	BS degrees in Physics, Electrical Engineering, Chemistry, Chemical Engineering, Rose-Hulman Institute of Technology
Christina Swan	Biology/Chemistry Teacher, Regents School of Austin, Austin Texas
	PhD, Molecular Pathology, University of California at San Diego
	BS, Biology, Westmont College
Christian J. Corley	Math-Science Department Chair, Regents School of Austin, Austin, Texas
	MS, Computer Science, Duke University
	BS, Geophysics, Texas A&M University

Contents

Preface
For Teachers — xii
    1. Student Audience and Preparedness — xii
    2. Our Emphasis on Wonder, Integration, Mastery, and Kingdom — xiii
    3. Recommendations for Teaching With This Text — xv
    4. Laboratory Work and Lab Reports — xvii
    5. Revisions in the Third Edition — xviii

Preface
For Students — xix

Introduction
What is Chemistry All About? — 2
  I.1 A Few Major Themes — 3
    I.1.1 Chemistry Is All About Electrons — 3
    I.1.2 Chemistry Is All About Electrical Forces — 5
        Hmm... Interesting.   *Why water forms beads* — 6
    I.1.3 Chemistry Is All About Minimizing Energy — 6
    I.1.4 Chemistry Is All About Whole-Number Ratios of Atoms — 9
    I.1.5 Chemistry Is All About Modeling — 10
  I.2 Conclusion — 12
  Introduction Study Questions — 12

Chapter 1
Measurements — 14
  1.1 Science and Measurements — 16
    1.1.1 No Measurements, No Science — 16
    1.1.2 Matter, Volume, and Mass — 16
    1.1.3 The US Customary System — 18
    1.1.4 The SI Unit System — 18
    1.1.5 Metric Prefixes — 20
  1.2 Converting Units of Measure — 22
    1.2.1 Basic Principles of Unit Conversion Factors — 22
    1.2.2 Tips for Converting Units of Measure — 24
    1.2.3 Converting Temperature Units — 27
  1.3 Accuracy and Precision — 28
    1.3.1 Distinguishing Between Accuracy and Precision — 28
    1.3.2 Significant Digits — 29
  1.4 Other Important Math Skills — 35
    1.4.1 Scientific Notation — 35
    1.4.2 Calculating Percent Difference — 37
  Chapter 1 Exercises — 37

Contents

Chapter 2
Atoms and Substances — 42
- 2.1 Atoms and Molecules — 44
  - 2.1.1 Atomic Facts — 44
  - 2.1.2 The History of Atomic Models — 45
- 2.2 Types of Substances — 49
  - 2.2.1 Pure Substances: Elements and Compounds — 49
  - 2.2.2 Mixtures — 53
    - *Hmm... Interesting.   Brownian motion* — 56
  - 2.2.3 Physical and Chemical Properties — 57
- 2.3 Isotopes and Atomic Masses — 59
  - 2.3.1 Isotopes — 59
  - 2.3.2 The Unified Atomic Mass Unit — 60
  - 2.3.3 Atomic Masses — 60
- 2.4 Density and Quantity of Substances — 62
  - 2.4.1 Density — 62
  - 2.4.2 The Mole and the Avogadro Constant — 64
  - 2.4.3 Molar Mass and Formula Mass — 65
  - 2.4.4 Gram Masses of Atoms and Molecules — 69
- Chapter 2 Exercises — 71

Chapter 3
Atomic Structure — 74
- 3.1 Atomic Spectra — 76
  - 3.1.1 The Electromagnetic Spectrum — 76
  - 3.1.2 Energy in Atoms — 78
    - *Hmm... Interesting.   Neon signs and phonons* — 79
  - 3.1.3 The Hydrogen Atom — 81
- 3.2 The Bohr Model of the Atom — 82
- 3.3 The Quantum Model of the Atom — 83
  - 3.3.1 Schrödinger and Pauli — 83
  - 3.3.2 Shells, Subshells, and Orbitals — 83
  - 3.3.3 The Aufbau Principle, the Madelung Rule, and Hund's Rule — 88
- 3.4 Electron Configurations — 90
  - 3.4.1 Electron Configurations — 90
  - 3.4.2 Condensed Electron Configurations — 92
- 3.5 Empirical Formulas — 93
  - 3.5.1 Percent Composition and Empirical Formulas — 93
  - 3.5.2 Determining a Molecular Formula from an Empirical Formula — 95
- Chapter 3 Exercises — 96

Chapter 4
The Periodic Law — 100
- 4.1 The Periodic Table of the Elements — 102
- 4.2 Periodic Table Nomenclature — 104
- 4.3 Periodic Physical Properties — 105
  - 4.3.1 Atomic Radius — 105
  - 4.3.2 Ionic Radius — 107
- 4.4 Periodic Chemical Properties — 108

	4.4.1 Core and Valence Electrons	108
	4.4.2 Ionization Energy	109
	4.4.3 Electron Affinity	113
	4.4.4 Electronegativity	115
4.5	A Few Notes About Hydrogen	116
	Chapter 4 Exercises	117
	Hmm... Interesting. Hydrogen in space	117

Chapter 5
Chemical Bonding — 120

5.1	Preliminaries	122
	5.1.1 Chemical Possibilities	122
	5.1.2 The Octet Rule	122
5.2	Ionic Bonding	123
	5.2.1 Ionic Bonds and Crystals	123
	5.2.2 Naming Ionic Compounds	125
	5.2.3 Energy in Ionic Bonds	127
	5.2.4 Hydrates	129
	5.2.5 Intensive and Extensive Properties	129
	5.2.6 Physical Properties of Ionically Bonded Substances	130
5.3	Covalent Bonding	131
	5.3.1 Covalent Bonds and Molecules	131
	5.3.2 Polyatomic Ions	133
	5.3.3 Ionic Compounds with Polyatomic Ions	134
	5.3.4 Polyatomic Ion Names	134
	5.3.5 Naming Acids	135
	5.3.6 Lewis Structures	136
	5.3.7 Exceptions to the Octet Rule	140
	5.3.8 Resonance Structures	141
	5.3.9 Naming Binary Covalent Compounds	142
	5.3.10 Energy in Covalent Bonds	143
	5.3.11 Physical Properties of Covalently Bonded Substances	144
5.4	Electronegativity, Polarity, and Bond Character	145
	5.4.1 Polarity and Dipoles	145
	5.4.2 The Nature of the Bond	146
	Chapter 5 Exercises	147
	Hmm... Interesting. The molecular structure of glass and quartz	148

Chapter 6
Molecular Theory and Metallic Bonding — 150

6.1	Molecular Structure	152
	6.1.1 Covalent Bond Theory	152
	6.1.2 Valence Shell Electron Pair Repulsion (VSEPR) Theory	152
	6.1.3 The Effect of Nonbonding Domains on Bond Angle	156
6.2	Metallic Bonding	157
	6.2.1 Metallic Lattices	157
	6.2.2 Physical Properties of Metals	158
6.3	Intermolecular Forces	160
	6.3.1 Bonding Forces	160

  6.3.2 Intermolecular Forces   160
   Hmm... Interesting. Tin pest   160
  6.3.3 Hydrogen Bonding   161
  6.3.4 Van der Waals Forces   162
 Chapter 6 Exercises   164

Chapter 7
Chemical Reactions and Stoichiometry   168
7.1 Introduction to Chemical Equations   170
  7.1.1 Fascinating Chemistry   170
  7.1.2 The Law of Conservation of Mass in Chemical Reactions   170
  7.1.3 Reaction Notation   171
  7.1.4 Balancing Chemical Equations   172
  7.1.5 Oxidation States   177
   Hmm... Interesting. Why nitrates and nitros blow up   179
7.2 General Types of Chemical Reactions   180
  7.2.1 Synthesis Reactions   180
  7.2.2 Decomposition Reactions   181
  7.2.3 The Activity Series of Metals   181
  7.2.4 Single Replacement Reactions   182
  7.2.5 Double Replacement Reactions   184
   Hmm... Interesting. A story about aqua regia   184
  7.2.6 Combustion Reactions   185
  7.2.7 Acid-Base Neutralization Reactions   185
  7.2.8 Oxidation-Reduction Reactions   186
7.3 Stoichiometry   187
  7.3.1 Stoichiometric Calculations   187
  7.3.2 Limiting Reactant   192
  7.3.3 Theoretical Yield and Percent Yield   194
 Chapter 7 Exercises   195

Chapter 8
Kinetic Theory and States of Matter   200
8.1 Temperature, Kinetic-Molecular Theory, and Pressure   202
  8.1.1 Temperature and Molecular Energy   202
  8.1.2 Velocity Distribution of Gases   202
  8.1.3 The Kinetic-Molecular Theory of Gases   203
  8.1.4 Gas Pressure   204
   Hmm... Interesting. How barometers work   206
8.2 States of Matter   208
  8.2.1 The Four Basic States of Matter   208
  8.2.2 Solids   208
  8.2.3 Liquids   209
  8.2.4 Gases   211
  8.2.5 Plasmas   212
   Hmm... Interesting. Gas diffusion   212
  8.2.6 Phase Transitions and Phase Diagrams   213
  8.2.7 Heat Capacity, Heat of Fusion, and Heat of Vaporization   216
  8.2.8 Evaporation   220

	8.2.9 Vapor Pressure	222
Chapter 8 Exercises		223

Chapter 9
The Gas Laws — 226

9.1	Early Formulations of the Gas Laws	228
	9.1.1 Boyle's Law	228
	9.1.2 Charles' Law	229
	9.1.3 Avogadro's Law	232
9.2	The Ideal Gas Law	233
	9.2.1 Standard Temperature and Pressure	233
	9.2.2 The Ideal Gas Law	233
	Hmm... Interesting. The gas laws as models	234
	9.2.3 Using the Ideal Gas Law to Find Molar Mass and Density	240
9.3	The Law of Partial Pressures	243
	9.3.1 Dalton's Law of Partial Pressures	243
	9.3.2 Collecting a Gas Over Water	246
9.4	Stoichiometry of Gases and Effusion	248
	9.4.1 Stoichiometry of Gases	248
	9.4.2 Gas Diffusion and Effusion	249
	Hmm... Interesting. Uranium enrichment	250
Chapter 9 Exercises		251

Chapter 10
Solution Chemistry — 256

10.1	Dissolution	258
	10.1.1 The Process of Dissolving	258
	10.1.2 Heat of Solution	261
	10.1.3 Entropy and Free Energy	261
	10.1.4 Electrolytes	263
10.2	Solubility	264
	10.2.1 Ionic Solids in Water	264
	10.2.2 Ionic Solids in Nonpolar Solvents	265
	10.2.3 Polar Liquids	266
	10.2.4 Nonpolar Liquids	267
	10.2.5 Solutions of Solids	267
	Hmm... Interesting. How soap works	268
	10.2.6 Gases in Liquid Solutions	270
	10.2.7 The Effect of Temperature on Solubility	271
10.3	Quantifying Solution Concentration	272
	10.3.1 Molarity	272
	10.3.2 Molality	274
10.4	Compounds in Aqueous Solution	275
	10.4.1 Ionic Equations and Precipitates	275
	10.4.2 Net Ionic Equations and Spectator Ions	277
10.5	Colligative Properties of Solutions	278
	10.5.1 Vapor Pressure Lowering	278
	10.5.2 Freezing Point Depression and Boiling Point Elevation	280
Chapter 10 Exercises		285

Contents

Chapter 11
Acids and Bases — 290

- 11.1 Properties and Nomenclature of Acids and Bases — 292
  - 11.1.1 Introduction — 292
  - 11.1.2 Properties of Acids and Bases — 292
  - 11.1.3 Acid Names and Formulas — 294
- 11.2 Acid-Base Theories — 295
  - 11.2.1 Arrhenius Acids and Bases — 296
  - 11.2.2 Brønsted-Lowry Acids and Bases — 298
    - *Hmm... Interesting. What is an alkali?* — 299
  - 11.2.3 Lewis Acids and Bases — 302
  - 11.2.4 Strength of Acids and Bases — 303
- 11.3 Aqueous Solutions and pH — 306
  - 11.3.1 The Self-ionization of Water — 306
  - 11.3.2 Calculating $[H_3O^+]$ and $[OH^-]$ — 306
  - 11.3.3 pH as a Measure of Ion Concentration and Acidity — 307
  - 11.3.4 pH Measurement, pH Indicators, and Titration — 312
  - 11.3.5 Titration Procedure — 316
  - 11.3.6 Determining $[H_3O^+]$ or $[OH^-]$ from Titration Data — 318
- Chapter 11 Exercises — 319

Chapter 12
Redox Chemistry — 324

- 12.1 Oxidation and Reduction — 326
  - 12.1.1 Introduction to Redox Reactions — 326
  - 12.1.2 Oxidation States — 326
  - 12.1.3 Strengths of Oxidizing and Reducing Agents — 330
- 12.2 Redox Reaction Equations — 332
  - 12.2.1 Redox Half-Reactions — 332
  - 12.2.2 Balancing Redox Equations — 336
- 12.3 Electrochemistry — 342
  - 12.3.1 Copper and Zinc Redox — 342
  - 12.3.2 Electricity Instead of Heat — 343
  - 12.3.3 Electrochemical Cells — 344
    - *Hmm... Interesting. How are salt bridges made?* — 348
  - 12.3.4 Electrode Potentials — 350
  - 12.3.5 Electrochemical Applications — 353
- Chapter 12 Exercises — 358

Glossary — 362

Answers to Selected Exercises — 378

Appendix A
Reference Data — 390

Appendix B
Scientists to Know About — 394

Appendix C
Memory Requirements   396

References and Citations   397

Image Credits   399

Index   401

Preface
For Teachers

Thank you for taking the time to begin here at the beginning. This chemistry text is quite different from other high school texts covering similar material. So to ensure that your experience with this text is successful, it is important to be aware of its unique features and the logic behind its structure.

Several important resources are described in the pages that follow. For your convenience, they are all listed here. Please visit our website (classicalsubjects.com) or contact us for more information.

- *Introductory Physics* (John D. Mays, 2016)
- *Accelerated Studies in Physics and Chemistry* (John D. Mays, 2018)
- *Chemistry for Accelerated Students* (John D. Mays, 2018)
- *From Wonder to Mastery: A Transformative Model for Science Education* (John D. Mays, 2021)
- *Chemistry Experiments for High School* (Christina Swan and John D. Mays, 2014)
- *Chemistry Experiments for High School at Home* (Christina Swan and John D. Mays, 2014)
- *Science for Every Teacher* (John D. Mays, 2013)
- *The Student Lab Report Handbook* (John D. Mays, 2014)

1. Student Audience and Preparedness

This text is designed for grade-level students. This means the typical student using this text is a junior in high school concurrently taking Algebra 2. In my view, students need to be at or above Algebra 2 in mathematics in order to study chemistry. Computations with pH involve logarithms and power functions, and computations with the gas laws involve a lot of algebraic manipulation. For these reasons, chemistry is usually undertaken when students are in Algebra 2 or later.

I promote a physics-first approach to secondary science programming. According to this program, students take an introductory physics course in 9th grade. The obvious choice for 10th grade science is then biology, followed by chemistry in 11th grade. My text *Introductory Physics* is specifically designed for this purpose. The benefits of using *Introductory Physics* in 9th grade followed two years later by *General Chemistry* in 11th grade are *significant*. Much of the content covered in the first two chapters of the present text is also addressed in *Introductory Physics*. Students who have had this background should find the first two chapters of *General Chemistry* to be a breeze—it will mostly be review for them. One of the results of this is that the study of chemistry, which can be heart-breakingly difficult for some students, becomes much more accessible.

Contrast this preparedness with the more typical scenario of students arriving in chemistry with very weak unit conversion skills and knowing virtually nothing about significant digits. In this situation, the challenge of chemistry is much greater simply because of students' lack of appropriate background in basic math skills. Like the present text, *Introductory Physics* is a mastery-oriented text. By using it, students get repeated, sustained practice at unit conversions throughout the year. With such extensive practice in 9th grade, brushing up those skills two years later in chemistry comes easily. *Introductory Physics* also gives students practice in the use of significant digits, an added bonus. (However, the addition rule for computations does not appear in *Introductory Physics*. That rule is introduced in Chapter 1 of the present text.)

Another important benefit of the physics-first approach is that students have a solid conceptual understanding of mass, potential energy, kinetic energy, electric charge, and the electromagnetic spectrum, all of which are hugely important in the study of chemistry. And we could add the topic of DC circuits as well, since the application section in the redox chapter is all about electrochemical cells and batteries.

It is important to mention here that we have a different set of texts for use by accelerated students. Accelerated students complete Algebra 1 in 8th grade. After Geometry in 9th grade they take Algebra 2 in 10th grade. For these students, we recommend *Accelerated Studies in Physics and Chemistry* for the 9th-grade course and *Chemistry for Accelerated Students* for the 10th-grade chemistry course. For details on these texts please visit classicalsubjects.com.

The benefits of the physics-first program and the logic behind the separation of science and math into separate pathways for grade-level and accelerated students are discussed in much more detail in two of the appendices in my book *From Wonder to Mastery: A Transformative Model for Science Education*.

2. Our Emphasis on Wonder, Integration, Mastery, and Kingdom

Wonder

The study of science should always begin with wonder. The world is a stunning place, full of surprises and jaw-dropping phenomena. But today there are many barriers standing in between our students and the world they might otherwise become fascinated with. Safety concerns interfere with kids playing and exploring outdoors. Liability concerns make it hard to find a decent chemistry set. And unfortunately, it is common today for young people grow up spending most of their time indoors with digital media. The natural draw of nature for developing youths is now commonly missed. Many people have never seen the night sky in an area that is completely dark and have no idea of the stunning beauty of the heavens at night. Most kids have not hiked or camped in the forests and have not learned to listen to the sounds made by animals, insects, and trees.

Additionally, the environmental challenges we face today from pollution, resource exploitation, and, especially, climate change require a new generation of people who care about the earth. But people usually do not care about what they do not love, and they do not love what they do not know about. Helping students to know the natural world has never been more important than it is today. Only if they know the world will students begin to love it, and only then will they be motivated to take care of it. To nurture this love, we begin with the natural wonder we feel when we encounter the natural world.

Integration

A second major aspect to the needed paradigm shift is that instruction must be integrative. The habit of compartmentalizing disciplines of learning must be eliminated. This habit currently pervades everything from problem assignments to lesson presentations to test design. Instead of isolating science content from everything else, critical points of effective integration must be developed. Some critical integration points include:

- frequent use of mathematical skills in science classes, and frequent science applications in math classes

Preface for Teachers

- maximizing opportunities to develop good written expression on exams, lab reports, and papers
- developing key historical connections that serve to enhance understanding of science as a process; and
- treating, in addition to basic skills, the nature of scientific and mathematical knowledge, and the roles these play in leading us toward truth, goodness, and beauty.

Naturally, for integration to be effective, specific learning objectives must be developed, explained to students, and incorporated into assessments. Novare Science texts include clear learning objectives in every chapter.

Mastery

Mastery essentially means proficiency and long-term retention of course content. The first step toward mastery-learning is to change how we define success. The broken default pattern is what I call the *Cram-Pass-Forget Cycle*: students cram for tests, pass them, and forget most of what they crammed in just a few weeks. Success in such an environment is a matter of jumping through assessment hoops. Students are not only cheated by this regimen, they are bored with it. And teachers are demoralized by the results.

By contrast, Novare Science advocates methods and curriculum designed to promote proficiency and long-term retention using a *Learn-Master-Retain Cycle*. This first involves culling the content scope to an amount that truly can be mastered in the course of a school year. Many educators unthinkingly prioritize quantity over quality. But we believe students should be presented with a right amount of material they can learn deeply rather than a bloated scope of content they will neither comprehend nor remember. Even with a reduced scope, students who study for mastery typically outperform their peers as they move to higher level classes.

Second, leading students to mastery and retention requires teaching methods designed to produce these results. The standard approach used today involves teaching a chapter and giving a test on the chapter. By contrast, pedagogy designed for mastery and retention involves continuous review, ongoing accountability for previously studied material, and embedding of basic skills into new material. Of course, an effective method includes innovative strategies to enable students to master course content.

Kingdom

Science and mathematics provide us with unique ways of seeing God's creative presence in the world. Bringing biblical faithfulness to science classes will not be accomplished simply by folding in a few Bible verses or prayers, nor by constructing random analogies between scientific content and biblical revelation. In fact, much more is involved. Science and math teachers need to think broadly about how we fulfill Christ's mandate to love God with all our mind, how we teach our students effectively to engage issues, and how we perceive God's fingerprints in creation.

We also need thoughtfully to engage with the scientific claims of our day. For example, it is not a scientific claim to say that the universe got here by itself; that is a metaphysical claim based on an atheistic worldview that Christians reject. But it is a scientific claim to say that the universe began with the Big Bang and is now 13.77 billion years old. The scientific evidence behind this claim is vast, and I believe an appropriate science text is one that teaches students how to productively engage such claims. I do not believe it is appropriate to teach students to be dismissive of claims like this simply because they do not line up with certain ways of interpreting Genesis. I write this as one who fully believes the Bible, who loves reading Genesis and the rest of Scripture, and who accepts as persuasive the strong evidence for an old universe. (I use the age of the earth as an example. The issue is not treated in this text.)

My ideas about all three parts of this core philosophy are described in more detail in *From Wonder to Mastery*. This book, along with a more detailed description of our textbook philosophy, may be found on our website.

3. Recommendations for Teaching With This Text

In this section, I make a few remarks pertaining to some specific chapters in this text, followed by some comments about teaching for Mastery and Integration.

Regarding the chapter content, there are some sections you may wish to skip, depending on the circumstances at your school. Details are as follows:

- Chapter 1   If your students were taught from my text *Introductory Physics* in a prior course, you may find you can move through this chapter fairly quickly. Feel free to speed through it if your students have mastered the content and are ready to move on. However, mastery of unit conversions and significant digits is so important that you should cover the chapter in depth if needed.

- Chapter 2   My comments about Chapter 1 apply here as well; most of this content is covered in *Introductory Physics*. Again, move through the chapter quickly if you find that your students remember this material well. However, I suggest that at a minimum you spend some time in Section 2.2 on the types of substances. Students often forget (or are confused about) this important information.

- Chapters 3–7   These chapters contain standard material and should be covered in full.

- Chapter 8   As before, students who have studied *Introductory Physics* can probably move quickly through this material, particularly the discussion of states of matter. Topics in this chapter that are not in *Introductory Physics* include calculations with molar heat capacity, molar heat of fusion/vaporization, and phase diagrams.

- Chapters 9–10   These chapters contain standard material and should be covered in full.

- Chapter 11   In this chapter (acids and bases), some teachers may prefer to stick to the Arrhenius model; others will consider Brønsted-Lowry theory indispensable. My advice is to think about your students and consider the merits of racing through material that students will not be able to master. If there is not sufficient time in your schedule to bring students to mastery, it would be better to skip the Brønsted-Lowry material and move on to pH and titrations.

- Chapter 12   This is the redox chapter. At a minimum, students should learn to balance equations using the half-reaction method, and understand what the electrode potentials represent—how they relate to the activity series of metals and redox reactions in general. However, if time is short at the time you are covering this final chapter, you could focus on a qualitative encounter with the electrochemistry.

Now for some remarks about teaching for Mastery and Integration, beginning with Mastery.

First, students need continually to be working with previously learned material to keep old skills fresh. Included in the Chapter Exercises at the end of each chapter is a set of General Review Exercises covering material from previous chapters. Students should always work through these exercises.

Preface for Teachers

Second, the teacher's assessment regimen should support the goal of students retaining previously learned material. For students at this level, I recommend the following four-part assessment regimen: (a) Do not award credit in students' grades for homework. (Do, however, require students to complete their assignments and hold them accountable for doing so.) The logic behind this principle is explained in detail in *From Wonder to Mastery*. (b) Between each chapter test, administer one or two quizzes covering material presented in class during the previous week. This motivates students to stay current with their studies instead cramming the night before the test. (It also gives them direct information about how prepared they are for that part of the chapter test.) Count the average of all the quizzes for a single semester together as equivalent to one chapter test in the semester grade. (c) Use the "Standard Problems List" technique described in *From Wonder to Mastery*. Inform students at the beginning of the year that each chapter test includes problems from the Standard Problems List on material from previously covered chapters. (d) On each chapter test, allocate about 20% of the exam to material on the Standard Problems List from previously covered chapters. I have placed a Standard Problems List for Chapters 1–11 in the Preface for Students. Individual teachers may wish to modify that list to suit what the teacher feels are the most important topics to emphasize.

Additionally, as described above, you should feel free to cut some of the material from any chapter if the students in a particular class simply can't take it in fast enough. Classes are different year to year, and some groups of students can handle more than others. Teaching according to mastery principles is quite different from conventional *Cram–Pass–Forget* methods, and it is more demanding for teacher and students alike. However, the rewards are huge, which is why I constantly promote mastery methods. But teachers need to administer course content with wisdom.

Now I will make a few points about how the principle of Integration should work with the course.

First, in addition to computations, I always promote questions on tests and quizzes requiring responses written in complete sentences. In their responses, students should be required to demonstrate competence with standard English. I discourage so-called objective items requiring true/false, matching, and multiple-choice responses.

Second, the incorporation of language skills into your course will be enhanced even further if you require students to write their lab reports from scratch. I address this more in the next section.

Third, a healthy epistemology of science should pervade all science courses. The *Cycle of Scientific Enterprise* model briefly described in the Introduction is the place to start. This topic is developed at length in *From Wonder to Mastery*. I strongly encourage all teachers to draw upon these resources and become fluent in the concepts and terminology about the nature of science and scientific knowledge. Then bring the subject up as often as possible in class. Science teachers are very busy, and there are constant pressures threatening to cause courses to lag behind in the curriculum. It is tempting to stick to the technical content and neglect teaching our students what science *is* and how science *works*. From neglect of this topic, only a small percentage of our adult population understands what the statement "science is modeling" means, or has any idea of the distinction between scientific claims and truth claims from Scripture. One of the results of this massive lack of understanding is fuel thrown on the fires of the "science-faith debate," as if one has to choose between godly faith and robust science. I argue that one does not have to make such a choice. A critical component of the science teacher's role is helping students learn to participate in a healthy, faithful, responsible dialog concerning what we know about the world from Scripture and what we know about the world from scientific inquiry.

Finally, as I state in Appendix B, Wolfgang Goethe once wrote that "the history of science *is* science." Science is about modeling nature. Our models—theories—are never perfect and never complete, and as a result they change over time. One of the best ways to help students grasp this is to get into the history of the subject—the scientists with their theories and discoveries. In my texts for younger students, I place learning objectives pertaining to the history prominently alongside those pertaining to the technical content. In this text, I do not do so. The reason is that there are quite a few historical references in this text, and I do not think it reasonable to require students to memorize a paragraph of information pertaining to every one of them. So instead of stipulating historical learning objectives, I leave it up to the individual teacher to decide how much history to include and which scientists to focus on. However, it is very important that some history is included in student learning objectives and assessments, and that the historical circumstances are related to the nature of scientific knowledge and the *Cycle of Scientific Enterprise* wherever possible.

To assist both teacher and student in managing historical content, all the scientists discussed in the text are listed in Appendix B. I know that some teachers and home school families using this text may wish for more prescriptive requirements when it comes to the history. Accordingly, I offer this proposal: have the students write a paper on the history of chemistry that summarizes the contributions to chemistry from at least six or eight scientists. Require that the contributions discussed be tied together in a unified narrative, rather than simply quoted one after the other from material in the text. Alternatively, have the students write papers in which they describe how several of the experiments or other contributions described in this text relate to the *Cycle of Scientific Enterprise* model discussed in the Introduction.

However you choose to go about it, the history of science is important. The historical material should be an important component of your course, not something students simply skip over.

4. Laboratory Work and Lab Reports

A laboratory component is essential for every high school chemistry class. Not only does a lab *practicum* give students direct knowledge and experience that are virtually impossible to obtain from a text, but the report writing component of the lab work provides a rich enhancement to the overall learning objectives for the course. As I state above, development of English language skills should be deeply integrated into our science courses. Requiring students to write lab reports from scratch—rather than by filling out blank spots in a workbook—provides a premier opportunity to do this.

High school science teachers should require that students write full-length lab reports from scratch, on a computer, five or six times per year. In *From Wonder to Mastery*, I outline assessment guidelines and learning objectives for lab reports at different grade levels throughout high school.

There is a lot that goes into writing a quality lab report, and teaching students how to do it without a guide to help is a difficult and time-consuming task. Thus, I commend to you and your students my manual, *The Student Lab Report Handbook*. I recommend that schools distribute copies at the beginning of the school year to each high school freshman. Let each student keep the book for use at home for the next four years. Students should begin learning how to write lab reports in their freshman science class and continue writing reports in all science classes throughout their high school years. Students trained this way astonish their lab instructors when they get to college, and are prepared for college in a way few students are.

Students in high school chemistry courses typically conduct at least 15 or 20 lab exercises during the year. But requiring students to write 15 or 20 lab reports from scratch in a year would be unduly burdensome. This is why I recommend that you require students to write reports from scratch six times during the year—three times in the fall and three times in the spring. For other experiments, use a short-form lab report in which students present data and interact with

Preface for Teachers

a few key questions. Our chemistry experiments books (described next) are designed to help with this.

To accompany this text in schools, Dr. Christina Swan and I produced *Chemistry Experiments for High School*. The experiments in the book are designed for use in a fully-equipped chemistry lab. (An alternative resource for those without a lab is discussed below.) At the end of each experiment in *Chemistry Experiments*, there are a few pages that may be used as a short-form lab report. The book is printed on perforated sheets that can be written on, removed from the book, and submitted. This provision is *not* intended to suggest that all the experiments should be documented this way. But for those that are, the book makes short-form reports easy.

For those that do not have access to a laboratory—and this includes most home-school students—we offer *Chemistry Experiments for High School at Home*. The experiments in this manual are adapted from *Chemistry Experiments for High School*, substituting alcohol burners for Bunsen burners, gravity filtration for vacuum filtration, etc. Some of the experiments do require special chemicals or apparatus—for students to have a legitimate chemistry laboratory experience there is just no way around it. But we have sought to keep these special items to a minimum.

5. Revisions in the Third Edition

In addition to error corrections and minor editorial revisions, the text has been updated to be consistent with the revisions to the SI unit system that went into effect in 2019. The affected sections are the opening page of Chapter 1, Sections 1.1.4, 2.4.2, 2.4.3, 3.1.1, 9.2.2, and Table A.2 in Appendix A. The changes to the mole and Avogadro's number are the most significant of these, requiring substantive changes in Sections 2.4.2 and 2.4.3. Section 2.3.2 was also revised to include mention of the dalton (Da).

Note also that a number of slight changes to the answers to exercises were necessary because of the change to Avogadro's number. These are only noticeable on exercises involving a large number of significant digits.

Preface

For Students

You probably don't normally read the Preface to Teachers in your textbooks. (You may not normally read the one to students either, but I am glad you are reading this one.) However, the books I write are quite different from other textbooks you may have used in the past. So in this case, I recommend that you read the Preface to Teachers so you learn about how this book is structured and what I recommend to teachers about how to use it. In particular, it is my view that students should master and *remember* what they are taught, instead of cramming for tests and forgetting everything a few weeks later.

For this to happen, your teacher needs to teach and test in ways you may not be accustomed to. My recommendation to your teacher is that all your chapter exams include problems and questions from prior chapters. Now, that doesn't mean you need to remember every detail from every chapter. But it does mean that there are certain questions and types of problems that are considered very foundational in any chemistry course. I call these Standard Problems. My advice to your teacher is that you use a list of standard problems in your course, and that your exams always include problems from the standard problems list. At the end of this Preface, you will find the Standard Problems List I recommend to your teacher for this chemistry course.

Naturally, your teacher is the person who decides whether to use the Standard Problems List. But if you do use it, then you need to study in a way that enables you to stay current with the material on the list. Simply doing your homework each night and cramming for tests won't cut it. You must have specific study strategies that help you remember definitions and concepts from previous chapters and how to solve previously learned types problems.

Here are my recommendations for how to do that.

1. Study the Objectives List at the beginning of each chapter carefully. Make it your goal to be able to do everything on the list (that is, for the objectives that have been covered so far in class) before your quizzes and tests occur.

2. Look over the Standard Problems List regularly. Identify any item that you cannot do or cannot remember how to do (assuming the topic has already been covered in class) and follow up on it.

3. Develop, maintain, and practice flash cards for each new chapter Objectives List and each item on the Standard Problems List.

4. Read every chapter in this text at least once, preferably twice. Ideally, every time your instructor covers new material you should read the sections in this book corresponding to that material within 24 hours.

5. Go back and read the chapters in this book again when you are a month or two down the road. You will be amazed at how much easier it is to remember things when you have reread a chapter. (Besides, reading is more fun than rehearsing flash cards.)

6. When you are working on exercises involving computations, check your answers against the answers in the back of the book. Every time you get an incorrect answer, dig in and stay with the problem until you identify your mistake and obtain the correct answer. If you can't figure out a problem after 10 or 15 minutes, raise the question in class.

Preface for Students

7. Every time you lose significant points on a quiz or test, follow up and fill in the gaps in your learning. If you didn't understand something, raise the question with your instructor. If you forgot something, rehearse it more thoroughly until you have it down. If you failed to commit something to memory, or didn't have it in your flash cards, then add it to the cards and commit it to memory. If you were not proficient enough at one or more of the computations, look up some similar problems from the exercises or from previous quizzes and tests and practice them thoroughly.

A complete list of constants, prefixes, and coversion factors students are expected to know from memory is found in Appendix C.

Finally, there is a fair bit of historical material scattered around in this text. It is my view that the history of science *matters* and that students should be held accountable for learning and remembering important historical information. In Appendix B you will find a list of all the scientists discussed in the text. This appendix is there to help you as you study. Your instructor will let you know which scientists to know about and what to know about them. I mention it here so that as you study and review you will know to use Appendix B as review a tool in addition to the Standard Problems List.

Standard Problems List

Given appropriate reference materials, students should be prepared to work the following problems on any chapter test throughout the course, assuming the material has already been covered in class.

FROM CHAPTER 1
1. Use the metric system fluently.
2. Use significant digits fluently.

FROM CHAPTER 2
3. Describe and define the various types of substances.
4. Calculate density.
5. Determine molar mass or formula mass and perform unit conversions between moles and grams.
6. Determine numbers of atoms in a given quantity of substance.

FROM CHAPTER 3
7. Describe the Bohr model and explain how it is able to explain emission spectra.
8. Write electron configurations.
9. Determine percent composition, determine an empirical formula from the percent composition, and determine the molecular formula from the empirical formula and the molar mass.

FROM CHAPTER 4
10. Describe the general structure and arrangement of the periodic table.
11. Describe general trends in the periodic table pertaining to atomic size, ionization energy, and electronegativity.
12. Identify the names of the major groups of elements and other significant regions in the periodic table.

FROM CHAPTER 5
13. Name binary ionic compounds and acids and write their formulas, including those incorporating polyatomic ions.

Preface for Students

14. Describe ionic and covalent bonding.
15. Draw Lewis structures.

FROM CHAPTER 6

16. Apply VSEPR theory to predict molecular shapes and bond angles.
17. Describe the different intermolecular forces, when they occur, and their relative strengths.

FROM CHAPTER 7

18. Balance chemical equations.
19. Perform stoichiometric calculations.
20. Determine oxidations states of pure elements and elements in compounds.
21. Use the activity series of metals to predict whether a single-replacement reaction will occur.

FROM CHAPTER 8

22. Describe the kinetic-molecular theory of gases and the Maxwell-Boltzmann velocity distribution in gases.
23. Explain the causes of surface tension and capillary action.
24. Describe the four basic states of matter.
25. Calculate the heat that must be added to or removed from a substance to change its temperature or effect a phase transition.
26. Explain the causes of evaporation and vapor pressure.

FROM CHAPTER 9

27. Describe Boyle's law, Charles' law, and Avogadro's law.
28. Solve problems using Boyle's law, Charles' law, the ideal gas law, and the density equation.
29. Distinguish between ideal gases and real gases.
30. Use Dalton's law of partial pressures to compute partial pressures, mole fractions, and the total pressure in a gas mixture.
31. Solve stoichiometric problems involving gas volumes.

FROM CHAPTER 10

32. Describe dissolution, electrolytes, and heat of solution.
33. Use solubility guidelines to classify compounds as soluble or insoluble in aqueous solution.
34. Use notions of ionization and polarity to explain the phrase "like dissolves like."
35. Calculate molarity and molality.
36. Use solubility guidelines to predict when precipitates will form.
37. Write ionic equations and net ionic equations describing precipitation reactions.
38. Describe three colligative properties of solutions.
39. Calculate boiling points and freezing points of solutions.

FROM CHAPTER 11

40. Describe two acid-base theories, including definitions for acids and bases in each.
41. Identify conjugate acid-base pairs.
42. Write ionic and net ionic equations for neutralization reactions.
43. Explain the concept of the self-ionization of water.

44. Describe the pH scale.
45. Compute pH, pOH, $[H_3O^+]$, and $[OH^-]$ from each other and from concentration data.
46. Explain the purpose of titration.
47. Determine $[H_3O^+]$ and $[OH^-]$ from titration data.

General Chemistry

A Mastery-Oriented Curriculum

Third Edition

Introduction

What is Chemistry All About?

This computer model depicts three of the *orbitals* available in atoms for holding electrons. Shown are one of the three *p* orbitals in each of three different *subshells*. The inner pair of orbitals can hold two of the highest energy electrons for elements 5 through 10 in the Periodic Table of the Elements. The middle pair is available to hold two of the highest energy electrons for elements 13 through 18 in the periodic table, and the outer pair can hold two of the highest energy electrons belonging to elements 31 through 36 in the periodic table.

In this Introduction, we touch briefly on electron orbitals. We treat the subject in more depth in Chapter 3.

Objectives for the Introduction

After studying this chapter and completing the exercises, you should be able to do each of the following tasks, using supporting terms and principles as necessary.

1. Briefly explain how electrons, electrical forces, minimizing energy, whole number ratios, and modeling can each be thought of as central to understanding what chemistry is all about.
2. State and explain examples illustrating a system moving to a lower energy state and a system experiencing an increase in entropy.
3. Briefly explain hydrogen bonding and why it plays such a large role in mixtures containing water.
4. Explain the relationship between energy and atomic orbitals.

I.1 A Few Major Themes

Chemistry is the study of the elements, how they combine to form mixtures and compounds, the properties of these substances, and the processes involved. One of the astonishing things about the physical world is that as complex as the details are, we can understand a lot about how it works in terms of just a few basic principles from physics. This striking situation is a direct result of the fact that nature is governed by an orderly, mathematical set of physical laws—the laws set in place by God according to his wisdom when he created the universe.

The existence of creation and of the laws of physics are two obvious clues to God's role in creating the universe: the universe is here because God made it, and it is governed in an orderly, mathematical way because it was God's pleasure to make it so. A third clue is that *we can understand it*.

Studying chemistry involves learning a great deal of terminology, and exploring quite a few different types of processes. The amount of information involved can be daunting! But one way to help organize all this information is to be alert to a few fundamental principles that turn up time and again. In this introductory chapter, we take a brief look at a few of these principles. As you read through the chapters ahead, you will see again and again that we can understand a lot about topics such as molecular structure, solubility, and chemical reactions in terms of a few basic concepts.

I.1.1 Chemistry Is All About Electrons

You probably already know that atoms consist of a tiny nucleus containing particles called protons and neutrons, and that the nucleus is surrounded by cloud-like regions containing the atoms' electrons. The protons and electrons carry electrical charge—protons are positively charged and electrons are negatively charged. The protons in an atom stay permanently in the atom's nucleus,[1] but atoms lose or gain electrons by interacting with other atoms. An atom that gains or loses one or more electrons is called an *ion*. Ions are charged particles. Gaining an electron means gaining negative charge, resulting in a negatively charged ion. Losing electrons means losing negative charge, and ending up with more protons (positive charges) than electrons. This results in a positively charged ion.

As it turns out, a lot of chemistry can be understood in terms of the atoms' electrons—where they are, how many there are, whether an atom has ionized by gaining or losing electrons, whether an atom is sharing electrons with another atom, and so on.

---

1 Except in the case of radioactive elements.

Introduction

The cloud-like regions containing an atom's electrons are called *orbitals*, and electrons reside in different orbitals according to how much energy they have. The arrangement of the orbitals is the same for all atoms, although the specific energies associated with each orbital vary from atom to atom, depending on the size of the nucleus and how many electrons an atom has. The orbitals in atoms are grouped into different energy groupings called *shells*. There are seven main shells containing the orbitals with the electrons of all the elements discovered so far. There are additional shells above these that high-energy electrons can move into when they absorb more energy.

In each shell, there is a specific number of orbitals, and each orbital and set of orbitals holds a specific number of electrons. One of the essential facts about atomic behavior is that atoms seek to gain, lose, or share electrons until they have just the right number of electrons so that they have only full shells, without any extra electrons and without electrons missing from any orbitals in the full shells. If only the first shell is full, an atom has two electrons. If the first two shells are full, 10 electrons. If the first three are full, 28 electrons, and so on. Significantly, these numbers relate to the numbers of elements in the rows of the Periodic Table of the Elements. With this one fact, we can understand a great deal about how atoms of one element bond with atoms of other elements to form compounds.

The position of the electrons within an atom also has a lot to do with how an atom behaves. One aspect of atoms that affects the position of electrons is the shapes of the different orbitals. Some orbitals are spherically shaped, some are shaped in pairs of protruding lobes often described as "dumbbells," and some are shaped as rings. There are other more complex shapes as well. Since all electrons repel each other due to their negative electrical charge, electrons located in the lobes of dumbbell-shaped orbitals push away from each other, resulting in molecules with very particular shapes. Examples are the water, ammonia, and methane molecules illustrated in Figure I.1.

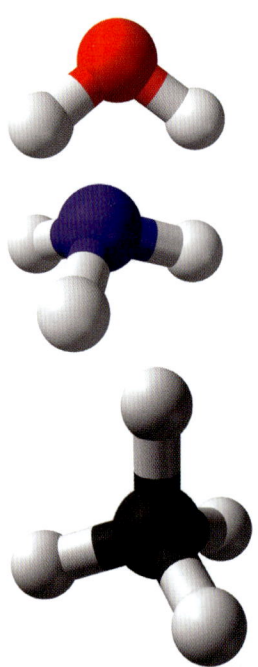

Figure I.1. Representations of the H_2O water molecule (top), the NH_3 ammonia molecule (middle), and the CH_4 methane molecule (bottom).

Electron position is also affected by the fact that within molecules some atoms attract electrons more strongly than others, an effect denoted by a value called the *electronegativity* of the atom. We discuss this in more detail later, but I will mention an important example here to illustrate this point. The electronegativity values for oxygen and hydrogen are 3.44 and 2.20, respectively. This means the oxygen atoms in water molecules attract electrons more strongly than the hydrogen atoms do. As a result, the four bonding electrons in the molecule crowd over toward the oxygen atom, making the oxygen region of the water molecule more electrically negative and the hydrogen regions more electrically positive. These differences make the water molecule electrically imbalanced—or *polar*, as we say—negative on one side and positive at the ends on the other side, as illustrated in Figure I.2. In this diagram, the arrows point from the positive regions of the molecule toward the negative region of the molecule.

The shapes and polarizations affect atomic behavior because of electrical attractions and repulsions, the basic theme we discuss in the next section.

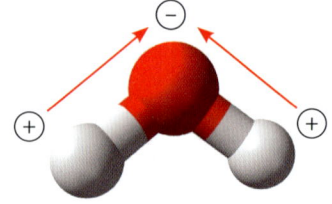

Figure I.2. The higher electronegativity of oxygen atoms compared to hydrogen atoms results in the polar water molecule.

I.1.2 Chemistry Is All About Electrical Forces

Consider again the two types of electrical charge, the positive protons and the negative electrons. You know that like charges repel each other (such as two positive charges) and opposite charges attract each other (positive and negative).

These electrical attractions and repulsions are highly important for chemistry because atoms and molecules are as prickly as porcupines with charges that repel or attract other charges. Some of these attractions and repulsions are stable and long-lasting, like the attractions between ions that hold together the atoms in a crystal of table salt. Other electrical interactions are sort of semi-stable, you might say, given the fact that molecules are moving around all the time. The world-class example of this is *hydrogen bonding*, which we examine in detail later. Hydrogen bonding takes us back to the polar water molecule described in the previous section. Since water molecules are polar, the positive regions of one water molecule are attracted toward the negative regions of other water molecules, as illustrated in Figure I.3.

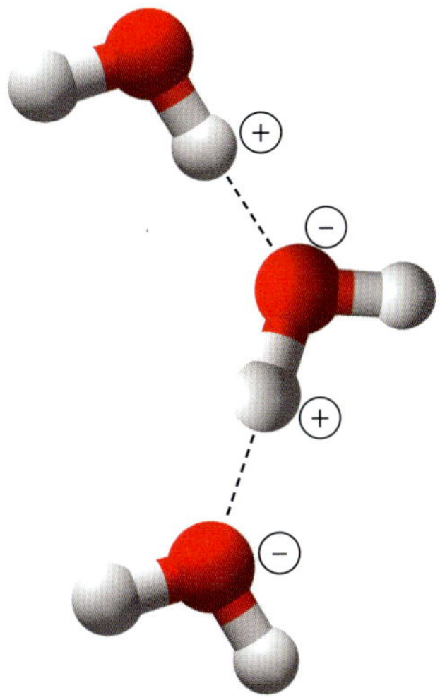

Figure I.3. Hydrogen bonding in water molecules. Dashed lines indicate hydrogen bonds between water molecules.

The importance of hydrogen bonding cannot be overstated. Water is everywhere, and thus so is hydrogen bonding. Hydrogen bonding explains why so many things dissolve in water. It explains why water travels upwards against the force of gravity when soaking into the fibers of a towel. And it explains why water gets less dense right before it freezes (which in turn explains why ice floats). Figure I.4 is a model of how the water molecules are arranged in ice. The dashed lines in the figure indicate the hydrogen bonds between water molecules. The result of these bonds is the three-dimensional, hexagonal structure of ice.

There are several other ways electrical forces between atoms and molecules are made manifest. In general, these different attractions and repulsions are called *intermolecular forces*. There is an electron cloud around every atom (except in the case of a hydrogen atom that has lost its only electron due to ionization). There is also an electron cloud around and between the atoms of every molecule. As the electrons swarm around in these clouds, there are moments when some regions in the molecule are more negatively charged because of electrons crowding together. There are other moments when regions are more positively charged because electrons have temporarily moved away and the positive charge on the protons in atomic nuclei are dominant in the area. These electron movements and crowding go on all the time and at extremely high speeds, giving rise to ever-changing patterns of intermolecular forces. Intermolecular forces are all caused by electrical attractions.

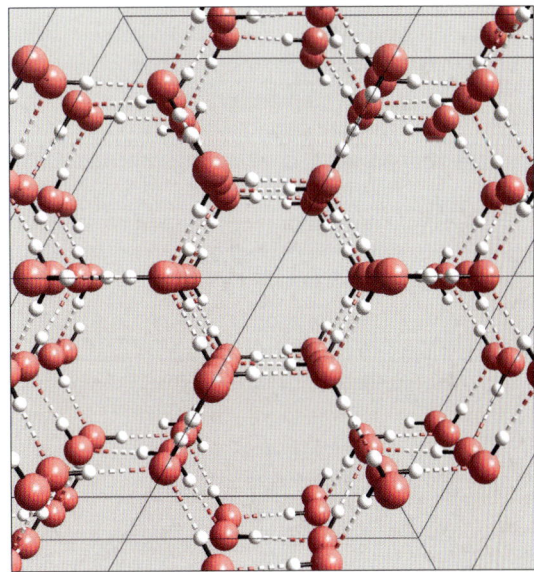

Figure I.4. The crystal structure in ordinary water ice. In this model, oxygen atoms are red and hydrogen atoms are white. Hydrogen bonds are shown as dashed lines. Thick black lines indicate the bonding of hydrogen to oxygen inside individual water molecules.

Introduction

> ### Hmm... Interesting.   ### Why water forms beads
>
> As described in this chapter, the polarity of water molecules makes them cling to one another. When nonpolar molecules are in contact with water, the water molecules are attracted to each other but not to the nonpolar molecules. The molecules in waxy leaves and oil-based wood finishing products are nonpolar. When water molecules rest on a surface of nonpolar molecules, they cling to each other but not to the surface,
>
>
>
> and the result is the formation of water drops. Small drops are nearly spherical because this shape minimizes the energy between the molecules. Larger drops flatten out due to their greater weight.
>
>
>
> Nonpolar molecules do not dissolve in water. The attractions between the polar water molecules squeeze out the nonpolar molecules, causing the two substances to separate. This is why oil and vinegar separate—oil molecules are nonpolar and vinegar is mostly water.

I.1.3 Chemistry Is All About Minimizing Energy

One of the primary drivers causing atoms to do what they do is the natural tendency of all things to minimize the energy associated with the state they are in. Minimizing energy is a concept that explains a great deal of chemical behavior. Here, we look at several examples of objects in different energy states. Then we apply the concept of minimizing energy to phenomena we see occurring in chemistry.

To begin, in a previous science course you may have studied different forms of potential energy. For example, gravitational potential energy is the energy an object has after being lifted up in a gravitational field. Figure I.5 shows a ball up on the side of the hill. The ball is trapped in a small valley or depression. The ball is located up above the ground, so it has gravitational potential energy. The ball always acts to reduce its potential energy if given a chance. If a tunnel opens up to a lower energy state, the ball goes there, releasing potential energy into some other form of energy (such as kinetic energy) as it goes. Another way for the ball to release potential energy and move to a lower state is for someone to hit it or kick it so that it has enough kinetic energy to get over the small hill where it is trapped. The point is that given the chance, the ball releases potential energy and moves to a lower energy state.

Another example of this idea is shown in Figure I.6. A cone held on its point has potential energy that is released if the cone is released and allowed to fall. In this case, the cone doesn't even need any kind of push or kick; it spontaneously moves to the lower energy state (laying down on its side) if released.

As a third example, consider the act of stretching a rubber band. To stretch out a rubber band, you have to supply energy.

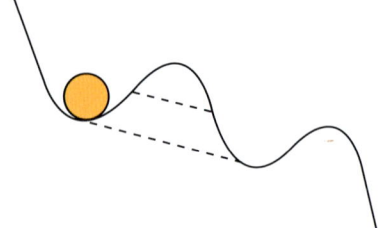

Figure I.5. The ball is trapped in the lowest-energy region in its vicinity. However, if a tunnel to a lower-energy region opens up, the ball goes there.

That is, you have to do mechanical work on the rubber band. If you release the stretched rubber band, it spontaneously contracts back to its lower energy (unstretched) state.

Let's now apply the idea of minimizing energy to a chemical reaction you may already be familiar with: the combustion of hydrogen to produce water. This reaction is represented by the following chemical equation:

$$2H_2 + O_2 \rightarrow 2H_2O$$

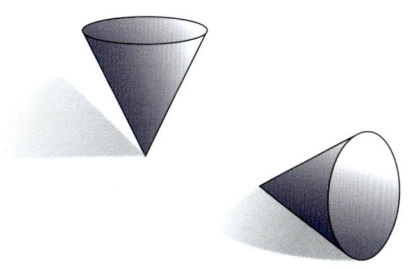

Figure I.6. The cone on the left is in a higher energy state. If released, it falls to the lower energy state shown on the right.

The left side of this equation indicates that the *reactants*—the substances taking part in the reaction—are molecules of hydrogen (H_2) and oxygen (O_2). Each of these gases exists as *diatomic* molecules, meaning that each hydrogen molecule is a pair of hydrogen atoms bonded together, and each oxygen molecule is a pair of oxygen atoms bonded together, as illustrated in Figure I.7. At room temperature, these gas molecules zoom around inside their container, colliding with one another several billion times per second, but otherwise nothing else happens.

In terms of the energy of these molecules, they are in a situation similar to the ball in Figure I.5: there are lower energy states the molecules can go to, releasing energy in the process, but they can't get there without a boost of energy to get the process started. Now, if a spark or flame is introduced to this gas mixture, the heat from the spark or flame excites the nearby molecules, causing them to move much faster and slam into each other with enough energy to break the bonds holding the molecules together. The result—which only lasts for a tiny fraction of a second—is a soup of unbonded gas atoms.

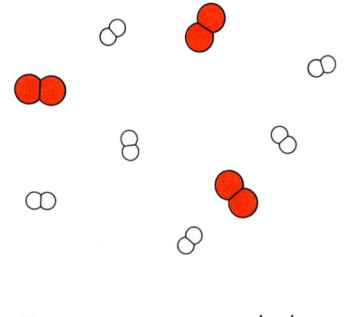

Figure I.7. Diatomic oxygen and hydrogen gas molecules.

At this point, we have a situation similar to the stretched rubber band the instant after being released, before it has had a chance to shrink. Electrical attractions between the protons and electrons in the isolated atoms of oxygen and hydrogen draw the atoms toward each other at an extremely high rate. Consider the collapse of the rubber band after it is released. It collapses to its unstretched state—a lower energy state—releasing energy in the process. The energy released might result in a snap (kinetic energy) that stings your hand and a sound wave (kinetic energy in moving air molecules) producing a snapping sound. Just as the relaxed rubber band is at a lower energy state and releases energy to get there, the hydrogen and oxygen atoms collapse together to the lowest energy state they can find, which is to form water molecules (H_2O). As they do so, they release a lot of energy in the form of light and heat and all this happens in an instant.

Figure I.8. Energy released as heat and light as hydrogen and oxygen atoms combine to form water molecules.

Introduction

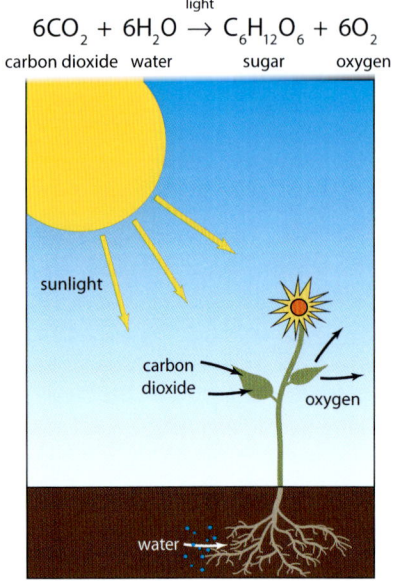

Figure I.9. The photosynthesis reaction is endothermic, as light from the sun is used by plants to convert water and carbon dioxide into sugar and oxygen.

This is the explosion of hydrogen, captured in the photograph of Figure I.8.

This release of heat indicates the reaction is *exothermic*—the reaction releases energy. When considering the way energy relates to various chemical processes, I have found it very helpful to remember the rubber band and to compare it in my mind to the way positive and negative ions are attracted to each other. If separated positive and negative ions are released and allowed to fly together, energy is released—the light and heat of the exothermic reaction—as the ions move to a lower energy state. To separate them, one has to pull them apart by putting in energy (doing work on them) and thus moving them to a higher energy state—just like stretching the rubber band. This is an *endothermic* process, where energy is being absorbed by the ions. The most well-known example of an endothermic chemical reaction is the photosynthesis reaction that occurs in plants, depicted in Figure I.9.

And I can't help pointing out in passing the exquisite elegance of the photosynthesis reaction, a process both simple and incredibly complex that happens automatically and continuously all over the world. Consider the care with which God placed oxygen-breathing creatures like ourselves on a planet covered with oxygen-producing vegetation. Of course, every school kid learns about photosynthesis, but do we also learn that the delicate balance displayed everywhere in the environment around us in creation is a tremendous gift? I encourage you, as a young student made in God's image, to give thanks and worship to our loving Creator for this most wonderful gift!

There are two more important concepts about the role of energy in chemistry to note here. The first involves a quantity called *entropy*. Entropy is a term that originated in the field of thermodynamics. Entropy is a measure of the *disorder* present in a system, and the second law of thermodynamics states that left to themselves, physical processes go in a direction that increases the entropy (disorder) in the system. As an example, consider a glass of water you may be holding in your hand versus a broken glass on the floor with water splashed everywhere. While the glass is intact with the water contained in it, the system of glass and water is in an orderly state. When you release the glass, disorder increases—the glass breaks and the water goes everywhere on the floor. If you leave the mess like this, the disorder continues to increase: the water evaporates and the water molecules are not even together any more at all. Instead, they are randomly distributed around in the atmosphere. And with time, the chunks of glass get trampled and broken more and more until the remnants of the glass are completely gone. You will never see this process occur in reverse!

For some chemical processes, the minimizing of energy and the increase of entropy both pull in the same direction. In other cases, they try to pull the system in opposite directions. This sets up a sort of tug of war, and the process goes in the most favorable direction. We discuss this in more detail later.

Finally, some detail is in order regarding the boundaries surrounding energy minimization. Try this little thought experiment: imagine a hydrogen ion, which is simply a proton with its positive charge. Nearby is an negatively charged electron, as illustrated in Figure I.10. Since these particles have opposite charges, they are strongly attracted to each other, and since the proton's mass is 1,836 times greater than the electron's mass, the electron dashes toward the proton while the proton essentially stays put, waiting for the electron to arrive. You might expect that the electron would crash right into the proton, bringing the potential energy between them right down to zero. But this is not what happens.

Figure I.10. Oppositely charged particles strongly attracted to each other.

In 1905, Albert Einstein theorized that energy is *quantized*—it comes in discrete chunks or packets. Since 1905, a host of scientists have explored the quantization of energy, confirming Einstein's proposal over and over and giving birth to the now well-developed theory of quantum mechanics. What quantum mechanics suggests for our proton-electron scenario is that an electron in an orbital of an atom cannot possess just any old amount of energy; it can only possess particular values of energy. In the context of dropping into one of the orbitals surrounding the proton, the electron can only possess an amount of energy corresponding to the one of the energies of the proton's orbitals. The bottom line is that instead of crashing into the proton and sticking to it like cat hair sticking to your pants, the electron instead pops into the lowest energy orbital available around the proton and stays there, captive, buzzing around furiously like a bee in a bottle. (But though this analogy may be suggestive, it is strictly metaphorical. Electrons are not at all like bees. For one thing, they don't have wings. And they don't make honey, either.)

I.1.4 Chemistry Is All About Whole-Number Ratios of Atoms

It is strange to think that even as recently as the beginning of the 20th century there was no consensus among scientists as to whether atoms even existed. In 1803, English scientist John Dalton put forward the first detailed, scientific atomic theory. Dalton proposed that all material substances are composed of atoms, and that the way different compounds are formed is by atoms combining together. Since various substances are composed of discrete, individual particles and not just a continuum of matter, there is always a whole number of each type of atom in the substance.

Although many scientists throughout the 19th century refused to accept the existence of atoms, we now agree that Dalton was correct. Compounds do form with whole-number ratios of the atoms involved. (Back then, those who accepted the existence of atoms were called "atomists." Today, everyone is an atomist, so we don't need a name for this view any more.) As an example, sulfuric acid, H_2SO_4, has two hydrogen atoms, one sulfur atom, and four oxygen atoms in every molecule, so the ratio of oxygen atoms to hydrogen atoms in the molecule is 2 to 1. The

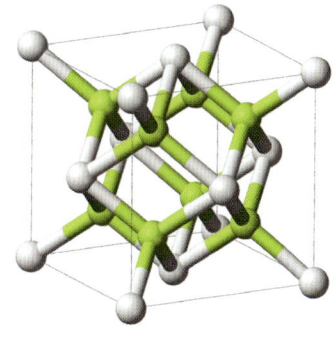

Figure I.11. The ratio of fluorine atoms (greenish-yellow) to calcium atoms (white) in fluorite is 2 to 1. To see this, note that there are 8 fluorine atoms completely within this crystalline cell. Each of the 8 calcium atoms at the corners is shared by 8 cells—the one shown and 7 other surrounding cells. Eight calcium atoms each shared 8 ways contributes a net of 1 calcium atom to the cell. Then there are 6 calcium atoms on the faces of the cell, each shared by the cell shown and the adjacent cell. Six atoms each shared 2 ways contributes a net of 3 atoms to the cell. In total then, this cell claims 8 fluorine atoms and 4 calcium atoms, a ratio of 2 to 1.

Introduction

ratio of oxygen atoms to sulfur atoms is 4 to 1. Of course, there are strange exceptions to every rule, including this one. Nevertheless, it is correct to say that just about every compound, regardless of how the atoms are structured, consists of atoms of different elements joined together in predictable whole-number ratios. Figure I.11 contains a photo of the mineral fluorite along with a computer model of the crystal structure of fluorite, or calcium fluoride, which has the formula CaF_2. In this crystal structure, the ratio of fluorine atoms to calcium atoms is 2 to 1. The caption in the figure explains this, using the computer model of the crystal structure as an aid.

The fact that atoms combine in whole-number ratios is a powerful computational tool. When we get into the math behind chemical reactions (*stoichiometry*, as it is called), we will appeal often to the whole-number ratios of atoms involved in order to compute how much of one compound reacts with a given quantity of another compound.

I.1.5 Chemistry Is All About Modeling

Chemical reactions are happening around us all the time. Just pour a can of soft drink into a glass and watch the carbonic acid (H_2CO_3) in the can convert into the carbon dioxide bubbles and water (CO_2 and H_2O). Light up the gas grill and watch propane (C_3H_8) reacting with the oxygen (O_2) in the air to produce carbon dioxide (CO_2) and water (H_2O). Heat up a pan of cake batter in an oven and the rather complicated molecules in the batter react and change into different complicated molecules in a cake.

In these examples, even though we see bulk materials going into a chemical reaction (the *reactants*) and resulting from the chemical reaction (the *products*), we are not able to see the actual atoms and molecules as they zoom around, combining with and separating from each other. Understanding the behavior of things we cannot see is tricky business.

As mentioned in the previous section, even in recent scientific history the existence of atoms was debated for a hundred years. The issue was finally resolved with experiments in the early 20th century that gave more and more support to the theory that material substances were composed of atoms. We certainly know a lot more today about atoms and their internal structure than we did just a few decades ago. We can even put this knowledge to use in designing amazing new engineering materials, specialty drugs, and even chemical delivery systems to get the drugs into our bodies. But there remains much we do not understand about atoms.

It is helpful to think of science as the process of building "mental models" of the natural world. These mental models are called *theories*. The information we use to build our mental models—scientific facts—comes from experiments, observations, and inferences from these.

Since chemistry deals so much with atoms and molecules, which we can't see, we are almost completely dependent on inferences to develop atomic models describing how the atomic world works. Knowing that the gunpowder in a firecracker explodes when ignited doesn't require a model. It is obvious to all of us that gunpowder is explosive. But *why* is it explosive? What are the rules governing how the atoms in those compounds behave? Understanding why gunpowder explodes does require a model. And the models we work with in chemistry come at us from two different directions.

First, there is the information we gather from experiments. Chemical experimentation has been going on for hundreds of years. In the early days of the scientific revolution, scientists were amazed to discover quantitative laws such as Dalton's whole number ratios and the inverse relationship between the pressure and volume of a gas, a relationship known as Boyle's Law. Second, there is the theoretical modeling that occurs when scientists attempt to apply physical principles from quantum mechanics, thermodynamics, and statistical mechanics to the solution of chemical problems. The shapes and sizes of the atomic orbitals, which we address in Chapter 3, are an example of this type of theoretical modeling.

What is Chemistry All About?

The theoretical models developed by scientists are the basis for our entire understanding of how the natural world functions. Successful theories are those that account for the facts we know and lead to new hypotheses (predictions) that can be put to the test. It is helpful to think about the relationship between facts, theories, hypotheses, and experiments as illustrated in Figure I.12. This diagram illustrates what I call the *Cycle of Scientific Enterprise*. It is important for every student to develop a correct understanding of the kind of knowledge scientific study provides for us. The *goal* of science is to uncover the truth about how nature works, but scientific theories are always works in progress. Even our best theories are provisional and subject to change. For this reason, science is not in the business of making truth claims about scientific knowledge. Science is in the business of modeling how nature works with theories based on research.

As our theories develop over time, our hope is that they get closer and closer to the truth—the amazing and profound truth about mysteries such as what protons and electrons are, why they have the properties they have, and how the two most successful theories of the 20th century—quantum mechanics and general relativity—can be reconciled with each other. But the truth about nature is always out in front of us somewhere, always outside our grasp. To know the truth about nature, we would have to understand nature as God understands it. We are nowhere close to that.

Here are some definitions to keep in mind as you consider the models we discuss in future chapters.

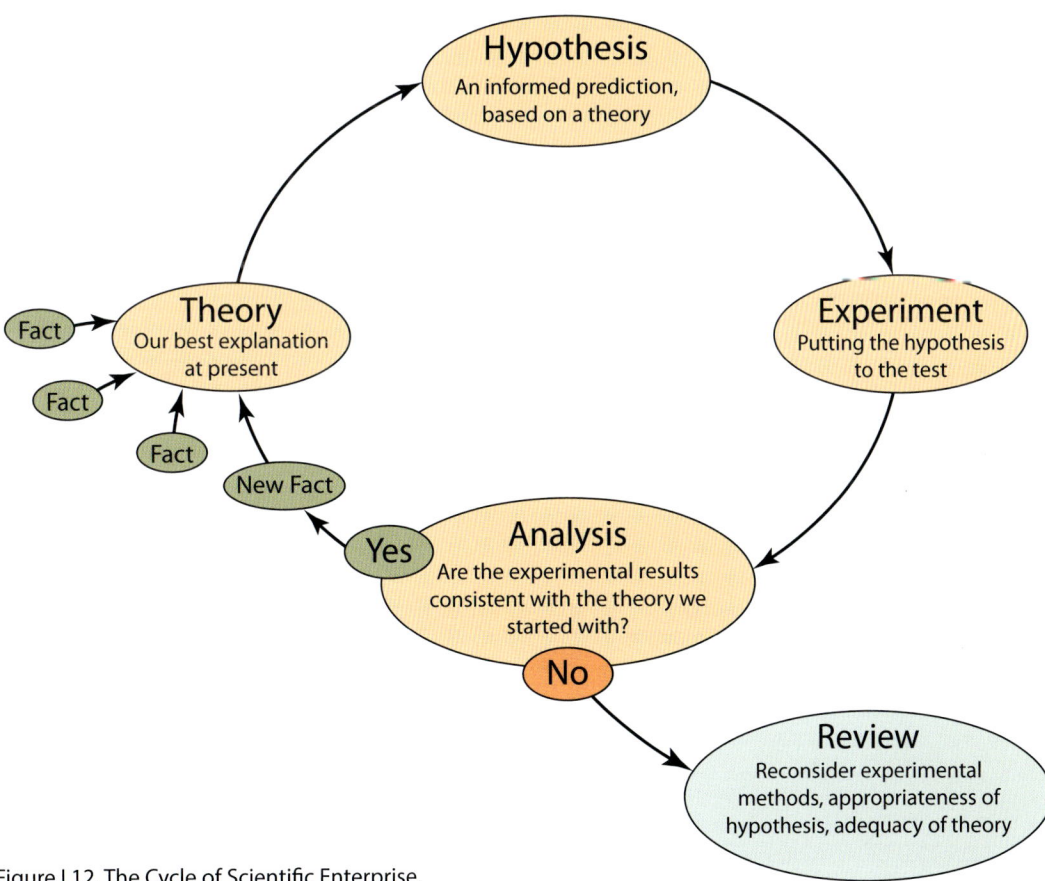

Figure I.12. The Cycle of Scientific Enterprise.

Introduction

Fact   A proposition based on a large amount of scientific data that is correct so far as we know. Facts are discovered by experiment, observation, and inferences from experiments and observations. Facts can and do change as new scientific knowledge—new data—is acquired. Since facts are always subject to change, we generally avoid terms like *true* or *proven*. Instead, we say a fact is correct so far as we know.

Theory   A mental model that accounts for the data (facts) in a certain field of research, and attempts to relate them together, interpret them, and explain them. Scientific theories are successful if they repeatedly allow scientists to form new hypotheses that can be put to experimental test. *Successful theories are the glory and goal of science.* Nevertheless, theories, like facts, are provisional and subject to change. Indeed, theories are almost constantly evolving as research continues. And as with facts, when referring to theories we avoid terms like *true* or *proven*. Instead, we speak in terms of how successful theories have been in generating hypotheses that are confirmed by experiments. A widely accepted scientific theory should be understood as our best explanation at present—our best *model* of how nature works.[2]

Hypothesis   An informed prediction about what will happen in certain circumstances. Every hypothesis is based on a particular theory. It is hypotheses that are tested in scientific experiments.

Experiment   A test designed to confirm or disconfirm a particular hypothesis. If a hypothesis is confirmed through experiment, and if other scientists are able to validate the confirmation by replicating the experiment, then the new facts gained from the experimental results become additional support for the theory the hypothesis came from.

Chemistry is a subject loaded with facts and heavily based on theories—models—that we know are incomplete descriptions of nature. That is why the research continues, as our models (hopefully) get nearer and nearer to the truth.

I.2 Conclusion

The goal of this introductory chapter is simply to alert you to some of the key concepts undergirding our understanding of chemical processes. Over and over in the coming chapters, you will find that thinking about the content in terms of one or more of these central ideas will help you develop a better grasp of the material.

Introduction Study Questions

1. Write five brief paragraphs summarizing the main ideas behind the titles of Sections I.1.1 through I.1.5.

2. Describe two examples, other than those in this Introduction, of a system of some kind spontaneously (without help) moving from a higher energy state to a lower energy state. (Hint: If energy is being released, it means the entities involved in the process are moving to a lower energy state.)

---

2 Note that the term *law* is simply an obsolete term for what we call a theory. For historical reasons, the term is still in use.

3. Describe two examples, other than those in this Introduction, of a system that will move to a lower energy state if allowed to, but which needs an initial boost of energy to get started (like the ball in Figure I.5 being kicked and then having enough energy to get out of the valley).

4. Describe two examples of processes in which entropy *decreases*. In each case, describe what source of energy and/or intelligence must be present for the decrease in entropy to occur. Here is an example to assist your thinking: an oxygen tank contains pressurized oxygen gas. The oxygen in this tank is more ordered than the oxygen in air because it has been separated from the air; there is a boundary (the tank) between the oxygen and the air. And if the valve on the tank is opened, the oxygen flows out into the air to increase the entropy (disorder). What we will never see: opening the tank valve and oxygen atoms from the atmosphere spontaneously flow into the tank. But the oxygen is put into the tank somehow, and the process that put it there decreases the entropy of that oxygen.

5. What is the ratio of nitrogen atoms to hydrogen atoms in ammonia molecules? What is the ratio of hydrogen atoms to carbon atoms in propane molecules?

6. Why are water molecules polar and what is the significance of this fact?

7. If oppositely charged objects attract, why can't a free electron and a free proton collide into one another and stick together because of their opposite charges?

8. A hydronium ion is a water molecule that has gained an extra proton. (A proton is identical to a hydrogen ion.) Hydronium ions form spontaneously in water, and are formed in greater quantities any time an acid is poured into water. What is the ratio of hydrogen atoms to oxygen atoms in hydronium ions?

9. What is hydrogen bonding?

10. Distinguish between endothermic and exothermic processes.

11. In a previous course, you may have learned about the "gold foil experiment" conducted by Ernest Rutherford in 1909. (I describe this experiment in Chapter 2.) This experiment led Rutherford to propose that the positive charge in atoms is concentrated in a tiny nucleus in the center of the atom. Think about this experiment and explain why Rutherford had to depend on inference as he interpreted his experimental data.

12. Why doesn't oil dissolve in water?

13. Distinguish between facts, theories, and hypotheses.

14. Explain why it is scientifically inappropriate to say, "no theory is true until it is proven."

Chapter 1
Measurements

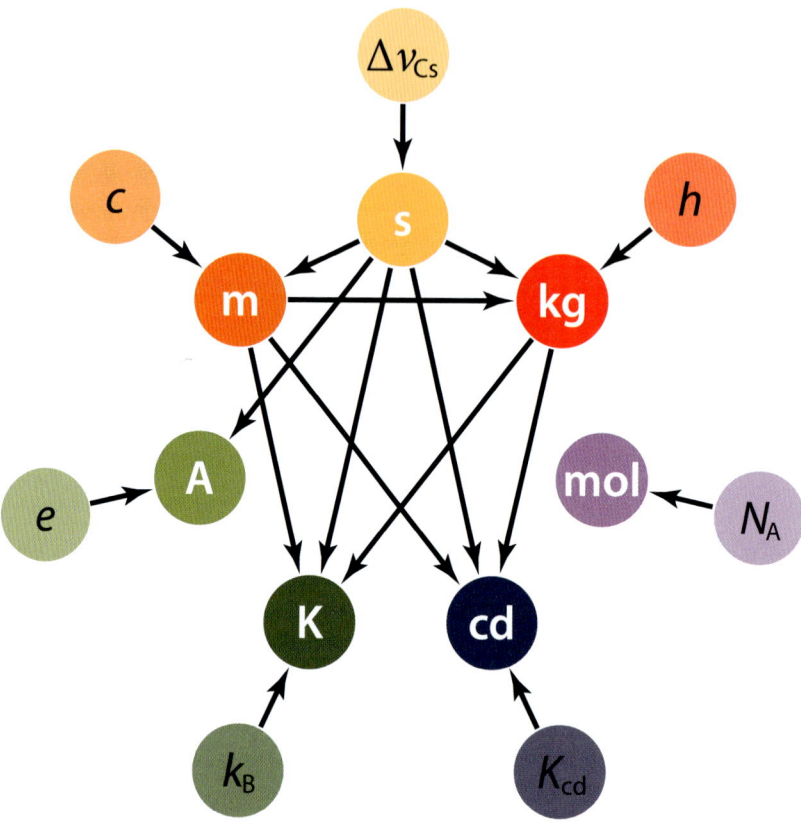

The SI unit system—or metric system—underwent a serious overhaul in 2019. Prior to the makeover, the kilogram was still defined by a physical object kept in a vault in France. Now, however, the seven base units in the SI system, represented by the circles in the inner ring in the graphic above, are defined in terms of physical constants (the outer ring) and each other. For example, the meter (m) is defined as the distance light travels in 1/299,792,458 seconds. That big number in the denominator is the speed of light in meters per second. So, the definition of the meter involves the speed of light (c) and the definition for the second (s). The arrows in the graphic indicate which units and constants affect others. As you see, the definition of the second affects every other unit definition except one.

Objectives for Chapter 1

After studying this chapter and completing the exercises, you should be able to do each of the following tasks, using supporting terms and principles as necessary.

SECTION 1.1
1. Define *matter* and *mass*.
2. Describe the advantages the SI system has over the USCS system for scientific work.
3. State the SI base units for length, mass, and time.
4. For the SI system of units, define the terms *base unit* and *derived unit*.
5. State several examples of base units and derived units in the SI system of units.
6. Use the metric prefixes listed in Table 1.4 from memory with various units of measure to perform unit conversions and solve problems.

SECTION 1.2
7. Given appropriate unit conversion factors, convert the units of measure for given quantities to different units of measure.
8. Given appropriate conversion factors, convert USCS units to SI units and vice versa.
9. Convert units of temperature measurements between °F, °C, and K.

SECTION 1.3
10. Define *accuracy* and *precision*.
11. Name several possibilities for sources of error in experimental measurements.
12. Explain how the significant digits in a measurement relate to the precision of the measurement.
13. Use significant digits correctly to record measurements from digital and analog measurement instruments.
14. Use significant digits correctly to perform computations, including multiplication, division, addition, subtraction, and combinations of these operations.

SECTION 1.4
15. Convert numerical values from standard notation into scientific notation and vice versa.
16. Use scientific notation when recording measurements and performing computations.
17. Calculate the percent difference between an experimental value and a theoretical value or accepted value.

Chapter 1

1.1 Science and Measurements

1.1.1 No Measurements, No Science

One of the things that distinguishes scientific research from other fields of study is the central role played in science by *measurement*. In every branch of science, researchers study the natural world, and they do it by making measurements. The measurements we make in science are the data we use to quantify the facts we have and to test new hypotheses. These data—our measurements—answer questions such as What is its volume?, How fast is it moving?, What is its mass?, How much time does it take?, What is its diameter?, What is its frequency and wavelength?, When will it occur again?, and many others. Without measurements, modern science would not exist.

Units of measure are crucial in science. Since science is so deeply involved in making measurements, you will work with measurements a lot in this course. The value of a measurement is always accompanied by the units of measure—a measurement without its units of measure is a meaningless number. For this reason, your answers to computations in scientific calculations must always show the units. In this chapter, we discuss units of measure at some length. The material in the chapter is very important—for the rest of this book, we will be engaged with calculations involving units of measure.

However, before we launch into our study of units, there is a topic of great importance we need to nail down—*mass*. Students often do not fully understand what mass is, so we will start by reviewing the topic.

1.1.2 Matter, Volume, and Mass

The best way to understand mass is to begin with *matter* and its properties. The term *matter* refers to anything composed of atoms or parts of atoms. Note that there are many things that are not material, that is, they are not matter, as illustrated in Figure 1.1. Your thoughts, your soul, and your favorite song are not matter. You can write down your thoughts in ink, which *is* matter, and your song can be recorded onto a CD, which is matter. But ideas and souls are not material and are not made of what we call matter. Another part of this world that is not matter

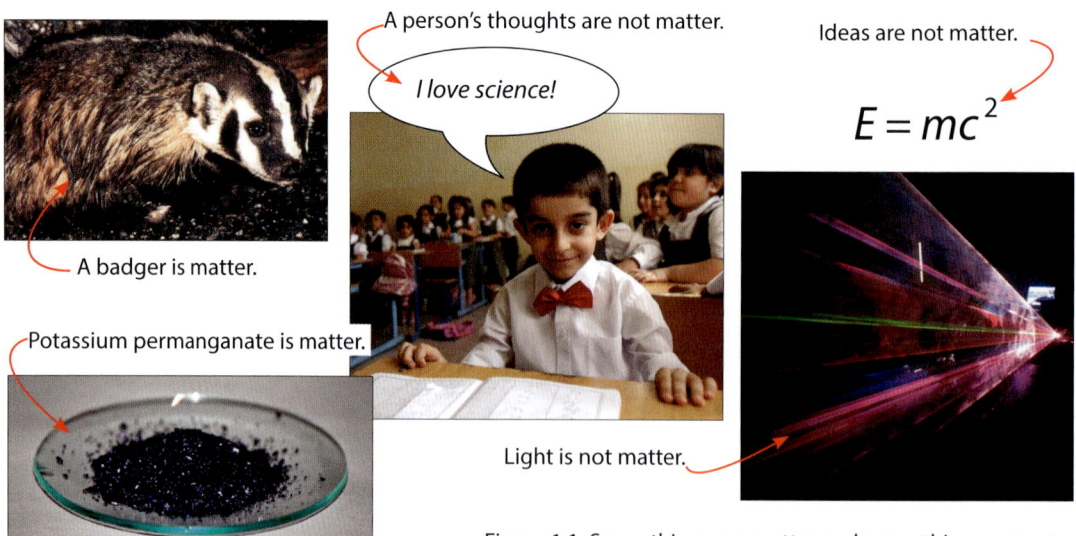

Figure 1.1. Some things are matter and some things are not.

is *electromagnetic radiation*—light, radio waves, X-rays, and all other forms of electromagnetic radiation. Light is pure energy; it is not matter and it has no mass.

We get into electromagnetic radiation a bit in Chapter 3. For now, we are going to focus on matter. A lot of what we discuss in this text is about different properties of matter. There are many different properties to discuss, but here we focus on just two properties that all matter possesses: all matter takes up space and all matter has inertia. Describing and comparing these two properties helps make clear what we mean by the term *mass*.

All matter takes up space. Even individual atoms and protons inside of atoms take up space. Now, how do we *quantify* how much space an object takes up? That is, how do we put a numerical measurement to it? The answer is, of course, by specifying its *volume*. Volume is the name of the variable we use to quantify how much space an object takes up. There are many different units of measure we use to specify an object's volume. Examples are gallons, liters, cubic meters, and pints. When we say that the volume of an object is 338 cubic centimeters, what we mean is that if we could hollow the object out and fill it up with little cubes, each with a volume of one cubic centimeter, 338 of them are needed to fill up the hollowed object.

All matter has inertia. The effect of this property is that objects resist being accelerated. The more inertia an object has, the more difficult it is to accelerate the object. For example, if the inertia of an object is small, as with, say, a golf ball, the object is easy to accelerate. Golf balls are easy to throw, and if you hit one with a golf club it accelerates at a high rate to a very high speed. But if the amount of inertia an object has is large, as with, say, a grand piano, the object is difficult to accelerate. Just try throwing a grand piano or hitting one with a golf club and you will see that it doesn't accelerate at all. This is because the piano has a great deal more inertia than a golf ball.

As with the property of taking up space, we need to quantify the property of inertia. The way we do this is with the variable we call *mass*. The mass of an object is a numerical measurement specifying the amount of inertia the object has. Since inertia is a property of matter, and since all matter is composed of atoms, it should be pretty obvious that the more atoms there are packed into an object, the more mass it has. And since the different types of atoms themselves have different masses, an object made of more massive atoms has more mass than an object made of an equal number of less massive atoms.

The main unit of measure we use to specify an object's mass is the *kilogram*. There are other units such as the gram and the microgram. The kilogram (kg) is one of the base units in the metric system, our topic in Section 1.1.4. On the earth, an object weighing 2.2 pounds (lb) has a mass of one kilogram. To give you an idea of what a kilogram mass feels like in your hand on the earth, the lantern battery pictured in Figure 1.2 weighs 2.2 lb, and thus has a mass of 1 kg.

We have established that the mass of an object is a measure of its inertia, which in turn depends on how many atoms it is composed of and how massive those atoms are. The implication of this is that an object's mass does not depend on where it is. A golf ball on the earth has the same mass as a golf ball at the bottom of the ocean, on the moon, or in outer space. Even where there is no gravity, the mass of the golf ball is the same. This is what distinguishes the *mass* of an object from its *weight*.

Figure 1.2. The mass of this battery is about one kilogram.

Weight is caused by the force of gravity acting on an object composed of matter (which we often simply refer to as *a mass*). The weight of an object depends on where it is. An object—or mass—on the moon weighs only about 1/6 its weight on earth, and in outer space, where there is no gravity, a mass has no weight at all. But the mass of an object does not depend on where it is. This is because an object's mass is based on the matter the object is made of. The lantern battery in the figure has a certain weight on the earth (2.2 lb). In outer space, it weighs nothing and floats right in front of you. But if you try to throw the battery, the force you feel on your hand is the same on the earth or in space. That's because the force you feel depends on the object's mass.

Here is a summary using slightly different terminology that may help even more. Inertia is a *quality* of all matter; mass is the *quantity* of a specific portion of matter. Inertia is a quality or property all matter possesses. Mass is a quantitative variable, and it specifies an amount of matter, a quantity of matter.

1.1.3 The US Customary System

The two major systems of units students should know about are the *International System of Units*, known as the *SI system* or the metric system, and the *U.S. Customary System*, or USCS. You have probably studied these systems before and should be already familiar with some of the SI units and prefixes. In this course, we do not make much use of the measurement system you are most familiar with—the USCS. For scientific work, the entire international scientific community uses the SI system. But here I address the USCS briefly before moving on.

Americans are generally comfortable with measurements in feet, miles, gallons, inches, and degrees Fahrenheit because they grow up using this system and are very familiar with it. But in fact, the USCS is rather cumbersome. One problem is that there are many different units of measure for every kind of physical quantity. Just for measuring length or distance, for example, we have the inch, foot, yard, and mile. The USCS is also full of random numbers such as 3, 12, and 5,280. A third problem is that there is no inherent connection between units for different types of quantities. Gallons have nothing whatsoever to do with feet, and quarts have nothing to do with miles.

The USCS may be familiar ground, and it may even feel patriotic to prefer it, but it is not the system of measurement scientists use. Scientists everywhere use the SI, and it is to that system we now turn.

Unit	Symbol	Quantity
meter	m	length
kilogram	kg	mass
second	s	time
ampere	A	electric current
kelvin	K	temperature
candela	Cd	luminous intensity
mole	mol	amount of substance

Table 1.1. The seven base units in the SI unit system.

1.1.4 The SI Unit System

In contrast to the USCS, the SI system is simple and has many advantages. There is usually only one basic unit for each kind of quantity, such as the meter for measuring length. Instead of having many unrelated units of measure for measuring quantities of different sizes, prefixes based on powers of ten are used on all the units to accommodate various sizes of quantities. And units for different types of quantities relate to one another in some way. Unlike the gallon and the foot, which have nothing to do with each other, the cubic meter is 1,000,000 cubic centimeters. For all these reasons, the USCS is not used much at all in scientific work. The SI system is the international standard.

There are seven *base units* in the SI System, listed in Table 1.1. All other SI units of measure,

Unit	Symbol	Quantity
joule	J	energy
newton	N	force
cubic meter	m^3	volume
watt	W	power
pascal	Pa	pressure

Table 1.2. Some SI System derived units.

such as the joule (J) for measuring quantities of energy and the newton (N) for measuring amounts of force, are based on these seven base units. Units based on combinations of the seven base units are called *derived units*. A few common derived units are listed in Table 1.2.

You are already familiar with the SI unit for time: the second. You may or may not be familiar with some of the other base units, so here are some facts and photos to help familiarize you with these. A meter is just a few inches longer than a yard (3 feet). Figure 1.3 shows a wooden measuring rule one meter long, commonly called a meter stick, along with a metal yardstick for comparison.

On earth, a mass of one kilogram weighs about 2.2 pounds. The six-volt lantern battery shown in Figure 1.2 weighs just under 2.2 pounds, so the mass of the battery is just about one kilogram, as I mentioned before.

I can't show you a picture of one ampere of electric current, but it may be helpful to know that a standard electrical receptacle (or "outlet") such as the one shown in Figure 1.4 is rated to carry 15 amperes of current. (However, the largest continuous current that the receptacle is allowed to supply is 80% of its rating, or 12 amperes. This is why vacuum cleaners are often advertised as having 12-amp motors. That's the upper limit of the current available to run them.)

Figure 1.3. A meter stick (left), with a yardstick for comparison.

Regarding temperature units, the Celsius scale is generally used for making scientific temperature measurements, but the Kelvin scale must be used for nearly all scientific calculations involving temperature. You must become familiar with both scales.

On the Celsius scale, water freezes at 0°C and boils at 100°C. Since Celsius temperature measurements can take on negative values, the Celsius scale (like the Fahrenheit scale) is not an absolute temperature scale. The Kelvin scale is an absolute scale, with 0 K (0 kelvins) being equal to absolute zero, theoretically the lower limit of possible temperatures. Note that the term "degrees" is not used when stating or writing values in kelvins.

Figure 1.4. A standard receptacle in American homes is rated for a current of 15 amperes.

A temperature change of one kelvin is the same as a temperature change of one degree Celsius, and both are almost double the change that a change of one degree Fahrenheit is. For reference, room temperature on the three scales is 72°F, 22.2°C, and 295.4 K.

All seven of the SI base units are defined in terms of physical constants, each one of which is defined with its own exact value. For example, the speed of light in a vacuum is defined to be 299,792,458 meters per second. The meter is defined from this: the distance light travels in 1/299,792,458 seconds. (And the second has its own definition.) Formerly, units such as the kilogram and the meter were defined by man-made physical objects (artifacts), such as the one shown in Figure 1.5. The bar shown in the figure was the standard in the U.S. for the meter from 1893 to 1960. But this method of definition was not at all convenient.

Figure 1.5. Standard meter bar number 27, owned by the U.S. and used as the standard meter from 1893 to 1960.

Chapter 1

	Prefix	deca–	hecto–	kilo–	mega–	giga–	tera–	peta–	exa–	zetta–	yotta–
Multiples	Symbol	da	h	k	M	G	T	P	E	Z	Y
	Factor	10	10^2	10^3	10^6	10^9	10^{12}	10^{15}	10^{18}	10^{21}	10^{24}
	Prefix	deci–	centi–	milli–	micro–	nano–	pico–	femto–	atto–	zetto–	yocto–
Fractions	Symbol	d	c	m	μ	n	p	f	a	z	y
	Factor	1/10	$1/10^2$	$1/10^3$	$1/10^6$	$1/10^9$	$1/10^{12}$	$1/10^{15}$	$1/10^{18}$	$1/10^{21}$	$1/10^{24}$

Table 1.3. The SI System prefixes.

In 1960, the definition of the meter was changed so that the meter was equal to a certain number of wavelengths of a certain color of light emitted by a certain isotope of the element krypton. In 1983, the definition of the meter was changed again to its present definition in terms of the speed of light and the second. The kilogram has a similar history, and its definition in terms of an artifact only ended in 2019.

The SI system includes not only the base and derived units, but all the other units that can be formed by adding metric prefixes to these units. We address the prefixes in the next section. But first I will mention a particular subset of the SI system known as the MKS system. MKS stands for meter-kilogram-second. The MKS system uses only the base and derived units without the prefixes (except for the kilogram, the only base unit with a prefix). The nice thing about the MKS system is that any calculation performed with MKS units produces a result in MKS units. For this reason, the MKS system is used almost exclusively in physics. However, in chemistry, it is common to use SI units that are not MKS units. Some commonly used non-MKS units are the gram (g), the centimeter (cm), the cubic centimeter (cm^3), the liter (L), and the milliliter (mL). The liter is not actually an official SI unit, but it is used all the time in chemistry anyway.

1.1.5 Metric Prefixes

In the system of units commonly used in the U.S., different units are used for different sizes of objects. For example, for short distances we might use the inch or the foot, whereas for longer distances we switch to the mile. For the small volumes used in cooking, we use the fluid ounce (or pint, quart, teaspoon, tablespoon, etc.), but for larger volumes like the gasoline in the gas tank of a car, we switch to the gallon. (That's six different volume units I just listed!)

The SI System is much simpler. Each type of quantity—such as length or volume—has one main unit of measure. Instead of using several different units for different sizes of quantities, the SI System uses multipliers on the units to multiply them for large quantities, or to scale them down for smaller quantities. We call these multipliers the *metric prefixes*. The complete list of the 20 metric prefixes is in Table 1.3. You do not need to memorize all these; some are rarely used. But you do need to memorize some of them. I recommend that all science students commit to memory the prefixes listed in Table 1.4.

Table 1.5 shows a few representative examples of how to use the prefixes to represent multiples (quantities larger than the SI base unit)

Fractions			Multiples		
Prefix	Symbol	Factor	Prefix	Symbol	Factor
centi–	c	$1/10^2$	kilo–	k	10^3
milli–	m	$1/10^3$	mega–	M	10^6
micro–	μ	$1/10^6$	giga–	G	10^9
nano–	n	$1/10^9$	tera–	T	10^{12}
pico–	p	$1/10^{12}$			

Table 1.4. SI System prefixes to commit to memory.

Measurements

	Prefix	Symbol	Meaning	Examples of usage
Multiples	kilo–	k	1,000	One kilojoule is 1,000 joules. There are 1,000 joules in one kilojoule, so 1,000 J = 1 kJ.
	mega–	M	1,000,000	One megawatt is 1,000,000 watts. There are 1,000,000 watts in one megawatt, so 1,000,000 W = 1 MW.
Fractions	centi–	c	1/100	One centimeter is 1/100 of a meter. There are 100 centimeters in one meter, so 100 cm = 1 m.
	milli–	m	1/1,000	One milligram is 1/1,000 of a gram. There are 1,000 milligrams in one gram, so 1,000 mg = 1 g.
	micro–	μ	1/1,000,000	One microliter is 1/1,000,000 of a liter. There are 1,000,000 microliters in one liter, so 1,000,000 μL = 1 L.

Table 1.5. Examples of correct usage of metric prefixes.

and fractions (quantities smaller than the SI base unit). As illustrations, let's look at a couple of these more closely.

The prefix *kilo–* is a *multiple*, and it means 1,000. One kilogram is 1,000 grams, one kilometer is 1,000 meters, and so on. Figure 1.6 illustrates with my favorite dairy product—chocolate chip cookie dough ice cream. One gram is only a fraction of a taste (and contains only two chocolate chips and no cookie dough). A kilogram of ice cream is 1,000 grams of ice cream, equivalent to two large bowls of ice cream.

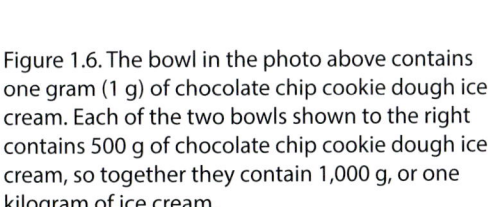

Figure 1.6. The bowl in the photo above contains one gram (1 g) of chocolate chip cookie dough ice cream. Each of the two bowls shown to the right contains 500 g of chocolate chip cookie dough ice cream, so together they contain 1,000 g, or one kilogram of ice cream.

The prefix *milli–* is a *fraction*, and it means one thousandth. One millimeter is one thousandth of a meter, and so on. The wooden rule in Figure 1.3 is one meter in length. A millimeter is one thousandth of this length, equal to the width of the line in Figure 1.7.

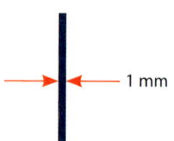

Figure 1.7. One millimeter, which is one thousandth of a meter.

We conclude this introduction to metric prefixes with a few brief notes. First, when using the prefixes for quantities of mass, prefixes are never added to the kilogram. Prefixes are only added to the gram, even though the kilogram is the base unit in the SI system, not the gram. Second, note that when writing the symbols for metric prefixes, the case of the letter matters: *kilo–* always takes a lower-case k, *mega–* always takes an upper-case M, and so on. Third, one of the prefix symbols is not an English letter. The prefix μ for *micro–* is the lower-case Greek letter *mu*, the m in the Greek alphabet. Finally, pay close attention to the difference between multiplier prefixes and fraction prefixes. Learning to use the fraction prefixes properly is the most challenging part of mastering the SI System of units, and using them incorrectly in unit conversion factors (our next topic) is a common student error.

1.2   Converting Units of Measure

1.2.1 Basic Principles of Unit Conversion Factors

For scientists and engineers, one of the most commonly used skills is re-expressing quantities into equivalent quantities with different units of measure. These calculations are called *unit conversions*. Mastery of this skill is essential for all students studying science. In this section, I describe what unit conversion factors are and how they are used.

Let's begin with the basic principles of how unit conversions work. First, we all know that multiplying any value by *unity* (one) leaves its value unchanged. Second, we also know that in any fraction, if the numerator and denominator are equivalent, the value of the fraction is unity (one). For example, the expression "12 bricks over 12 bricks" is equal to one:

$$\frac{12 \text{ bricks}}{12 \text{ bricks}} = 1$$

This is because the numerator and denominator are equivalent, and any time this is the case, the value of the fraction is one, or unity.

A *unit conversion factor* is simply a fractional expression in which the numerator and denominator are equivalent ways of writing the same physical quantity with different units of measure. This means a conversion factor is just a special way of writing unity (one).

The third basic principle is that when multiplying fractions, factors that appear in both the numerator and denominator may be "cancelled out." So when performing ordinary unit conversions, what we are doing is repeatedly multiplying a given quantity by unity so that cancellations alter the units of measure until they are expressed the way we wish. Since all we are doing is multiplying by one, the value of our original quantity is unchanged; it simply looks different because it is expressed with different units of measure.

There are many different units of measure and there are many different conversion factors used for performing unit conversions. Table A.3 in Appendix A lists a number of important ones.

Let me elaborate a bit more on the idea of unity I mentioned above, using one common conversion factor as an example. American school kids all learn that there are 5,280 feet in one mile, which means 5,280 ft = 1 mi. One mile and 5,280 feet are equivalent ways of writing the same length. If we place these two expressions into a fraction, the numerator and denominator are equivalent, so the value of the fraction is unity, regardless of the way we write it. The equation 5,280 ft = 1 mi can be written as a conversion factor two different ways, and the fraction equals unity either way:

$$\frac{5280 \text{ ft}}{1 \text{ mi}} = \frac{1 \text{ mi}}{5280 \text{ ft}} = 1 \qquad (1.1)$$

Now, I need to make an important clarification about the use of the equal sign in the expressions I just wrote. In mathematics, the equal sign means *identity*. An expression such as 3 miles = 3 miles is a mathematical *identity*; the first expression, "3 miles" is *identical* to the second expression, also "3 miles." But when we are dealing with converting units of measure from one set of units to a different set of units, we don't use the equal sign to mean identity. Obviously, the expression

$$\frac{5280 \text{ ft}}{1 \text{ mi}}$$

is not *identical* to the expression

$$\frac{1 \text{ mi}}{5280 \text{ ft}}$$

One of these has units of ft/mi and the other has units of mi/ft. But even though these expressions are not identical, they are *equivalent*. When we are dealing with converting units of measure, this is the sense in which we interpret the equal sign. We are using it to mean *equivalent*. This is why I can write Equation (1.1) using equal signs. The three terms in Equation (1.1) are not identical, but they are equivalent.

Suppose you have a measurement such as 43,000 feet that you wish to re-express in miles. To convert the units from feet to miles, first write down the quantity you are given, with its units of measure:

43,000 ft

Next, select a unit conversion factor containing the units you presently have and the ones you want to convert to. (This is not always possible. Sometimes more than one conversion factor is required, as Example 1.1 below illustrates.) In this case, those units are feet and miles, and the conversion factors containing these units are the two written in Equation (1.1). As I explain below, the one we need for converting 43,000 ft into miles is the second one. So to perform the conversion, you multiply your given quantify by the conversion factor. Then you cancel any units that appear both in the numerator and the denominator, as follows:

$$43,000 \text{ ft} \cdot \frac{1 \text{ mi}}{5280 \text{ ft}} = 8.1 \text{ mi}$$

There are two important comments to make here. First, since any conversion factor can be written two ways (depending on which quantity is placed in the numerator), how do we know which way to write the conversion factor? Well, we know from algebra that when we have quantities in the numerator of a fraction that are multiplied (factors), and quantities in the denominator of the fraction that are multiplied (factors), any quantities that appear in both the numerator and denominator can be cancelled out. In the example above, we want to cancel out the "feet" in the given quantity (which is in the numerator), so the conversion factor needs to be written with feet in the denominator and miles in the numerator.

Second, if you perform the calculation above (43,000 ÷ 5,280), the result that appears on your calculator screen is 8.143939394. So why didn't I write down all those digits in my result? Why did I round my answer off to simply 8.1 miles? The answer to that question has to do with the significant digits in the value 43,000 ft that we started with. We address the issue of significant digits later in this chapter, but in the examples that follow I always write the results with the correct number of significant digits for the values involved in the problem.

The following example illustrates the use of conversion factors based on metric prefixes. This example also illustrates how to perform a conversion when more than one conversion factor is required.

▼ Example 1.1

Convert the value 2,953,000 µg into kilograms.

Referring to Tables 1.4 and 1.5, you see that the symbol µ means *micro–*, which means one millionth. Thus, µg means millionths of a gram. We use this information to make conversion factors. Since it takes 1,000,000 millionths of a gram to make one gram, 1,000,000 µg = 1 g, and thus

$$\frac{1,000,000 \text{ µg}}{1 \text{ g}} = \frac{1 \text{ g}}{1,000,000 \text{ µg}} = 1$$

Note that converting from µg to g only gets us part of the way toward the solution. We need another conversion factor to get from g to kg. Looking again at Tables 1.4 and 1.5, we see that there are 1,000 grams in one kilogram, or 1,000 g = 1 kg. From this we can make additional conversion factors:

$$\frac{1000 \text{ g}}{1 \text{ kg}} = \frac{1 \text{ kg}}{1000 \text{ g}} = 1$$

Now to perform the conversion, first convert micrograms into grams. Then convert grams into kilograms. You can do this two-step conversion at the same time by simply multiplying both conversion factors at the same time as follows:

$$2,953,000 \text{ µg} \cdot \frac{1 \text{ g}}{1,000,000 \text{ µg}} \cdot \frac{1 \text{ kg}}{1000 \text{ g}} = 0.002953 \text{ kg}$$

The µg in the given quantity cancels with the µg in the denominator of the first conversion factor. The g in the first conversion factor cancels with the g in the denominator of the second conversion factor. The units we are left with are the kg in the numerator of the second conversion factor. The kg did not cancel out with anything, so these are the units of our result.

1.2.2 Tips for Converting Units of Measure

There are several important points you must remember in order to perform unit conversions correctly. I illustrate them below with examples. You should rework each of the examples on your own paper as practice to make sure you can do them correctly. The conversion factors used in the examples below are all listed in Table A.3 in Appendix A.

Point 1 *Never use slant bars in your unit fractions. Use only horizontal bars.*

In printed materials, we often sees values written with a slant fraction bar in the units, as in the value 35 m/s. Although writing the units this way is fine for a printed document, you should not write values this way when you are performing unit conversions. This is because it is easy to get confused and not notice that one of the units is in the denominator in such an expression (s, or seconds, in my example), and the conversion factors used must take this into account.

▼ Example 1.2

Convert 57.66 mi/hr into m/s.

Writing the given quantity with a horizontal bar makes it clear that the "hours" is in the denominator. This helps you write the hours-to-seconds factor correctly.

To perform this conversion, we must convert the miles in the given quantity into meters, and we must convert the hours into seconds. From Table A.3, we select the two conversion factors we need. Then we multiply the given quantity by them to convert the mi/hr into m/s. When doing the multiplying, we write all the unit fractions with horizontal bars.

$$57.66 \, \frac{\text{mi}}{\text{hr}} \cdot \frac{1609 \, \text{m}}{1 \, \text{mi}} \cdot \frac{1 \, \text{hr}}{3600 \, \text{s}} = 25.77 \, \frac{\text{m}}{\text{s}}$$

As you see, the miles cancel and the hours cancel, leaving meters in the numerator and seconds in the denominator. Now that you have your result, you may write it as 25.77 m/s if you wish, but do not use slant fraction bars in the units when you are working out the unit conversion.

Point 2 *The term "per," abbreviated p, implies a fraction.*

Some units of measure are commonly written with a "p" for "per," such as mph for miles per hour, or gps for gallons per second. Change these expressions to fractions with horizontal bars when you work out the unit conversion.

▼ Example 1.3

Convert 472.2 gps to L/hr.

When you write down the given quantity, change the gps to gal/s and write these units with a horizontal bar:

$$472.2 \, \frac{\text{gal}}{\text{s}} \cdot \frac{3.785 \, \text{L}}{1 \, \text{gal}} \cdot \frac{3600 \, \text{s}}{\text{hr}} = 6{,}434{,}000 \, \frac{\text{L}}{\text{hr}}$$

Point 3 *Use the* ⊠ *and* ⊟ *keys correctly when entering values into your calculator.*

When dealing with several numerator terms and several denominator terms, multiply all the numerator terms together first, hitting the ⊠ key between each, then hit the ⊟ key and enter

Chapter 1

all the denominator terms, hitting the ÷ key between each. This way you do not need to write down intermediate results, and you do not need to use any parentheses.

▼ Example 1.4

Convert 43.2 mm/hr into km/yr.

The setup with all the conversion factors is as follows:

$$43.2 \frac{\text{mm}}{\text{hr}} \cdot \frac{1 \text{ m}}{1000 \text{ mm}} \cdot \frac{1 \text{ km}}{1000 \text{ m}} \cdot \frac{24 \text{ hr}}{1 \text{ dy}} \cdot \frac{365 \text{ dy}}{1 \text{ yr}} = 0.378 \frac{\text{km}}{\text{yr}}$$

To execute this calculation in your calculator, you enter the values and operations in this sequence:

$43.2 \times 24 \times 365 \div 1000 \div 1000 =$

If you do so, you get 0.37843200. (Again, significant digits rules require us to round to 0.378.)

Point 4 *When converting units for area and volume such as cm^2 or m^3, you must use the appropriate length conversion factor twice for areas and three times for volumes.*

The units "cm^2" for an area mean the same thing as "cm × cm." Likewise, "m^3" means "m × m × m." So when you use a length conversion factor such as 100 cm = 1 m or 1 in = 2.54 cm, you must use it twice to get squared units (areas) or three times to get cubed units (volumes).

▼ Example 1.5

Convert 3,550 cm^3 to m^3.

$$3550 \text{ cm}^3 \cdot \frac{1 \text{ m}}{100 \text{ cm}} \cdot \frac{1 \text{ m}}{100 \text{ cm}} \cdot \frac{1 \text{ m}}{100 \text{ cm}} = 0.00355 \text{ m}^3$$

Notice in this example that the unit cm occurs three times in the denominator, giving us cm^3 when they are all multiplied together. This cm^3 term in the denominator cancels with the cm^3 term in the numerator. And since the m unit occurs three times in the numerator, they multiply together to give us m^3 for the units in our result.

The issue of needing to repeat conversion factors only arises when you are using a unit raised to a power, such as a when a length unit is used to represent an area or a volume. When using a conversion factor such as 3.785 L = 1 gal, the units of measure are written using units that are strictly volumetric (liters and gallons), and are not obtained from lengths the way they are with in^2, ft^2, cm^3, and m^3. Another common unit that uses an exponent is acceleration, which has units of m/s^2 in the MKS unit system.

▼ Example 1.6

Convert 5.85 mi/hr^2 into MKS units.

The MKS unit for length is meters (m), so we must convert miles to meters. The MKS units for time is seconds (s), so we must convert hr² into s².

$$5.85 \frac{mi}{hr^2} \cdot \frac{1609 \text{ m}}{1 \text{ mi}} \cdot \frac{1 \text{ hr}}{3600 \text{ s}} \cdot \frac{1 \text{ hr}}{3600 \text{ s}} = 0.000726 \frac{m}{s^2}$$

With this example, you see that since the "hours" unit is squared in the given quantity, the conversion factor converting the hours to seconds must appear twice in the conversion calculation. The "miles" unit in the given quantity has no exponent, so the conversion factor used to convert miles to meters only appears once in the calculation.

▲

1.2.3 Converting Temperature Units

Converting temperature values from one scale to another requires the use of equations rather than conversion factors. This is due to the fact that the Fahrenheit and Celsius scales are not absolute temperature scales. If all temperature scales were absolute scales like the Kelvin scale is, temperature conversions could be performed with conversion factors just as other conversions are. To convert a temperature in degrees Fahrenheit (T_F) into degrees Celsius (T_C), use this equation:

$$T_C = \frac{5}{9}(T_F - 32°)$$

Using a bit of algebra, we can work this around to give us an equation that can be used to convert Celsius temperatures to Fahrenheit values:

$$T_F = \frac{9}{5}T_C + 32°$$

To convert a temperature in degrees Celsius into kelvins (T_K), use this equation:

$$T_K = T_C + 273.15$$

Again, some algebra gives us the equation the other way around.

$$T_C = T_K - 273.15$$

All four of the temperature conversion equations above are exact. This is important to know later when we discuss significant digits. These equations are listed in Table A.3 in Appendix A.

▼ Example 1.7

The normal temperature of the human body is 98.6°F. Express this value in degrees Celsius and kelvins.

Since the given value is in degrees Fahrenheit, write down the equation that converts values from °F to °C.

$$T_C = \frac{5}{9}(T_F - 32°)$$

Chapter 1

Now insert the Fahrenheit value and calculate the Celsius value.

$$T_C = \frac{5}{9}(98.6° - 32°) = 37.0°C$$

Now we are able to use the Celsius value to compute the Kelvin value.

$$T_K = T_C + 273.15 = 37.0 + 273.15 = 310.2 \text{ K}$$

The reason the answer is 310.2 K instead of 310.15 K is due to the significant digits rule for addition. Again, the topic of significant digits is coming up next.

▼ Example 1.8

The melting point of aluminum is 933.5 K. Express this temperature in degrees Celsius and degrees Fahrenheit.

Write down the equation that converts Kelvin values to Celsius values.

$$T_C = T_K - 273.15$$

From this we calculate the Celsius value as

$$T_C = T_K - 273.15 = 933.5 - 273.15 = 660.3°C$$

Next, write down the equation for converting a Celsius value to a Fahrenheit value.

$$T_F = \frac{9}{5}T_C + 32°$$

Insert the Celsius value into this equation and calculate the Fahrenheit value.

$$T_F = \frac{9}{5}T_C + 32° = \frac{9}{5} \cdot 660.3°C + 32° = 1220.5°F$$

1.3   Accuracy and Precision

1.3.1 Distinguishing Between Accuracy and Precision

The terms *accuracy* and *precision* refer to the practical limitations inherent in making measurements. Science is all about investigating nature, and to do that we must make measurements.

Accuracy relates to error—that is, to the lack of it. Error is the difference between a measured value and the true value. The lower the error is in a measurement, the better the accuracy. Error arises from many different sources, including human mistakes, malfunctioning equipment, incorrectly calibrated instruments, vibrations, changes in temperature or humidity, or unknown causes that are influencing a measurement without the knowledge of the experimenter. All measurements contain error, because (alas!) perfection is simply not a thing we have access to in this world.

Precision refers to the resolution or degree of "fine-ness" in a measurement. The limit to the precision that can be obtained in a measurement is ultimately dependent on the instrument being used to make the measurement. If you want greater precision, you must use a more precise instrument. The degree of precision in every measurement is signified by the measurement value itself because the precision is a built-in part of the measurement. *The precision of a measurement is indicated by the number of significant digits (or significant figures) included in the measurement value when the measurement is written down* (see below).

Here is an example that illustrates the idea of precision and also helps distinguish between precision and accuracy. The photograph in Figure 1.8 shows a machinist's rule and an architect's scale set one above the other. Since the marks on the two scales line up consistently, *these two scales are equally accurate*. But the machinist's rule (on top) is more precise. The architect's scale is marked in 1/16-inch increments, but the machinist's rule is marked in 1/64-inch increments. Thus, *the machinist's rule is more precise*.

It is important that you are able to distinguish between accuracy and precision. Here is another example to help illustrate the difference. Let's say Shana and Marius each buy digital thermometers for their homes. The thermometer Shana buys costs $10 and measures to the nearest 1°F. Marius pays $40 and gets one that reads to the nearest 0.1°F. Shana reads the directions and properly installs the sensor for her new thermometer in the shade. Marius doesn't read the directions and mounts his sensor in the direct sunlight, which causes a significant error in the thermometer reading when the sun is shining on it; thus Marius' measurements are not accurate. The result is that Shana has lower-precision, higher-accuracy measurements!

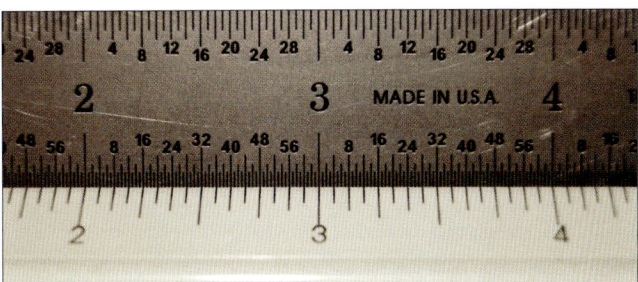

Figure 1.8. The accuracy of these two scales is the same, but the machinist's rule (above) is more precise than the architect's scale (below).

1.3.2 Significant Digits

The precision in any measurement is indicated by the number of *significant digits* it contains. Thus, the number of digits we write in any measurement we deal with in science is very important. The number of digits is meaningful because it shows the precision inherent in the instrument used to make the measurement.

Let's say you are working a computational exercise in a science book. The problem tells you that a person drives a distance of 110 miles at an average speed of 55 miles per hour and wants you to calculate how long the trip takes. The correct answer to this problem *is different* from the correct answer to a similar problem with given values of 110.0 miles and 55.0 miles per hour. And if the given values are 110.0 miles and 55.00 miles per hour, the correct answer is different yet again. Mathematically, of course, all three answers are the same. If you drive 110 miles at 55 miles per hour, the trip takes two hours. But scientifically, the correct answers to these three problems are different: 2.0 hours, 2.00 hours, and 2.000 hours, respectively. The difference between these cases is in the precision indicated by the given data, which are *measurements*. (Even though this is just a made-up problem in a book and not an actual measurement someone made in an experiment, the given data are still measurements. There is no way to talk about distances or speeds without talking about measurements, even if the measurements are only imaginary or hypothetical.)

Chapter 1

So when you perform a calculation with physical quantities (measurements), you can't simply write down all the digits shown by your calculator. The precision inherent in the measurements used in a computation governs the precision in any result you might calculate from those measurements. And since the precision in a measurement is indicated by the number of significant digits, data and calculations must be written with the correct numbers of significant digits. To do this, you need to know how to count significant digits, and you must use the correct number of significant digits in all your calculations and experimental data.

Correctly counting significant digits involves four different cases:

1. Rules for determining how many significant digits there are in a given measurement.
2. Rules for writing down the correct number of significant digits in a measurement you are making and recording.
3. Rules for computations you perform with measurements—multiplication and division.
4. Rules for computations you perform with measurements—addition and subtraction.

We address each of these cases below, in order.

Case 1   We begin with the rule for determining how many significant digits there are in a given measurement value. The rule is as follows:

- *The number of significant digits in a number is found by counting all the digits from left to right beginning with the first nonzero digit on the left. When no decimal is present, trailing zeros are not considered significant.*

Let's apply this rule to the following values to see how it works.

15,679   This value has five significant digits.

21.0005   This value has six significant digits.

37,000   This value has only two significant digits because when there is no decimal, trailing zeros are not significant. Notice that the word *significant* here is a reference to the precision of the measurement, which in this case is rounded to the nearest thousand. The zeros in this value are certainly *important*, but they are not *significant* in the context of precision.

0.0105   This value has three significant digits because we start counting digits with the first nonzero digit on the left.

0.001350   This value has four significant digits. Trailing zeros count when there is a decimal.

The significant digit rules enable us to tell the difference between two measurements such as 13.05 m and 13.0500 m. Again, these values are obviously equivalent *mathematically*. But they are different in what they tell us about the process of how the measurements were made—and science deals in measurements. The first measurement has four significant digits. The second measurement is more precise—it has six significant digits and was made with a more precise instrument.

Now, just in case you are bothered by the zeros at the end of 37,000 that are not significant, here is one more way to think about significant digits that may help. The precision in a measurement depends on the instrument used to make the measurement. If we express the measurement in different units, this cannot change the precision of the value. A measurement of 37,000 grams is equivalent to 37 kilograms, as shown in the following calculation:

$$37,000 \text{ g} \cdot \frac{1 \text{ kg}}{1000 \text{ g}} = 37 \text{ kg}$$

Whether we express this value in grams or kilograms, it still has two significant digits.

Case 2 The second case addresses the rules that apply when you are recording a measurement yourself, rather than reading a measurement someone else has made. When you make measurements yourself, as when conducting the laboratory experiments in this course, you must know the rules for which digits are significant in the reading you are making on the measurement instrument. The rule for making measurements depends on whether the instrument you are using is a digital instrument or an analog instrument. Here are the rules for these two possibilities:

- *Rule 1 for digital instruments*   For the digital instruments commonly found in introductory science labs, assume all the digits in the reading are significant except leading zeros.

- *Rule 2 for analog instruments*   The significant digits in a measurement include all the digits known with certainty, plus one digit at the end that is estimated between the finest marks on the scale of your instrument.

The first of these rules is illustrated in Figure 1.9. The reading on the left has leading zeros, which do not count as significant. Thus, the first reading has three significant digits. The second reading also has three significant digits. The third reading has five significant digits.

The fourth reading also has five significant digits because with a digital display the only zeros that don't count are the leading zeros. Trailing zeros are significant with a digital instrument. *However*, when you write this measurement down, you must write it in a way that shows those zeros to be significant. The way to do this is by using scientific notation. When a value is written in scientific notation, *the digits that are written down in front of the power of 10* (the stem, also called the mantissa) *are the significant digits.* Thus, the right-hand value in Figure 1.9 must be written as 4.2000×10^4. We address scientific notation in more detail in the next section.

Dealing with digital instruments is actually more involved than the simple rule above implies, but the issues involved go way beyond what we can deal with in introductory science classes. So, simply make your readings and assume that all the digits in the reading except leading zeros are significant.

0042.0   42.0   42.000   42,000

Figure 1.9. With digital instruments, all digits are significant except leading zeros. Thus, the numbers of significant digits in these readings are, from left to right, three, three, five, and five.

Now let's look at some examples illustrating the rule for analog instruments. Figure 1.10 shows a machinist's rule being used to measure the length in millimeters (mm) of a brass block. We know the first two digits of the length with certainty; the block is clearly between 31 mm and 32 mm long. We have to estimate the third significant digit. The scale on the rule is marked in increments of 0.5 mm. Comparing the edge of the block with these marks, I estimate the next digit to be a 6, giving a measurement of 31.6 mm. Others might estimate the last digit to be a 5 or a 7; these small differences in the last digit are unavoidable because the last digit is estimated. Whatever you estimate the last digit to be, two digits of this measurement are known with certainty, the third digit is estimated, and the measurement has three significant digits.

Chapter 1

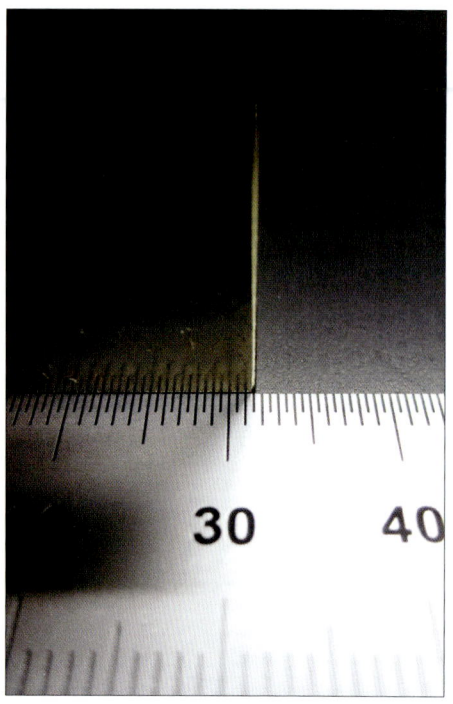

Figure 1.10. Reading the significant digits with a machinist's rule.

The photograph in Figure 1.11 shows a liquid volume measurement in milliliters (mL) being made with an article of apparatus called a *buret*. Notice in this figure that when measuring liquid volume the surface of the liquid curls up at the edge of the cylinder, forming a bowl-shaped surface on the liquid. This curved surface is called a *meniscus. For most liquids, liquid measurement readings are taken at the bottom of the meniscus.* Liquid mercury is the major exception, because the meniscus in liquid mercury is inverted—liquid mercury curves down at the edges. In that case, the measurement is read at the top of the meniscus. But that is an unusual case. For most liquids, the reading is made at the bottom of the meniscus.

For the buret in the figure, you can see that the scale is marked in increments of 0.1 milliliters (mL). This means we are to estimate to the nearest 0.01 mL. To one person, it may look like the bottom of the meniscus (where the black curve touches the bottom of the silver bowl) is just above 2.2 mL, so that person would call this measurement 2.19 mL. To someone else, it may seem that the bottom of the meniscus is right on 2.2, in which case that person would call the reading 2.20 mL. Either way, the reading has three significant digits and the last digit is estimated to be either 9 or 0.

The third example involves a liquid volume measurement with an article of apparatus called a *graduated cylinder*. The scales on small graduated cylinders like this one are marked in increments of 1 mL. In the photo of Figure 1.12, the entire meniscus appears silvery in color with a black curve at the bottom. For the liquid shown in the figure, we know the first two digits of the volume measurement with certainty because the reading at the bottom of the meniscus is clearly between 82 mL and 83 mL. We have to estimate the third digit, and I estimate the edge of the meniscus to be at 60% of the distance between 82 and 83, giving a reading of 82.6 mL. Others may prefer a different value for that third digit.

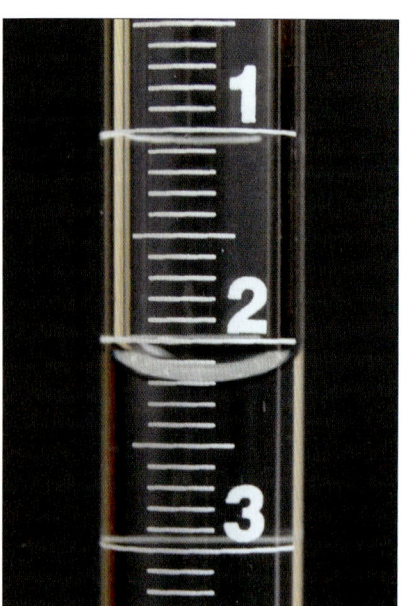

Figure 1.11. Reading the significant digits on a buret.

Figure 1.12. Reading the significant digits on a graduated cylinder.

It is important for you to keep the significant digits rules in mind when you are making measurements and entering data for your lab reports. The data in your lab journal and the values you use in your calculations and report should correctly reflect the use of the significant digits rules as they apply to the actual instruments you use to make your measurements.

Case 3   The third and fourth cases of rules for significant digits apply to the calculations you perform with measurements. In Case 3, we deal with multiplication and division. The main idea behind the rule for multiplying and dividing is that the precision you report in your result cannot be higher than the precision that is in the measurements you start with. The precision in a measurement depends on the instrument used to make the measurement, nothing else. Multiplying and dividing things cannot increase that precision, and thus your results can be no more precise than the measurements used in the calculations. In fact, your result can be no more precise than the *least precise value* used in the calculation. The least precise value is, so to speak, the "weak link" in the chain, and a chain is no stronger than its weakest link.

Here are the two rules for using significant digits in calculations involving multiplication and division:

- *Rule 1*   When multiplying or dividing, count the significant digits in each of the values you are using in the calculation, including any conversion factors involved. (However, note: Conversion factors that are exact are not considered.) Determine how many significant digits there are in the least precise of these values. The result of your calculation must have this same number of significant digits.

- *Rule 2*   When performing a multi-step calculation, keep at least one extra digit during intermediate calculations, and round off to the final number of significant digits you need at the very end. This practice ensures that small round-off errors don't accumulate during a multi-step calculation. This extra digit rule also applies to unit conversions performed as part of the computation.

I illustrate the two rules above, along with some more unit conversions, in the following example problem and calculation.

▼ Example 1.9

At a chemical research lab, a stream of a reactant solution is flowing into a reaction vessel at a rate of 56.75 µL per second. A volume of 1.0 ft^3 of this solution is required in the vessel for the reaction. Determine the amount of time needed for the required volume to be collected. State your result in hours.

First note that the value of the flow rate has four significant digits, and the required volume has two significant digits. The two-digit value is the least precise of these, so our result must be rounded to two significant digits. But to avoid rounding error, we must work with values having at least three significant digits (one more than we need) until the very end.

One of the volumes in this problem is in µL and the other is in ft^3. I begin by converting the required volume from ft^3 to µL so our volumes all have the same units. We have no conversion factor that goes directly from ft^3 to µL, so we must use a chain of conversion factors that we know or have available. Since we are dealing with relatively small volumes based on length units, the main conversion from USCS units to SI units is the inch to centimeter factor of 1 in = 2.54 cm. This factor is exact and should be committed to memory. If we were starting with gallons instead of ft^3, we might use the conversion 1 gal = 3.785 L, although this factor is not exact.

We first convert ft^3 to in^3, then from in^3 to cm^3, then from cm^3 to L, and finally from L to µL.

$$1.0 \text{ ft}^3 \cdot \frac{12 \text{ in}}{1 \text{ ft}} \cdot \frac{12 \text{ in}}{1 \text{ ft}} \cdot \frac{12 \text{ in}}{1 \text{ ft}} \cdot \frac{2.54 \text{ cm}}{1 \text{ in}} \cdot \frac{2.54 \text{ cm}}{1 \text{ in}} \cdot \frac{2.54 \text{ cm}}{1 \text{ in}} \cdot \frac{1 \text{ L}}{1000 \text{ cm}^3} \cdot \frac{1 \times 10^6 \text{ μL}}{1 \text{ L}} = 28{,}300{,}000 \text{ μL}$$

This result has three significant digits. We are keeping one extra digit during the intermediate calculations. It is not incorrect to write down all the digits your calculator shows. But it *is* pointless. A person who writes all the digits regardless of whether they are needed simply shows that he or she doesn't understand significant digits. Those extra digits are meaningless.

Notice that all the conversion factors used in the calculation above are exact; none of them are approximations. Since they are all exact, they play no role in limiting the significant digits in our result. Note that if you must use a conversion factor that is approximate, you should make sure the precision of the value in the conversion factor is at least as high as the precision in your data (two significant digits, in this case). That way your conversion factor does not limit the precision of your result. If, for some reason, you do not have a conversion factor with as many significant digits as your least precise measurement, then the precision of your result must match the precision of the conversion factor. The least precise value in the entire calculation always governs the precision in your result.

Now we compute the time required by dividing the required volume by the flow rate.

$$t = \frac{28{,}300{,}000 \text{ μL}}{56.75 \frac{\text{μL}}{\text{s}}} = 499{,}000 \text{ s}$$

This value also has three significant digits—one more than we need. We now convert this value from seconds to hours as the problem statement requires.

$$499{,}000 \text{ s} \cdot \frac{1 \text{ hr}}{3600 \text{ s}} = 139 \text{ hr}$$

Finally, we need to round the result to the required two significant digits. The second non-zero digit from the left (3) is in the tens place, so we round to the nearest ten.

$t = 140$ hr

Case 4 The fourth case of rules for significant digits also applies to the calculations you perform with measurements. In Case 4, we deal with addition and subtraction.

The rule for addition and subtraction is completely different from the rule for multiplication and division. When performing addition, it is not the number of significant digits that governs the precision of the result. Instead, it is the *place value of the last digit that is farthest to the left in the numbers being added* that governs the precision of the result. This rule is quite wordy and is best illustrated by an example. Consider the following addition problem:

```
    13.65
     1.9017
 + 1,387.069
  1,402.62
```

Of the three values being added, 13.65 has digits out to the hundredths place, 1.9017 has digits out to the ten thousandths place, and 1,387.069 has digits out to the thousandths place. Looking at the final digits of these three, you can see that the final digit farthest to the left is the 5 in 13.65, which is in the hundredths place. This is the digit that governs the final digit of the result. There can be no digits to the right of the hundredths place in the result. The justification for this rule is that one of our measurements is precise only to the nearest hundredth, even though the other two are precise to the nearest thousandth or ten thousandth. We are going to add these values together, and one of them is precise only to the nearest hundredth. It makes no sense to have a result that is precise to a place more precise than that, so hundredths are the limit of the precision in the result.

Correctly performing addition problems in science (where nearly everything is a measurement) requires that you determine the place value governing the precision of your result, perform the addition, then round the result. In the above example, the sum is 1,4602.6207. Rounding this value to the hundredths place gives 1,4602.62.

Going back to Example 1.7, we saw the following equation for converting a temperature from kelvins to degrees Celsius:

$$T_K = T_C + 273.15 = 37.0 + 273.15 = 310.2 \text{ K}$$

The two values we are adding are 37.0, which has digits out to the tenths place, and 273.15, which has digits out to the hundredths place. The final digit in 37.0 is the one farther to the left, so it governs the final digit we can have in the sum. The final digit in 37.0 is in the tenths place, so the final digit in the sum must also be in the tenths place. Adding the values gives 310.15 K. Rounding this to the tenths place gives us 310.2 K for our result.

1.4   Other Important Math Skills

1.4.1 Scientific Notation

No doubt you have studied *scientific notation* in your math classes. However, beginning in high school, scientific notation is used all the time in scientific study. Knowing how to use scientific notation correctly—including the use of the special key found on scientific calculators for working with values in scientific notation—is very important.

Mathematical Principles

Scientific notation is a way of expressing very large or very small numbers without all the zeros, unless the zeros are *significant*. This is of enormous benefit when one is dealing with a value such as 0.0000000000001 cm (the approximate diameter of an atomic nucleus). The basic idea will be clear from a few examples.

Let's say we have the value 3,750,000. This number is the same as 3.75 million, which can be written as $3.75 \times 1,000,000$. Now, 1,000,000 itself can be written as 10^6 (which means one followed by six zeros), so our original number can be expressed equivalently as 3.75×10^6. This expression is in scientific notation. The number in front, the stem, is always written with one digit followed by the decimal and the other digits. The multiplied 10 raised to a power has the effect of moving the decimal over as many places as necessary to recreate our original number.

As a second example, the current population of earth is about 7,200,000,000, or 7.2 billion. One billion has nine zeros, so it can be written as 10^9. So we can express the population of earth in scientific notation as 7.2×10^9.

Chapter 1

When dealing with extremely small numbers such as 0.000000016, the process is the same, except the power on the 10 is negative. The easiest way to think of it is to visually count how many places the decimal in the value has to be moved over to get 1.6. To get 1.6, the decimal has to be moved to the right 8 places, so we write our original value in scientific notation as 1.6×10^{-8}.

Using Scientific Notation with a Scientific Calculator

All scientific calculators have a key for entering values in scientific notation. This key is labeled $\boxed{\text{EE}}$ or $\boxed{\text{EXP}}$ on most calculators, but others use a different label.[1] It is very common for those new to scientific calculators to use this key incorrectly and obtain incorrect results. So read carefully as I outline the procedure.

The whole point of using the $\boxed{\text{EE}}$ key is to make keying in the value as quick and error free as possible. *When using the scientific notation key to enter a value, you do not press the $\boxed{\times}$ key, nor do you enter the 10.* The scientific calculator is designed to reduce all this key entry, and the potential for error, by use of the scientific notation key. You only enter the stem of the value and the power on the ten and let the calculator do the rest.

Here's how. To enter a value, simply enter the digits and decimal in the stem of the number. Then hit the $\boxed{\text{EE}}$ key, and then enter the power on the ten. The value is then in the calculator and you may do with it whatever you need to. As an example, to multiply the value 7.2×10^9 by 25 using a standard scientific calculator, the sequence of key strokes is as follows:

7.2 $\boxed{\text{EE}}$ 9 $\boxed{\times}$ 25 $\boxed{=}$

Notice that between the stem and the power, the only key pushed is the $\boxed{\text{EE}}$ key.

When entering values in scientific notation with negative powers on the 10, the $\boxed{+/-}$ key is used before the power to make the power negative. Thus, to divide 1.6×10^{-8} by 36.17, the sequence of key strokes is:

1.6 $\boxed{\text{EE}}$ $\boxed{+/-}$ 8 $\boxed{\div}$ 36.17 $\boxed{=}$

Again, neither the 10 nor the × sign that comes before it are keyed in. The $\boxed{\text{EE}}$ key has these built in.

Students sometimes wonder why it is *incorrect* to use the $\boxed{10^x}$ key for scientific notation. To calculate 7.2×10^9 times 25, they are tempted to enter the following:

7.2 $\boxed{\times}$ $\boxed{10^x}$ 9 $\boxed{\times}$ 25 $\boxed{=}$

The problem with this approach is that sometimes it works and sometimes it doesn't, and calculator users need to use key entries that *always* work. The scientific notation key ($\boxed{\text{EE}}$) keeps all the parts of a value in scientific notation together as one number. That is, when the $\boxed{\text{EE}}$ key is used, a value such as 7.2×10^9 is not two separate numbers to the calculator; it is a single numerical value. But when the $\boxed{\times}$ key is manually inserted, the calculator treats the numbers separated by the $\boxed{\times}$ key as two separate values, and this can caused the calculator to use a different order of operations than you intend. For example, using the $\boxed{10^x}$ key causes the calculator to render an incorrect answer for a calculation such as this:

$$\frac{3.0 \times 10^6}{1.5 \times 10^6}$$

---

[1] One infuriating model uses the extremely unfortunate label $\boxed{\times 10^x}$ which looks a *lot* like $\boxed{10^x}$, a different key with a completely different function.

The denominator of this expression is exactly half the numerator, so the value of this fraction is obviously 2.0. But when using the $\boxed{10^x}$ key, the 1.5 and the 10^6 in the denominator are separated and treated as separate values. The calculator then performs the following calculation:

$$\frac{3.0 \times 10^6}{1.5} \times 10^6$$

This comes out to 2,000,000,000,000 (2×10^{12}), which is not the same as 2.0!

The bottom line is that the $\boxed{\text{EE}}$ key, however it may be labeled, is the correct key to use for scientific notation.

Finally, when writing a result in scientific notation, it is not acceptable to write it using the EE notation your calculator uses. For example, your calculator might display a result as 3.14EE8 or 3.14E8, but you must write this as 3.14×10^8.

1.4.2 Calculating Percent Difference

One of the conventional calculations in science experiments is the so-called "experimental error." Experimental error is typically defined as the difference between a predicted value and an experimental value, expressed as a percentage of the predicted value, or

$$\text{experimental error} = \frac{|\text{predicted or accepted value} - \text{experimental value}|}{\text{predicted or accepted value}} \times 100\%$$

Although the term "experimental error" is widely used, it is a poor choice of words. When there is a mismatch between theory and experiment, the experiment may not be the source of the error. Often, it is the theory that is found wanting—this is how science advances.

I now prefer to use the phrase *percent difference* to describe the value computed by the above equation. When quantitative results are compared to quantitative predictions or accepted values, students should compute the percent difference as

$$\text{percent difference} = \frac{|\text{predicted or accepted value} - \text{experimental value}|}{\text{predicted or accepted value}} \times 100\%$$

Chapter 1 Exercises

For all exercises, note that physical constants and unit conversion factors are found in Tables A.2 and A.3 of Appendix A.

SECTION 1.1
1. Write a paragraph distinguishing between matter and mass.
2. Distinguish between base units and derived units in the SI system of units and give three examples of each.
3. Describe the advantages the SI system has over the USCS system for scientific work.
4. Why does the SI system use prefixes on the units of measure?

Chapter 1

5. Re-express the quantities in the following table using only a single numerical digit followed by an SI unit symbol, with a metric prefix if necessary. Example: 5 thousand liters = 5 kL

   a. 8 pascals
   b. 5 hundredths of a meter
   c. 3 million amperes
   d. 2 thousand meters
   e. 4 thousandths of a second
   f. 6 thousand newtons
   g. 8 thousand grams
   h. 7 millionths of a liter
   i. 1 thousandth of a joule

6. Re-write the quantities in the following table by writing out the unit names without symbols. Example: 5 km = 5 kilometers

   a. 14 m^3
   b. 164.1 kg
   c. 250 MPa
   d. 16.533 ms
   e. 160 kA
   f. 19.55 cL
   g. 31.11 µJ
   h. 2300 K
   i. 13.0 mmol

SECTION 1.2

7. Why must equations be used instead of conversion factors for most temperature unit conversions?

8. Perform the USCS unit conversions required in the following table. (Note: The answers in the back of the book are given with the correct number of significant digits.)

Convert This Quantity	Into These Units	Convert This Quantity	Into These Units
a. 12.55 ft	yd	b. 0.44556 mi	ft
c. 147.55 in	ft	d. 55.08 gal	ft^3
e. 934 ft^3	in^3	f. 739.22 ft^3/s	gal/hr
g. 12.4 yr	hr	h. 51,083 in	mi
i. 14,560.77 gal/hr	qt/s	j. 15.90 mi/dy	in/hr

9. Perform the SI/metric unit conversions required in the following table.

Convert This Quantity	Into These Units	Convert This Quantity	Into These Units
a. 35.4 mm	m	b. 76.991 mL	µL
c. 34.44 cm^3	L	d. 6.33 g/cm^2	kg/m^2
e. 9.35 m/s^2	mm/ms^2	f. 542.2 mJ/s	J/s
g. 56.6 µs	ms	h. 44.19 mL	cm^3
i. 532 nm	µm	j. 96,963,000 mL/ms	m^3/s
k. 295.6 cL	µL	l. 0.007873 m^3	mL
m. 8,750 mm^2	m^2	n. 87.1 cm/s^2	m/s^2
o. 15.75 kg/m^3	g/cm^3	p. 0.875 km	m
q. 16,056 MPa	kPa	r. 7,845 µA	mA

Measurements

10. Reproduce the following table on your own paper and fill in the empty cells.

	°F	°C	K		°F	°C	K
a.	431.1			b.		−56.1	
c.			16.0	d.	0.0 (exact)		
e.		−77.0		f.			4,002
g.	−32.0			h.		65.25	
i.		1,958		j.			998.0

SECTION 1.3

11. Distinguish between accuracy and precision.
12. Describe the measurements you would obtain from an instrument that was very precise but not very accurate.
13. Which is more important on the speedometer of a car—accuracy or precision?
14. Explain why accuracy is important on a heart rate monitor but precision is not.
15. Sometimes we want high accuracy in a measurement, but are not too concerned about high precision. Sometimes we want both high accuracy and high precision. Explain why no one wants low accuracy and high precision.
16. On the package of a digital stopwatch I once purchased was the phrase: "1/100th second accuracy." The stopwatch readings in seconds contained two decimal places, but the values the stopwatch actually displayed were spaced 0.03 seconds apart. Thus, it could read 12.31 s, 12.34 s, 12.37 s, etc. Comment on the accuracy and precision of this stopwatch with respect to the claim on the package.
17. Using the correct number of significant digits and the correct units of measure, record the measurements represented by the following instruments.

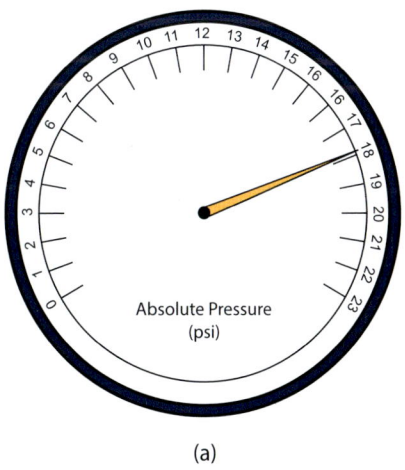

(a)

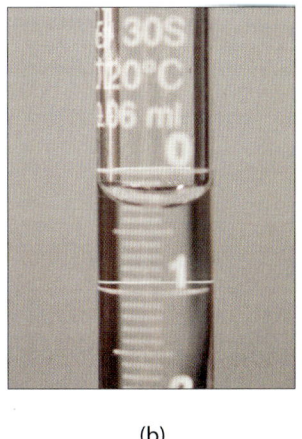

(b)

(c) gallons

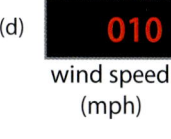

(d) wind speed (mph)

SECTION 1.4

18. Using the correct number of significant digits, compute the percent difference for the experimental results in each of the following cases:

39

a. A scientist measures the masses of three compounds resulting from a certain chemical reaction. Her measurements are 0.234 g, 1.678 g, and 4.446 g. Her calculations predict that the reaction results in masses of 0.239 g, 1.688 g, and 4.678 g, respectively. Determine the percent difference for each of the three compounds.

b. A student measures the density of aluminum and finds it to be 2.81 g/cm^3. The accepted density value for this alloy is 2.72 g/cm^3.

c. According to the Periodic Table of the Elements, the atomic mass of carbon is 12.011 g/mol. A calculation from experimental data results in a figure of 12.0117 g/mol.

d. The predicted yields for the products of certain chemical reaction are 23.4 kg of compound A and 2.21 kg of compound B. Careful measurements of the masses of the compounds produced indicate masses of 21.610 kg for compound A and 1.995 kg of compound B.

19. Perform each of the unit conversions indicated in the table below. Express each result using the correct number of significant digits. Where possible and appropriate, express your result in both standard notation and scientific notation. (Note: By possible, I refer to the fact that sometimes a result can only be expressed with the correct number of significant digits if it is written in scientific notation, such as a value of 100 with two or three significant digits. By appropriate, I refer to the fact that it is silly to write a value with a very large number of zeros. Such values should always be expressed in scientific notation. It is also silly to use scientific notation to express a value such as 3 or 4.1. Such values should only be expressed in standard notation.)

Convert This Quantity	Into These Units
a. 1,737 km (radius of the earth's moon)	ft
b. 2.20 g (mass of a single peanut m&m)	kg
c. 591 mL (volume of a typical water bottle)	µL
d. 7 × 10^8 m (radius of the sun)	mi
e. 1.616 × 10^{-35} m (Planck length, a fundamentally small length)	ft
f. 750 cm^3 (size of the engine in my old motorcycle)	m^3
g. 2.9979 × 10^8 m/s (speed of light in a vacuum)	mi/hr
h. 168 hr (one week)	s
i. 5,570 kg/m^3 (average density of the earth)	g/cm^3
j. 45 gps (gal/sec, flow rate of Mississippi River at the source)	m^3/min
k. 600,000 ft^3/s (flow rate of Mississippi River at New Orleans)	L/hr
l. 5,200 mL (volume of blood in a typical man's body)	m^3
m. 5.65 × 10^2 mm^2 (area of a postage stamp)	in^2
n. 32.16 ft/s^2 (acceleration of gravity, or one "g")	m/s^2
o. 10.6 µm (wavelength of light from a CO$_2$ laser)	in
p. 1.1056 g/mL (density of heavy water)	kg/m^3

Convert This Quantity	Into These Units
q. 13.6 g/cm^3 (density of liquid mercury metal)	mg/m^3
r. 93,000,000 mi (distance from earth to the sun)	cm
s. 65 mph (typical highway speed limit)	m/s
t. 633 nm (wavelength of light from a red laser)	in
u. 5.015% of the speed of light (see item g, or Table A.2)	mph
v. 6.01 kJ/mol (molar heat of fusion of water)	J/mol
w. 32.1 bar (pressure in saltwater at 318 m, free diving record depth)	psi
x. 0.116 nm (radius of a sodium atom)	cm
y. 6.54 × 10^{-24} cm^3 (volume of a sodium atom)	in^3
z. 0.385 J/(g·K) (specific heat capacity of copper)	J/(mg·K)
aa. 370 mL (volume of a soft drink can)	ft^3
ab. 268,581 mi^2 (land area of Texas)	mm^2
ac. 50,200 mi^2/yr (current rate of global deforestation)[2]	ft^2/s

---

[2] As a science educator who believes that we should take care of our planet, I note that items (ab) and (ac) above indicate that every 5.4 years, we lose an area of forest the size of Texas. Texas is a big place. Think about it.

Chapter 2
Atoms and Substances

Some substances pose a minimal threat to human health and safety. Others need to be labeled and classified so we know about the risks they pose. If you examine the packaging on a shipment of chemicals or look on the back of a tank truck on the highway, you may find the Fire Diamond, a symbolic representation of the various hazards associated with the substance inside. The numerical values in each of the three colored zones range from 0 to 4, with 4 representing the most extreme hazard. The blue region pertains to health, the red to flammability, the yellow to instability or reactivity, and the white to special notices. In the sample symbol above, the numerical codes specify the following:

Blue: Health 3—Short exposure could cause serious temporary or moderate residual injury. (Example: chlorine)

Red: Flammability 2—Must be moderately heated before ignition can occur (Example: diesel fuel)

Yellow: Instability/reactivity 1—Normally stable, but can become unstable at elevated temperature and pressure (Example: alcohol)

White: Special notices "W bar"—Reacts with water in an unusual or dangerous manner (Example: sodium)

Objectives for Chapter 2

After studying this chapter and completing the exercises, you should be able to do each of the following tasks, using supporting terms and principles as necessary.

SECTION 2.1
1. Define and describe *atom* and *molecule*.
2. State the five points of John Dalton's atomic model.
3. Write brief descriptions of J.J. Thomson's cathode ray tube experiment, Robert Millikan's oil drop experiment, and Ernest Rutherford's gold foil experiment. In your descriptions, include definitions for the terms *cathode ray* and *alpha particle*.
4. Describe the atomic models proposed by J.J. Thomson and Ernest Rutherford.

SECTION 2.2
5. Define *pure substance, element, compound, mixture, heterogeneous mixture,* and *homogeneous mixture* and give several examples of each.
6. Define *suspension* and *colloidal dispersion* and give several examples of each.
7. Explain the Tyndall effect and how it can be used to identify a mixture as colloidal.
8. Describe the two basic types of structures atoms form when bonding together.
9. Distinguish between compounds and mixtures.
10. Define and distinguish between *physical properties, chemical properties, physical changes,* and *chemical changes*. Give several examples of each.
11. Define and give examples of the terms *malleable* and *ductile*.

SECTION 2.3
12. Define *isotope, nuclide,* and *atomic mass*.
13. Given isotope mass and abundance data, calculate the atomic mass of an element.
14. Given the periodic table, determine the number of protons, electrons, and neutrons in the atoms of a given nuclide.
15. Define the *unified atomic mass unit*, u.

SECTION 2.4
16. Use the density equation to calculate the density, volume, or mass of a substance.
17. Define the *mole*.
18. State the Avogadro constant to four digits of precision.
19. Define *molar mass, formula mass,* and *molecular mass*.
20. Calculate the molar mass, formula mass, or molecular mass of a compound or molecule.
21. Calculate the mass in grams of a given mole quantity of a compound or molecule, or vice versa.
22. Calculate the number of atoms or molecules in a given quantity of substance.
23. Calculate the gram masses of an atom or molecule of a given pure substance.

Chapter 2

2.1 Atoms and Molecules

2.1.1 Atomic Facts

We begin this chapter with a summary of the basic facts about atoms and molecules. Much of this information you probably already know.

As we saw in the previous chapter, all matter is made of atoms, the smallest basic units matter is composed of. An atom of a given element is the smallest unit of matter that possesses all the properties of that element.

Atoms are almost entirely empty space. Each atom has an incredibly tiny nucleus in the center containing all the atom's protons and neutrons. Since the protons and the neutrons are in the nucleus, they are collectively called *nucleons*. The masses of protons and neutrons are very nearly the same, although the neutron mass is slightly greater. Each proton and neutron has nearly 2,000 times the mass of an electron, so the nucleus of an atom contains practically all the atom's mass. Outside the nucleus is a weird sort of cloud surrounding the nucleus containing the atom's electrons.

We address the details about electrons in the next chapter, but here is a brief preview. The electron cloud consists of different *orbitals* where the electrons are contained. Electrons are sorted into the atomic orbitals according to the amount of energy they have. For an electron to be in a specific orbital means the electron has a certain amount of energy—no more, no less.

I wrote above that atoms are almost entirely empty space because the nucleus is incredibly small compared to the overall size of the atom with its electron cloud. It's quite easy for us to pass over that remark without pausing to consider what it means. To help visualize the meaning, consider the athletic stadium pictured in Figure 2.1. Using this stadium as an enlarged atomic model, the electrons in their orbitals would be zipping around in the region where the red seating sections are in the stadium. Each electron in this enormous atomic model is far smaller than the period at the end of this sentence. The atomic nucleus containing the protons and neutrons is located at the center of the playing field, and is the size of a pinhead. And what fills all the vast space inside the atom? Nothing, not even air, since air, of course, is also made of atoms. The inside of an atom is empty space.

Figure 2.1. The head of a pin at center field in a stadium is analogous to the nucleus in the center of an atom.

Returning to our discussion of atomic facts, one of the fundamental physical properties of the subatomic particles is *electric charge*. Neutrons have no electric charge. They are electrically neutral, hence their name. Protons and electrons each contain exactly the same amount of charge, but the charge on protons is positive and the charge on electrons is negative. If an atom or molecule has no net electric charge, it contains equal numbers of protons and electrons.

Atoms are significantly smaller than the wavelengths of light, which means light does not reflect off atoms and there is no way to see them. The same is true of *molecules*. Molecules are clusters of atoms chemically bonded together. When atoms of different elements are bonded together in a molecule they form a compound, which we discuss later in this chapter. But sometimes atoms of the same element bond together in molecules, as illustrated in Figure 2.2. Oxygen and chlorine are two of the *diatomic gases* that form molecules consisting of a pair of atoms chemically bound together. Hydrogen, nitrogen, and fluorine also exist naturally as diatomic gases.

2.1.2 The History of Atomic Models

The story of atomic theory starts back with the ancient Greeks. As we look at how the contemporary model of the atom developed, we also hit on some of the great milestones in the history of chemistry and physics along the way.

In the 5th century BC, the Greek philosopher Democritus proposed that everything was made of tiny, indivisible particles. Our word atom comes from the Greek word *atomos*, meaning "indivisible." Democritus' idea was that the properties of substances were due to characteristics of the atoms they are made from. So atoms of metals were supposedly hard and strong, atoms of water were assumed to be wet and slippery, and so on. At this same time, there were various views about what the most basic substances—that is, the elements—were. One of the most common views was that there are four elements—earth, air, water, and fire—and that everything is composed of these.

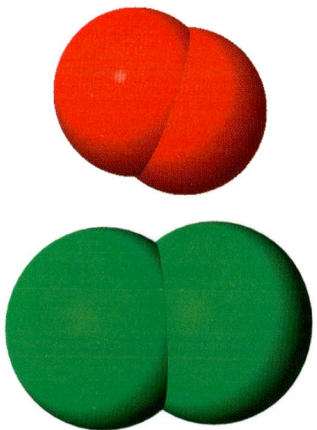

Figure 2.2. Space-filling models of the diatomic oxygen (top) and chlorine molecules.

Not much real chemistry went on for a very long time. During the medieval period, of course, there were the alchemists, who sought to transform lead and other materials into gold. But this cannot be done by the methods available to them, so their efforts were not successful.

But in the 17th century, things started changing as scientists became interested in experimental research. The goal of the scientists described here was to figure out what the fundamental constituents of matter are. This meant figuring out how atoms are put together, what the basic elements are, and understanding what is going on when various chemical reactions take place. The nature of earth, air, fire, and water was under intense scrutiny over the next 200 years.

In 1803, English scientist John Dalton (Figure 2.3) produced the first scientific model of the atom. Dalton's atomic model is based on five main points, listed in Table 2.1.

Figure 2.3. English scientist John Dalton (1766–1844).

The impressive thing about Dalton's atomic theory is that even today the last three of these points are regarded as correct, and the first two are at least partially correct. On the first point, it is still scientifically factual that all substances are made of atoms, but we now know that atoms are not indivisible. This is now obvious, since atoms themselves are composed of protons, neutrons, and electrons. The second point is correct in every respect but one. Except for the number of neutrons in the nucleus, every atom of a given element is identical. However, we now know that

1. All substances are composed of tiny, indivisible substances called atoms.
2. All atoms of the same substance are identical.
3. Atoms of different elements have different weights.
4. Atoms combine in whole-number ratios to form compounds.
5. Atoms are neither created nor destroyed in chemical reactions.

Table 2.1. The five tenets of Dalton's 1803 atomic model.

Chapter 2

Figure 2.4. English scientist Joseph John (J.J.) Thomson (1856–1940).

atoms of the same element can vary in the number of neutrons they have in the nucleus. These varieties of nuclei are called *isotopes*, a topic we return to soon.

After Dalton, the next breakthrough in our understanding of atomic structure came from English scientist J.J. Thomson (Figure 2.4). Thomson worked at the Cavendish Laboratory in Cambridge, England. In 1897, he conducted a series of landmark experiments that revealed the existence of electrons. Because of his work, he won the Nobel Prize in Physics in 1906. A photograph of the *cathode ray tube* Thomson used for his work is shown in Figure 2.5.

Thomson placed electrodes from a high-voltage electrical source inside a very elegantly made, sealed-glass vacuum tube. This apparatus can generate a so-called *cathode ray* from the negative electrode (1), called the *cathode*, to the positive one, called the *anode* (2). A cathode ray is simply a beam of electrons, but this was not known at the time. The anode inside Thomson's vacuum tube had a hole in it for some of the electrons to escape through, which created a beam of cathode rays heading toward the other end of the tube (5).

Thomson placed the electrodes of another voltage source in-

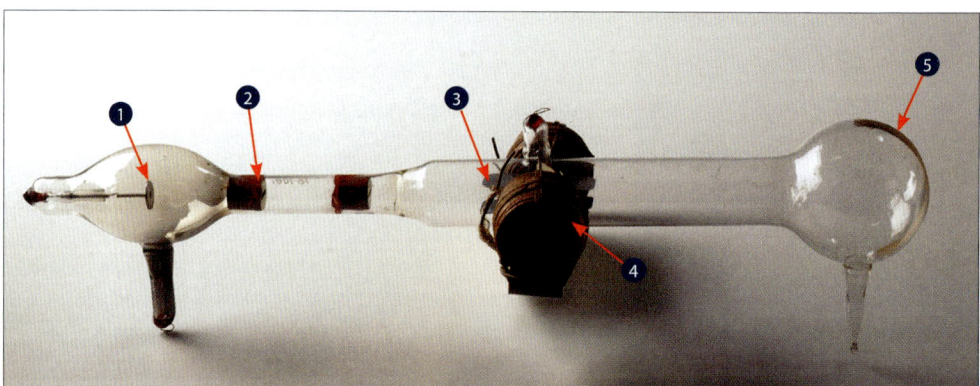

Figure 2.5. J.J. Thomson's cathode ray tube.

side the tube (3), above and below the cathode ray, and discovered that the beam of electrons deflected when this voltage was turned on. He also placed magnetic coils on the sides of the tube (4) and discovered that the electrons also deflected as they passed through the magnetic field produced by the coils. The deflection of the beam toward the positive electrode led Thomson to theorize that the beam was composed of negatively charged particles, which he called "corpuscles." (The name *electron* was first used a few years later by a different scientist.) By trying out many different arrangements of cathode ray tubes, Thomson confirmed that the ray was negatively charged. Then using the scale

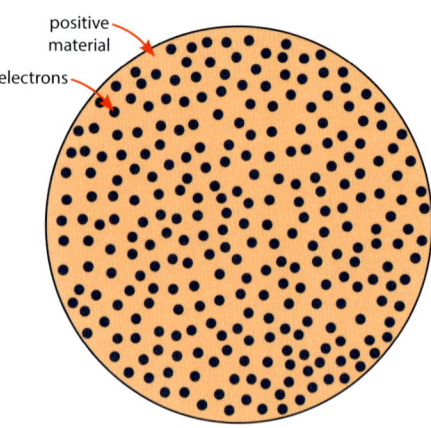

Figure 2.6. Thomson's plum pudding model.

46

Atoms and Substances

Figure 2.7. American scientist Robert Millikan (1868–1953).

on the end of the tube to measure the deflection angle (5), he was able to determine the charge-to-mass ratio of the individual electrons he had discovered. This value is 1.8×10^{11} C/kg, where C stands for coulomb, the SI unit of electric charge.

Thomson went on to theorize that electrons came from inside atoms. He developed a new atomic model that envisions atoms as tiny clouds of massless, positive charge sprinkled with thousands of the negatively charged electrons, as illustrated in Figure 2.6. Thomson's model is usually called the *plum pudding model*.

In 1911, American scientist Robert Millikan (Figure 2.7) devised his famous *oil drop experiment*, an extraordinary procedure that allowed him to determine the charge on individual electrons, 1.6×10^{-19} C. Once this value was known, Millikan used Thomson's charge-to-mass ratio and calculated the mass of the electron, 9.1×10^{-31} kg. Millikan's apparatus is pictured in Figure 2.8, and his sketch of the system is shown in Figure 2.9.

Inside a heavy metal drum (1) about the size of a 5-gallon bucket, Millikan placed a pair of horizontal metal plates (2) connected to an adjustable high-voltage source. The upper plate had a hole in the center and was connected to the positive voltage, the lower plate to the negative. He used an atomizer spray pump (3) to spray in a fine mist of watchmaker's oil above the positive plate. Some of the oil droplets would fall through the hole in the upper plate and move into the region between the plates. Connected through the side of the drum between the two plates was a telescope eyepiece (4) and lamp (5) so that Millikan could see the oil droplets between the plates.

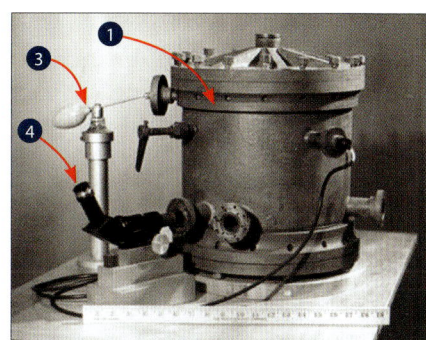

Figure 2.8. Apparatus for the oil drop experiment.

The process of squirting in the oil droplets with the atomizer sprayer caused some of the droplets to acquire a charge of static electricity. This means the droplets had excess electrons on them and carried a net negative charge. They picked up these extra electrons by friction as the droplets squirted through the rubber sprayer tube. As Millikan looked at an oil droplet through the eye-

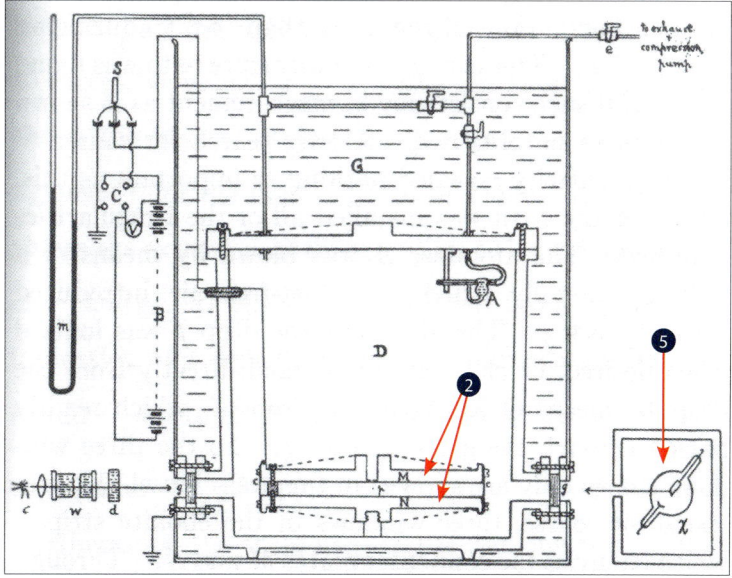

Figure 2.9. Millikan's schematic of his oil drop apparatus.

Figure 2.10. New Zealander and physicist Ernest Rutherford (1871–1937).

piece and adjusted the voltage between the plates, he could make the charged oil droplet hover when the voltage was just right. Millikan took into account the weight of the droplets and the viscosity of the air as the droplets fell and was able to determine that every droplet had a charge on it that was a multiple of 1.6×10^{-19} C. From this he deduced that this must be the charge on a single electron, which it is. Millikan won the Nobel Prize in Physics in 1923 for this work.

The last famous experiment in this basic history of atomic models was initiated in 1909 by one of Thomson's students at Cambridge, New Zealander Lord Ernest Rutherford (Figure 2.10). Rutherford was already famous when this experiment occurred, having just won the Nobel Prize in Chemistry the previous year. Rutherford's *gold foil experiment* resulted in the discovery of the atomic nucleus. To understand this experiment you need to know that an alpha particle, or α-particle (using the Greek letter alpha, α) is a particle composed of two protons and two neutrons. Alpha particles are naturally emitted by some radioactive materials in a process called *nuclear decay*.

Rutherford created a beam of α-particles by placing some radioactive material (radium bromide) inside a lead box with a hole in one end. The α-particles from the decaying radium atoms streamed out the hole at very high speed (15,000,000 m/s!). Rutherford aimed the α-particles at an extremely thin sheet of gold foil only a few hundred atoms thick. Surrounding the gold foil was a ring-shaped screen coated with a material that glows when hit by α-particles. Rutherford could then determine where the α-particles went after encountering the gold foil.

When Rutherford began taking data, he had Thomson's plum pudding model in mind and was expecting results consistent with that atomic model. This scenario is depicted in the upper part of Figure 2.11. The atom's positive charge, shown in the light orange color, is spread throughout the atom and the negative electrons are embedded in the positive material. Rutherford expected the massive and positively charged α-particles to blow right through the gold foil.

What Rutherford found was astonishing. Most of the α-particles passed straight through the foil and struck the screen on the other side, just as Rutherford expected. However, occasionally an α-particle (one particle out of every several thousand) deflected with a small angle. And sometimes the deflected particles bounced almost straight back. This situation is depicted in the lower part of Figure 2.11.

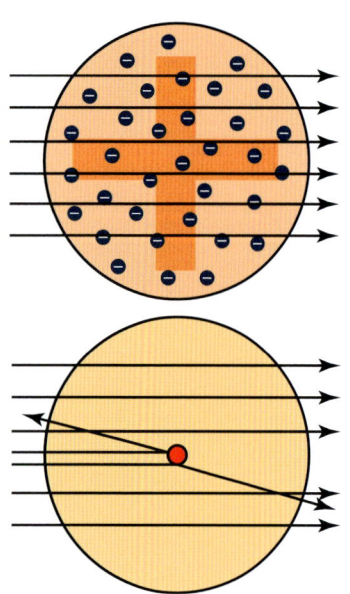

Figure 2.11. Alpha-particle pathways through the gold atoms expected from Thomson's plum pudding model (top) and the pathways Rutherford observed (bottom).

The astonished Rutherford commented that it was like firing a huge artillery shell into a piece of tissue paper and having it bounce back and hit you! Rutherford's work led to his new proposal in 1911 for a model of the atom. Rutherford's model included the key points listed in Table 2.2.

In 1917, Rutherford became the first to "split the atom." In this experiment he used α-particles again, this time strik-

Atoms and Substances

ing nitrogen atoms. His work led to the discovery of the positively-charged particles in the atomic nucleus, which he named protons.

It took another twenty years before James Chadwick (Figure 2.12), another Englishman, discovered the neutron. Before World War I, Chadwick studied under Rutherford (at Cambridge, of course). Then the war began. Not only did the war interrupt the progress of the research in general, but Chadwick was a prisoner of war in Germany. Working back in England after the war, he discovered the neutron in 1932 and received the Nobel Prize in Physics for his discovery in 1935.

Chadwick's discovery of the neutron enabled physicists to fill in a lot of blanks in their understanding of the basic structure of atoms. But years before Chadwick made his discovery, Rutherford's atomic model was already being taken to another level through the work of Niels Bohr. We explore Bohr's atomic model, and the quantum model to which it led, in the next chapter.

1. The positive charge in atoms is concentrated in a tiny region in the center of the atom, which Rutherford called the *nucleus*.
2. Atoms are mostly empty space.
3. The electrons, which contain the atoms' negative charge, are outside the nucleus.

Table 2.2. The main ideas in Ernest Rutherford's 1911 atomic model.

Figure 2.12. English physicist James Chadwick (1891–1974).

2.2 Types of Substances

A *substance* is anything that contains matter. There are several major classifications of substances, but as shown in Figure 2.13, they all fall into two major categories, *pure substances* and *mixtures*.

2.2.1 Pure Substances: Elements and Compounds

There are two kinds of pure substances, *elements* and *compounds*. We will discuss elements first. In previous science classes you may have seen or studied the Periodic Table of the Elements, which lists all the known elements. This famous table plays a major role in the study of chemistry and is shown in Figure 2.14 on the next page. We dive into the periodic table in detail in Chapter 4, but I bring up the table here to assist in our discussion of elements.

Elements   The characteristic that defines each element in the periodic table is the number of protons the element has in each of its atoms, a number called the *atomic number*. The elements are ordered in the periodic table by atomic number. For example, carbon is element number six in the periodic table. This means that an atom of carbon has six

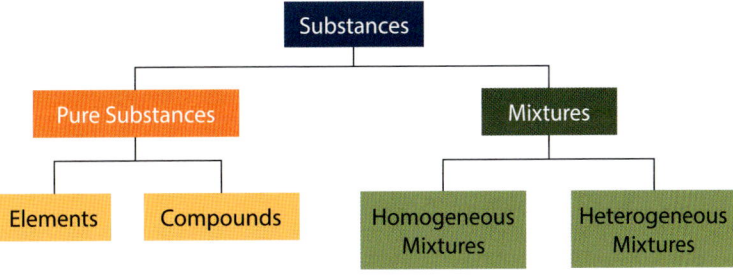

Figure 2.13. Classifications of substances.

49

Figure 2.14. The Periodic Table of the Elements.

	1 1A	2 2A	3 3B	4 4B	5 5B	6 6B	7 7B	8	9 8B	10	11 1B	12 2B	13 3A	14 4A	15 5A	16 6A	17 7A	18 8A
1	1 **H** Hydrogen 1.0079																	2 **He** Helium 4.0026
2	3 **Li** Lithium 6.941	4 **Be** Beryllium 9.0122											5 **B** Boron 10.811	6 **C** Carbon 12.011	7 **N** Nitrogen 14.0067	8 **O** Oxygen 15.9994	9 **F** Fluorine 18.9984	10 **Ne** Neon 20.1797
3	11 **Na** Sodium 22.9898	12 **Mg** Magnesium 24.3050											13 **Al** Aluminum 26.9815	14 **Si** Silicon 28.0855	15 **P** Phosphorus 30.9738	16 **S** Sulfur 32.066	17 **Cl** Chlorine 35.4527	18 **Ar** Argon 39.948
4	19 **K** Potassium 39.098	20 **Ca** Calcium 40.078	21 **Sc** Scandium 44.9559	22 **Ti** Titanium 47.88	23 **V** Vanadium 50.9415	24 **Cr** Chromium 51.9961	25 **Mn** Manganese 54.9380	26 **Fe** Iron 55.847	27 **Co** Cobalt 58.9332	28 **Ni** Nickel 58.6934	29 **Cu** Copper 63.546	30 **Zn** Zinc 65.39	31 **Ga** Gallium 69.723	32 **Ge** Germanium 72.61	33 **As** Arsenic 74.9216	34 **Se** Selenium 78.96	35 **Br** Bromine 79.904	36 **Kr** Krypton 83.80
5	37 **Rb** Rubidium 85.468	38 **Sr** Strontium 87.62	39 **Y** Yttrium 88.9059	40 **Zr** Zirconium 91.224	41 **Nb** Niobium 92.9064	42 **Mo** Molybdenum 95.94	43 **Tc** Technetium 98.9072	44 **Ru** Ruthenium 101.07	45 **Rh** Rhodium 102.9055	46 **Pd** Palladium 106.42	47 **Ag** Silver 107.8682	48 **Cd** Cadmium 112.411	49 **In** Indium 114.82	50 **Sn** Tin 118.710	51 **Sb** Antimony 121.76	52 **Te** Tellurium 127.60	53 **I** Iodine 126.9045	54 **Xe** Xenon 131.29
6	55 **Cs** Cesium 132.905	56 **Ba** Barium 137.327	71 **Lu** Lutetium 174.967	72 **Hf** Hafnium 178.49	73 **Ta** Tantalum 180.9479	74 **W** Tungsten 183.85	75 **Re** Rhenium 186.207	76 **Os** Osmium 190.2	77 **Ir** Iridium 192.22	78 **Pt** Platinum 195.08	79 **Au** Gold 196.9665	80 **Hg** Mercury 200.59	81 **Tl** Thallium 204.3833	82 **Pb** Lead 207.2	83 **Bi** Bismuth 208.9804	84 **Po** Polonium 208.9824	85 **At** Astatine 209.9871	86 **Rn** Radon 222.0176
7	87 **Fr** Francium 223.0197	88 **Ra** Radium 226.0254	103 **Lr** Lawrencium 262.11	104 **Rf** Rutherfordium 261.11	105 **Db** Dubnium 262.114	106 **Sg** Seaborgium 263.118	107 **Bh** Bohrium 262.12	108 **Hs** Hassium (265)	109 **Mt** Meitnerium (266)	110 **Ds** Darmstadtium (281)	111 **Rg** Roentgenium (281)	112 **Cn** Copernicium (285)	113 **Nh** Nihonium (284)	114 **Fl** Flerovium (289)	115 **Mc** Moscovium (288)	116 **Lv** Livermorium (293)	117 **Ts** Tennessine (294)	118 **Og** Oganesson (294)

57 **La** Lanthanum 138.9055	58 **Ce** Cerium 140.115	59 **Pr** Praseodymium 140.9077	60 **Nd** Neodymium 144.24	61 **Pm** Promethium 144.9127	62 **Sm** Samarium 150.36	63 **Eu** Europium 151.965	64 **Gd** Gadolinium 157.25	65 **Tb** Terbium 158.9253	66 **Dy** Dysprosium 162.50	67 **Ho** Holmium 164.9303	68 **Er** Erbium 167.26	69 **Tm** Thulium 168.9342	70 **Yb** Ytterbium 173.04
89 **Ac** Actinium 227.0278	90 **Th** Thorium 232.0381	91 **Pa** Protactinium 231.0359	92 **U** Uranium 238.0289	93 **Np** Neptunium 237.0482	94 **Pu** Plutonium 244.0642	95 **Am** Americium 243.0614	96 **Cm** Curium 247.0703	97 **Bk** Berkelium 247.0703	98 **Cf** Californium 251.0796	99 **Es** Einsteinium 252.083	100 **Fm** Fermium 257.0951	101 **Md** Mendelevium 258.10	102 **No** Nobelium 259.1009

■ liquid at room temperature

■ radioactive

Note: The upper set of group numbers has been recommended by the International Union of Pure and Applied Chemistry (IUPAC) and is now in wide use. The lower set of numbers is still in common use in America.

Atoms and Substances

protons—*all* carbon atoms have six protons. If an atom does not have six protons, it is not a carbon atom, and if an atom does have six protons, it is a carbon atom. An element is therefore a type of atom, classified according to the number of protons the atom has. A lump of elemental carbon—which could be graphite, diamond, coal, or several other varieties of pure carbon—is any lump of atoms that contain only six protons apiece. Oxygen (element 8) is another example of an element. Pure oxygen is a gas (ordinarily) that contains only atoms with eight protons each, because oxygen is element number eight. Other examples of elements you have heard of are iron, gold, silver, neon, copper, nitrogen, lead, and many others.

For every element, there is a *chemical symbol* that is used in the periodic table and in the chemical formulas for compounds. For a few elements, a single upper-case letter is used, such as N (7) for nitrogen and C for carbon. But for most elements, an upper-case letter is followed by one lower-case letter, such as Na (11) for sodium and Mg (12) for magnesium. The three-letter symbols for elements 113, 115, 117, and 118 are placeholders until official names and two-letter symbols are selected by the appropriate governing officials.

Some of the chemical symbols are based on the Latin names of elements, such as Ag for silver, from its Latin name *argentum*. Other examples are Au for gold, from the Latin *aurum*, and Pb for lead, from the Latin *plumbum* (everyone's favorite Latin name).

The most common representations of the periodic table show four pieces of information for each element, indicated in Figure 2.15. At the top of the cell is the atomic number, symbolically represented as Z. Again, this number indicates the number of protons in each atom of the element and is the number used to order the elements in the periodic table. Below the atomic number are the element's chemical symbol and name. At the bottom of the cell is the atomic mass. We look at the atomic mass more closely in Section 2.3.

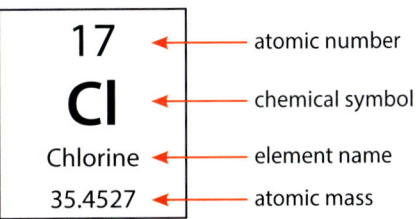

Figure 2.15. The basic information in each cell of the periodic table.

Compounds   As the name implies, a compound is formed when two or more different elements are chemically bonded together. This bonding is always the result of a chemical reaction. A chemical reaction is any process in which connecting bonds between atoms are formed or broken. Once bonded together chemically, the elements in a compound can only be separated by chemical means. In other words, it takes a different chemical reaction to break atoms apart.

The elements or compounds that go into a chemical reaction are called the *reactants*. The compounds formed by the reaction are called the *products*. The physical and chemical properties of a compound are completely different from the properties of any of the elements in the compound. For example, consider oxygen, hydrogen, and water. Hydrogen and oxygen react with a boom (Figure I.8) to form water, according to this chemical equation:

$$2H_2 + O_2 \rightarrow 2H_2O$$

Oxygen is an invisible gas that we breathe in the air and that supports combustion. Hydrogen is an invisible, flammable gas. Water is composed of oxygen atoms bonded to hydrogen atoms, but one cannot breathe water, nor does water combust or support combustion. Or consider the sodium and chlorine in sodium chloride. These elements combine according to this chemical equation:

$$Na + Cl \rightarrow NaCl$$

Chapter 2

We all require salt in our diets, and we find it tasty. But both sodium and chlorine, the two elements of which sodium chloride is composed, are deadly dangerous in their pure, elemental forms. Sodium is a shiny, peach-colored metal, pictured in Figure 2.16. Sodium slices just like cheddar cheese and the photo shows me slicing a chunk of sodium open to show its beautiful, pinkish shiny color. Chlorine is a greenish gas, pictured in a glass bottle in Figure 2.17. As you can see, the properties of these substances are nothing at all like the properties of the salt they form when they react together. We discuss physical and chemical properties in more detail a bit later.

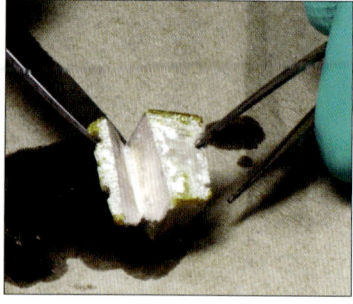

Figure 2.16. Sodium metal.

Figure 2.17. Chlorine gas.

When atoms bond together to form a compound, the atoms in the compound can be arranged in either of two basic types of structures. In many cases, the atoms join together to form molecules. In every molecule of a given substance, the atoms bond together in the same whole-number ratio—a perfect example of the important general principle that we first encountered in the Introduction. In the Introduction are shown computer models of water (H_2O), ammonia (NH_3), and methane (CH_4) molecules. A few other well-known molecular substances are carbon dioxide (CO_2), propane (C_3H_8), and ozone (O_3), all represented by space-filling models in Figure 2.18. The standard color coding used in computer models is white for hydrogen, black or charcoal gray for carbon, and red for oxygen. It is important to note that in any chemical formula for a molecular substance, the formula indicates the number of each type of atom in the molecule. Propane, C_3H_8, has three carbon atoms and eight hydrogen atoms in each molecule. The subscripts are only shown when the quantity of atoms of an element is greater than one. The formula for sucrose—table sugar—is $C_{12}H_{22}O_{11}$. Each molecule of sucrose contains 12 carbon atoms, 22 hydrogen atoms, and 11 oxygen atoms.

The other common way atoms combine is by forming a continuous geometric arrangement. These compounds are called *crystals*, and the structure the atoms in the compound make when they join together is called a *crystal lattice*. The number of different arrangements atoms can make in a lattice is endless, and these arrangements are responsible for many of the unusual properties crystals possess. But what all lattices have in common is the regular arrangement of the atoms into

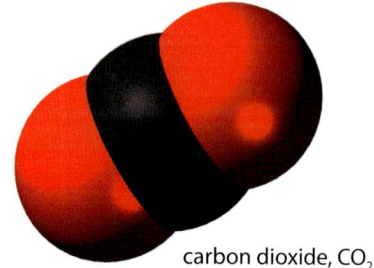

carbon dioxide, CO_2

propane, C_3H_8

ozone, O_3

Figure 2.18. More common molecular substances. Hydrogen atoms are white, carbon atoms are black, and oxygen atoms are red.

52

repeating, geometrical patterns. A space-filling model of the very simple crystal structure for sodium chloride, NaCl (table salt), is shown in Figure 2.19. The formula tells us that the ratio of sodium atoms to chlorine atoms in the crystal is 1 : 1. The model shows that the atoms are bonded together in a simple alternating arrangement.

We have seen several space-filling models so far. Another common type of computer model is the ball-and-stick model. Figure 2.20 shows a ball-and-stick model of the somewhat more complex crystal structure of copper(II) chloride, $CuCl_2$. As the formula indicates, in the lattice structure there are two chlorine atoms for each copper atom.

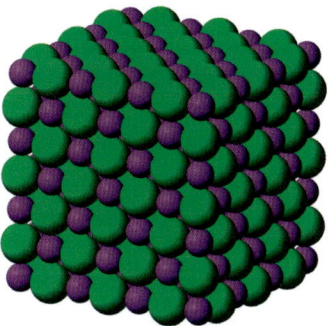

Figure 2.19. A space-filling model of the sodium chloride lattice structure. Sodium atoms are purple and chlorine atoms are green.

2.2.2 Mixtures

So far we have been discussing one major category of substances—pure substances. Elements and compounds are pure substances. The other major category is *mixtures*. Anytime substances are mixed together without a chemical reaction occurring, a mixture is formed. Remember—if a chemical reaction occurs, compounds are formed, not mixtures. If you toss vegetables in a salad, you've made a mixture. If you put sugar in your tea or milk in your coffee, you've made a mixture. If you mix up a batch of chocolate chip cookie dough, a bowl of party mix, or the batter for a vanilla cake, you've made a mixture.

In contrast to compounds, when a mixture is formed the individual substances in the mixture retain their physical and chemical properties. If you mix salt in water, you've made a mixture. The salt is still there and tastes salty. The water is still there too, and tastes watery. Also in contrast to compounds, the substances in a mixture can be separated by physical means such as filtering, boiling, freezing, or settling.

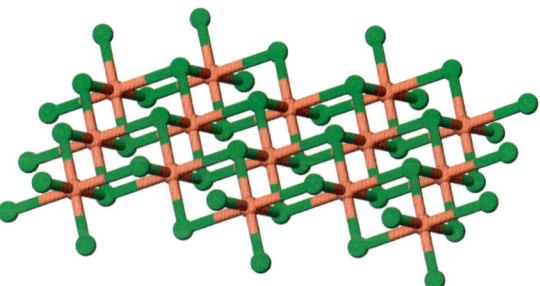

Figure 2.20. A ball-and-stick model of the copper(II) chloride crystal structure. Copper atoms are copper colored and chlorine atoms are green.

Again, in compounds, the original properties of the elements in the compound are chemically changed into the properties of the new chemical substance that is formed by means of a chemical reaction. Further, the elements in a compound cannot be separated by physical means. They can only be separated by the same means that brought them together in the first place—a chemical reaction. The distinguishing features of mixtures and compounds are summarized in Table 2.3. There are two classes of mixtures—*homogeneous mixtures* and *heterogeneous mixtures*. We examine these next.

Homogeneous Mixtures Homogeneous mixtures have uniform composition down to, but not including, the groups of atoms at the atomic level. The individual particles of the different substances in a homogeneous mixture cannot be seen with the eye, not even with the most powerful microscope. Particles at the atomic level—atoms, molecules, ions—are too small to reflect visible light, and so cannot be seen in the ordinary way, regardless of the magnification.

Chapter 2

Mixtures	Compounds
• Formed when substances combine without a chemical reaction occurring.	• Elements combine chemically to form a new substance—a compound.
• The individual substances in the mixture retain their physical and chemical properties.	• The physical and chemical properties of the compound are completely different from those of the reactants that formed the compound.
• The substances in mixtures can be separated by physical means such as filtering, boiling, freezing, or settling.	• The elements in a compound can only be separated by chemical means.

Table 2.3. Summary of the distinctions between mixtures and compounds.

The implication of this definition is that homogeneous mixtures are identical with *solutions*, mixtures in which one pure substance (or more than one) is dissolved in another pure substance. Solutions are so important in chemistry that we devote an entire chapter to the topic later on. For now, we move on to the other types of mixtures.

Heterogeneous Mixtures
In contrast to homogeneous mixtures, in a heterogeneous mixture there are lumps of different substances mixed together. You might be able to see the lumps with the naked eye, as with the mixture of spices in meat seasoning. Or the different substances may be visible only under a microscope, such as microscopic particulates in well water. Either way, if the different substances can be seen, the mixture is heterogeneous. In addition to the examples of salads and so on I mentioned above, there are two classes of heterogeneous mixtures that we encounter every day: *suspensions* and *colloidal dispersions*.

Suspensions
A *suspension* is formed when particles of size approximately 1 micrometer (1 μm) or larger are dispersed in a fluid (liquid or gas) medium. Particles this large do not remain in suspension indefinitely; they eventually settle out due to gravity.

Figure 2.21 shows an example of a suspension. At room temperature, corn starch is not soluble in water; that is, it does not dissolve in water. But if corn starch is stirred into water, a suspension is formed, as shown in the upper photo in the figure. After a few minutes, it is evident that the starch does not dissolve in the water, and after a few hours the starch particles are all at the bottom of the container, as shown in the lower photo.

Figure 2.21. Corn starch mixed in water forms a suspension (top). The starch particles eventually settle out of the liquid and fall to the bottom of the beaker (bottom).

Muddy water is another example of a suspension. If left to stand, the particles of sand, soil, and other organic matter eventually settle out, leaving the water quite transparent.

Colloidal Dispersions
Another class of heterogeneous mixtures is the *colloidal dispersions*, usually referred to simply as *colloids*. A colloid is formed when microscopic particles, ranging in

Atoms and Substances

		Dispersing Medium State		
		Solid	Liquid	Gas
Dispersed Particle State	Solid	**solid sol** cobalt glass, cranberry glass	**sol** blood, paint	**solid emulsion** butter, cheese	**solid aerosol** smoke, fine airborne dust
	Liquid	**gel** gelatin, hair gel	**liquid emulsion** mayonnaise, milk	**liquid aerosol** fog, hair spray, insect repellent
	Gas	**solid foam** Styrofoam, foam rubber, marshmallow	**foam** shaving cream, whipped cream	**(none)**

Table 2.4. Types of colloids with examples.

size from 1 nm to 1,000 nm, are dispersed throughout a dispersing medium. Unlike the particles in suspensions, these particles do not settle out. Forces from *Brownian motion* (see the box on the next page) keep the particles dispersed in the dispersing medium.

Table 2.4 lists the different types of colloids, distinguished by the states of the dispersed particles and the dispersing medium.[1] Since most of these substances are probably familiar to you, although perhaps not by name, I will just point out a few things about the information in the table.

You may or may not have seen items made of cobalt glass or cranberry glass, both examples of *solid sols* and shown in Figure 2.22. Glass is silicon dioxide, SiO_2, just like sand. Cobalt glass is silicon dioxide with cobalt salts such as cobalt(II) aluminate, $CoAl_2O_4$, dispersed within the molten glass. Cranberry glass contains particles of gold(III) oxide in the molten glass. Once the glass hardens, the colloid that remains consists of solid particles dispersed within a solid medium.

Two different terms are used for colloids consisting of solid particles dispersed within a liquid medium, *sol* and *solid emulsion*. As the examples in the table show, these two classes of colloids are rather different.

Figure 2.22. Cobalt glass (left) and cranberry glass (right), also known as "gold ruby."

In a *gel*, the solid particles link together in long molecules known as *polymers*, and these linked solids form a network or matrix throughout the substance. Finally, there are no colloids consisting of gas particles dispersed in other gas particles. This is because when no reaction occurs, all gases are *miscible* in each other. That is, the particles of the gases mix together in all proportions. Thus, mixed gases form solutions, not colloids.

We conclude this section with two more interesting facts about colloids. First, although the dispersed particles in a colloid are smaller than the particles in a suspension, they are still not

---

1 Sources differ to a frustrating degree on the definitions of some of the terms and on the identifications of some of the substances. If you go digging around in different texts or on-line sources, prepare to get confused.

Chapter 2

> ### Hmm... Interesting.   Brownian motion
>
>
>
> In 1827, Scottish botanist Robert Brown (1773–1858) was using a microscope to examine tiny particles from pollen grains. The particles were suspended in water, and Brown noticed that they jittered back and forth and moved through the water in an erratic and apparently random path. The motion of those particles is now known as *Brownian motion*, although no one knew what caused it until 1905. That was the year Albert Einstein published a paper demonstrating that the motion was caused by a particle being bombarded on all sides by moving water molecules. On this topic, still pictures just don't cut it. There are lots of animations online, but nothing beats the real thing. Go to commons.wikimedia.org, search under Brownian motion, and watch the clip with the file name *Brownianmotion beads in water spim video.gif* (sic).

small enough to pass through a filtration medium. In a true solution, the individual particles are ions or molecules, and their diameters are down in the sub-one nanometer range. Particles this small can pass right through a filtration medium such as filter paper. (So don't bother trying to separate salt from water by passing saltwater through a filter.) Colloidal particles are much larger than this and cannot pass through a filter. This is why people can wear dust masks to filter out pollens and dust (but not water vapor) from breathing air.

Figure 2.23. A salt water solution (top) allows the flashlight beam to pass through without scattering the light. A colloidal dispersion (bottom) scatters the light, making the beam visible, illuminating the surrounding liquid and the markings on the beaker from behind.

For colloids in fluid media, there is a test that distinguishes between solutions and colloids. Figure 2.23 illustrates the so-called *Tyndall effect* in a colloid. In the upper photograph, a flashlight beam is passed from right to left through a solution of saltwater. The light from the beam cannot be seen passing through the solution because the particles in the solution are so small they do not scatter light. In the lower photo, the beam is passed through a mixture of gelatin and water. The particles of gelatin form a colloidal dispersion, and are large enough that they do scatter the light. This is demonstrated by the fact that the beam is visible, by the illumination of the liquid, and by the brightness of the white markings on the beaker. The two photos were taken under the same lighting conditions. The light in the room was very dim except for the flashlight.

The Tyndall effect is what makes the beams from a car's headlights visible in foggy conditions. Fog is a colloid of liquid particles of water in air (a liquid aerosol), and the liquid particles scatter the light. And by the way, true water vapor—which is what steam is—is an invisible gas, just as the humidity in the air is invisible. If you can see the moisture, you are seeing light scattered by liquid water particles in air, and you are looking at a colloid, a mist of tiny condensed water droplets, not at steam.

2.2.3 Physical and Chemical Properties

All substances have certain properties. We divide up the different properties substances can possess into two broad classes. Some properties have to do with the physical characteristics of the substance, such as color, shape, size, phase, boiling point, texture, thermal conductivity, electrical conductivity, opacity, and density. These properties are called *physical properties*. Below is a list of example statements about the physical properties of substances. Consider how each one relates to the definition of physical properties just given.

- Iron is gray in color.
- Copper has a high electrical conductivity.
- Mica is shiny.
- Glass is smooth, but has sharp edges.
- Aluminum has a high thermal conductivity.
- Ethyl alcohol is transparent and colorless.
- Chlorine is greenish-yellow gas at atmospheric pressure and room temperature.
- Helium is a gas at atmospheric pressure and room temperature.
- At standard pressure, water freezes at 0°C.
- Milk is opaque.
- Oil is slippery.
- Clay brick is ochre in color.
- At 4.0°C, the density of water is 1.0 g/cm^3.
- Gold is malleable and ductile.
- Cast iron is not malleable.
- Glass is not malleable.
- Play-Doh is malleable, but not ductile.
- Jello is not ductile.

You may have noticed a couple of unfamiliar terms in the above list. The terms *malleable* and *ductile* are used to describe two important properties possessed by many metals. A substance is malleable if it can be hammered into different shapes, or hammered flat into sheets. A substance is ductile if it can be *drawn* into a wire. Wire drawing is a process of making wire by pulling the metal through a small hole in a metal block called a *die*. Usually the metal is already formed into a wire of larger diameter. The end of this larger diameter wire is hammered down or filed to get it through the hole in the die, and then a machine pulls the wire through the die to make the new, smaller diameter wire. Substances that can be drawn through a die like this without simply snapping are said to be ductile.

Notice from the above examples that a good student of science should be careful when describing physical properties. We need to make sure our statements are accurate in cases where temperature or pressure affect the property in question. For example, it is inaccurate to say that H$_2$O is a liquid. A more accurate statement is to say that one of the physical properties of water is that it is a liquid at temperatures between 0°C and 100°C. The statement is made even more accurate by specifying that the preceding sentence is correct at atmospheric pressure because at other pressures the boiling and freezing points of water are different.

The second broad class of properties has to do with the kinds of chemical bonds a substance forms, that is, the chemical reactions a substance does or does not participate in. These properties are called *chemical properties*. We have not yet studied chemical reactions, so you may not know that much about them. However, there are two common chemical reactions that you are quite familiar with—burning and rusting. Both of these are chemical reactions in which a substance combines with oxygen, and both are examples of a type of reaction called *oxidation*. Fiery explosions are simply combustions that happen very rapidly. But whether the combustion happens slowly, as with a log on a fire, or rapidly, as with a firecracker, combustion is a chemical reaction with oxygen. Substances that react with oxygen in this way are said to be *flammable* or *combustible*. (Oddly, *inflammable* means the same thing!)

Iron oxidizes to form compounds known as *oxides*. There are several different forms of iron oxide, colored red, yellow, brown, and black. Other metals oxidize as well. When copper oxidizes, it can form two different oxides, one red and one black. This is why copper objects exposed to the air turns dark brown or black. (Over a longer period of time the copper oxide forms other compounds, such as copper carbonate, which give the copper its pretty blue-green color. The Statue of Liberty is made of copper, and has been there for a long time. It is essentially covered with a layer of copper carbonate and other copper compounds.) Aluminum also oxidizes. Aluminum oxide is dark gray, and anyone who has done a lot of hand work with aluminum parts has noticed his or her hands blackened by the particles of aluminum oxide building up. Figure 2.24 shows a few different oxides.

Here are some examples describing chemical properties of substances:

- Hydrogen is combustible.
- Aluminum oxidizes to form aluminum oxide.
- Water is not flammable.
- Platinum does not oxidize. (This is why it is so valuable. It stays shiny as a pure element.)
- Baking soda (sodium bicarbonate, $NaHCO_3$) reacts with vinegar (acetic acid, CH_3COOH).
- Iron oxidizes to form iron oxide, or rust.
- Sodium reacts violently with water.
- Hydrogen reacts with a number of different polyatomic ions to form acids.
- Dynamite is explosive.
- Sodium hydroxide reacts with aluminum.
- Sulfuric acid reacts with many metals.

Figure 2.24. Oxides: red iron oxide (top), yellow iron oxide (center), and black copper oxide (bottom).

We discuss the physical and chemical properties of substances further in the coming chapters.

The two broad classes of properties we have been discussing, physical properties and chemical properties, are related to two broad classes of changes that substances undergo. If a substance experiences a change with respect to one of its physical properties, we call this a *physical change*. When a physical change occurs, the substance is still the same substance, it just looks or feels different. If a chemical reaction occurs to a substance, this is called a *chemical change*. Chemical properties basically describe the kinds of chemical changes a substance can undergo. When a chemical change occurs, the original substances that go into the reaction—the reactants—are converted into new substances—the products—with totally different physical and chemical properties. When asked to describe a given change as physical or chemical, ask yourself if the substance is still the same substance, or if it has actually gone through a chemical reaction to become a different substance.

Table 2.5 lists some examples of physical and chemical changes, with comments explaining why the type of change is physical or chemical.

Process	Change	Comments
glass breaking	physical	The broken glass is still glass, it just changes shape.
firecracker exploding	chemical	This explosion is a combustion. All combustions are chemical reactions. The substances in the firecracker react to form new substances such as ash and various gases.
mercury boiling	physical	The mercury is still mercury, it just changes from the liquid state to the vapor state.
copper turning dark brown or black	chemical	This occurs because the copper is oxidizing and forming copper oxide, a new substance. This is a chemical reaction.
iron pipes corroding	chemical	Corrosion is a chemical reaction. In this case, the iron reacts with the substances surrounding the pipes to form a new substance.
water evaporating	physical	The substance is still H_2O, it simply changes state from liquid to vapor.
mixing cake batter	physical	The eggs and flour and so on form a mixture, but no chemical change (reaction) occurs.
baking cookies	chemical	The heat causes a chemical reaction to occur in the dough. The substance is no longer cookie dough. It is cookie.
spilled pancake batter drying out	physical	No chemical reaction occurs. Dried batter is still batter. (If you want a pancake you have to cook it, which causes a chemical change.)
molten lead hardening	physical	The lead is still lead, it just changes state from liquid to solid.
balloon popping	physical	The balloon material is still the same material, it is just in shreds now. The air inside the balloon is at a lower pressure and is not contained in the balloon any longer, but it is still air.

Table 2.5. Examples of physical and chemical changes.

2.3 Isotopes and Atomic Masses

2.3.1 Isotopes

As you know, the atomic number (Z) of an element designates the number of protons in the nucleus of an atom of that element. For a given element, the atomic number is fixed: if an atom has a different number of protons, it is an atom of a different element. But the number of neutrons that may be present in the nucleus is not always the same for atoms of a given element. For most elements, there are variations in the number of neutrons that can be present in the nucleus. These varieties are called *isotopes*. For most elements, there is one isotope that is the most abundant in nature and several other isotopes that are also present but in smaller quantities. The general term for any isotope of any element is *nuclide*.

Isotopes are designated by writing the name of the element followed by the number of nucleons (protons and neutrons) in the isotope. The number of nucleons in a nucleus is called

Chapter 2

the *mass number*. For example, the most common isotope of carbon is carbon-12, accounting for about 98.9% of all the naturally occurring carbon. In the nucleus of an atom of carbon-12 there are six protons and six neutrons. There are two other naturally occurring carbon isotopes. Carbon-13 with seven neutrons accounts for about 1.1% of natural carbon. Atoms of carbon-14, of which only a trace exists in nature, have eight neutrons in the nucleus.

2.3.2 The Unified Atomic Mass Unit

The mass of a single atom is an extremely small number. But so much of our work in chemistry depends on atomic masses that scientists having been using units of *relative* atomic mass for a long time—all the way back to John Dalton, before actual masses of atoms were even known. Prior to the discovery of isotopes in 1912, the so-called *atomic mass unit* (amu) was defined as 1/16 the mass of an oxygen atom. After the discovery of isotopes, physicists defined the amu as 1/16 the mass of an atom of oxygen-16, but the definition used by chemists was 1/16 the average mass of naturally occurring oxygen, which is composed of several isotopes. To eliminate the confusion resulting from these conflicting definitions, the new *unified atomic mass unit* (u) was adopted in 1961 to replace them. Many texts continue to use the amu as a unit, but they define it as the u is defined. Strictly speaking, the amu is an obsolete unit that has been replaced by the u, now also called the dalton (Da). The u and the Da are alternative names (and symbols) for the same unit. The use of the dalton has increased in recent years, particularly in molecular biology.

The unified atomic mass unit, u, is defined as exactly 1/12 the mass of an atom of carbon-12. Table 2.6 lists a few nuclides and their atomic masses using the u as a unit of mass. All the elements listed exist as other isotopes in addition to those shown, but as you see from the percentage abundances, the ones shown are the major ones for the elements represented in the table.

2.3.3 Atomic Masses

In addition to the atomic number, the Periodic Table of the Elements lists the atomic mass in unified atomic mass units (u) for each element. But since there are multiple isotopes for just about every element, *the atomic mass values in the periodic table represent the weighted average of the masses of naturally occurring isotopes.*

An example of a weighted average is the average age of the students in the junior class at your school. Let's say there are 47 juniors, 40 of whom are 16 years old and 7 of whom are 17 years old at the beginning of the school year. To determine the average age of these students, let's first determine the proportion of the students at each age.

$$\frac{40}{47} = 0.851 \ (85.1\%)$$

$$\frac{7}{47} = 0.149 \ (14.9\%)$$

Z	isotope	mass (u)	abundance (%)
1	hydrogen-1	1.0078	99.9885
1	hydrogen-2	2.0141	0.0115
6	carbon-12	12.0000	98.93
6	carbon-13	13.0034	1.078
14	silicon-28	27.9769	92.223
14	silicon-29	28.9765	4.685
14	silicon-30	29.9738	3.092
17	chlorine-35	34.9689	75.76
17	chlorine-37	36.9659	24.24
20	calcium-40	39.9626	96.941
20	calcium-42	41.9586	0.647
26	iron-54	53.9396	5.845
26	iron-56	55.9349	91.754
26	iron-57	56.9354	2.119
26	iron-58	57.9333	0.282
29	copper-63	62.9296	69.15
29	copper-65	64.9278	30.85
92	uranium-235	235.0439	0.7204
92	uranium-238	238.0508	99.2742

Table 2.6. Major isotopes for a few elements.

To calculate the average age, we first multiply each student age by the proportion of students of that age to find the contribution to the average from each age group. Then we add the contributions together to find the weighted average age for the junior class.

$$\begin{array}{r} 16 \text{ years} \cdot 0.851 = 13.6 \text{ years} \\ +17 \text{ years} \cdot 0.149 = 2.53 \text{ years} \\ \hline = 16.1 \text{ years} \end{array}$$

We perform a similar calculation when computing the average atomic mass of an element from the masses of its isotopes, as shown in the following example.

▼ Example 2.1

Given the isotope masses and abundances for copper-63 and copper-65 in Table 2.6, determine the atomic mass for naturally occurring copper.

We need to multiply each isotope's mass by its abundance to get the isotope's contribution to the average mass of the element, which is the atomic mass. Then we add together the contributions from each isotope. The data from the table are:

copper-63: mass = 62.9296 u, abundance = 69.15%

copper-65: mass = 64.9278 u, abundance = 30.85%

$$\begin{array}{r} 62.9296 \text{ u} \cdot 0.6915 = 43.51 \text{ u} \\ + \ 64.9278 \text{ u} \cdot 0.3085 = 20.03 \text{ u} \\ \hline = 63.55 \text{ u} \end{array}$$

Compare this value to the value shown in the periodic table in Figure 2.14 or inside the back cover of the text.

The unified atomic mass unit, u, is defined as 1/12 the mass of an atom of carbon-12. Although the value of this mass is quite close to the masses of the proton and neutron, it is not exact because of the mass of the electrons in atoms of carbon-12, and also because of the mass-energy involved in binding the nucleus of the atom together. (The mass of nucleons bound together in a nucleus does not equal the sum of their individual masses.) Table 2.7 shows the masses of the three basic subatomic particles in unified atomic mass units.

Still, the proton and neutron masses are very close to unity (one) and the electron mass is extremely small. This means that for elements with a very large abundance of one isotope we can use the atomic mass and atomic number in the periodic table to determine the numbers of protons and neutrons in the nucleus of the most common isotope. For example, from Table 2.6, the mass of uranium-238 is very close to 238 u. Since an atom of uranium-238 has 92 protons, the balance of the mass is essentially all neutrons. Thus, there are 238 − 92 = 146 neutrons in uranium-238.

particle	mass
proton	1.007277 u
neutron	1.008665 u
electron	0.0005486 u

Table 2.7. Masses in u of the three basic subatomic particles.

Chapter 2

2.4 Density and Quantity of Substances

2.4.1 Density

Density is a physical property of substances. Density is a measure of how much matter is packed into a given volume for different substances. No doubt you are already familiar with the concept of density. You know that if you hold equally sized balloons in each hand, one filled with water and one filled with air, the water balloon weighs more because water is denser than air. You know that for equal weights of sand and Styrofoam packing peanuts the volume of the packing material is much larger because the packing material is much less dense. And you probably also know that objects less dense than water float, while objects denser than water sink.

The equation for density is

$$\rho = \frac{m}{V} \qquad (2.1)$$

where the Greek letter ρ (spelled rho and pronounced "row," which rhymes with snow) is the density in kg/m^3, m is the mass in kg, and V is the volume in m^3. These are the variables and units in the MKS unit system. However, since laboratory work typically involves only small quantities of substances, it is more common in chemistry for densities to be expressed in g/cm^3 (for solids) or g/mL (for liquids). In the examples that follow, I illustrate the use of g/cm^3 and kg/m^3. Since 1 mL = 1 cm^3 (see Table A.3 in Appendix A), calculations of densities in g/mL are essentially the same as those solving for densities in g/cm^3. If all you are doing is using the density equation, then any of these units of measure is fine. One final item for you to note is that the density of water at room temperature is

$$\rho_w = 0.998 \; \frac{\text{g}}{\text{cm}^3} \qquad (22.0°C)$$

This value is useful to know because water comes up in many different applications. The densities at other temperatures, along with other properties of water, are listed in Table A.5 of Appendix A.

▼ Example 2.2

The density of germanium is 5.323 g/cm^3. A small sample of germanium has a mass of 17.615 g. Determine the volume of this sample.

Begin by writing the given information.

$$\rho = 5.323 \; \frac{\text{g}}{\text{cm}^3}$$

$$m = 17.615 \; \text{g}$$

$$V = ?$$

Now write Equation (2.1) and solve for the volume.

$$\rho = \frac{m}{V}$$

$$\rho \cdot V = m$$

$$V = \frac{m}{\rho}$$

Next, insert the values and compute the result.

$$V = \frac{m}{\rho} = \frac{17.615 \text{ g}}{5.323 \frac{\text{g}}{\text{cm}^3}} = 3.309 \text{ cm}^3$$

This value has four significant digits, as it should based on the given information.

▼ Example 2.3

Determine the density of a block of plastic that has a mass of 1,860 g and dimensions 4.0 in × 2.5 in × 9.50 in. State your result in kg/m³.

To solve this problem, we use the given dimensions to calculate the volume of the block. Then we use Equation (2.1) to calculate the density.

Always begin your problem solutions by writing down the given information and performing any necessary unit conversions. Since the units of measure required for the result are kg/m³, we convert the mass to kilograms and the lengths to meters. When solving problems requiring unit conversions like this, write down the given information on separate lines down the left side of your page. Then perform the unit conversions by multiplying the conversion factors out to the right.

From the given information, our result must have two significant digits. This means we must perform the unit conversions and volume calculation with three significant digits (one more than we need), and round to two digits when we get our final result.

$$m = 1860 \text{ g} \cdot \frac{1 \text{ kg}}{1000 \text{ g}} = 1.86 \text{ kg}$$

$$l = 4.0 \text{ in} \cdot \frac{2.54 \text{ cm}}{1 \text{ in}} \cdot \frac{1 \text{ m}}{100 \text{ cm}} = 0.102 \text{ m}$$

$$w = 2.5 \text{ in} \cdot \frac{2.54 \text{ cm}}{1 \text{ in}} \cdot \frac{1 \text{ m}}{100 \text{ cm}} = 0.0635 \text{ m}$$

$$h = 9.5 \text{ in} \cdot \frac{2.54 \text{ cm}}{1 \text{ in}} \cdot \frac{1 \text{ m}}{100 \text{ cm}} = 0.241 \text{ m}$$

Now that the units are squared away, let's determine the volume of the block.

$$V = l \cdot w \cdot h = 0.102 \text{ m} \cdot 0.0635 \text{ m} \cdot 0.241 \text{ m} = 0.00156 \text{ m}^3$$

Finally, using Equation (2.1) the density is

$$\rho = \frac{m}{V} = \frac{1.86 \text{ kg}}{0.00156 \text{ m}^3} = 1192 \frac{\text{kg}}{\text{m}^3}$$

Rounding to two significant digits we have

$$1200 \frac{\text{kg}}{\text{m}^3}$$

2.4.2 The Mole and the Avogadro Constant

When solving problems in chemistry, we are generally working with chemical reactions in which huge numbers of atoms are involved, including all the naturally occurring isotopes, so performing reaction calculations with the masses of individual atoms is not practical. However, the average mass of a given multiple of some kind of atom is simply that multiple times the atomic mass. The mass of one million atoms of aluminum is 1,000,000 times the atomic mass of aluminum.

In chemistry, the standard bulk quantity of substance used in calculations is the *mole* (mol), the SI base unit for quantity of substance (Table 1.1). The mole is a particular number of particles of a substance, just as the terms *dozen*, *score*, and *gross* refer to specific numbers of things (12, 20, and 144, respectively). A mole is exactly $6.02214076 \times 10^{23}$ particles of a substance.

This value is known today as the *Avogadro constant*, N_A. More formally, the Avogadro constant is defined as exactly:

$$N_A = 6.02214076 \times 10^{23} \text{ mol}^{-1} \tag{2.2}$$

Usually, we just round this value to $6.022 \times 10^{23} \text{ mol}^{-1}$. In the next section, I'll describe why this value is what it is, instead of being a more convenient round number. For the moment, let's focus on what it means. Now, don't freak out over the unit of measure. Allow me to explain. Raising a unit of measure to the power -1 is mathematically equivalent to placing the unit in a denominator because $x^{-1} = \frac{1}{x}$. In other words, Equation (2.2) is the same thing as saying "$6.02214076 \times 10^{23}$ per mole." To make things even clearer, it's okay to say it this way: "N_A is about 6.022×10^{23} particles per mole." This is the way I like to think of it when performing unit conversions, as we do quite a lot in coming chapters. Without the units of measure, the value $6.02214076 \times 10^{23}$ is called *Avogadro's number*. With the units, it is called the Avogadro constant. Using this terminology, the mole can be defined this way:

A mole is the amount of a pure substance (element or compound) that contains Avogadro's number of particles of the substance.

Let's now consider what we mean when we refer to *particles* of a substance. For substances that exist as molecules, the particles are the molecules. Examples of these are the molecular substances we encountered back in Figure 2.18. For substances that exist as individual atoms, the particles are the individual atoms. Metals are like this, since a pure metal is composed of individual atoms of the same element joined together in a crystal lattice. The noble gases are also like this. The noble gases are located in the far right-hand column of the Periodic Table of the Elements. As I discuss more in coming chapters, atoms of noble gases are almost completely unreactive—they don't bond with other atoms at all. At ordinary temperature and pressure, the noble gases are gases composed of individual atoms.

For crystalline compounds, the "particles" in a mole of the substance are the *formula units* in the crystal lattice. A formula unit is one set of the atoms represented by the chemical formula of the compound. For example, the chemical formula for calcium carbonate is $CaCO_3$. One formula unit of calcium carbonate includes one calcium atom, one carbon atom, and three oxygen atoms.

The value of the Avogadro constant was determined approximately by French Physicist Jean Perrin (Figure 2.25) in the early 20th century. Perrin determined the value of the constant through several different experimental methods. In the 19th century, many scientists did not yet accept the existence of atoms as a scientific fact and Perrin's research put the atomic nature of matter beyond dispute. For this work, he received the Nobel Prize in Physics in 1926. Perrin proposed naming the constant after Amedeo Avogadro, a 19th-century Italian scientist who was the first to propose that the volume of a gas at a given temperature and pressure is proportional to the number of particles of the gas (atoms or molecules), regardless of the identity of the gas. In fact, at 0°C and atmospheric pressure, one mole of any gas occupies a volume of 22.4 L.

Figure 2.25. French physicist Jean Perrin (1870–1942).

2.4.3 Molar Mass and Formula Mass

Since 2019, the value of the Avogadro constant in Equation (2.2) is exact by definition. But the number has the value it does because it was originally chosen so that the average atomic mass in u of a molecule of a compound, as computed from the mass values in the periodic table, would be numerically equivalent to the mass of one mole of the compound in grams per mole. (We address these calculations below.) Now, recall that the definition of the unified atomic mass unit (or dalton) is such that an atom of carbon-12 has a mass of exactly 12 u. According to the original definition of Avogadro's number, there were also exactly 12 grams of carbon-12 in one mole of carbon-12. So according to these definitions, an atom of carbon-12 has a mass of exactly 12 u, and a mole of carbon-12 had a mass of exactly 12 grams. This quantity, the mass of one mole of a substance, is called the *molar mass*. Because of the way the molar and atomic masses were defined, the molar mass for an atom was numerically equivalent to the atomic mass. As a result of the 2019 redefinition of Avogadro's number, the atomic mass in u and the molecular mass in g/mol are no longer exactly equivalent. However, they are extremely close and may still be treated as equal for practical purposes. (The difference is a factor of only about 4×10^{-10}.)

Even though the exact equivalence ended in 2019, these are still very handy definitions! For example, from the periodic table we find that the average mass of one atom of silicon ($Z = 14$) is 28.0855 u. This also tells us that the mass of one mole of silicon is 28.0855 g, so the molar mass of silicon is 28.0855 g/mol. Likewise, from the periodic table we find that the average mass of one atom of copper ($Z = 29$) is 63.546 u. This also tells us that the mass of one mole of copper is 63.546 g, so the molar mass of copper is 63.546 g/mol. For the elements that exist as single atoms, the molar mass in g/mol and the atomic mass in u are numerically equivalent (almost).

From the periodic table, we can also determine the molar mass of compounds—the mass of one mole of the compound. We simply add up the molar masses for the elements in the chemical formula, taking into account any subscripts present in the formula, and we have the molar mass for the compound in g/mol. If we add up the element atomic masses in unified atomic mass

Chapter 2

Quantity	Units	Definition
molar mass	g/mol	The mass of one mole of a substance, approximately equal to the sum of the atomic masses of the elements in a chemical formula, taking into account the subscripts indicating atomic ratios in the compound.
formula mass	u	The mass of one formula unit of a substance. Numerically nearly equivalent to the molar mass.
molecular mass	u	The average mass of a single molecule of a molecular substance. Numerically equivalent to the formula mass. (May also be converted to grams and expressed in grams, see Section 2.4.4.)

Table 2.8. Definitions and units for molar mass, formula mass, and molecular mass.

units, we obtain what is called the *formula mass* of the compound in u (or Da). If the compound is molecular, then the formula mass may also be referred to as the *molecular mass*, the average mass of a single molecule of the substance.

The details of these three different mass terms are summarized in Table 2.8.

▼ Example 2.4

Determine the molar mass and formula mass for water, H_2O. Note that since water is composed of molecules, the formula mass may also be called the molecular mass.

From the periodic table, the atomic masses of hydrogen (H) and oxygen (O) are:

H: 1.0079 u

O: 15.9994 u

Written as molar masses, these are:

H: 1.0079 $\frac{g}{mol}$

O: 15.9994 $\frac{g}{mol}$

The formula for water, H_2O, tells us there are two hydrogen atoms and one oxygen atom in each molecule of water, so we multiply these numbers by the molar masses and add them up to get the molar mass of H_2O.

$$\left(2 \times 1.0079 \ \frac{g}{mol}\right) + \left(1 \times 15.9994 \ \frac{g}{mol}\right) = 18.0152 \ \frac{g}{mol}$$

The calculation of the formula mass is identical, except we use units of u instead of g/mol. From the periodic table, the atomic masses of hydrogen (H) and oxygen (O) are:

H: 1.0079 u

O: 15.9994 u

There are two hydrogen atoms and one oxygen atom in each molecule, so we multiply these numbers by the element atomic masses and add them up to get the formula mass of H_2O.

$(2\times 1.0079 \text{ u}) + (1\times 15.9994 \text{ u}) = 18.0152 \text{ u}$

Thus, the formula mass for water is 18.0152 u. This value is also the molecular mass of water.

▼ Example 2.5

Nitrogen is one of several elements that exist in nature as diatomic gases. (The others include hydrogen, oxygen, fluorine, and chlorine.) Determine the molar mass of nitrogen gas, N_2.

From the periodic table, the atomic mass of nitrogen (N), written as a molar mass for individual nitrogen atoms, is:

N: $14.0067 \dfrac{\text{g}}{\text{mol}}$

There are two nitrogen atoms in each molecule of N_2, so we multiply the molar mass of N by two to get the molar mass of N_2.

$\left(2\times 14.0067 \dfrac{\text{g}}{\text{mol}}\right) = 28.0134 \dfrac{\text{g}}{\text{mol}}$

▼ Example 2.6

Determine the mass in grams of 2.5 mol sodium bicarbonate, $NaHCO_3$ (baking soda).

In any problem like this, we first find the molar mass of the given compound. Then we simply use that molar mass to compute the mass of the given quantity. From the periodic table, the molar masses of the elements in the compound are:

Na: $22.9898 \dfrac{\text{g}}{\text{mol}}$

H: $1.0079 \dfrac{\text{g}}{\text{mol}}$

C: $12.011 \dfrac{\text{g}}{\text{mol}}$

O: $15.9994 \dfrac{\text{g}}{\text{mol}}$

The oxygen appears three times in the formula, so its mass must be multiplied by three and added to the others.

$22.9898 \dfrac{\text{g}}{\text{mol}} + 1.0079 \dfrac{\text{g}}{\text{mol}} + 12.011 \dfrac{\text{g}}{\text{mol}} + \left(3\times 15.9994 \dfrac{\text{g}}{\text{mol}}\right) = 84.007 \dfrac{\text{g}}{\text{mol}}$

This value is the molar mass of $NaHCO_3$. To find the mass of 2.5 mol, we multiply:

Chapter 2

$$2.5 \text{ mol} \cdot 84.007 \frac{g}{mol} = 210 \text{ g}$$

Notice that the significant digits in the molar mass are obtained by the addition rule. The atomic mass of carbon, 12.011 u, only goes to three decimal places, so the molar mass of the compound goes to three decimal places. The final result is obtained by multiplying the compound's molar mass by the quantity 2.5 mol, which only has two significant digits. Thus, the result must have two significant digits as well.

▼ Example 2.7

A scientist measures out 125 g of potassium chloride (KCl). How many moles of KCl does this quantity represent?

First, determine the molar mass of KCl. From the periodic table:

K: $39.098 \frac{g}{mol}$

Cl: $35.4527 \frac{g}{mol}$

The formula includes one atom of each, so we add them to obtain the molar mass:

$$39.098 \frac{g}{mol} + 35.4527 \frac{g}{mol} = 74.551 \frac{g}{mol}$$

Beginning now, *always think of the molar mass of any substance as a conversion factor* that can be written right side up or upside down to convert grams to moles or vice versa. For KCl, 74.551 g is equivalent to 1 mol, so these quantities can be written as conversion factors, like this:

$$\frac{74.551 \text{ g}}{1 \text{ mol}} = \frac{1 \text{ mol}}{74.551 \text{ g}}$$

This makes the last step of this problem easy. Just select the way of writing the molar mass conversion factor that cancels out the given units (g) and gives the units required (mol). This is nothing but a unit conversion.

$$125 \text{ g} \cdot \frac{1 \text{ mol}}{74.551 \text{ g}} = 1.68 \text{ mol}$$

The photograph in Figure 2.26 shows one mole of each of four substances. The first is one mole of copper, equal to 63.5 g. The second is a 250-mL beaker containing one mole of water. As you can see, this is not much water—only 18 mL. At the upper right is a weigh tray

Figure 2.26. Clockwise from left are shown 1 mole of copper, 1 mole of water, 1 mole of table salt, and 1 mole of baking soda.

Atoms and Substances

containing one mole of sodium chloride, 40.0 g. (This is just under 1/4 cup.) Finally, one mole of baking soda, 84.1 g. (This is right at 1/3 cup.)

▼ Example 2.8

Calculate the number of water molecules in a 1.00-liter bottle of water.

The logic of this problem, in reverse, is as follows: To calculate a number of molecules, we must use the Avogadro constant. To use the Avogadro constant, we need to know the number of moles of water we have. To determine the number of moles, we need to know both the molar mass and the mass of the water. To determine the mass from a volume, we use the density equation.

So we begin with the given information and the density equation to determine the mass of water we have. The given information and unit conversions are as follows:

$$V = 1.00 \text{ L} \cdot \frac{1000 \text{ cm}^3}{1 \text{ L}} = 1.00 \times 10^3 \text{ cm}^3$$

$$\rho = 0.998 \frac{\text{g}}{\text{cm}^3}$$

$$m = ?$$

Now we write down Equation (2.1) and solve for the mass:

$$\rho = \frac{m}{V}$$

$$m = \rho \cdot V = 0.998 \frac{\text{g}}{\text{cm}^3} \cdot 1.00 \times 10^3 \text{ cm}^3 = 998 \text{ g}$$

Next, we need the molar mass of water. We calculated this in Example 2.4 and obtained 18.0152 g/mol. We use this molar mass as a conversion factor to convert the mass of water into a number of moles of water:

$$998 \text{ g} \cdot \frac{1 \text{ mol}}{18.0152 \text{ g}} = 55.40 \text{ mol}$$

This intermediate result has four significant digits—one more than we need in the final result. Finally, with the number of moles in hand we use the Avogadro constant to determine how many particles of water this is, which is identical to the number of water molecules.

$$55.40 \text{ mol} \cdot \frac{6.022 \times 10^{23} \text{ particles}}{\text{mol}} = 3.34 \times 10^{25} \text{ particles}$$

▲

2.4.4 Gram Masses of Atoms and Molecules

The atomic mass from the periodic table and the Avogadro constant can be used to calculate the mass in grams of an individual atom. Recall that the atomic mass value in the periodic table gives both the average atomic mass in u, and the molar mass in g/mol. Knowing the molar mass in g/mol we can simply divide by the number of atoms there are in one mole to find the mass of

Chapter 2

one atom in grams. Although this kind of calculation is quite simple, I have found that it is *very* easy for students to get confused and not be able to determine whether one should multiply or divide or what. So here's a problem solving tip: let the units of measure help you figure out what to do. If you include the units of measure in your work and pay attention to how the units cancel out or don't cancel out, these calculations are pretty straightforward. *Keep this principle firmly in mind throughout your study of chemistry!* Units of measure are not an annoying burden; they are the student's friend.

▼ Example 2.9

Determine the average mass in grams of an atom of boron.

From the periodic table, we find that the molar mass of boron is 10.811 g/mol. One mole consists of Avogadro's number of atoms of boron, so if we divide the molar mass by the Avogadro constant, we have the mass of a single atom of boron. Let's begin by setting up the division I just described, and then use the old invert-and-multiply trick for fraction division to help with the unit cancellations.

$$\frac{10.811 \frac{g}{mol}}{6.0221 \times 10^{23} \frac{particles}{mol}} = 10.811 \frac{g}{mol} \cdot \frac{1}{6.0221 \times 10^{23}} \frac{mol}{particles}$$

$$= \frac{10.811}{6.0221 \times 10^{23}} \frac{g}{particle} = 1.7952 \times 10^{-23} \frac{g}{particle}$$

So the average mass of one boron atom is 1.7952×10^{-23} g. Note that I use five digits in the value of the Avogadro constant to preserve the precision we have in the molar mass.

For molecular substances, the molar mass is used to compute the molecular mass in grams—the average mass of one molecule. This is done by first computing the molar mass of the compound, just as we did before. Then we simply divide by the Avogadro constant to obtain the mass of a single molecule.

Like the atomic mass, the molecular mass is an average mass, since the atomic masses used in calculating the molar mass are all based on the average mass of different isotopes with their abundances taken into account. The molecular mass for a *specific* molecule must be calculated using the specific masses of the nuclides in the molecule.

▼ Example 2.10

Determine the mass in grams of one molecule of carbon tetrachloride, CCl_4.

From the periodic table we find that the molar masses of carbon and chlorine are 12.011 g/mol and 35.4527 g/mol, respectively. From this we calculate the molar mass of CCl_4:

$$\left(1 \times 12.011 \frac{g}{mol}\right) + \left(4 \times 35.4527 \frac{g}{mol}\right) = 153.822 \frac{g}{mol}$$

With this molar mass we can use the Avogadro constant to get the molecular mass in grams. This time, instead of writing the Avogadro constant in the denominator of a big fraction, I simply treat it as a conversion factor and write it in the equation such that the mole units cancel

Atoms and Substances

out. (This is the way I always perform such calculations.) I also use six digits in the Avogadro constant to preserve the precision we have in the molar mass.

$$153.822 \; \frac{g}{mol} \cdot \frac{1 \; mol}{6.02214 \times 10^{23} \; particles} = 2.55427 \times 10^{-22} \; \frac{g}{particle}$$

Chapter 2 Exercises

SECTION 2.1

1. Write paragraphs describing the experiments performed by J.J. Thomson, Robert Millikan, and Ernest Rutherford.

2. Describe the main points or features in the atomic models proposed by John Dalton, J.J. Thomson, and Ernest Rutherford.

3. Explain why Ernest Rutherford found the reflection of alpha particles off gold foil so astonishing.

SECTION 2.2

4. Write paragraphs distinguishing between these pairs of terms:
   a. compounds and elements
   b. mixtures and compounds
   c. heterogeneous mixtures and homogeneous mixtures
   d. suspensions and colloids

5. Classify each of the following as element, compound, homogeneous mixture, or heterogeneous mixture.

a. water	b. cesium chloride	c. pond water	d. methane
e. a soft drink	f. nitric acid	g. black coffee	h. argon
i. air	j. hydrogen nitrate	k. exhaust fumes	l. quartz
m. brass	n. hydrogen gas	o. hydrogen cyanide	p. mouthwash
q. platinum	r. dirt	s. radon	t. a smoothie

6. Explain why salt water and sugar water are homogeneous mixtures while automotive paint, which contains invisible particulates, is not.

7. Write a paragraph describing the two basic types of structures atoms can take when bonding together.

8. Select three pure substances not mentioned in the chapter. For each substance, list at least eight physical properties and three chemical properties.

9. Explain why colloids reflect light. What is this effect called?

Chapter 2

10. Identify each of the following as a physical change or a chemical change. For each, explain your choice.

   a. an avalanche
   b. a cigar burning
   c. spilling a glass of milk
   d. digesting your food
   e. swatting a fly
   f. stirring cream into coffee
   g. firing a pop gun
   h. firing a real gun
   i. boiling mercury
   j. welding steel
   k. filling a helium balloon
   l. allowing molten iron to harden
   m. frying chicken
   n. snow melting
   o. a car exhaust pipe rusting
   p. paint "drying"
   q. wood rotting
   r. a ball rolling down a hill

SECTION 2.3

11. How is the unified atomic mass unit, u, defined?

12. Referring to Table 2.6, calculate the atomic mass for silicon, calcium, and uranium. Compare your results to the values shown in the periodic table.

13. Which two nuclides in Table 2.6 have 20 neutrons?

14. In Table 2.6, how many protons, neutrons, and electrons are there in the heaviest nuclide listed? How many protons, neutrons, and electrons are there in the lightest nuclide listed?

15. As mentioned in the chapter, the sum of the masses of the particles in an atom does not equal the mass of the atom. Some of the mass of the individual particles is converted to energy, and the atom weighs less than the sum of the weights of its parts. How much mass is converted into energy when the individual protons, neutrons, and electrons are assembled to form an atom of uranium-238?

SECTION 2.4

16. What is the density of carbon dioxide gas if 0.196 g of the gas occupies a volume of 100.1 mL?

17. Oil floats because its density is less than that of water. Determine the volume of 550 g of a particular oil with a density of 955 kg/m^3. State your answer in mL.

18. A factory orders 15.7 kg of germanium. The density of germanium is 5.32 g/cm^3. Calculate the volume of this material and state your answer both in m^3 and cm^3.

19. A graduated cylinder contains 23.35 mL of water. An irregularly shaped stone is placed into the cylinder, raising the volume to 27.79 mL. If the mass of the stone is 32.1 g, what is the density of the stone?

20. A standard 55-gallon drum is 34.5 inches tall and 24 inches in diameter. Consider a 55-gallon drum filled with kerosene. Using the dimensions in inches to calculate the volume, determine the mass of kerosene that fills this drum, given that the density of kerosene is 810 kg/m^3.

21. Iron has a density of 7,830 kg/m^3. An iron block is 2.1 cm by 3.5 cm at the base and has a mass of 94.5 g. How tall is the block?

22. A student measures out 22.5 mL of mercury and finds the mass to be 306 g. Determine the density of mercury, and state your answer in kg/m^3.

23. A large contemporary water tower holds over 3 million gallons of water. Determine the mass in kilograms of 3.0×10^6 gallons of water.

24. The famous Kon-Tiki was a raft sailed by Norwegian explorers in 1947 from South America to the Polynesian islands in the South Pacific. (Thor Heyerdahl's book about it is a great read.) The trip took three and a half months and covered 4,300 miles in the Pacific Ocean. The crew of six men built the main section of the raft out of 9 massive balsa trees, each 2.0 ft

in diameter. Assume for simplicity that the balsa trees had a typical balsa wood density of 160 kg/m³, and the logs were each 45 ft in length. Determine the total mass in kilograms of these logs used to build the raft.

25. How is the mole defined?

26. Determine the number of atoms in each of the following.

    a. 73.2 g Cu
    b. 1.35 mol Na
    c. 1.5000 kg W

27. Determine the mass in grams for each of the following.

    a. 6.022×10^{23} atoms K
    b. 100 atoms Au
    c. 0.00100 mol Xe
    d. 2.0 mol Li
    e. 4.2120 mol Br
    f. 7.422×10^{22} atoms Pt

28. Determine the number of moles present in each of the following.

    a. 25 g $Ca(OH)_2$
    b. 286.25 g $Al_2(CrO_4)_3$
    c. 2.111 kg KCl
    d. 47.50 g $LiClO_3$
    e. 10.0 g O_2
    f. 1.00 mg $C_{14}H_{18}N_2O_5$

29. Calculate the molar mass for each of the following compounds or molecules.

    a. ammonia, NH_3
    b. carbon dioxide, CO_2
    c. chlorine gas, Cl_2
    d. copper(II) sulfate, $CuSO_4$
    e. calcium nitrite, $Ca(NO_2)_2$
    f. sucrose, $C_{12}H_{22}O_{11}$
    g. ethanol, C_2H_5OH
    h. propane, C_3H_8
    i. glass, SiO_2

30. Determine the formula masses for these compounds:

    a. $MgCl_2$
    b. $Ca(NO_3)_2$
    c. $(SO_4)^{2-}$ (The 2– indicates this is an ion with an electrical charge of –2. The charge does not affect your calculation.)
    d. $CuSO_4$
    e. BF_3
    f. CCl_4

31. Determine the mass in grams of 2.25 mol silver nitrate, $AgNO_3$.

32. Given 2.25 kg CCl_4, answer these questions:

    a. How many moles CCl_4 are present?
    b. How many carbon atoms are present?
    c. Approximately how many carbon-13 atoms are present?

33. Given 1.00 gal H_2O at 4°C, answer the questions below. (Hint: You must use the appropriate volume conversion and the density of water to determine the mass of 1.00 gal H_2O. See the information in Tables A.3 and A.5 in Appendix A.)

    a. How many moles H_2O are present?
    b. How many hydrogen atoms are present?
    c. Approximately how many deuterium (hydrogen-2) atoms are present?

Chapter 3
Atomic Structure

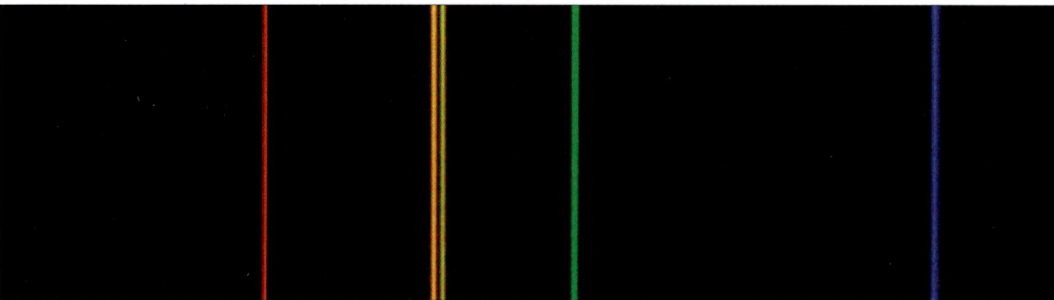

Hg

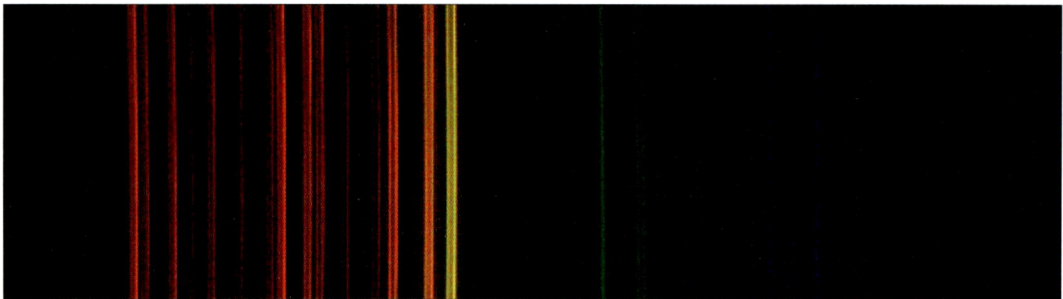

Ne

When excited by heat or electricity, the atoms of each element emit a specific, unique set of wavelengths of light—the *atomic spectrum* for that element. The visible spectra emitted by mercury and neon are shown above. Atomic spectra were known and studied in the 19th century, but there was no theory at that time that could explain the source of the colors different elements emit. Then in 1913, Niels Bohr published his new model of the atom, locating the electrons in atoms in specific energy levels. Bohr theorized that when excited, electrons jump to higher energy levels, and that to drop back down to a lower energy level an electron emits a packet of electromagnetic energy—what we now call a photon. In 1901, Max Planck had published the equation relating specific amounts of energy to specific wavelengths (colors) of light. Bohr's successful explanation for atomic spectra opened the door for detailed study of the internal structure of atoms.

The two spectra shown above were imaged in the Laser Optics Lab at Regents School of Austin in Austin, Texas.

Atomic Structure

Objectives for Chapter 3

After studying this chapter and completing the exercises, you should be able to do each of the following tasks, using supporting terms and principles as necessary.

SECTION 3.1
1. Describe the electromagnetic spectrum and state approximate wavelengths for the ends of the visible spectrum.

2. Define *quantum* and explain what it means for energy to be quantized.

3. Given the Planck relation and Planck's constant, determine the energy of a photon of a given wavelength and vice versa.

SECTION 3.2
4. Describe the Bohr model of the atom and how the model explained the phenomenon of atomic spectra.

SECTION 3.3
5. Distinguish qualitatively between the orbital energies in the hydrogen atom and those of other atoms.

6. Describe the two main ways that atoms can possess energy.

7. State and describe the four quantum numbers required to describe the quantum state of an electron.

8. For the first three principal quantum numbers, describe the orbitals available for electrons.

9. Describe the principles governing the quantum states of electrons in atoms.

10. State the Pauli exclusion principle and explain its relationship to the placement of electrons in atoms.

SECTION 3.4
11. Given the periodic table, write electron configurations (full and condensed) for elements in the first four periods.

SECTION 3.5
12. Given mass data for an unknown compound, determine the percent composition and empirical formula of the compound.

13. Use the percent composition along with the molar mass or molecular mass of an unknown compound to determine its molecular formula.

Chapter 3

3.1 Atomic Spectra

3.1.1 The Electromagnetic Spectrum

Understanding our present theory of atomic structure and the story of how it unfolded requires a basic understanding of the electromagnetic spectrum. We will thus begin this chapter with a brief presentation about the electromagnetic spectrum, light, and the relationship between energy and wavelength in light.

The spectrum of visible light is shown in Figure 3.1. The *visible spectrum* runs through the colors of the rainbow—red, orange, yellow, green, blue, violet—and includes wavelengths from about 750 nm (red) down to about 400 nm (violet). This is the range of wavelengths our eyes can see.

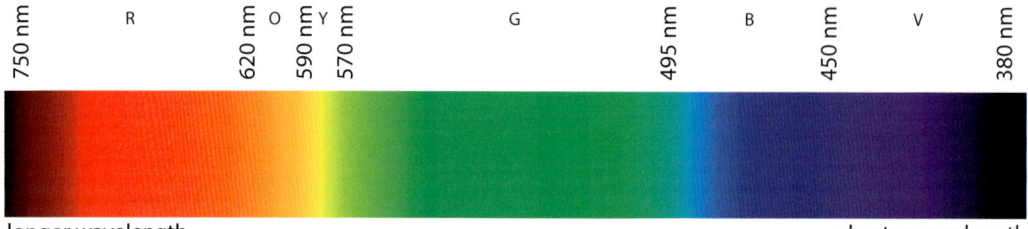

Figure 3.1. Colors and approximate wavelength ranges in the visible portion of the electromagnetic spectrum.

Visible light is just a small portion of a vast spectrum of electromagnetic radiation that occurs in nature. Figure 3.2 shows the most important regions of the *electromagnetic spectrum*, from radiation with wavelengths in the range of 1 km, the region of AM radio waves, down to the high-energy gamma rays, with wavelengths in the range of 1 picometer (pm). As you see from the figure, the solar emission spectrum runs from wavelengths of about 1 mm down to

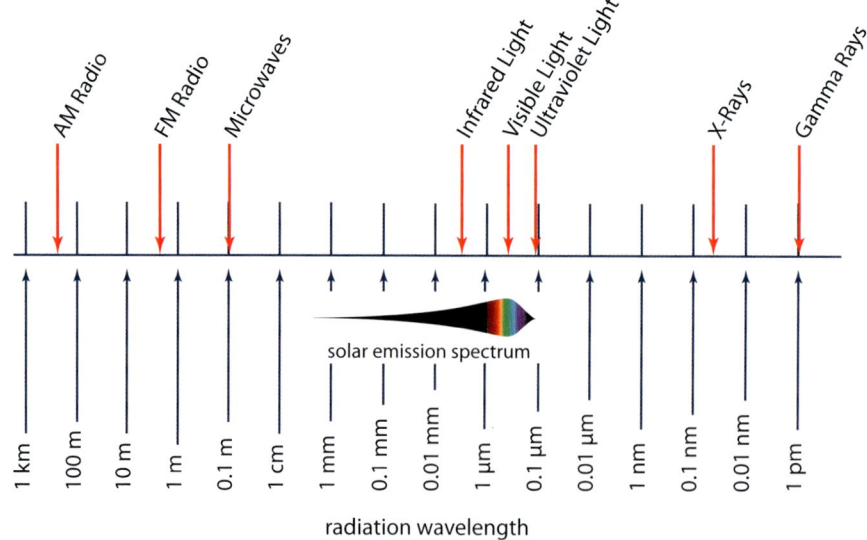

Figure 3.2. The electromagnetic spectrum.

wavelengths of about 0.1 μm. The solar spectrum includes the infrared, visible, and ultraviolet regions, and is strongest in the middle of the visible spectrum.

The contemporary theory of light (a shorter term for electromagnetic radiation in general) holds that light exhibits both wave-like properties and particle-like properties. Since light behaves like waves, we can refer to the wavelengths of particular colors. But light also behaves like particles. We call these particles *photons*, and each photon represents a single packet of energy. The packet of energy in a photon is also called a *quantum* of energy. (The plural is *quanta*.)

When thinking of light as waves, we characterize those waves by the wavelength. When considering light as discrete packets of energy, we tend to think of the amount of energy in each packet (each photon). It turns out there is a simple equation relating these together.

Figure 3.3. German physicist Max Planck (1858–1947).

It was in 1901 that German physicist Max Planck (Figure 3.3) conceived of treating energy as if it were quantized. He was working on a different problem at the time (the so-called blackbody radiation problem), and he did not imagine that energy really *is* quantized. However, he introduced what he thought was a mathematical trick—the quantization of energy—and in the process, quantum theory was born. Planck won the Nobel Prize in Physics for this work in 1918. (It's ironic, isn't it, to win the Nobel Prize for a major discovery that the scientist thinks is just a mathematical trick?) Four years later in 1905, German physicist Albert Einstein (Figure 3.4) proposed that energy really is quantized and used this idea to solve another problem (explaining the photoelectric effect). For this, Einstein won the Nobel Prize in Physics in 1921.

Figure 3.4. German physicist Albert Einstein (1879–1955).

The equation Planck introduced is called the *Planck relation*. This important equation is:

$$E = hf \tag{3.1}$$

In this equation, E is the energy in the photon in joules (J). The next term in the equation, h, is a constant known as the *Planck constant*. The value of h is defined as exactly

$$h = 6.62607015 \times 10^{-34} \text{ J} \cdot \text{s} \tag{3.2}$$

The last term in the Planck relation, f, is the frequency of the wave. The frequency and wavelength of a wave are related by the equation

$$v = \lambda f \tag{3.3}$$

In this equation, v is the velocity of the wave, which is the speed of light for all wavelengths of electromagnetic radiation (2.9979×10^8 m/s). The wavelength is represented by the Greek letter λ (lambda, the Greek lower-case letter l). If we solve Equation (3.3) for the frequency, we get

$$v = \lambda f$$

$$f = \frac{v}{\lambda}$$

Chapter 3

Now we will insert v/λ into the Planck relation in place of f, giving

$$E = \frac{hv}{\lambda} \tag{3.4}$$

With this equation, we can compute the energy in a single photon of light at any wavelength, or vice versa. The following example illustrates such a calculation.

▼ Example 3.1

The bright blue line in the mercury vapor spectrum (see the upper image on the opening page of this chapter) has a wavelength of 435.8 nm. Determine the energy contained in a single photon of this blue light.

We begin by writing down the given information and the unknown we seek to find.

$\lambda = 435.8$ nm

$E = ?$

Next, we convert the given wavelength into the MKS length unit, meters.

$$\lambda = 435.8 \text{ nm} \cdot \frac{1 \text{ m}}{10^9 \text{ nm}} = 4.358 \times 10^{-7} \text{ m}$$

From this value, along with the Planck constant and the speed of light, we calculate the energy of a photon with this wavelength.

$$E = \frac{hv}{\lambda} = \frac{(6.626 \times 10^{-34} \text{ J·s})\left(2.9979 \times 10^8 \frac{\text{m}}{\text{s}}\right)}{4.358 \times 10^{-7} \text{ m}} = 4.558 \times 10^{-19} \text{ J}$$

This is an extremely small amount of energy, less than a billionth of a billionth of a joule. The given wavelength and the value for the Planck constant each have four significant digits. Thus, the result is stated with a precision of four significant digits.

▲

3.1.2 Energy in Atoms

As mentioned in the caption on the opening page of this chapter, the atoms of every element emit a specific set of colors when excited. In the context of atomic theory, the term *excitation* refers to the absorption of energy by atoms, either from electromagnetic radiation (light) or from collisions with other particles.

Let's spend a moment considering the ways an individual atom can possess energy. There are two basic mechanisms by which atoms can possess energy. First, all atoms possess *kinetic energy*, the energy associated with motion. Kinetic energy in atoms is illustrated in Figure 3.5. In solids, the atoms are fixed in place and are not free to move around, so the kinetic energy is manifest in the atoms' vibrations. In liquids and gases (fluids), atoms are free to move around, so the energy possessed by atoms in fluids is in their translational kinetic energy. Also, when atoms in fluids are bound together in molecules, the molecules can tumble and rotate, so some of their kinetic energy is in the energy of rotation. Atoms in molecules also vibrate, just as balls attached to one another by springs can wiggle back and forth. In all these cases, the kinetic energy in atoms and

Atomic Structure

molecules correlates directly to their temperature. The hotter they are, the more vigorously they vibrate and the faster they move.

The second basic way an atom can possess energy is in the energies of the atom's electrons. As mentioned in the Introduction and Chapter 1 (and discussed in detail later in this chapter), the electrons in atoms are located in various orbitals, and different orbitals are associated with different amounts of electron energy. Atoms can absorb quanta of energy from the photons of electromagnetic radiation and from collisions with other particles, such as ions and free electrons. When an atom absorbs energy in this way, the quantum of energy absorbed by the atom is manifest in one or more of the atom's electrons moving into higher-energy orbitals. This is atomic excitation.

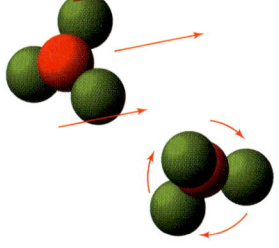

When an atom's electrons are all in their lowest-energy orbitals, the atom is said to be in the *ground state*. Excitation occurs when an atom absorbs a quantum of energy causing an electron to move to a higher-energy orbital. When this happens, the atom is said to be in an *excited state*. Atoms tend not to remain in excited states. Instead, after becoming excited an atom typically heads straight back to the ground state, generally by emitting the energy it absorbed in the form of one or more new photons.[1]

Figure 3.5. The atoms in the solid crystal at the top are blurred to illustrate their vibrations. The gas molecules at the bottom are translating and tumbling.

A newly emitted photon may not have the same amount of energy as the original quantum of energy the atom absorbed. To explain this, we need to introduce a commonly used graphical representation of the different energies electrons can possess. For now, let's call these *energy levels*. We will relate these more carefully to the energies of electrons in atoms in the next few sections.

---

1 In some substances, electrons in atoms can remain in excited states for an extended period of time. As the atoms in such a substance return to the ground state over time the substance gradually radiates the energy away. This is the way *phosphorescence* (glowing in the dark) works.

Hmm... Interesting.   Neon signs and phonons

When excited atoms in gases return to the ground state they do so by emitting photons. Neon signs are tubes of gas excited by high-voltage electricity. Their glowing colors are caused by the atoms returning to the ground state. When excited atoms in solids (and some liquids) return to the ground state, they can do so by emitting photons as gases do, but they can also emit *phonons*, packets of vibrational energy. Phonons can travel as waves through the crystal lattice in a solid, displacing the atoms from their equilibrium positions. In the image on the right, the wavelength of the emitted energy is shown in red (and the displacement of the atoms is greatly exaggerated).

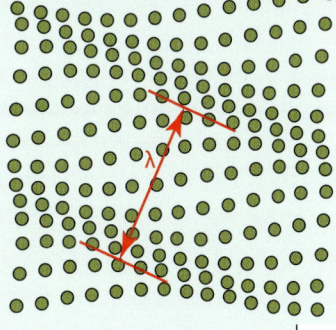

Chapter 3

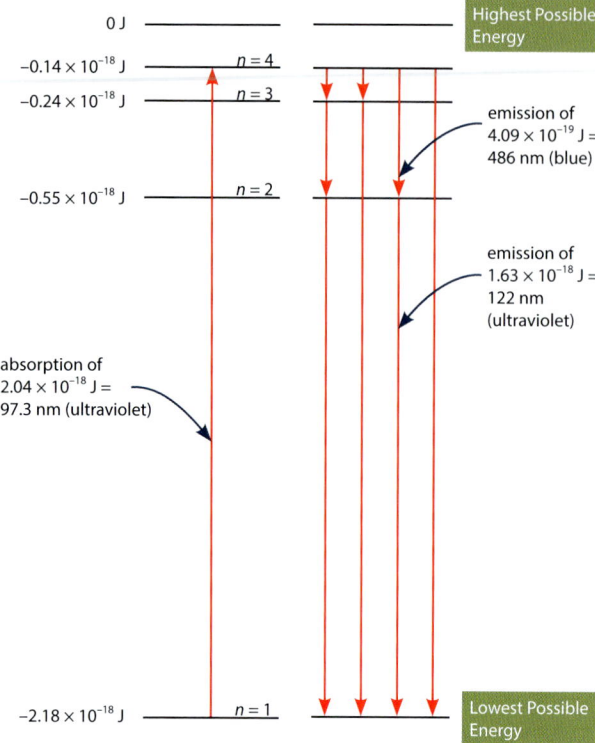

Figure 3.6. An energy-level diagram illustrating how quanta of energy are absorbed and released by electrons in hydrogen atoms.

Figure 3.6 is a diagram representing four different energy levels actually available to electrons in hydrogen atoms, labeled in the figure $n = 1$, $n = 2$, and so on. I explain this diagram carefully below, but first we need to pause here to revisit one of the points made in the Introduction. In Section I.3, I note that when an electron is held at a certain distance away from the positive nucleus, it has a high potential energy. As the electron is allowed to move closer to the nucleus, its potential energy decreases just as the gravitational potential energy of an object above the ground decreases with decreasing height. Since the electron's potential energy decreases as the electron moves closer to the nucleus, the electron releases energy as it gets closer and closer to the nucleus.

In discussions of the energies of electrons in atoms, it is customary to assign a reference value of 0 joules to the energy an electron has when it is completely free from the nucleus—in other words, when it is very far away. Then, since the electron's potential energy decreases as it gets closer to the nucleus, the energy an electron has is expressed as a negative value. This happens because instead of setting the potential energy to be zero at the nucleus, we set the zero energy reference to be when the electron is far away from the nucleus. So don't let the negative energy values bother you. It makes sense that the electron's energy is set to zero when it is far away from the nucleus because at that point the electron really has nothing to do with that nucleus. But to be consistent with what we know about potential energy, the electron's energy must decrease as it enters the area near the nucleus, so its energy takes on increasingly negative values relative to the zero-energy reference.

Going back now to Figure 3.6, let's assume that an electron in a hydrogen atom is in the ground state, which means it has the lowest possible energy. Since hydrogen atoms only have one electron, this places the electron at the bottom of the figure at the first energy level, $n = 1$. An energy of -2.18×10^{-18} J is the energy an electron has when it is in this first energy level. Assume now that this electron absorbs a quantum of energy equal to 2.04×10^{-18} J from an incoming photon, indicated by the arrow pointing upward on the left side of the figure. From the Planck relation, you can verify that this corresponds to a wavelength of 97.3 nm, placing this photon in the ultraviolet region of the electromagnetic spectrum. When this photon strikes our hydrogen atom, the atom absorbs it. The photon ceases to exist and the energy it had is added to the atom's electron. The result is that the atom's electron is now in the fourth energy level. If you subtract the energy of $n = 1$ from that of $n = 4$, the difference is the amount of energy the electron absorbed, 2.04×10^{-18} J.

Remember, energy in atoms is quantized. Electrons can only have certain specific values of energy, and the permissible values of energy an electron in a hydrogen atom can have (for the first four energy levels) are the energies listed down the left side of Figure 3.6. Very quickly the atom emits this energy in the form of new photons and the electron drops back down to the ground state. But as you can see from the right side of the figure, the electron in the hydrogen atom has four different ways of doing this.

First, the electron can release the smallest permissible amount of energy each time it emits a photon. This causes it to drop down one energy level at a time, emitting three separate photons on its way back to the ground state, as shown by the sequence of three downward pointing arrows leading from $n = 4$ to $n = 1$. Second, the electron can first drop from $n = 4$ to $n = 3$ and then drop to $n = 1$. Third, the electron can first drop from $n = 4$ to $n = 2$ and then drop to $n = 1$. For this possibility, the emitted amounts of energy are shown in the figure. The energy emitted when dropping from $n = 4$ to $n = 2$ is 4.09×10^{-19} J. Using the Planck relation, you can calculate the wavelength of a photon with this energy. Doing so gives a wavelength of 486 nm, which is in the visible portion of the electromagnetic spectrum. A check of Figure 3.1 indicates that this wavelength corresponds to blue light. The drop from $n = 2$ to $n = 4$ is a much larger energy drop, 1.63×10^{-18} J. This energy corresponds to a wavelength of 122 nm, which is in the ultraviolet region and is not visible.

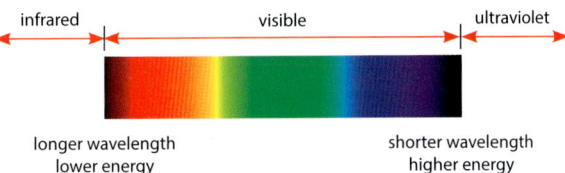

Figure 3.7. Wavelengths and energies in and near the visible spectrum.

Finally, the electron can drop back to $n = 1$ by emitting 2.04×10^{-18} J, the same amount of energy it absorbed in the first place. This energy produces a new ultraviolet photon with the same wavelength as the photon originally absorbed by the atom.

Note from Equation (3.4) that energy and wavelength are inversely proportional. Longer wavelengths represent lower energies; shorter wavelengths represent higher energies. This relationship is illustrated in Figure 3.7.

The spectrum of wavelengths emitted by each element is unique, which means that light spectra can be used to identify the element's presence in a gas or solution. The science of such identifications is called *spectroscopy*.

3.1.3 The Hydrogen Atom

The hydrogen atom is the simplest atom, with only one electron, and thus it has been studied extensively. The wavelengths for the possible electron energy transitions in the first six energy levels of the hydrogen atom are shown in Figure 3.8. All the arrows in this diagram are shown pointing in both directions because the wavelengths shown can represent either the absorption or emission of energy. These are the energies hydrogen atoms can absorb and emit.

In 1885, Swiss mathematician and physicist Johann Balmer discovered the formula that predicts the lines in the visible hydrogen spectrum.[2] This series of lines is now called the Balmer series. In 1888, Swedish physicist Johannes Rydberg worked out the more general formula for all the hydrogen wavelengths. The ultraviolet and infrared lines in the hydrogen spectrum were not known initially (because they are invisible). But in 1906, American physicist Theodore Lyman observed the ultraviolet series that bears his name, and in 1908, German physicist Friedrich Paschen observed the infrared series of lines in the hydrogen spectrum.

---

2 The wavelengths in an atomic spectrum are called *lines* because when we image them they become lines of color, as in the spectra shown on the opening page of the chapter.

Chapter 3

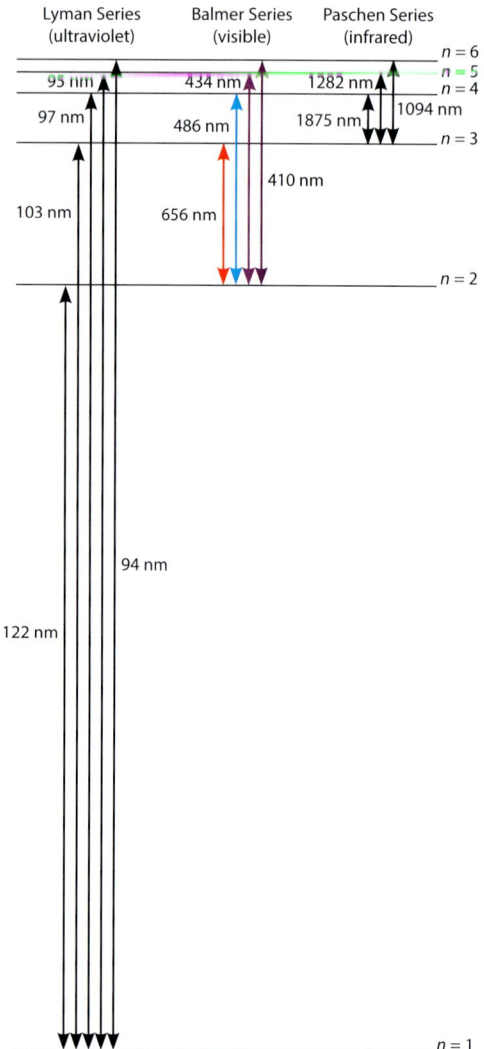

Figure 3.8. Wavelengths for electron transitions in the first six energy levels of the hydrogen atom.

The Rydberg formula that predicts all these wavelengths has an interesting mathematical structure. It is worth showing here, even though we will not be doing any calculations with it:

$$\frac{1}{\lambda} = R\left(\frac{1}{n_1^2} - \frac{1}{n_2^2}\right) \qquad (3.5)$$

The R in this equation is the so-called Rydberg constant (1.097×10^7 m^{-1}), and n_1 and n_2 represent the numbers for the two energy levels in question. Whenever I see mathematical patterns like this in nature, I am always reminded that the mathematical structure found everywhere in nature could not have arisen apart from the hand of an intelligent Creator. The mathematics embedded in creation is strong evidence that the physical universe we are studying in our science courses is a great and beautiful gift, and as Nicolaus Copernicus said, was "built for us by the Best and most Orderly Workman of all."

3.2   The Bohr Model of the Atom

As mentioned on the opening page of this chapter, Danish physicist Niels Bohr (Figure 3.9) introduced his new model of the atom in 1913. This new atomic model was of tremendous importance in the development of atomic theory. Rydberg's formula predicting the lines in the hydrogen spectrum had been known since 1888, but until Bohr's model there was no theoretical basis for the observed spectrum.

In Bohr's atomic model, the electrons orbit the nucleus like planets orbiting the sun. In the model, the electrons have fixed energies—the same energies as those shown in Figures 3.6 and 3.8. These different energy levels correspond to different orbits around the nucleus.

Bohr correctly described the cause of the specific lines in the emission spectra of atoms—electrons absorbing energy and moving to higher energy levels and then releasing photons at specific energies as they move back to lower energy levels.

Another significant feature of the Bohr model is the number of electrons that he permitted at each energy level. These numbers are shown in Figure 3.10. If you compare these numbers to the Periodic Table of the Elements, you see that the number of electrons in each energy level corresponds to the number of elements in each period

Figure 3.9. Danish physicist Niels Bohr (1885–1962).

Atomic Structure

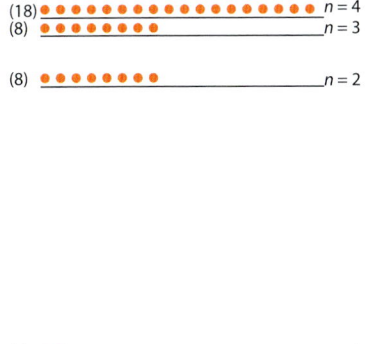

Figure 3.10. Number of permissible electrons in the first four energy levels of the Bohr model of the atom.

(row) of the table: two in the first, eight in the second, eight in the third, and 18 in the fourth, etc.

As powerful as it is, the Bohr model was known to have weaknesses from the start. For one thing, there is no explanation for why electrons are able to stay in their orbits. Electrons moving in circles radiate energy, so one would think electrons in orbits would gradually lose energy, slow down, and spiral in to the nucleus. Another issue is that for atoms other than hydrogen, the energies between the energy levels do not match up precisely with the observed wavelengths in the emission spectra of elements.

3.3 The Quantum Model of the Atom

3.3.1 Schrödinger and Pauli

In 1926, Austrian physicist Erwin Schrödinger (Figure 3.11) published what is now called the *Schrödinger equation*. This was a landmark achievement and one of the hallmarks of 20th-century physics. For this work, Schrödinger received the Nobel Prize in Physics in 1933.

Solutions to the Schrödinger equation are now understood to provide us with the details of the internal structure of energy levels in atoms. With the arrival of the Schrödinger equation, the *quantum model* of the atom began to unfold. The history of quantum physics is still being written. There are many mysteries associated with the behavior of electrons as described by the quantum model. But quantum theory has a colossally impressive string of achievements, and its success in predicting atomic behavior is undeniable. We know the quantum model will continue to evolve, and may some day even be replaced. But the details we will consider in this section are now generally accepted as correct. Remember, chemistry is all about modeling—developing theories. Theories are explanations, and the quantum model is widely accepted as our best explanation of how atoms are structured.

Figure 3.11. Austrian physicist Erwin Schrödinger (1887–1961).

In addition to Schrödinger's equation, there is one other theoretical milestone that we need to have in hand to understand the details to follow. In 1925, while Schrödinger was working on his equation, another Austrian physicist, Wolfgang Pauli (Figure 3.12) formulated what is now known as the *Pauli exclusion principle*. In short, *the Pauli exclusion principle holds that no two electrons in the same atom can occupy the same quantum state*. We will unpack this further as we go along. For this important contribution to quantum theory, Pauli won the Nobel Prize in Physics in 1945.

3.3.2 Shells, Subshells, and Orbitals

The quantum state of an electron in an atom—its unique address, we might say, within the atomic quantum realm—is specified by four different *quantum numbers*. According to the Pauli exclusion principle, every electron in an atom has a unique quantum state. This is one of the laws of nature governing the way atoms are structured. This situation in atoms is analogous to postal addresses.

Figure 3.12. Austrian physicist Wolfgang Pauli (1900–1958).

Every postal customer in the U.S. has a unique address. For a house, this unique address requires four pieces of information—the street number, street name, city, and state. For an apartment complex, an apartment number is also required. (The zip code doesn't contain any additional location information; it just helps speed things up. That's why it's called a *zip* code.)

The physics behind these quantum numbers is quite complicated, and in this introductory course we will not be getting much into that. However, introductory chemistry classes do generally now require students to learn the arrangement of shells, subshells, and orbitals in atoms for "energy levels" $n = 1$ through $n = 4$ because knowing this structure allows us to specify where the electrons are in an atom. And as you recall from the Introduction, chemistry is all about electrons! So here we go. There is a *lot* of detail in this section, and most of it is important. Read it slowly and carefully, and read it more than once.

The phrase *energy levels* is in quotes just above for an important reason. We are transitioning now from the energy levels in Bohr's atomic model to those of the far more accurate quantum model. In Bohr's model, and in the hydrogen atom as we still understand it, there is only one energy level for each value of n. The quantum model is quite different, as we will see.

Recall from Figure 3.10 that the numbers of electrons permitted in the first four levels of Bohr's model are 2, 8, 8, and 18. These numbers correspond to the number of elements in the first four periods of the periodic table. Bohr was on the right track, but did not initially perceive the correct pattern. We now refer to n as the *principal quantum number*, and in every atom except hydrogen there are multiple energy levels associated with each value of n. As explained in detail below, the number of electrons allowed for each value of n is actually $2n^2$. This gives us 2, 8, 18, and 32 electrons in the various energy levels associated with $n = 1$ through $n = 4$.

The clusters of energy levels associated with each value of n are commonly called *shells*. As I state just above, the quantum state of an electron in an atom, including its energy, is specified by four quantum numbers; the principal quantum number—the shell number—is the first of them. So beginning with the principal quantum number, here is a list of the names and other details for the four quantum numbers:

1. Principal Quantum Number, n Values for n are the integers 1, 2, 3, 4, 5, ... These are the main clusters of energy levels in the atom, usually called *shells*. So far as we know, there is no highest value for n.

2. Azimuthal Quantum Number, l Within each shell except the first one ($n = 1$), there are *subshells*. The number of subshells in a shell is equal to the principal quantum number. For example, for $n = 3$ there are three subshells. Values for l are integers ranging from 0 to $(n - 1)$. Typically, these subshells are referred to by the letters s, p, d, f, and g rather than by the values of l. These common letter designations are shown Table 3.1. The azimuthal quantum numbers describe specific types of subshell configurations. So for example, within any shell the s subshell is always structured the same way. Likewise, the p subshell has the same general structure in every shell except $n = 1$ (since $n = 1$ doesn't have a p subshell). Again, the number of subshells in a given shell is equal to the principal quantum number. So, in the first shell there is one subshell, denoted as $1s$. In the $n = 2$ shell, there are two subshells, denoted as $2s$ and $2p$, and so on. (Note: The azimuthal quantum number is also sometimes called the *angular momentum quantum number*.)

l value	Common Letter Designation
0	s
1	p
2	d
3	f
4	g

Table 3.1. Letters used to designate values of the azimuthal quantum number, l.

3. Magnetic Quantum Number, m_l Within each subshell (numbered l), the possible values for m_l are the integers ranging from $-l$ to l. So, in a subshell with $l = 2$, the values for m_l are $-2, -1, 0, 1$, and 2. The magnetic quantum number is associated with specific shapes and orientations of orbitals within a subshell. A important point to note is that *any orbital in an atom can hold at most two electrons.*

4. Spin Projection Quantum Number, m_s As you recall, the Pauli exclusion principle requires every electron in an atom to be in a unique quantum state. That is, the state of each electron is described by a unique set of quantum numbers. And since each orbital can hold two electrons, we need one more piece of information to distinguish from one another the quantum states of the two electrons in an orbital. This characteristic is called *spin*. Unfortunately, it's a very misleading term because electrons aren't really spinning. In fact, it's pretty hard to say exactly *what* they are doing. But anyway, accepting spin as a real property analogous in some way to spinning, any two electrons in the same orbital have opposite spins. The two possible values for electron spin are $m_s = +1/2$ and $m_s = -1/2$, and we call these "spin up" and "spin down." At this point in your career you really don't need to worry about what these strange names and numbers mean. The fact is, when there are two electrons in the same orbital (and there can be at most two) one has spin up and one has spin down. This final quantum specification allows each electron in every atom to inhabit a unique quantum state.

All this information pertaining to the first three quantum numbers is summarized in Table 3.2. Hopefully your understanding of all these shells, subshells, and orbitals will be enhanced by looking at images of computer models of the orbitals. Let's be a bit clearer about what these orbitals are: they represent the solutions to the Schrödinger equation for electrons in atoms with different energies. Table 3.3 depicts the orbitals in the various subshells associated with the first three shells, $n = 1$ through $n = 3$. Note first that in every shell there is an s orbital. These are spherical in shape. The models shown depict the sphere cut in half so you can see the relative sizes. If you look carefully at the $2s$ orbital, you can see the tiny $1s$ orbital inside it. And inside the $3s$ orbital you can see both the $1s$ and $2s$ orbitals inside it. We are coming back to electron energy soon, but for now note that within any shell, the s orbital is the lowest energy orbital in that shell. Note also that for all these orbital arrangements, the atomic nucleus is at the center. All orbitals are symmetric about the nucleus.

Beginning with $n = 2$, there is a p subshell in each shell, and beginning with $n = 3$ there is also a d subshell in each shell. The orbitals in the p subshell are usually described as resembling

n	Possible Values of l	Subshell Name	Possible Values of m_l (Each value corresponds to one orbital.)	Number of Orbitals in the Subshell	Total Number of Orbitals in the Shell (= n^2)
1	0	1s	0	1	1
2	0	2s	0	1	4
	1	2p	$-1, 0, 1$	3	
3	0	3s	0	1	9
	1	3p	$-1, 0, 1$	3	
	2	3d	$-2, -1, 0, 1, 2$	5	
4	0	4s	0	1	16
	1	4p	$-1, 0, 1$	3	
	2	4d	$-2, -1, 0, 1, 2$	5	
	3	4f	$-3, -2, -1, 0, 1, 2, 3$	7	

Table 3.2. Subshells and orbitals for the $n = 1$ through $n = 4$ shells.

Chapter 3

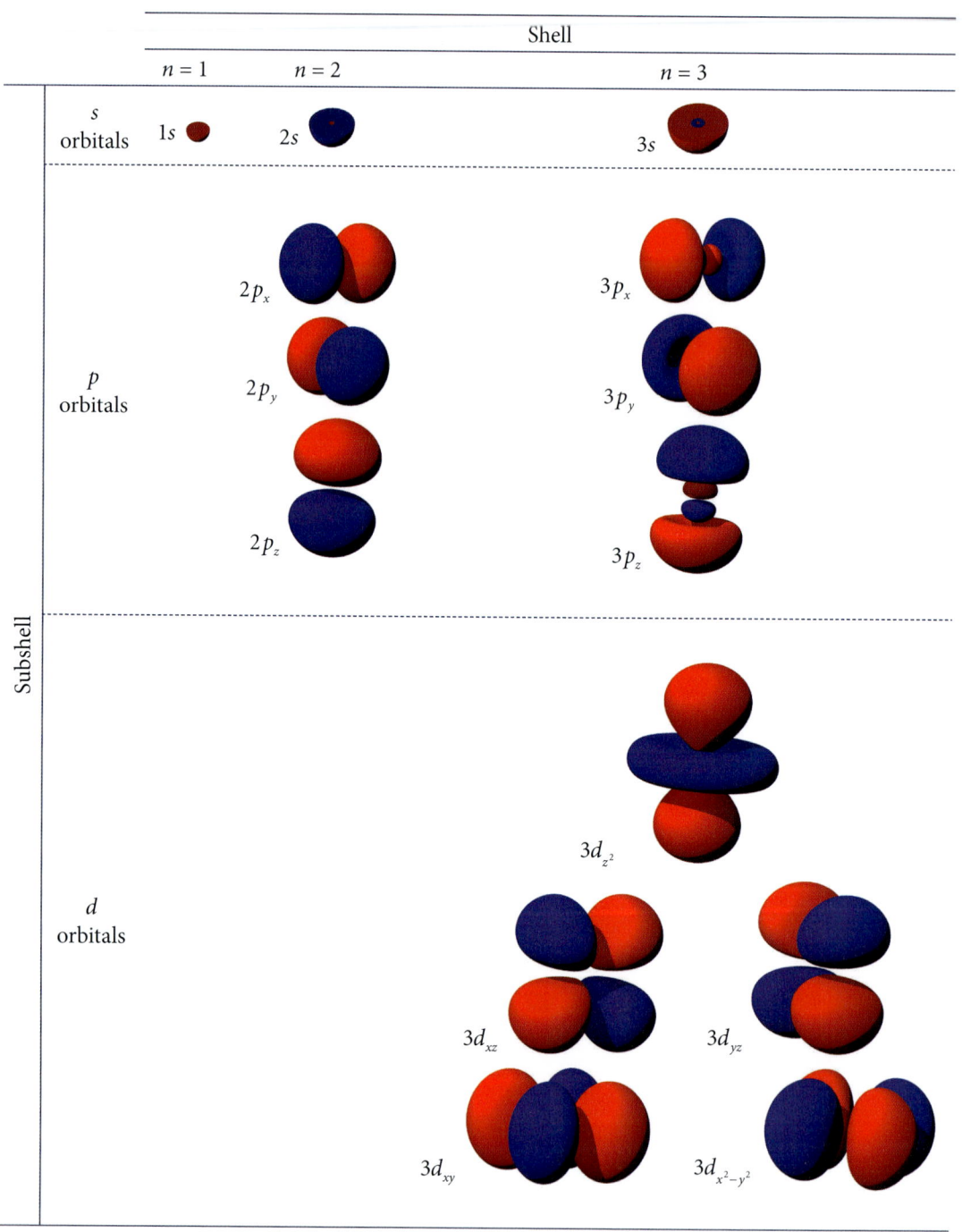

Table 3.3. Shapes of *s*, *p*, and *d* orbitals for the $n = 1$, $n = 2$, and $n = 3$ shells.

"dumbbells" because of their twin lobes. There are three of these twin-lobed orbitals in each p subshell, each oriented at right angles to the other two. For this reason, they are designated the p_x, p_y, and p_z orbitals—they can be thought of as lined up along the x, y, and z axes in a three-dimensional Cartesian coordinate system as depicted in Figure 3.13. (I explain the elongated appearance of the orbitals shown in Figure 3.13 shortly.) In Table 3.3, in order to make the s orbitals large enough to see and still have room to fit the p and d orbitals on the page, the p and d orbitals are shown much smaller than their actual size relative to the s orbitals.

Looking now at the $n = 3$ orbitals in Table 3.3, note that the $3p$ orbitals are shown surrounding the $2p$ orbitals. The three $3p$ orbitals are superimposed on each other just as the $2p$ ones are (Figure 3.13). Finally, you can see that shapes of the five $3d$ orbitals are pretty bizarre. These orbitals are also superimposed on each other, and the whole bunch of them is superimposed on top of all the other orbitals in the table. Then, of course, there are all the orbitals for higher principal quantum numbers superimposed on top of them. Beginning with the $n = 4$ shell (not shown in the table) there is an f subshell in each shell. There are seven orbitals in each f subshell, and they sort of resemble the d orbitals, only with six or eight lobes instead of four.

Recall that each orbital can house a maximum of two electrons (with opposite spins). For example, just to be clear, the $2p_x$ orbital with its two lobes is a single orbital, (even though Table 3.3 shows the two lobes in different colors). Likewise, the $3d_{z^2}$ orbital with its two lobes and doughnut around the middle is also a single orbital. With a maximum of two electrons in each orbital, you can see that the first shell, $n = 1$, can hold at most two electrons, both in the $1s$ orbital. The $n = 2$ shell can hold a maximum of eight electrons: two in the $2s$ orbital, and two in each of the three $2p$ orbitals. The $n = 3$ shell can hold up to 18 electrons: two in the $3s$ orbital, six in the $3p$ orbitals, and a total of 10 in the $3d$ orbitals.

Before we move on and get back to talking about energy, one more important point should be made about orbitals. As noted above, the orbitals shown in Table 3.3 are the solutions to the Schrödinger equation. However, it is not correct to think of these shapes as locating where the electrons *are*. (Remember, the world of quantum mechanics is weird.) But it turns out that if we *square* the solutions to the Schrödinger equation we get shapes indicating *probabilities* of where the electrons are. This is what is depicted in Figure 3.13. Squaring the $2p$ solutions elongates the shapes of the orbitals. These orbitals that come from squaring solutions to the Schrödinger equation are called *probability distributions*. They should be envisioned as fuzzy at the edges and denser in the middle, indicating a lower probability that an electron is at the edge of the orbital and a higher probability that an electron is in the center part of the orbital.

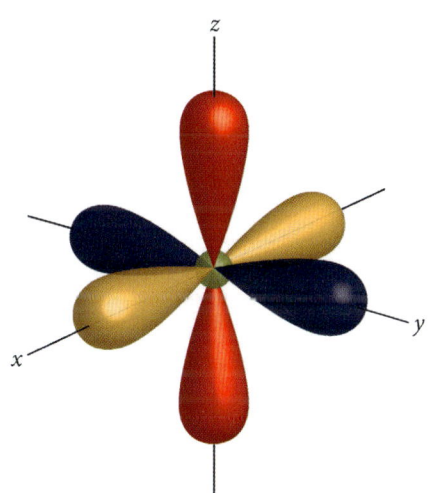

Figure 3.13. The elongated "probability distributions" of the 2s and 2p orbitals shown together. (2s = green; $2p_x$ = yellow, $2p_y$ = blue, $2p_z$ = red)

Note just one more feature of the orbital arrangements: just because an orbital has more than one part—like the two lobes of a p orbital—does not mean that one electron is in one lobe and the other electron is in the other lobe. Instead, both electrons inhabit both lobes. Even stranger, to pass from one lobe to the other the electron somehow passes right through (around?) the atomic nucleus. (Don't hurt your brain by trying too hard to understand this. No one else understands it either! Electrons are very strange.)

3.3.3 The Aufbau Principle, the Madelung Rule, and Hund's Rule

Now that you know how the orbitals are arranged, we return to the topic of electron energies. Let's begin by recalling how one knows how many electrons an atom has. Unless it has ionized, an atom has the same number of electrons as protons, and the number of protons is given by the *atomic number* (Z). For example, if you check the periodic table you see that iron is element 26. This means an atom of iron has 26 protons and 26 electrons. The protons are all in the nucleus with the neutrons. The electrons are distributed around in various orbitals.

Figure 3.14 is another type of energy level diagram and illustrates the energies associated with the different orbitals. In this diagram, each little square represents an orbital, and each string of connected squares represents a subshell. On the left are the orbital energies for the hydrogen atom. As you see, all orbitals associated with a given principal quantum number (1, 2, 3, ..., etc.) have the same energy. These are the energies shown in Figures 3.6 and 3.8.

On the right side of Figure 3.14 is a general arrangement depicting the energies for atoms other than hydrogen. Here the energies go up with each subshell. For example, subshell 4f has a higher energy than 4d, which has a higher energy than 4p, which has a higher energy than 4s. Also, note especially that the energies associated with different principal quantum numbers (shells) overlap. Thus, subshell 4s has a lower energy than subshell 3d.

An important point to note about the right side of Figure 3.14 is that the exact energies associated with various subshells are different for every element. With only one electron, the orbitals in an energy level of a hydrogen atom are basically all the same. But with multiple electrons repelling each other in an atom, the subshells begin spreading upward and each subshell is at a different energy. The amount of spread—and thus the exact energy associated with each subshell—is different for every element. The important consequence of this for what we have covered so far in this chapter pertains to atomic spectra. The amount of energy released by an electron transition from, say, a 5d orbital to a 4p orbital depends on the atom—that is, the element—involved. As you know, the energy in an emitted photon determines its wavelength and color (the Planck relation). The fact that the energies for the different orbitals depend on the element is the reason why spectroscopy can be used to identify the presence of elements in a sample. Each atom emits its own spectrum of wavelengths corresponding to the unique energy differences between the orbitals in that particular kind of atom.

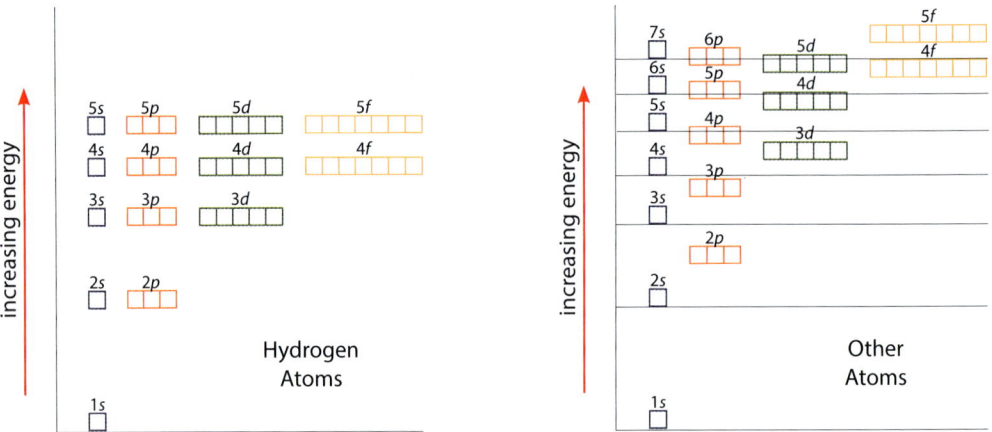

Figure 3.14. In hydrogen atoms, all orbitals within a given shell are at the same energy level. In atoms of other elements, orbital energy increases with increasing azimuthal quantum number, and the sequence of energies follows the Madelung rule.

The colors emitted by excited atoms in two metals are illustrated in Figure 3.15. The images show lithium and copper wires heated in a stove-top gas flame, causing electrons in the metal atoms to absorb photons of heat energy (electromagnetic radiation in the infrared region). As the electrons return to the ground state, they emit photons of visible light, and the colors produced depend on the energies of the subshells in the atoms of the respective metals. If the flames are observed through a prism, the colors in the flames are separated into a line spectrum and the individual color wavelengths can be identified. A test like this that uses flame as the energy source for exciting the metal is called a *flame test*.

With the energy picture under our belts, we are finally ready to describe the principles governing how electrons are arranged in atoms. We are describing here the electron positions when atoms are in the ground state. You already know that when atoms are excited, electrons jump from ground state energies up to higher energies.

There are three principles involved in determining electron arrangement in ground state atoms. The first is the *Aufbau principle*, named after a German word meaning "building up." The Aufbau principle states that *electrons fill places in orbitals in order of increasing energy, starting from the lowest energy orbital and going up from there*. Remember: chemistry is all about minimizing energy. Electrons in a ground-state atom go into the lowest energy orbitals available.

Figure 3.15. Lithium (top) and copper flame tests.

The second principle is the *Madelung rule*. This principle specifies the order of the shells and orbitals for increasing energy. On the right side of Figure 3.14, the sequence the orbitals are in as energy increases follows the Madelung rule. Another common way of depicting the sequence of energies according to the Madelung rule is shown in Figure 3.16. If you start at the top and follow the arrows in descending order, you get the same sequence of orbitals as shown on the right side of Figure 3.14.[3]

The third principle involved in electron arrangements is *Hund's rule*, which applies to the case of subshells that are only partially filled. Hund's rule states that *if orbitals of equal energy are available within a subshell, electrons fill them all up singly before they begin doubling up in orbitals*. For example, as you can see from the right side of Figure 3.14, the 3d subshell contains five orbitals of equal energy. According to Hund's rule, if there are electrons in this subshell, but not enough electrons to fill the subshell, the electrons go into the orbitals as one electron per orbital until each of the five orbitals has one electron in it. After that, any remaining electrons go

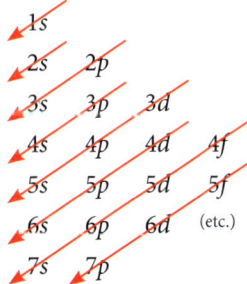

Figure 3.16. Since each red arrow represents a particular value of $n + l$, this diagram shows the order in which the subshells fill with electrons, according to the Madelung rule.

---

3 There is a mathematical principle involved in this ordering: each arrow in Figure 3.16 represents a particular value of the sum of the principal quantum number and the azimuthal quantum number, $n + l$. For example, look at the arrow starting at the 3d orbital. For 3d, $n = 3$ and $l = 2$, so $n + l = 5$. (The values of l are shown in Table 3.1.) For 4p, $n = 4$ and $l = 1$, and $n + l = 5$. For 5s, $n = 5$ and $l = 0$, so $n + l = 5$. So the subshells fill up in the sequence shown in Figure 3.16.

Chapter 3

in as the second electron in each orbital until each electron has a place. And again, remember that each orbital can hold at most two electrons.

The principle at work behind Hund's rule is again energy minimization. Spreading single electrons in the orbitals of unfilled subshells is a lower energy configuration than putting pairs of electrons together when other orbitals remain empty. Minimizing the energy this way also makes the atom more stable, just as the cone on its side in Figure I.6 is more stable than the cone on its point.

3.4 Electron Configurations

3.4.1 Electron Configurations

You may be pleased to know that the ocean of information described in the previous section will be a lot easier to remember after you have had a bit of practice writing *electron configurations* to indicate where all the electrons are in an atom of a given element. The electron configuration for a given element is a list, written in a particular format, of all the subshells in use in an atom and how many electrons are in each one. As an example, consider iron, atomic number 26 ($Z = 26$). There are 26 electrons in an atom of iron. The subshells required to hold them all, in order of increasing energy according to the Madelung rule, are as follows:

1s	2s	2p	3s	3p	4s	3d
holds 2 electrons	holds 2 electrons	holds 6 electrons	holds 2 electrons	holds 6 electrons	holds 2 electrons	holds 6 electrons in 10 places

The electron configuration is formed simply by chaining these together, placing the numbers of electrons as superscripts on the subshells they go with, without any punctuation. In front of the electron configuration, it is customary to place the element's chemical symbol followed by a colon. So, the electron configuration for iron is written as follows:

Fe: $1s^2 2s^2 2p^6 3s^2 3p^6 4s^2 3d^6$

Electron configurations only indicate subshells; they do not indicate which orbitals electrons are in inside the subshells. But we can use an *orbital diagram* similar to Figure 3.14 to show more precisely where the electrons are. Remember, Hund's rule comes into play, requiring that orbitals of equal energy each receive one spin-up electron before any of them take a second spin-down electron. A great metaphor for this was first used by Wolfgang Pauli, who formulated the Pauli exclusion principle. Pauli said that when filling up the orbitals in a subshell, electrons are like passengers filling a bus. Each takes a seat by himself until every seat has one person in it. After that, people start doubling up.

Figure 3.17 shows the electron arrangement for phosphorus, $Z = 15$. Each of the little arrows represents one electron, with upward arrows representing spin up and downward arrows representing spin down. Notice that the three electrons in the 3p subshell are placed so that each orbital contains one spin-up electron, as Hund's rule requires.

To make an orbital diagram, you simply show the orbitals in order, side by side, and put in the arrows representing the electrons. Thus, the orbital diagram and electron configuration for phosphorus are as follows:

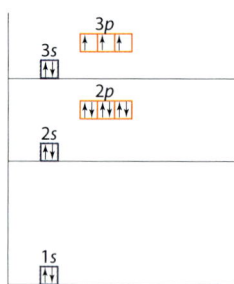

Figure 3.17. The electron arrangement for phosphorus, with 15 electrons.

Orbital Diagram	Electron Configuration
1s 2s 2p 3s 3p	P: $1s^2 2s^2 2p^6 3s^2 3p^3$

Here are three more examples: sodium (Na) with 11 electrons, chlorine (Cl) with 17 electrons, and nickel (Ni) with 28 electrons:

Z	Orbital Diagram	Electron Configuration
11	1s 2s 2p 3s	Na: $1s^2 2s^2 2p^6 3s^1$
17	1s 2s 2p 3s 3p	Cl: $1s^2 2s^2 2p^6 3s^2 3p^5$
28	1s 2s 2p 3s 3p 4s 3d	Ni: $1s^2 2s^2 2p^6 3s^2 3p^6 4s^2 3d^8$

Note in each case that the superscripts add up to the number of electrons being represented in the notation: 11, 17, and 28 in the three examples above.

To write an electron configuration, you need to know the atomic number (Z) to get the number of electrons, the number of electrons that can reside in each of the four types of subshells (s, p, d, and f; we won't deal with g subshells in this course), and the energy sequence according to the Madelung rule. The periodic table itself is the best aid to writing an electron configuration. Figure 3.18 is a depiction of the periodic table in its full, long form. The rare-earth elements, normally shown separated out beneath the main table, are shown in their rightful place.[4] Notice the captions on the different blocks of elements in the table. There are two groups (columns) in the s block, because s subshells can only hold two electrons. The p subshells can hold six electrons because there are three orbitals in each p subshell, and the p block is six groups wide. In the same way, d subshells have five orbitals, so the d block is 10 groups wide, and f subshells have seven orbitals, so the f block is 14 groups wide. (See Table 3.2 to confirm the number of orbitals in a subshell, keeping in mind that each orbital can hold up to two electrons.)

Now look what happens as we move in order through elements in the table:

- Hydrogen (H) has one electron, helium (He) has two. In the element at the end of the 1st period (row)—helium—the 1s subshell is full.
- Lithium (Li) has three electrons, beryllium (Be) has four. With Be, the 1s and 2s subshells are full. The 5th electron possessed by boron (B) goes into the 2p subshell. So do all the additional electrons added in elements 6 through 10. Neon (Ne) has 10 electrons, so in the element at the end of the 2nd period—neon—subshells 1s, 2s, and 2p are all full.
- Sodium (Na) has 11 electrons, and the 11th one goes in the 3s subshell. Magnesium (Mg) has 12, and the 12th one fills the 3s subshell. Aluminum (Al) has 13, and the 13th one goes in the 3p subshell. In the element at the end of the third period—argon (Ar)—the 1s, 2s, 2p, 3s, and 3p subshells are full.
- The 4th period begins with potassium (K) and calcium (Ca), in which new electrons are placed in the 4s subshell. Then look what happens next: the next element, scandium (Sc), has 21 electrons and the 21st one goes into a d subshell, the 3d subshell, in fact. Each new

---

4 The f-block elements shown are also called the *inner transition metals* or *rare-earth elements*. They are usually removed from the table and shown beneath it for the simple reason that with them in place, the table is inconveniently wide. Since most of our work in chemistry is with s-, p-, and d-block elements, the removal of the f-block elements in the standard representation of the periodic table doesn't cause much trouble.

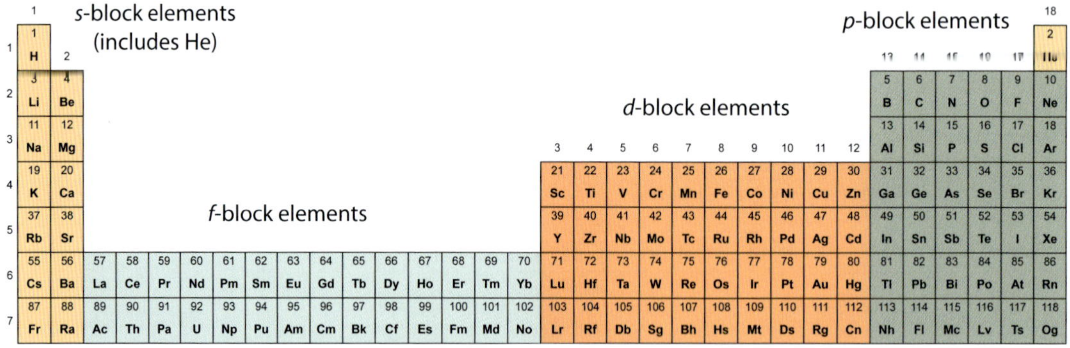

Chapter 3

Figure 3.18. Common terms applied to groups of elements based on the type of suborbital that is being filled as we move through the block in a given period (row). Group numbers shown are the numbers typically shown on the standard table with the f-block elements taken out.

electron for elements 21 through 30 goes into the 3d subshell. Then the 31st electron in gallium (Ga) is placed in the 4p subshell. In the element at the end of the 4th period—krypton (Kr)—the 1s, 2s, 2p, 3s, 3p, 4s, 3d, and 4p subshells are full.

This pattern continues, and the order of new shells coming into use continues to follow the Madelung rule. Now you can see why the blocks of elements in Figure 3.18 are identified the way they are. Moving from left to right across any period in the table, the additional electron for atoms of the next element goes into a subshell of the type indicated by the name of the block the element is in.

Be careful when writing electron configurations for elements at the beginning of the d and f blocks. The first d subshell that occurs in the sequence of shell filling is the 3d subshell, even though the elements that fill it are in the 4th period. Table 3.2 and Figure 3.16 will help remind you that the number of subshells for a particular principal quantum number n is equal to n.

3.4.2 Condensed Electron Configurations

You may already know that the elements in Group 18, the noble gases, are very unreactive. In fact, these elements eluded discovery by researchers for a long time. Since they don't form compounds, scientists didn't even know they existed!

The reason the noble gases are so nonreactive is that their electron arrangements are very stable, low-energy configurations. Obviously, each of the noble gases is at the end of a period in the periodic table (Group 18). Figure 3.19 illustrates the pattern that occurs in the orbital filling of these elements. In each case, all orbitals are filled up to but not including the s orbital of the next principal quantum number.

Since the noble gases are so stable, occurring as they do at the end of each period, the chemical symbols of the noble gases are used to form the so-called *condensed electron configurations*. The condensed electron configuration is a shorter, more convenient form.

Here's an example to show how this works. A glance at the periodic table shows that the only difference in the electron configurations of, say, titanium (Ti, $Z = 22$) and argon (Ar, $Z = 18$), is that titanium has four extra electrons. The electron configurations for argon and titanium are:

Ar: $1s^2 2s^2 2p^6 3s^2 3p^6$

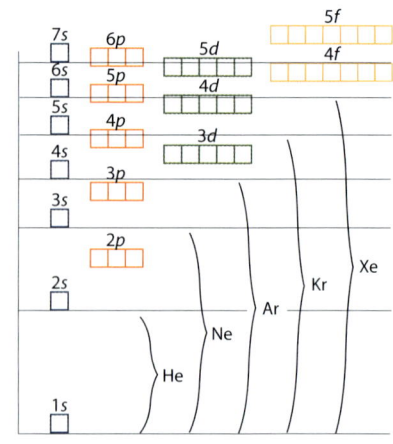

Figure 3.19. The orbital filling pattern of the noble gases.

Ti: $1s^2 2s^2 2p^6 3s^2 3p^6 4s^2 3d^2$

The condensed electron configuration for any element is written by using the chemical symbol of the noble gas in the previous period to represent all the filled orbitals up to that point, and then just adding on the orbitals to hold the additional electrons due to the period where the element is. The noble gas chemical symbol is always written in square brackets. For titanium, argon is the noble gas at the end of the previous period. So, the condensed electron configuration for titanium is written as

Ti: $[Ar]4s^2 3d^2$

As one more example, neon is at the end of the period prior to the period phosphorus is in, so the condensed electron configuration for phosphorus ($Z = 15$) is written as

P: $[Ne]3s^2 3p^3$

In the condensed electron configuration, the inner-shell electrons lumped together under the noble gas symbol are called the *core electrons*.

To wrap up this section, I will mention briefly that there are a few elements (19 in all) whose electron configurations do not completely follow the sequence specified by the Madelung rule. In these exceptions, one or two of the higher energy electrons in the atom occupies an orbital in a subshell one level higher than we would expect. For example, we would expect the condensed electron configuration for chromium ($Z = 24$) to be

Cr: $[Ar]4s^2 3d^4$

However, it is not! One of the 4s electrons goes into the 3d subshell instead, giving

Cr: $[Ar]4s^1 3d^5$

Copper is another element that has an anomalous electron configuration. We are not going to go into the details about these anomalies, except to mention that the cause has to do with one of the themes we reviewed in the Introduction—minimizing energy. All the anomalies are *f*-block and *d*-block metals. These metals have lots of subshells and the energy differences between them are typically very slight. In these anomalous cases, it turns out that the way to minimize the total energy in the atom is for one of the electrons to jump up a notch. If you look again at the chromium example above, you see that with the extra electron in the 3d subshell the 3d subshell is exactly half full. This turns out to represent a lower total energy configuration for the atom than having the 4s subshell full.

3.5   Empirical Formulas

3.5.1 Percent Composition and Empirical Formulas

Laboratory chemical analysis of a substance enables a chemist to determine the *percent composition* of the substance. When a new compound is discovered, chemists place a high priority on determining the percentages, by mass, of each element in the substance. This is the percent composition. From the percent composition, a so-called *empirical formula* for the substance can be worked out—a formula that represents the ratios of the elements in the substance.

For example, suppose laboratory analysis of a 221.6-g sample of ascorbic acid (vitamin C) results in the following mass data:

H: 10.15 g

C: 90.68 g
O: 120.8 g

If we divide each of the mass values by the total mass of the sample, we have the percent composition of the sample:

H: $\dfrac{10.15 \text{ g}}{221.6 \text{ g}} = 0.04580$

C: $\dfrac{90.68 \text{ g}}{221.6 \text{ g}} = 0.4092$

O: $\dfrac{120.8 \text{ g}}{221.6 \text{ g}} = 0.5451$

Thus, the percent composition is 4.58% hydrogen, 40.92% carbon, and 54.51% oxygen. Note that we expect these percentages to add up to 100%, but due to limits on the precision of the data they add to 100.01%.

Given either the percent composition or the actual masses from a sample, we can determine the empirical formula for a substance. The empirical formula may differ from the actual *molecular formula* of the substance. An empirical formula represents the smallest whole number ratios of the elements in the substance, while the molecular formula represents the actual numbers of each element in the molecule.

For example, hydrogen peroxide, H_2O_2, is a common household disinfectant. Molecules of hydrogen peroxide contain two atoms of hydrogen and two atoms of oxygen, so the molecular formula for this substance is H_2O_2. But the empirical formula is HO, because the empirical formula contains the smallest whole number values that can represent the ratios in the compound. Since each molecule of H_2O_2 contains two atoms of H and two atoms of O, the ratio of H to O in the molecule is 2 : 2, which is the same ratio as 1 : 1, giving an empirical formula of HO.

In many cases, the empirical and molecular formulas are identical. The molecular formula for methane, for example, is CH_4. This formula indicates a ratio of carbon to hydrogen atoms in the molecule of 1 : 4. This same formula is the empirical formula because 1 and 4 are the smallest whole numbers that can represent this ratio.

To determine the empirical formula from percent composition, begin by assuming you have a sample of the substance with a mass of exactly 100 g. Use the percent composition to determine the masses of each element in the 100-g sample, then use the mass data from the periodic table to convert each of these masses to numbers of moles. Finally, divide each of the mole values by the smallest number of moles to determine the whole number ratios in the formula. The following example illustrates this calculation.

▼ Example 3.2

Given percent composition data for ascorbic acid (see above), determine the empirical formula for this substance.

We assume a sample with a mass of exactly 100 g. We begin by using the percent composition to obtain masses in grams for each element in the substance. Assuming a 100-g sample just makes this easy. Since hydrogen is 4.58% of the 100-g sample, the mass of the hydrogen in the sample is 4.58 g. Similarly, the masses of the carbon and oxygen are 40.92 g and 54.51 g, respectively.

Next, we use the molar masses for each element to convert each of these masses to number of moles. We use the molar mass as a conversion factor, just as we have before.

$$4.58 \text{ g H} \cdot \frac{1 \text{ mol}}{1.0079 \text{ g}} = 4.54 \text{ mol H}$$

$$40.92 \text{ g C} \cdot \frac{1 \text{ mol}}{12.011 \text{ g}} = 3.407 \text{ mol C}$$

$$54.51 \text{ g O} \cdot \frac{1 \text{ mol}}{15.9994 \text{ g}} = 3.408 \text{ mol O}$$

Next, to determine the ratios of elements in the substance, divide each of these mole amounts by the smallest of them.

$$\frac{4.54 \text{ mol}}{3.407 \text{ mol}} = 1.33$$

$$\frac{3.407 \text{ mol}}{3.407 \text{ mol}} = 1.00$$

$$\frac{3.408 \text{ mol}}{3.407 \text{ mol}} = 1.00$$

These values tell us that the ratio of hydrogen to carbon to oxygen in ascorbic acid is 1.33 : 1.00 : 1.00. Now, we need the smallest whole numbers that preserve this same ratio. Noting that the value 1.33 is very close to 4/3, we multiply all the values by 3 to get whole number ratios of 4 : 3 : 3 for hydrogen : carbon : oxygen. Finally, we use these ratios to write the empirical formula. In formulas containing these three elements, it is traditional to write the elements in the formula in the order C—H—O. Doing so gives us

$C_3H_4O_3$.

3.5.2 Determining a Molecular Formula from an Empirical Formula

The empirical formula determined in the previous example relates to the molecular formula by some simple multiple. Recall that the subscripts in the molecular formula of hydrogen peroxide, H_2O_2, are simply double the subscripts in the empirical formula, HO. We can determine the molecular formula for a compound from the empirical formula if we have access to the molecular mass of the compound. We do this by computing the formula mass for the empirical formula and comparing this to the molecular mass to see what the multiple is between the empirical formula mass and the molecular mass. Then we can multiply the subscripts in the empirical formula by the same multiple to get the molecular formula. In other words,

$$\text{whole number multiple} = \frac{\text{molecular mass}}{\text{empirical formula mass}}$$

This calculation is illustrated in the following example. Note that although this example uses atomic masses and molecular mass in u, the same computation can be performed using molar masses in g/mol.

▼ Example 3.3

The experimentally determined molecular mass for ascorbic acid is 176.1 u. Use this value and the empirical formula from Example 3.2 to determine the molecular formula for ascorbic acid.

We begin by determining the formula mass for the empirical formula, $C_3H_4O_3$.

C: 12.011 u

H: 1.0079 u

O: 15.9994 u

$(3 \times 12.011 \text{ u}) + (4 \times 1.0079 \text{ u}) + (3 \times 15.9994 \text{ u}) = 88.063 \text{ u}$

Next we calculate the whole number ratio by dividing the molecular mass by the empirical formula mass:

$$\text{whole number multiple} = \frac{176.1 \text{ u}}{88.063 \text{ u}} = 2.000$$

Finally, we multiply all the subscripts in the empirical formula by this multiple to obtain the molecular formula:

$C_6H_8O_6$.

Chapter 3 Exercises

SECTION 3.1

1. What does it mean for an atom to be in the ground state?
2. What does it mean for an atom to be in an excited state, and in what ways can it occur?
3. What conditions have to be met for a photon to ionize an atom?
4. Determine the energy in a photon of light from a green laser with a wavelength of 543 nm.
5. An atom absorbs a photon, causing one of its electrons to move to an orbital associated with 2.2718×10^{-19} J higher energy. Determine the wavelength of the absorbed photon and state what region of the electromagnetic spectrum it is in.
6. For a single photon to ionize a ground-state hydrogen atom, its energy has to raise the energy of the atom's electron to 0 J. What wavelength of light does this and what part of the electromagnetic spectrum is it in?
7. Calculate the energies for the four lines in the visible spectrum of the hydrogen atom (see Figure 3.8).

SECTION 3.2

8. What are two of the limitations of the Bohr model of the atom?

9. In the Bohr model, how many electrons would you expect the 5th energy level to be able to hold? Explain your response.

SECTION 3.3

10. Describe the difference between the orbital energies in the hydrogen atom and those of other atoms.

11. A certain atom is in the ground state. The 3p subshell of this atom is 2/3 full.
    a. Identify the element this atom represents.
    b. How many unpaired electrons are there in the atom? (A paired electron is one in an orbital with another one possessing opposite spin.)

12. In a certain ground-state atom, the 4d subshell has two electrons in it.
    a. Identify the element this atom represents.
    b. How many unpaired electrons are there in the atom?

13. Generally speaking, what is the explanation for an atom's electron configuration not following the sequence described by the Madelung rule?

14. Describe the two main ways that atoms can possess energy.

15. State and describe the four quantum numbers required to describe the quantum state of an electron.

16. For the first three principal quantum numbers, describe the orbitals available for electrons.

17. Write a description of the principles governing the quantum state of electrons in atoms.

18. State the Pauli exclusion principle, and explain its relationship to the placement of electrons in atoms.

SECTION 3.4

19. Write the full-length electron configuration for each of the following elements.
    a. chlorine
    b. oxygen
    c. ruthenium
    d. potassium
    e. vanadium
    f. bromine

20. For each of the following elements, write the condensed electron configuration.
    a. chlorine
    b. nitrogen
    c. aluminum
    d. yttrium
    e. strontium
    f. tungsten
    g. cesium
    h. iodine
    i. neodymium

21. Compare the electron configurations for beryllium, magnesium, and calcium. Formulate a general rule for the condensed electron configuration of a Group 2 element.

22. For which group of elements does the electron configuration always end with np^2? Explain how you know.

23. Write the condensed electron configurations for ytterbium, einsteinium, and nobelium.

SECTION 3.5

24. Automobile antifreeze is composed of ethylene glycol. The composition of this green liquid is 38.7% C, 9.7% H, and 51.6% O by mass. The molecular mass is 62.1 u. Determine the empirical formula and the molecular formula for ethylene glycol.

25. A scientist isolates 47.593 g of a new, unidentified substance. The scientist also determines the following masses for the elements in the substance: carbon: 43.910 g; hydrogen: 3.683 g.

Chapter 3

Finally, the scientist is also able to determine the molecular mass of the substance to be 78.11 u. From these data, determine:

a. the percent composition
b. the empirical formula
c. the molecular formula

26. Determine the percentage composition of these compounds:

a. sodium bicarbonate, $NaHCO_3$
b. sodium oxide, Na_2O
c. iron(III) oxide, Fe_2O_3
d. silver nitrate, $AgNO_3$
e. calcium acetate, $Ca(CH_3COO)_2$
f. aspirin, $C_9H_8O_4$

27. A *hydrate* is a compound with water molecules trapped in the crystal lattice. Determine the mass percentage of water in zinc sulfate septahydrate, $ZnSO_4 \cdot 7H_2O$. (The coefficient on the H_2O indicates the number of water molecules present for each unit of $ZnSO_4$.)

28. The results of quantitative analysis show that a compound contains 22.65% sulfur, 32.38% sodium, and 44.99% oxygen. Determine the empirical formula for this compound.

29. A compound has an empirical formula of CH_2O and a molar mass of 120.12 g/mol. Determine the molecular formula for this compound.

30. Determine the empirical and molecular formulas for each of the following:

a. caffeine, which contains 49.5% C, 5.15% H, 28.9% N, and 16.5% O by mass, and has a molecular mass of 195 u.

b. ibuprofen, which contains 75.69% C, 8.80% H, and 15.51% O by mass, and has a molar mass of 206 g/mol.

c. propane, which contains 81.71% C and 18.29% H by mass, and has a molar mass of 44.096 g/mol.

d. aspartame, a sugar-free sweetener, which contains 57.14% C, 6.16% H, 9.52% N, and 27.18% O, and has a molecular mass of 294.302 u.

e. acetylene, a gas used in cutting torches, which contains 92.26% C and 7.74% H, and has a molar mass of 26.038 g/mol.

31. Toluene is a solvent commonly found in chemistry labs. An analysis of a 10.5-g sample shows that the sample contains 9.581 g carbon and 0.919 g hydrogen. If the molar mass of toluene is 92.140 g/mol, determine the percent composition, empirical formula, and molecular formula.

GENERAL REVIEW EXERCISES

32. Determine the molar mass and formula mass of sulfuric acid, H_2SO_4.

33. Hydrogen chlorate, $HClO_3$, is a molecular substance that becomes chloric acid when dissolved in water. Determine the number of molecules present in 125.0 g $HClO_3$.

34. A standard Olympic competition swimming pool is 50.0 m long, 25.0 m wide, and 2.00 m deep. Determine the mass in kg of the water needed to fill this pool at 25°C.

35. Determine the mass of exactly one million atoms of lead. Express your result in µg.

36. How many molecules are there in 1.000 pt of acetic acid, CH_3COOH? The density of acetic acid (a liquid) is 1.049 g/cm^3.

37. The propane in a typical propane tank connected to a gas grill weighs 20.0 lb when the tank is full. Determine the number of propane molecules contained in a full tank. (Hint: Determine the mass from the equation $F_w = mg$, where F_w is the weight in newtons and g is the acceleration due to gravity, $g = 9.80$ m/s^2.)

38. The density of gold (Au) is 19.30 g/cm^3. Determine the volume of 3.000 mol Au.

39. Why is it important to show the correct number of significant digits in measurements?

40. Why is it important to show the correct number of significant digits in calculations performed on measurements?

41. What is the definition for the unified atomic mass unit, u?

42. Distinguish between theories and hypotheses.

43. Distinguish between theories and truth.

44. Distinguish between matter and mass.

45. Why is the Kelvin temperature scale referred to as an "absolute" scale?

46. Perform the following unit conversions:

Convert This Quantity	Into These Units	Convert This Quantity	Into These Units
a. 13.96 mL	gal	b. −62.7°F	K
c. 2.2901 × 10^{-3} m^3	in^3	d. 1.2509 × 10^4 nm	μm
e. 130,005 kPa	bar	f. 60,000 mi	cm
g. 65.7°C	°F	h. 10,600 mL	L
i. 23.17 μs	s	j. 98.6°F	°C
k. 1.600076 × 10^{-9} km	cm	l. 1.002 × 10^{-14} kg	ng

Chapter 4
The Periodic Law

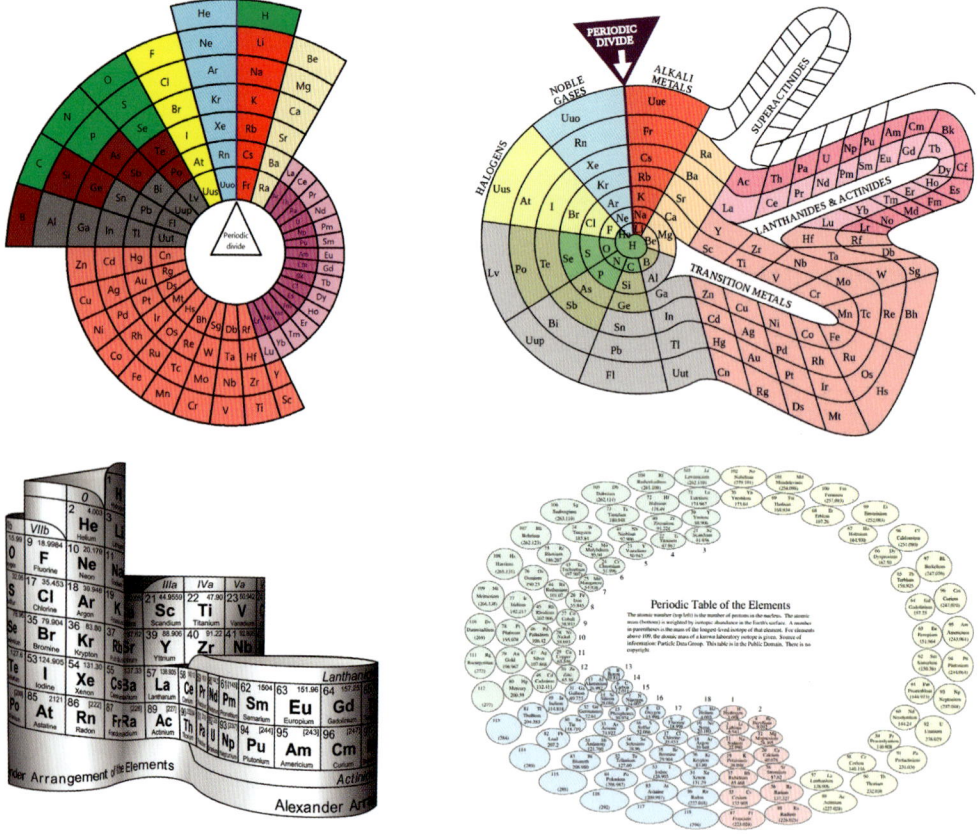

The patterns among the elements have tempted many scientists to try their hands at developing different forms of the Periodic Table of the Elements. At the top is the original periodic table, published by Russian scientist Dmitri Mendeleev in 1869.

Objectives for Chapter 4

After studying this chapter and completing the exercises, you should be able to do each of the following tasks, using supporting terms and principles as necessary.

SECTION 4.1
1. Describe the general structure and arrangement of the Periodic Table of the Elements.

SECTION 4.2
2. Identify the names and locations of the major regions in the periodic table.
3. Identify the names and locations of the elements in Groups 1, 2, 3–12, 16, 17, and 18. Also identify the groups known as the rare-earth elements and the names of the two rows of these elements.
4. State the two collective names for Groups 1–2 and 13–18.
5. State the chief chemical property that distinguishes the metals from the nonmetals.

SECTION 4.3
6. Define the length unit known as the *angstrom* (Å).
7. Describe the trends in atomic radii in the periodic table across periods and down groups.
8. Explain three factors influencing the trends in atomic radii in the periodic table.
9. Compare ionic radii to atomic radii for metals and nonmetals.
10. Use your knowledge of atomic and ionic size trends in the periodic table to arrange lists of elements and ions in order by size.

SECTION 4.4
11. Distinguish between core electrons and valence electrons.
12. Define *ionization energy*, describe the trends for ionization energy across periods and down groups in the periodic table, and describe causes for these trends.
13. Write regular and condensed electron configurations for an ion in a given oxidation state.
14. Use the shield effect to explain the large difference in ionization energy between an atom's core electrons and its valence electrons.
15. Predict oxidation states for metals in Groups 1–4 and nonmetals in Groups 15–17.
16. Define *electron affinity*, account for the high electron affinity of elements in Groups 16 and 17, and explain why electron affinity values are not available for the noble gases.
17. Define *electronegativity*.
18. Describe the trends for electronegativity across periods and down groups in the periodic table. Name the elements with the lowest and highest electronegativity values.

SECTION 4.5
19. Explain why hydrogen is located in Group 1 and why it acts like the Group 17 elements.

Chapter 4

4.1 The Periodic Table of the Elements

The contemporary Periodic Table of the Elements is shown again in Figure 4.1 and in the inside the rear cover of the text. Discovering new elements and figuring how they relate to one another was one of the hottest issues in science in the 19th century. By the 1860s, several scientists had noticed *periodicities* in the properties of the known elements. A periodicity is a regular, cyclic variation of some kind. These scientists noticed, for example, that when elements were listed in order according to atomic weight, the physical property of density increased and decreased in a cyclic fashion. In 1864, German chemist Lothar Meyer published a paper describing cyclic variation in the chemical property known as *valence*, an important property related to the number of bonds an atom makes with other atoms to form compounds.

Credit for the discovery of the *periodic law* and the development of the first Periodic Table of the Elements is generally given to Russian scientist Dmitri Mendeleev, who published his table of the elements in 1869. Mendeleev (Figure 4.2) had not only noticed the periodicities, but he also arranged some 67 elements into a table and predicted the existence of several unknown elements based on gaps in the table as he had organized it. The elements Mendeleev predicted included those now known as gallium, germanium, technetium, and others. The properties Mendeleev predicted for these elements included valence, density, atomic weight, and color. His predictions of not only the existence of these elements but also their properties is the reason for the general credit Mendeleev gets for discovering the periodic law.

Figure 4.2. Russian scientist Dmitri Mendeleev (1834–1907).

As Mendeleev organized the elements, he ordered them by atomic weight and aligned them into rows and columns based on their chemical and physical properties. Today, just as back then, one of the most important things to know about the periodic table is that *elements in the same group (column) exhibit very similar chemical properties.*

Mendeleev's original table is shown on the opening page of this chapter. In that image, just above the center and a bit to the left you can see gaps at the atomic mass values 68 and 72. These are the positions now occupied by gallium and germanium, discovered in 1875 and 1886, respectively. Mendeleev was a brilliant scientist. His fields of expertise included physics, chemistry, and a host of areas of technology. Mendeleev taught in St. Petersburg at several different institutions, and because of his work there St. Petersburg became internationally known for prominence in chemical research.

There were some debates in the 19th century regarding proper placement for four of the elements. If you look at tellurium ($Z = 52$) and iodine ($Z = 53$) in the periodic table to the right, you see that the atomic mass of tellurium is the larger of the two. Because of this, many scientists felt that iodine should come before tellurium in the table. But iodine exhibits all the properties of the Group 17 elements and Mendeleev argued that it should come after tellurium. This problem was resolved when scientists realized that the atomic number was the correct parameter to use for ordering the elements in the table. Mendeleev had ordered them by atomic mass (known then as atomic weight); today the atomic number governs the order. A similar debate over placement surrounded the elements cobalt ($Z = 27$) and nickel ($Z = 28$).

The noble gases were not known in Mendeleev's time and he provided no place for them in his table. Interestingly, when they were eventually discovered Mendeleev was resistant to accept the discovery because the new elements didn't fit into his table. The solution to this little problem was simply to add another column for them, now known as Group 18.

Figure 4.1. The Periodic Table of the Elements.

■ liquid at room temperature
■ radioactive

	1 1A	2 2A	3 3B	4 4B	5 5B	6 6B	7 7B	8 8B	9 8B	10 8B	11 1B	12 2B	13 3A	14 4A	15 5A	16 6A	17 7A	18 8A
1	1 H Hydrogen 1.0079																	2 He Helium 4.0026
2	3 Li Lithium 6.941	4 Be Beryllium 9.0122											5 B Boron 10.811	6 C Carbon 12.011	7 N Nitrogen 14.0067	8 O Oxygen 15.9994	9 F Fluorine 18.9984	10 Ne Neon 20.1797
3	11 Na Sodium 22.9898	12 Mg Magnesium 24.3050											13 Al Aluminum 26.9815	14 Si Silicon 28.0855	15 P Phosphorus 30.9738	16 S Sulfur 32.066	17 Cl Chlorine 35.4527	18 Ar Argon 39.948
4	19 K Potassium 39.098	20 Ca Calcium 40.078	21 Sc Scandium 44.9559	22 Ti Titanium 47.88	23 V Vanadium 50.9415	24 Cr Chromium 51.9961	25 Mn Manganese 54.9380	26 Fe Iron 55.847	27 Co Cobalt 58.9332	28 Ni Nickel 58.6934	29 Cu Copper 63.546	30 Zn Zinc 65.39	31 Ga Gallium 69.723	32 Ge Germanium 72.61	33 As Arsenic 74.9216	34 Se Selenium 78.96	35 Br Bromine 79.904	36 Kr Krypton 83.80
5	37 Rb Rubidium 85.468	38 Sr Strontium 87.62	39 Y Yttrium 88.9059	40 Zr Zirconium 91.224	41 Nb Niobium 92.9064	42 Mo Molybdenum 95.94	43 Tc Technetium 98.9072	44 Ru Ruthenium 101.07	45 Rh Rhodium 102.9055	46 Pd Palladium 106.42	47 Ag Silver 107.8682	48 Cd Cadmium 112.411	49 In Indium 114.82	50 Sn Tin 118.710	51 Sb Antimony 121.76	52 Te Tellurium 127.60	53 I Iodine 126.9045	54 Xe Xenon 131.29
6	55 Cs Cesium 132.905	56 Ba Barium 137.327	71 Lu Lutetium 174.967	72 Hf Hafnium 178.49	73 Ta Tantalum 180.9479	74 W Tungsten 183.85	75 Re Rhenium 186.207	76 Os Osmium 190.2	77 Ir Iridium 192.22	78 Pt Platinum 195.08	79 Au Gold 196.9665	80 Hg Mercury 200.59	81 Tl Thallium 204.3833	82 Pb Lead 207.2	83 Bi Bismuth 208.9804	84 Po Polonium 208.9824	85 At Astatine 209.9871	86 Rn Radon 222.0176
7	87 Fr Francium 223.0197	88 Ra Radium 226.0254	103 Lr Lawrencium 262.11	104 Rf Rutherfordium 261.11	105 Db Dubnium 262.114	106 Sg Seaborgium 263.118	107 Bh Bohrium 262.12	108 Hs Hassium (265)	109 Mt Meitnerium (266)	110 Ds Darmstadtium (281)	111 Rg Roentgenium (281)	112 Cn Copernicium (285)	113 Nh Nihonium (284)	114 Fl Flerovium (289)	115 Mc Moscovium (288)	116 Lv Livermorium (293)	117 Ts Tennessine (294)	118 Og Oganesson (294)

57 La Lanthanum 138.9055	58 Ce Cerium 140.115	59 Pr Praseodymium 140.9077	60 Nd Neodymium 144.24	61 Pm Promethium 144.9127	62 Sm Samarium 150.36	63 Eu Europium 151.965	64 Gd Gadolinium 157.25	65 Tb Terbium 158.9253	66 Dy Dysprosium 162.50	67 Ho Holmium 164.9303	68 Er Erbium 167.26	69 Tm Thulium 168.9342	70 Yb Ytterbium 173.04
89 Ac Actinium 227.0278	90 Th Thorium 232.0381	91 Pa Protactinium 231.0359	92 U Uranium 238.0289	93 Np Neptunium 237.0482	94 Pu Plutonium 244.0642	95 Am Americium 243.0614	96 Cm Curium 247.0703	97 Bk Berkelium 247.0703	98 Cf Californium 251.0796	99 Es Einsteinium 252.083	100 Fm Fermium 257.0951	101 Md Mendelevium 258.10	102 No Nobelium 259.1009

Note: The upper set of group numbers has been recommended by the International Union of Pure and Applied Chemistry (IUPAC) and is now in wide use. The lower set of numbers is still in common use in America.

Chapter 4

The first 92 elements in the periodic table are found in nature; elements 93–118 have been synthesized in laboratories. The "discovery" (by synthesis) of elements 114 and 116 was confirmed in 2011, and in January 2016 official confirmation of elements 113, 115, 117, and 118 was announced. The reason confirmations for the last few elements took so long is that once the nucleus of one of these heavy elements is assembled it doesn't stay around very long—far less than one second.

4.2 Periodic Table Nomenclature

The columns in the periodic table are called *groups* and the rows are called *periods*. The images in Figures 4.3 and 4.4 identify several different specific regions of elements in the periodic table. In the long form of the table shown in Figure 4.3, the elements are classified as *metals*, *nonmetals*, and *metalloids*. Note that hydrogen (H) is classified as a nonmetal, even though it is positioned with the metals in Group 1. Hydrogen's location in Group 1 is due to the fact that hydrogen has one valence electron, which I address in more detail later in the chapter.

As you probably know, metals possess a number of properties in common. Common physical properties include high electrical and thermal conductivity, malleability, ductility, and shininess or *luster*. Chemically, the metals are known for ionizing by losing electrons to form positive ions, known as *cations* (pronounced cat-ion). As positive ions, they bond with negative ions to form ionic compounds. (Again, we address compounds in detail later.) People commonly think of metals as shiny conductors of electricity. Chemists think of them as elements that form positive ions.

The metalloids possess properties that are neither clearly metallic nor clearly nonmetallic. For example, under some conditions they conduct electricity and under other conditions they don't. This property is the reason why some of the metalloids are the elements used to manufacture computer "semiconductors."

The nonmetals have their own distinguishing properties, such as ionization by gaining electrons to form negative ions, called *anions*. They also bond with each other—something metals almost never do.

In the standard form periodic table of Figure 4.4 are shown the common names for several specific groups (columns) of elements. All these names are used frequently in scientific discourse and you need to commit them to memory. Groups 1–2 and 13–18 are also collectively referred to as the *main group elements* or *representative elements*.

You probably noticed that in both the figures elements 113–118 are not included with the rest of the elements in the different classes or groups. This is due to the fact that they have only existed for extremely short periods of time in laboratories and little is known about their properties. By the way, you don't need to feel sorry for Groups 13–15 not having nicknames. They do,

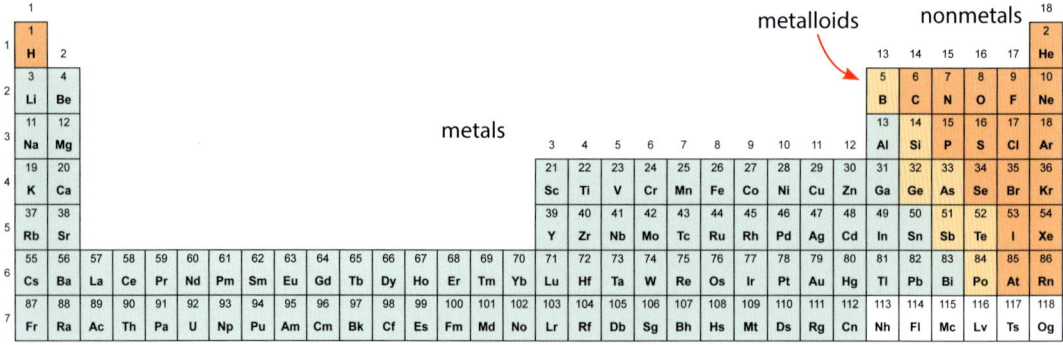

Figure 4.3. The long form of the periodic table indicating the three major classes of elements.

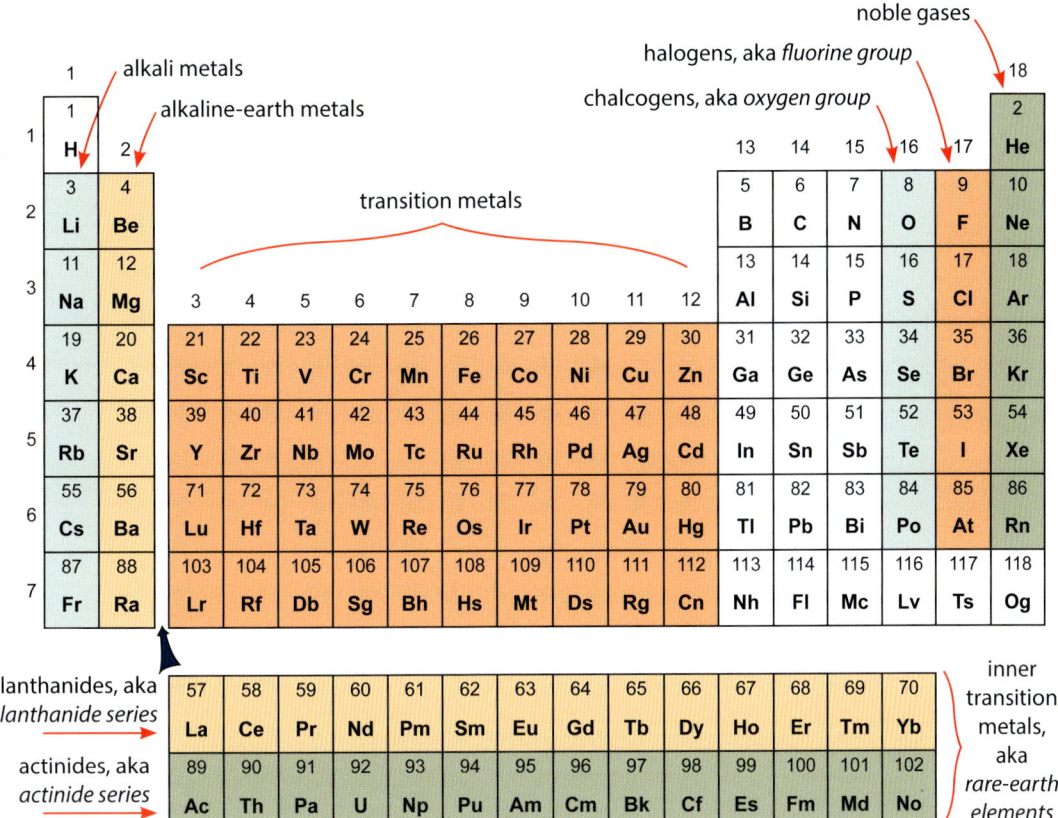

Figure 4.4. The standard form table indicating many of the common names for particular groups of elements.

but the names are seldom used. (Okay, Group 13 is called the *icosagens*, Group 14 the *crystallogens*, and Group 15 the *pnictogens*. See? Well, now you know.)

4.3 Periodic Physical Properties

Many of the physical properties of elements exhibit strong periodicity—cyclic rises and falls with cycles that correspond to the periods in the periodic table. In this section, we look at one of the most important of these—atomic size. As we look at periodic trends in this section and the next, you should closely compare the charts presented with the periodic table in Figure 4.1.

4.3.1 Atomic Radius

Defining the size of atoms is a bit tricky. The nucleus of the atom is extremely small, and virtually all the space an atom takes up is defined by the electrons in their orbitals surrounding the nucleus. The size of the orbitals themselves is defined by the probability distribution of where electrons may be found, and this probability does not drop cleanly to zero at the edge of the orbital. Instead, it fades toward zero, making the atomic radius fuzzy. Still, the electrons in the orbitals create a negatively charged shell around the atom that strongly resists penetration by the shells around other atoms. Thus scientists can measure atomic radius by firing two atoms at one another and examining how closely the two atoms come together before bouncing apart. The electron clouds repel each other so stiffly that atoms bounce apart as if they were steel spheres.

When discussing atomic size, a convenient length unit is the *angstrom* (Å), which is equal to 10^{-10} m, which is equivalent to one tenth of a nanometer (0.1 nm or 100 pm). Figure 4.5 shows

Chapter 4

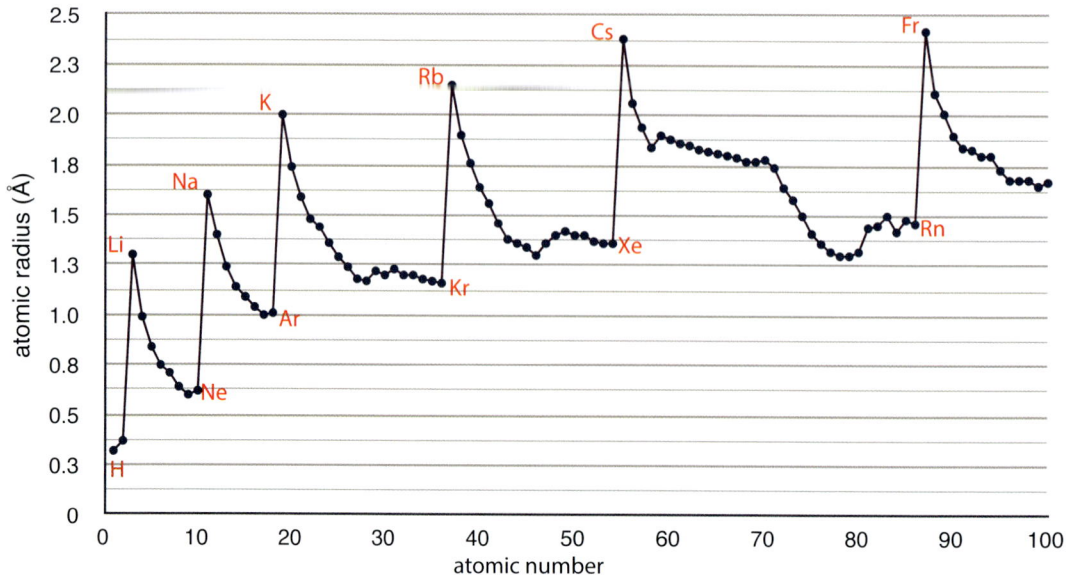

Figure 4.5. Atomic radius values. The six peaks are the alkali metals in Periods 2–7.

the presently accepted values for the atomic radii of the first 100 elements. As you see, typical radii are in the range of 1–2 Å, or 0.1–0.2 nm. This means that typical atomic diameters are in the range of 0.2–0.4 nm.

The most striking things about the graph are the strong peak that occurs at the beginning of each new period and the gradual decrease in atomic size from one element to the next within the period. There are at least three different effects present governing the size of atoms and leading to these patterns.

The first is the number of subshells in use holding the atoms' electrons. At the beginning of each period, an electron appears with a new, higher principle quantum number, n, and a new s subshell associated with that principle quantum number. This new shell allows for electrons to be much farther from the nucleus and is the major cause of the peak in atomic radius at the beginning of each period, as well as the overall upward trend in atomic size.

The second effect is the increasing attraction between the nucleus and the electrons as we move from left to right across the periodic table in any given period, or from top to bottom in a given group. Moving left to right across the periodic table, with each new element comes a new proton in the nucleus. The increasing positive electrical charge at the center of the atom tends to pull the atom's electrons in tighter and tighter. Thus, with a few exceptions, atomic size decreases from left to right in a period. In the chart of Figure 4.5, this decrease in size after the start of a new period is quite pronounced.

The third effect is called the *shield effect* or *atomic shielding*. The electrons in shells with lower values of n effectively form an electrical screen around the nucleus and to some extent shield off the attraction of the positive nucleus for electrons in higher shells. Because of the shield effect, the electrons in higher-energy shells are not attracted to the nucleus as strongly as we might expect. This is another reason for the large increase in atomic radius that occurs at the beginning of each new period.

A second feature to note from Figure 4.5 is that the radius of each of the alkali metals is larger than the one just above it in the periodic table. The same holds for each of the noble gases. In fact, this same trend is present in every group in the table. Down every group, the atomic radius of each element is greater than that of the element above it, and this results in the overall upward trend in atomic size shown in the chart. In summary, atomic size generally decreases from left to right in a period and always increases from top to bottom in a group.

4.3.2 Ionic Radius

Figure 4.6 depicts the sizes of atoms and their ions in four of the groups of the representative elements. The blue disks are the atomic radii and the yellow disks are the ionic radii, both in angstroms.

Metals ionize by losing electrons to form cations (positive ions). The loss of an electron leaves an atom with more protons than electrons, and thus the atom possesses a net positive charge. As we will discuss more in Section 4.4, Group 1 metals always ionize by losing one electron to become ions with a charge of +1. With sodium, for example, we write this ion as Na^+. Group 2 metals always ionize by losing two electrons to become ions with a charge of +2. Thus, for the calcium ion we write Ca^{2+}.[1] Nonmetals ionize by gaining electrons to become anions (negative ions). Group 17 elements ionize by gaining one electron to become ions with a net charge of –1. Group 16 elements ionize by gaining two electrons to become ions with a net charge of –2.

The blue circles in Figure 4.6 represent the atomic sizes, and yellow circles represent ionic sizes. Since the metals always ionize by losing the electrons in their highest *s* orbital, their diameters decrease considerably when they ionize. The opposite happens to the nonmetals. Gaining electrons adds to the mutual electron repulsion in the highest-energy orbitals, increasing atomic size.

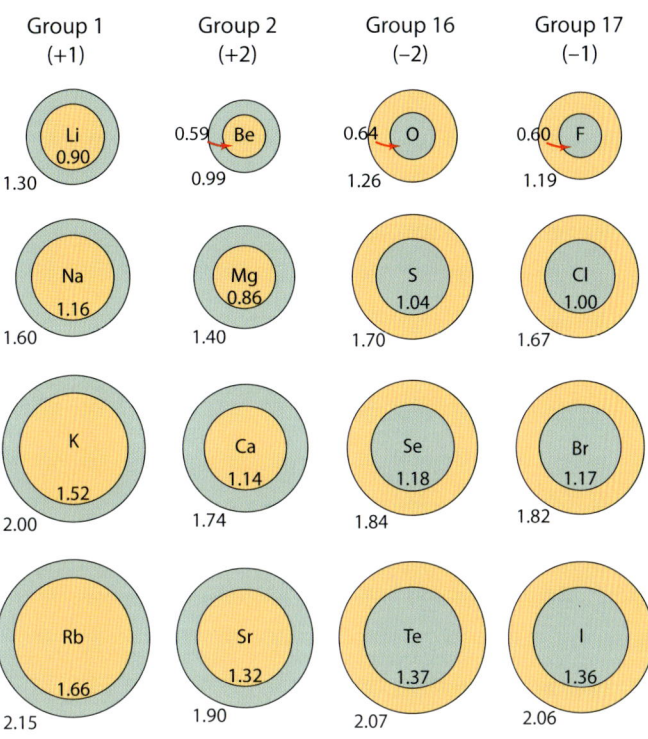

Figure 4.6. Atomic radii (blue) and ionic radii (yellow) for the Group 1 and 2 positive ions and Group 16 and 17 negative ions. All values are in angstroms.

Just as with neutral atoms, ionic sizes decrease in a period from left to right in the periodic table and increase going down a group. Further, we can summarize the paragraphs above by saying that cations are smaller than their neutral atoms and anions are larger than their neutral atoms. Refer to the periodic table while studying the following examples.

▼ Example 4.1

Based on your knowledge of trends in the periodic table, place the following atoms and ions in order of decreasing size: S, Se^{2-}, O, and S^{2-}.

---

1 The most common convention is to write the sign of the charge after the value of the charge on the chemical symbol for an ion.

Anions are larger than their neutral atoms, so $S^{2-} > S$. Elements and ions farther down in a group are larger than those above, so $S > O$. Also, Se^{2-} is larger than S^{2-}, so

$Se^{2-} > S^{2-} > S > O$.

▼ Example 4.2

Arrange the following atoms and ions in order of size from largest to smallest: Ca, Mg, Sr, Mg^{2+}, and Rb.

Cations are smaller than their neutral atoms, so $Mg > Mg^{2+}$. Going down a group, atoms increase in size, which means $Sr > Ca > Mg$. Going left to right in a period atoms decrease in size, so $Rb > Sr$. Putting these facts together, we have

$Rb > Sr > Ca > Mg > Mg^{2+}$.

4.4   Periodic Chemical Properties

4.4.1 Core and Valence Electrons

Chemistry is all about electrons, and when it comes to the chemical bonding that occurs in chemical reactions, the *valence electrons* of the elements involved determine what kinds of compounds form. To illustrate, we consider the elements sulfur ($Z = 16$), cobalt ($Z = 27$), and bromine ($Z = 35$) located in the partial periodic tables of Figure 4.7.

The electron configurations for sulfur and cobalt are as follows:

Z	Electron Configuration	Condensed Electron Configuration
16	S: $1s^2 2s^2 2p^6 3s^2 3p^4$	S: $[Ne]3s^2 3p^4$
27	Co: $1s^2 2s^2 2p^6 3s^2 3p^6 4s^2 3d^7$	Co: $[Ar]4s^2 3d^7$

As we saw in the previous chapter, when moving from one element to the next in the periodic table, we add one electron and one proton for each new element. As we look at an element's position in the periodic table, we can think of each of the elements before it as representing the position of one of the element's electrons, arranged according to the Madelung rule and Hund's rule.

Now consider Figure 4.7 and the electron configuration for sulfur. The first 10 electrons in a sulfur atom go to completely filling the $n = 1$ and $n = 2$ shells. The next six electrons go into the $n = 3$ shell. This shell can hold eight electrons, so it is only partially full. The electrons in the full shells are called *core electrons*. The electrons in the partially filled shell are called *valence electrons*, and the shell they are in is called the *valence shell*. The core electrons are the ones involved in the shield effect I mentioned a couple of pages back. The core electrons have more negative energies than the valence electrons, which means they are more tightly bound to the nucleus. As a result, the core electrons are not involved in the electron swapping and sharing that takes place in chemical reactions. That involvement is limited to the valence electrons.

When you look at sulfur's position in the periodic table, just a quick glance indicates that sulfur has six valence electrons, all of them in the third shell, $n = 3$. It's as simple as observing that sulfur is in the third period and counting columns from the left over to where sulfur is.

The Periodic Law

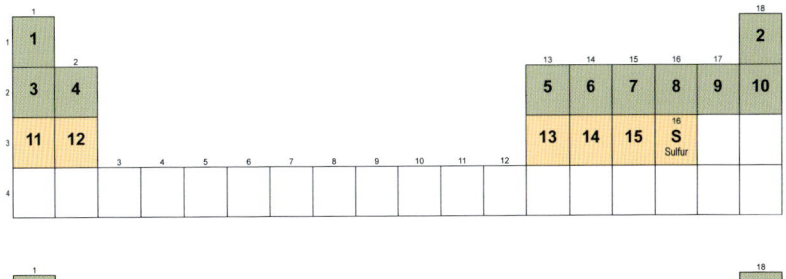

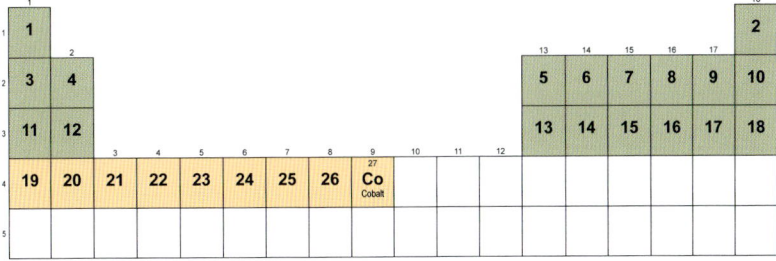

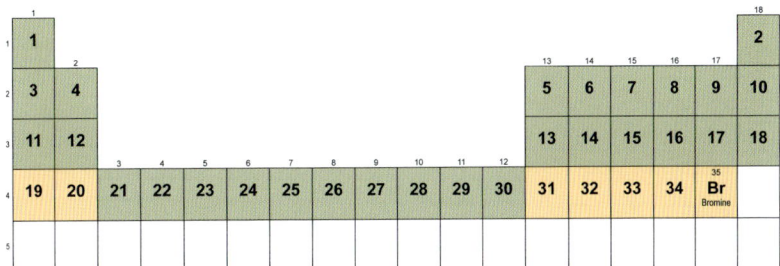

Figure 4.7. Core electrons (green) and valence electrons (yellow) for sulfur, cobalt, and bromine.

Notice that in the condensed electron configuration notation, all of sulfur's core electrons are represented by [Ne]. The valence electrons are still shown explicitly as $3s^2 3p^4$.

Looking now at cobalt, the element positions colored in green represent cobalt's core electrons. The first 18 electrons are the core electrons, represented by [Ar] in the condensed electron configuration. The next nine electrons are the valence electrons, easily seen from the fact that cobalt is in the fourth period and the ninth column from the left side of the periodic table.

Finally, note that for *p*-block elements in Periods 4–7, the electrons in filled, lower *d* subshells do not act as valence electrons. The only valence electrons in these atoms are those in the *s* and *p* subshells of the unfilled shell. This is illustrated by the diagram for bromine at the bottom of Figure 4.7.

4.4.2 Ionization Energy

Ionization energy is defined as the amount of energy required to remove a ground-state electron from an isolated, gaseous atom. Recall from our discussion in the previous chapters that adding an electron to an atom releases energy because the electron is going into a lower energy state. Conversely, removing an electron from an atom requires an input of energy, which is the work required to pull the electron from its negative energy state up to zero energy, where it is free from the nuclear attraction of the atom.

Using hydrogen as an example, the removal of the electron from an atom of hydrogen is modeled by the following *ionization equation*:

$$H(g) \rightarrow H^+ + e^- \qquad \Delta E = 2.18 \times 10^{-18} \text{ J} \qquad (4.1)$$

Chapter 4

This expression shows the neutral hydrogen atom on the left, with (g) indicating that the atom is in the gaseous state. On the right, H^+ indicates a hydrogen ion with a charge of +1, and e^- indicates a free electron. In an equation like this, it is customary to write the energy change (ΔE) that occurred in the atomic system (the atom and its electron) during the process. Notice that ΔE is positive, meaning that a certain amount of energy has to be added in order to accomplish the ionization. This amount of energy is the ionization energy.

Because the amounts of energy involved in ionization are so small, it is more convenient in discussions of this sort to use an energy unit called the *electron volt* (eV). The electron volt is defined as

$$1 \text{ eV} = 1.60218 \times 10^{-19} \text{ J} \qquad (4.2)$$

For the ionization of hydrogen, converting the energy above into eV gives

$$\Delta E = 2.18 \times 10^{-18} \text{ J} \cdot \frac{1 \text{ eV}}{1.602 \times 10^{-19} \text{ J}} = 13.6 \text{ eV}$$

The energy required to remove one electron from a neutral atom is called the *first ionization energy*. The additional energy required to remove a second electron is called the *second ionization energy*, and so on.

Figure 4.8 charts the first ionization energies of the elements. The strong upward trend in each period is easily accounted for by the decreasing efficacy of the shield effect that occurs moving from left to right in the period. Moving from left to right, the strength of the shielding provided by the atom's core electrons remains the same, but the positive charge of the nucleus increases with each additional proton. This means the shield effect becomes less effective and the attraction between the nucleus and the valence electrons increases from left to right. The higher the nuclear attraction is, the stronger the nucleus attracts the outermost electrons and the greater the energy required to remove one of them. Notice from the Group 1 and Group 18 elements labeled in the figure that the trend down a group is for the ionization energy to decrease, even though the attraction of the nucleus increases down the group because of the increasing

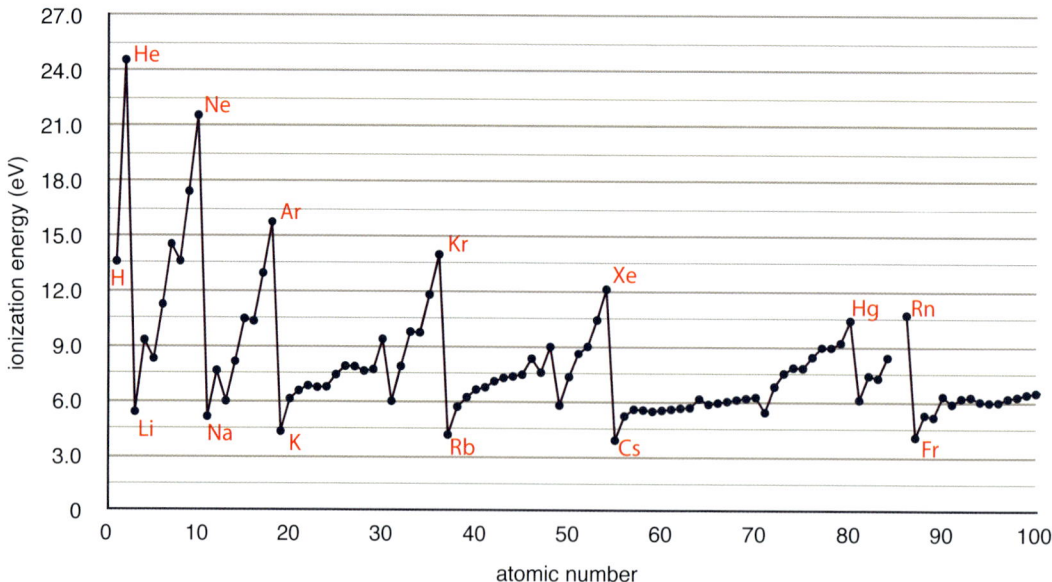

Figure 4.8. First ionization energies.

number of protons in the nucleus. This is accounted for by the fact that the elements in each new period have their valence electrons in a shell with a higher principle quantum number. These electrons are thus farther from the nucleus, have a less negative energy, and are easier to remove.

Writing electron configurations for ions requires you to keep a somewhat surprising rule in mind: *the electrons an element loses during ionization come from the orbital with the highest principle quantum number, n. Further, among the orbitals associated with the highest principle quantum number, the electrons come from the orbitals with the highest value of the azimuthal quantum number, l.* This means, for example, that sodium (Z = 11, Group 1) ionizes as we would expect by losing its 3s electron. However, scandium (Z = 21, Group 3) ionizes by losing a 4s electron, not a 3d electron. The following example illustrates further.

▼ Example 4.3

Write the electron configurations for Sn^{4+} and Mn^{4+}.

Sn is a p-block metal in Period 5 and Group 14 with configuration

Sn: $[Kr]5s^2 4d^{10} 5p^2$

Its valence electrons are therefore in the 5s and 5p subshells. It first loses its two 5p electrons, because these have $n = 5$ and $l = 1$. The next two to go are the 5s electrons which have $n = 5$ and $l = 0$. The electron configuration of the ion is:

Sn^{4+}: $[Kr]4d^{10}$

Mn is a transition metal with the electron configuration:

Mn: $[Ar]4s^2 3d^5$

To ionize to Mn^{4+}, the atom first loses the 4s electrons, which have $n = 4$ and $l = 0$. Then it loses two of its 3d electrons, which have $n = 3$ and $l = 2$. Thus, the configuration of the ion is:

Mn^{4+}: $[Ar]3d^3$

Table 4.1 lists the ionization energies up through the 7th ionization energy for the Period 2 elements. Values without shading represent the energies required to remove valence electrons. The yellow shading indicates where core electrons are being removed. Notice the whopping

Element	1st	2nd	3rd	4th	5th	6th	7th
Li	5.38	75.64	122.45				
Be	9.32	18.21	153.90	217.72			
B	8.30	25.15	37.93	259.38	340.23		
C	11.26	24.38	47.89	64.49	392.09	489.99	
N	14.53	29.60	47.45	77.47	97.89	552.07	667.05
O	13.62	35.12	54.94	77.41	113.90	138.12	739.29
F	17.42	34.97	62.71	87.14	114.24	157.17	185.19
Ne	21.56	40.96	63.45	97.12	126.21	157.93	207.28

Table 4.1. Ionization energies (eV) for Period 2 elements.

increase in the ionization energy once we start getting into an atom's core electrons. This is explained by the shield effect we discussed previously. Core electrons shield the nucleus from exerting its full attraction on the valence electrons. But inside the core, the full attraction of the nucleus is felt and the energy required to remove an electron is significantly greater.

This discussion about ionization is an appropriate time to show representative values for ionizations exhibited by the main group elements and transition metals. These values, displayed in Figure 4.9, are called *oxidation states*. I address the origin and use of the term "oxidation" in a later chapter. In the present context, it just means that when atoms of a given element ionize, these are the values of charge they typically acquire. The first and most important thing to notice is that on both ends of the periodic table, elements ionize so as to end up with either an empty valence shell or a full one, depending on whether the valence shell was closer to being full or empty to start with. All Group 1 metals and hydrogen ionize by losing their only valence electron and retaining their core electrons. After this ionization, the Group 1 metals are left with one more proton than electron, so they are cations with a charge of +1. Similarly, Group 2 metals lose two electrons and Group 3 metals lose three electrons to end up as cations with charges of +2 or +3, respectively.

On the other end of the table, the noble gases don't readily ionize, so no oxidation states are shown. The nonmetals ionize by taking on electrons to become anions. The halogens all have p subshells containing five electrons, so they ionize by gaining one electron to fill up the p subshell, becoming anions with a charge of –1 in the process. Similarly, Group 16 nonmetals gain two electrons and Group 15 nonmetals gain three electrons to become anions with charges of –2 or –3, respectively.

All the transition metals have multiple oxidation states, and the oxidation state they assume depends on what other elements are around to swap electrons with. The values shown are the most common states. This chart makes it easy to visualize the metalloids as the boundary between the cations and the anions. In the cases of lutetium (Lu) and astatine (At), these elements have multiple oxidation states with none preferred, and for this reason none are shown.

It should be clear now that although ionization energy is defined as the energy required to *remove* an electron from an atom, which is what happens to metals when they ionize, the nonmetals ionize by *gaining* electrons to form anions. The energy involved in a neutral atom gaining an electron is called the *electron affinity*, and is our next topic.

	1	2	3	4	5	6	7	8	9	10	11	12	13	14	15	16	17	18
1	H^+																	He
2	Li^+	Be^{2+}											B^{3+}	C^{4+} C^{4-}	N^{3-}	O^{2-}	F^-	Ne
3	Na^+	Mg^{2+}											Al^{3+}	Si^{4+}	P^{3-}	S^{2-}	Cl^-	Ar
4	K^+	Ca^{2+}	Sc^{3+}	Ti^{4+}	V^{5+} V^{4+}	Cr^{6+} Cr^{3+}	Mn^{4+} Mn^{2+}	Fe^{3+} Fe^{2+}	Co^{3+} Co^{2+}	Ni^{2+}	Cu^{2+} Cu^+	Zn^{2+}	Ga^{3+}	Ge^{4+} Ge^{2+}	As^{3+} As^{5+}	Se^{2-}	Br^-	Kr
5	Rb^+	Sr^{2+}	Y^{3+}	Zr^{4+}	Nb^{5+}	Mo^{6+} Mo^{4+}	Tc^{7+} Tc^{4+}	Ru^{4+} Ru^{3+}	Rh^{3+}	Pd^{4+} Pd^{2+}	Ag^+	Cd^{2+}	In^{3+}	Sn^{4+} Sn^{2+}	Sb^{5+} Sb^{3+}	Te^{2-}	I^-	Xe
6	Cs^+	Ba^{2+}	Lu	Hf^{4+}	Ta^{5+}	W^{6+}	Re^{7+} Re^{6+}	Os^{4+}	Ir^{4+} Ir^{3+}	Pt^{4+} Pt^{2+}	Au^{3+} Au^+	Hg^{4+} Hg^{2+}	Tl^+	Pb^{4+} Pb^{2+}	Bi^{5+} Bi^{3+}	Po^{4+}	At	Rn

Figure 4.9. Representative oxidation states.

4.4.3 Electron Affinity

Electron affinity is defined as the amount of energy *released* when adding an electron to a ground-state, isolated, gaseous atom. The term "affinity" indicates that electron affinity is a measure of how eager the atoms of a given element are to take on another electron, an important measure when trying to understand or predict chemical reactions.

Here is yet another instance in which energy relationships are crucial for understanding chemical behavior. Most atoms release energy when an electron is added, and accept the additional electron into an orbital to become an anion with a charge of –1. The major exceptions to know about are the noble gases. These elements do not release energy when an electron is added. Instead, energy is required to attach the electron and as a result the atom is not stable. Such an ion immediately rejects the electron and returns to its neutral state. This is because the valence shells in the noble gas atoms are full—the most stable electron configuration there is.

Using sulfur as an example, the gaining of an electron is modeled with an ionization equation as follows:

$$S(g) + e^- \rightarrow S^-(g) \qquad \Delta E = 2.077 \text{ eV} \qquad (4.3)$$

As with ionization Equation (4.1), the (g) indicates that the sulfur is in the gaseous state, and the energy change is stated to the right of the equation.

It is important for you to distinguish clearly in your mind between ionization energy and electron affinity. Ionization energy is the energy required to *remove* an electron from a neutral atom (creating a cation). This energy is positive for every element. Electron affinity is the energy released when an electron is *added* to a neutral atom (creating an anion). Defined this way, this energy is also positive for every element except for the noble gases and a few other elements.

Figure 4.10 shows the electron affinities for the stable atoms. As with other properties we have addressed, there is an obvious periodicity in the affinity values corresponding to the periods in the periodic table. We can again explain the large values associated with the Group 16 and Group 17 elements in terms of the shield effect. In these two groups, the positive charge in the nucleus is as high as it gets, so the nuclei of these elements exert the highest possible attraction

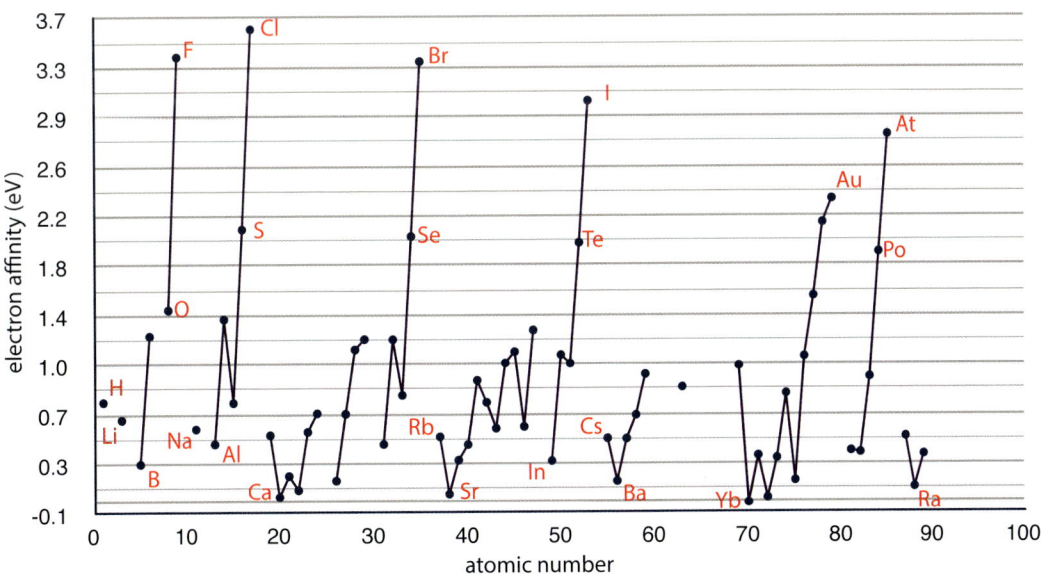

Figure 4.10. Electron affinities.

on a free electron, causing the atom/electron system to release as much energy as possible when the electron drops down into the lower energy state in one of the atom's orbitals.

The breaks between connected dots in Figure 4.10 indicate that several other elements are missing besides the noble gases. These are Be, N, Mg, Mn, and Zn. All these elements exhibit the same unstable behavior the noble gases do when an electron is added to them. Most of these can be at least partially explained by referring to the periodic table. Beryllium has two electrons in the $2s$ subshell. An added electron would go into the $2p$ subshell, where its energy is too high for the small nucleus to hold it. The same is the case for magnesium. Nitrogen has three electrons in the $2p$ subshell—a stable, low-energy state. Adding a fourth electron results in repulsion between two electrons in the p_x orbital, and once again the small nucleus does not have a strong enough attraction to wrangle those electrons and keep them both there. Similar considerations apply for manganese and zinc. Manganese has five electrons in the $3d$ subshell, so the $3d$ subshell is exactly half full with one electron in each orbital. This puts manganese in a situation similar to that of nitrogen. Zinc has 10 electrons in the $3d$ subshell and none in the $4p$ subshell, so zinc is in a situation similar to that of beryllium and magnesium.

There is one element with a very slightly negative value for electron affinity that will still accept an electron. This element is ytterbium (Yb, $Z = 70$). Ytterbium is at the far right of the rare-earth elements, in Period 6. Its $4f$ subshell is full and an additional electron goes into the $5d$ subshell.

There is one final point to make before we move on. I have used care in defining ionization energy and electron affinity so that it is clear which way the energy is going when energy values are positive or negative. But you should be aware that some texts and other sources use definitions that may not be as clear, possibly leading to confusion. Instead of using the terms *required* or *released* to define the direction of positive energy flow, some sources use the term *involved*, and define positive energy flow in terms of energy going into or out of the atomic system. Both definitions are valid, as long as one specifies the direction of energy flow that constitutes a positive value of energy.

The definitions we have seen in the last two sections are fairly standard. But to help you avoid confusion, Table 4.2 spells out my definitions and the alternative definitions side by side.

Term	Definition Used In This Text	If the energy quantity ΔE is positive, it means:
ionization energy	energy *required* to *remove* an electron	energy flows *into* the atomic system from outside
electron affinity	energy *released* when *adding* an electron	energy flows *out of* the atomic system

Term	Alternative Definition	If the energy quantity ΔE is positive, it means:
ionization energy	energy *involved* when removing an electron	energy flows *into* the atomic system from outside
electron affinity	energy *involved* when adding an electron	energy flows *into* the atomic system from outside (thus, ΔE and the electron affinity values are nearly always negative)

Table 4.2. Definitions used here and elsewhere for ionization energy and electron affinity.

The Periodic Law

4.4.4 Electronegativity

American chemist Linus Pauling (Figure 4.11) was one of the most important chemists of the 20th century. His work in the 1930s, 1940s, and 1950s on the nature of chemical bonds remains foundational for our understanding of chemistry to this day. For his work, Pauling won the Nobel Prize in Chemistry in 1954. Beginning in the mid-1940s, Pauling showed a deep concern for the negative health effects due to nuclear fallout from nuclear weapons testing, and he presented a petition signed by 11,000 scientists to the United Nations in 1958 to urge a ban on nuclear weapons testing. A scientific study that came out in 1961 showed that radioactivity contamination was indeed widespread in the population, leading to nuclear test ban treaties with the Soviet Union. All this led to Pauling winning the Nobel Peace Prize in 1962. To this day, Pauling remains the only person ever to win two unshared Nobel Prizes.

Figure 4.11. American chemist Linus Pauling (1901–1994).

In 1932, as part of his efforts to understand chemical bonding, Pauling introduced what is now called the *Pauling electronegativity scale*. The electronegativity scale uses a dimensionless quantity called *electronegativity*, with values running from 0.7 (francium) to 3.98 (fluorine). Electronegativity is a measure of how strongly atoms attract the electrons shared between atoms inside molecules. The higher an element's electronegativity relative to the other elements in a molecule, the more an atom attracts the shared electrons in the molecule toward itself.

Once again, the importance of electrical attraction in the atomic world is crucial. Recall from the Introduction that many of the important properties of water are due to the difference between the electronegativities of oxygen (3.44) and hydrogen (2.20) in the water molecule. The oxygen atom attracts shared electrons more strongly than the hydrogen atoms do, and as a result the four bonding electrons in the molecule crowd over toward the oxygen atom. The result is that the oxygen region of the water molecule is more electrically negative and the hydrogen regions are more electrically positive, so water molecules are *polar*—negative at the elbow and positive at the ends, as shown in Figure 4.12. In this diagram, the arrows point from the positive region of the molecule toward the negative region of the molecule. We consider electronegativity further in the next chapter in the context of covalent bonding.

Figure 4.13 shows the periodicity of electronegativity and the trend of increasing electronegativity from left to right in the periodic table. There are gaps between the halogens and the alkali metals because the noble gases are not included. This is because electronegativity measures atomic attraction within molecules, and the noble gases don't form bonds with other elements, molecular or otherwise, except under the extreme and unusual conditions that can be produced in a specialized laboratory.

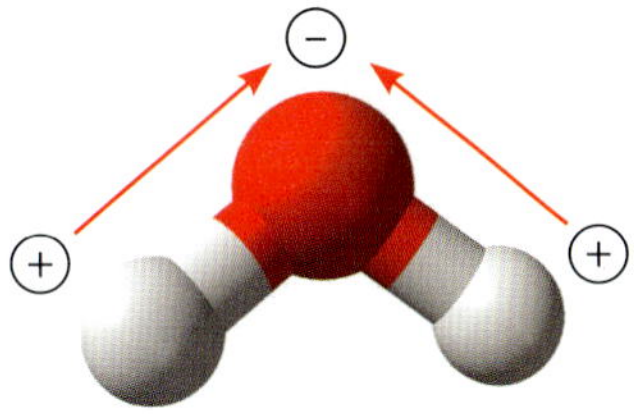

As with electron affinity and ionization energy, the trends in electronegativity can be understood in terms of the shield effect, which decreases in strength from left to right as the number of protons in the nucleus increases while the number of core electrons screening the nucleus remains constant. Decreased shielding efficacy means not only that an atom attracts its own valence electrons more tightly. It also means the atom attracts the valence electrons shared with neighboring atoms more strongly relative to the attraction of neighboring atoms.

Figure 4.12. The higher electronegativity of oxygen atoms compared to hydrogen atoms results in the polar water molecule.

The diagram in Figure 4.14 summarizes the general trend of electronegativity values over the periodic table from the lowest val-

115

Chapter 4

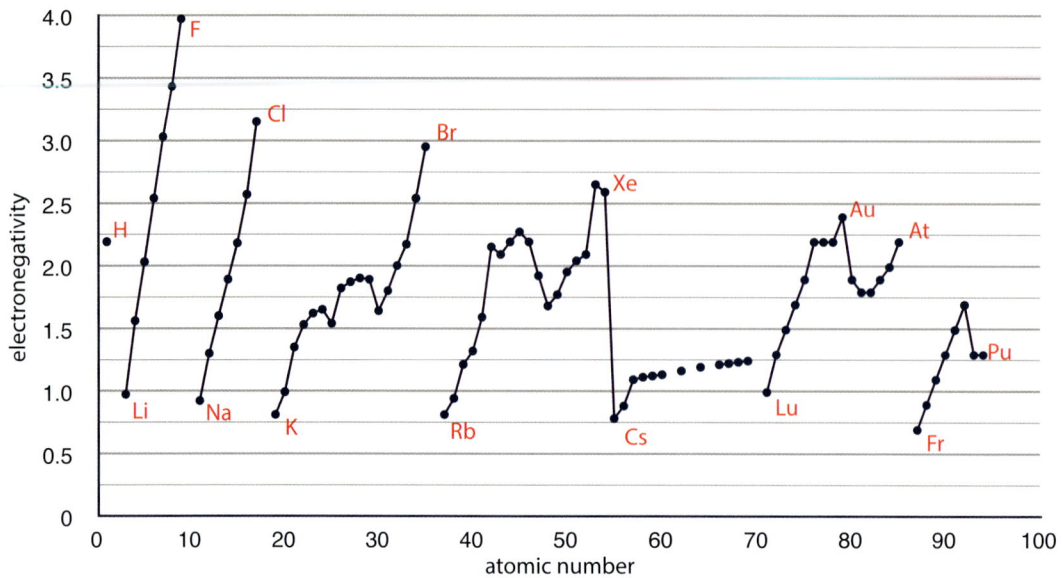

Figure 4.13. Electronegativity values, according to the Pauling electronegativity scale.

ue held by francium, to the highest value held by fluorine. You may need various electronegativity values for exercises in this and later chapters, so they are all shown in Figure 4.15 and inside the back cover. If you look at the values toward the right end of the transition metals, you see that around Groups 10–13 there is a small decline in the electronegativities, going against the general upward trend in values from lower left to upper right.

4.5   A Few Notes About Hydrogen

Hydrogen is located in Group 1 for a couple of reasons. First, it is an *s*-block element. Its lone electron is in a partially filled *s* subshell, just like the alkali metals. Second, hydrogen usually ionizes by losing an electron to form a cation, like all metals. In fact hydrogen forms the most basic of all cations—a lone proton. Third, in an aqueous solution (a solution with water as the solvent), acids, which are formed with hydrogen, dissociate (come apart) just as ionic compounds do. And just as the metal in a soluble ionic compound is a cation, so is the hydrogen from a dissolved acid. Hydrogen's atomic structure and its normal ionization as a cation with a charge of +1 indicate that hydrogen belongs at the top of Group 1.

But though hydrogen's *structure* assures that its place in Group 1 is probably not going to change, hydrogen's chemical *behavior* is more like a halogen than an alkali metal. For one thing, hydrogen shares electrons in covalent bonds with other elements to form molecules, just as nonmetals do. In contrast, when metals bond they form crystals. A second factor is that hydrogen atoms bond to *themselves* to

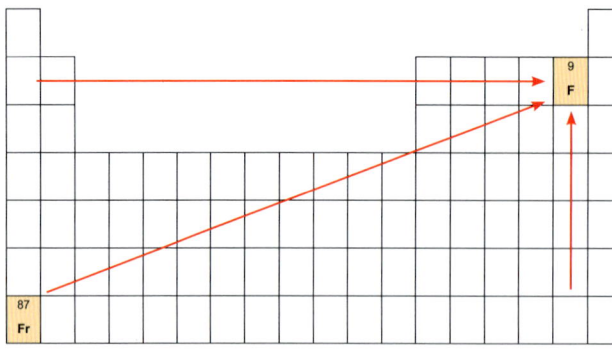

Figure 4.14. The arrows show the trends of increasing electronegativity from the lowest value at francium to the highest value at fluorine.

116

The Periodic Law

1																	18
1 **H** 2.20	2											13	14	15	16	17	2 **He**
3 **Li** 0.98	4 **Be** 1.57											5 **B** 2.04	6 **C** 2.55	7 **N** 3.04	8 **O** 3.44	9 **F** 3.98	10 **Ne**
11 **Na** 0.93	12 **Mg** 1.31	3	4	5	6	7	8	9	10	11	12	13 **Al** 1.61	14 **Si** 1.90	15 **P** 2.19	16 **S** 2.58	17 **Cl** 3.16	18 **Ar**
19 **K** 0.82	20 **Ca** 1.00	21 **Sc** 1.36	22 **Ti** 1.54	23 **V** 1.63	24 **Cr** 1.66	25 **Mn** 1.55	26 **Fe** 1.83	27 **Co** 1.88	28 **Ni** 1.91	29 **Cu** 1.90	30 **Zn** 1.65	31 **Ga** 1.81	32 **Ge** 2.01	33 **As** 2.18	34 **Se** 2.55	35 **Br** 2.96	36 **Kr**
37 **Rb** 0.82	38 **Sr** 0.95	39 **Y** 1.22	40 **Zr** 1.33	41 **Nb** 1.6	42 **Mo** 2.16	43 **Tc** 2.10	44 **Ru** 2.2	45 **Rh** 2.28	46 **Pd** 2.20	47 **Ag** 1.93	48 **Cd** 1.69	49 **In** 1.78	50 **Sn** 1.96	51 **Sb** 2.05	52 **Te** 2.1	53 **I** 2.66	54 **Xe**
55 **Cs** 0.79	56 **Ba** 0.89	71 **Lu** 1.0	72 **Hf** 1.3	73 **Ta** 1.5	74 **W** 1.7	75 **Re** 1.9	76 **Os** 2.2	77 **Ir** 2.2	78 **Pt** 2.2	79 **Au** 2.4	80 **Hg** 1.9	81 **Tl** 1.8	82 **Pb** 1.8	83 **Bi** 1.9	84 **Po** 2.0	85 **At** 2.2	86 **Rn**
87 **Fr** 0.7	88 **Ra** 0.9	103 **Lr**	104 **Rf**	105 **Db**	106 **Sg**	107 **Bh**	108 **Hs**	109 **Mt**	110 **Ds**	111 **Rg**	112 **Cn**	113 **Nh**	114 **Fl**	115 **Mc**	116 **Lv**	117 **Ts**	118 **Og**

Figure 4.15. Electronegativity values, according to the Pauling electronegativity scale.

form molecules of H_2, something metals almost never do. Third, hydrogen can ionize by *gaining* an electron to fill the 1s subshell and become H^-, an anion known as *hydride*.

With its only subshell half full with one electron, easily emptied by losing one electron and easily filled by gaining one electron or sharing one pair of electrons (details next chapter), hydrogen is unique among the elements in the periodic table.

Chapter 4 Exercises

SECTION 4.1
1. What entire group of elements did not appear in Mendeleev's original periodic table? Why were they left out and how were they put in?

2. Write a paragraph explaining the general structure and arrangement of the periodic table.

SECTION 4.2
3. State the chief chemical property that distinguishes the metals from the nonmetals.

Hmm... Interesting.   Hydrogen in space

Hydrogen is the most abundant element in the universe, making up about 75% of the mass of all matter. The image below, taken by the Hubble Space Telescope, is of the NGC 604 nebula, an enormous region of ionized hydrogen gas (a plasma) in the constellation *Triangulum*. The gas cloud is about 1,500 light years across, and has been called a "nursery of new stars" because of all the new stars formed within it. At the center are over 200 hot stars, each 10–15 times the size of our sun. These hot stars excite the hydrogen atoms, causing them to fluoresce, and heat the nebula to 10,000 kelvins—about twice the temperature at the surface of our sun.

Chapter 4

4. Distinguish between cations and anions.
5. What is the "long form" of the periodic table, and why are there two forms (see Section 3.4.1).

SECTION 4.3

6. Describe the trend in atomic radius going down a group and across a period.
7. Using the concept of the shield effect, write a description accounting for the trends in atomic size in the periodic table.
8. From your knowledge of the periodic table, put the elements rubidium (Rb), silver (Ag), xenon (Xe), and yttrium (Y) in order of increasing atomic radius. Explain your order by referring to trends in the periodic table.
9. From your knowledge of the periodic table, put the elements sodium (Na), barium (Ba), cesium (Cs), and magnesium (Mg) in order of increasing atomic radius. Explain your order by referring to trends in the periodic table.
10. Based on your knowledge of trends in the periodic table, place the following atoms and ions in order of decreasing size: Be^{2+}, Mg, Ca, and Mg^{2+}.

SECTION 4.4

11. Aluminum and scandium both ionize to +3, even though scandium is in Group 3 and aluminum is in Group 13. Explain why this is.
12. Define ionization energy and describe the trends for ionization energy in the periodic table across periods and down groups.
13. Referring again to Table 4.1, explain the large increase in ionization energy that occurs in the yellow shaded region of that table.
14. Write a description accounting for the trends in ionization energy in terms of the shield effect and other factors.
15. Write the condensed electron configurations for Cu^{2+}, As^{5+}, Ag^+, and Au^{3+}.
16. Why are ionization energies so much higher when core electrons are involved than they are when only valence electrons are involved?
17. Based on your knowledge of trends in the periodic table, place the following atoms in order of increasing ionization energy: Ar, Sr, P, Mg, and Ba.
18. Distinguish between ionization energy and electron affinity.
19. Based on your knowledge of trends in the periodic table, place the following atoms in order of increasing electron affinity: Br, Rb, and S.
20. Explain what the electronegativity scale is used for and how it arose.
21. Of the following cations, which is least likely to form: Ca^{3+}, Mg^{2+}, K^+? Explain your response.
22. Why do the chalcogens form ions with a charge of −2?
23. Distinguish between electron affinity and electronegativity.
24. Referring to the periodic table, describe the chemical properties of potassium (K), sulfur (S), xenon (Xe), iodine, (I), and manganese (Mn).
25. Which of the following is likely to have the greatest difference between the third and fourth ionization energies: Cl, Sc, Na, C?
26. Using only the periodic table as a reference (without electronegativity data), predict the relative electronegativities of these elements and put them in order from least to greatest: Ni, Ta, Se, F, Cs, Cl.

27. Develop an explanation for why the electron affinity values for the chalcogens are each significantly lower than those of their halogen neighbors.

28. Why don't the noble gases have electronegativity values listed in Figure 4.15?

SECTION 4.5

29. Describe the chemical properties that place hydrogen in Group 1, and the chemical properties hydrogen shares with Group 17 elements.

GENERAL REVIEW EXERCISES

30. Determine the energy released by the largest electron transition in the Lyman series (see Figure 3.8). State your answer in eV.

31. If a beam of laser light consists of photons with energies of 5.09×10^{-19} J, is the light visible? Explain your response.

32. Identify the block, period, and group for the elements represented by each of the following condensed electron configurations:

    a. $[Ne]3s^2 3p^3$
    b. $[Xe]6s^2 4f^{14} 5d^{10} 6p^1$
    c. $[Kr]5s^1 4d^5$
    d. $[Ar]4s^2 3d^3$

33. How many orbitals are there in the shell associated with $n = 4$? How many electrons can this shell hold?

34. Determine the number of carbon atoms present in 112 g CO_2.

35. Given 35.0 g H_2SO_4 (hydrogen sulfate, which is called sulfuric acid in aqueous solution), determine the percent composition. Then determine numbers of hydrogen, sulfur, and oxygen atoms present.

36. Identify some specific differences between the chemical properties of the alkali metals and those of the transition metals.

37. How many grams of calcium are there in 3.00 mol $CaBr_2$?

38. Analysis of a certain sample finds that the sample consists of 53.64% chlorine and 46.36% tungsten. Determine the empirical formula for this compound.

39. What is meant by the phrase, "chemistry is all about modeling"?

40. Why is it that scientists, when they are being accurate in their speech, avoid using the term *truth*? What are they likely to say instead?

41. Is there a difference between scientific facts and historical facts? If so, what is it?

42. What is the difference between molar mass and molecular mass?

43. Why must a calculation of molecular mass based on periodic table data necessarily be an average mass and not the mass of a specific molecule?

Chapter 5
Chemical Bonding

Above are pictured the famous white cliffs of Dover at the southeast corner of England. The cliffs are chalk, one of the many naturally occurring forms of calcium carbonate, $CaCO_3$. Limestone, marble, and sea shells are also made of $CaCO_3$. Marine animals combine the calcium Ca^{2+} ions and carbonate CO_3^{2-} ions dissolved in the water to form their shells. In fact, the limestone and marble were themselves formed by ancient layers of tiny sea creatures with their calcium carbonate shells depositing on the ocean floor as they died.

The space-filling model to the right is of the mineral calcite, also made of $CaCO_3$. The green spheres represent the calcium ions and the black and red spheres are the CO_3^{2-} ions.

As shown later in Figure 5.13, the crystal form of calcium carbonate, calcite, possesses an interesting optical property.

Objectives for Chapter 5

After studying this chapter and completing the exercises, you should be able to do each of the following tasks, using supporting terms and principles as necessary.

SECTION 5.1
1. Explain the octet rule.

SECTION 5.2
2. Describe the details of ionic bonding and write formulas for ionic compounds formed from given elements.
3. Given a chart with common oxidation states for transition metals, give names for ionic compounds formed from given elements using the Stock system and the older system based on Latin names.
4. Explain the terms *lattice energy* and *hydrate*.
5. Explain why ionically bonded compounds tend to be brittle and dense and have high melting points.

SECTION 5.3
6. Draw Lewis structures for atoms, molecules, ions, and polyatomic ions.
7. State the names of ionic compounds formed with polyatomic ions.
8. State the names of acids formed from given atoms and polyatomic ions.
9. Describe and explain the three general cases of exceptions to the octet rule.
10. Explain what resonance structures are and depict them in Lewis structures.
11. Give the names for binary molecular compounds.
12. Describe some physical properties of molecular substances and compare the properties of covalently bonded substances to those of ionically bonded substances.

SECTION 5.4
13. Use the electronegativity scale to explain molecular polarity.
14. Given electronegativity values, predict the polarity and the percent ionic character of a given molecule, and characterize bonds as nonpolar covalent, polar covalent, or ionic.
15. Use percent ionic character to give a general account for the colors of aqueous solutions.

Chapter 5

5.1 Preliminaries

5.1.1 Chemical Possibilities

As with the atomic orbital structures we have seen in previous chapters, much of our understanding about chemical bonding derives from theoretical calculations. When it comes to chemical bonding, some of those calculations involve the Schrödinger equation; others involve ionization energy, electron affinity, and electronegativity. For all but the last of these, our considerations in this text are primarily qualitative. In other words, we will not go into the calculations themselves; we simply look at what the results tell us about how atomic bonding works. The calculations involving electronegativity are simple, so we will have a look at those.

Much of chemistry is about understanding the mechanisms that bond atoms to other atoms. *And guess what that means*—electrical attractions, electrons, minimizing energy, whole-number ratios, and modeling! Everything from the Introduction is on center stage here.

In some cases, atomic bonds hold atoms together in the rigid, extended, regular structures of crystal lattices. In other cases, atoms are bonded together in molecules. And a third type of structure again involves regular arrangement of atoms in a strong crystal lattice, but with much less rigidity and much less brittleness. These three bonding types—named *ionic bonds*, *covalent bonds*, and *metallic bonding*—are each caused by a particular combination of electrical attraction and electron location that minimizes the energy in a given structure. We explore the first two in this chapter and save metallic bonding for the next chapter.

5.1.2 The Octet Rule

For main group elements, the electrons in the *s* and *p* subshells are the valence electrons. Adding the two possible electrons in an *s* subshell to the six possible electrons in a *p* subshell gives eight electrons—an *octet*. The noble gases (except helium) each have an octet without ionizing. This is the most stable of possible electron configurations, where all electrons are in a stable state that minimizes their energies, and for this reason, atoms of noble gas elements have high ionization energies and do not readily bond with other atoms. Other main group elements (and many of the transition metals) tend to ionize or share electrons with other atoms in such a way that they acquire a noble gas electron configuration, the stablest way to be. In other words, *atoms tend to gain or lose electrons to achieve an octet of electrons in the highest-energy occupied shell*. This is such a common characteristic that it has been dubbed the *octet rule*. People say that atoms try to "surround themselves" with an octet of electrons. The obvious main group exceptions to the octet rule are hydrogen and helium. These elements are each satisfied with a single pair of electrons in the 1*s* subshell. There are some other noteworthy exceptions (such as boron), which we explore later in the chapter.

The octet rule is an easy way to think about what happens with the electrons in both ionic bonding and covalent bonding. When ionic bonding occurs, atoms ionize, and the oxidation states they take on when they ionize usually involve achieving an octet. In covalent bonding, atoms share electrons in pairs, and each atom involved in the sharing acts as if all shared electrons are its own. In covalent bonds, atoms generally try to share the correct number of electron pairs so that they get their octets. In fact, we even use diagrams that symbolize this, such as the diagram of the hydrogen and oxygen atoms in a water molecule shown in Figure 5.1. We elaborate on this type of diagram when we get to covalent bonding. But for now, you see that each hydrogen is adjacent to a pair of electrons and the oxygen is surrounded by its octet.

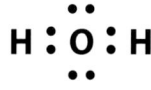

Figure 5.1. Life is good. Each hydrogen has its pair and oxygen has its octet.

Chemical Bonding

5.2　Ionic Bonding

5.2.1 Ionic Bonds and Crystals

As its name implies, ionic bonds are bonds between ions—the symmetrical electrical attraction between cations and anions is what bonds the atoms together. As you know, the minimum energy arrangement for objects with opposite electrical charge is for them to get as close together as possible. Minimum energy for like charges is to be as far away as possible. To minimize the energy of the attractions and repulsions between the ions, the ions arrange themselves into a crystal lattice, as illustrated in Figure 5.2. The figure shows four cells of the cubic crystal lattice formed by +1 cations (alkali metals) and −1 anions (halogens). Sodium chloride is the usual example given. In this crystal structure, the ratio of cations to anions is one to one. Figure I.11 shows the more complicated structure of calcium fluoride, in which there are two fluorine atoms for each calcium atom.

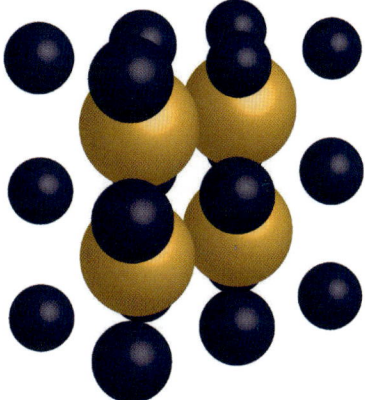

Figure 5.2. The cubic crystal structure formed by ionic bonds between +1 cations and −1 anions. In the lattice, the ratio of cations to anions is 1 : 1.

In the previous chapter, we saw that metals ionize to become cations and nonmetals ionize to become anions. It follows then that ionic bonds form between metals and nonmetals. More generally, ionic bonds form between any type of positive and negative ions, even if the ions are molecules instead of single atoms.

As illustrated in Figure 5.3, we usually think of ionic bonding as a process of electron *transfer*. A metal ionizes by the removal of one or more electrons, and a nonmetal ionizes by acquiring one or more electrons. In the process, electrons are transferred from the metal atoms to the nonmetal atoms. The figure illustrates the electron transfer that occurs between sodium and chlorine. A glance at the periodic table shows that sodium ionizes by the removal of its one valence electron to become a +1 cation. Chlorine, being a Group 17 nonmetal, needs only one additional electron for its valence shell to be full—an attractive, low-energy state—so it becomes a −1 anion.

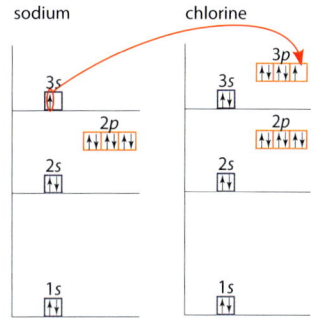

Figure 5.3. Electron transfer between sodium and chlorine atoms. The sodium atom becomes a Na^+ ion and the chlorine atom becomes a Cl^- ion. In the process, both sodium and chlorine atoms acquire octets.

The $3s$ electron from the sodium atom moves into the $3p$ subshell of the chlorine atom. In the process, both sodium and chlorine atoms acquire octets.

Figure 5.4 illustrates the electron transfer that occurs between magnesium and chlorine. Magnesium is an alkaline-earth metal. As such, it has two valence electrons and ionizes to become a +2 cation. To accommodate both these electrons, two atoms of chlorine are required, since each takes on only one electron to fill the $3p$ subshell. In the process, all three atoms acquire octets.

When writing the chemical formula for an ionic compound, the convention is to place the chemical symbol for the cation first followed by the chemical

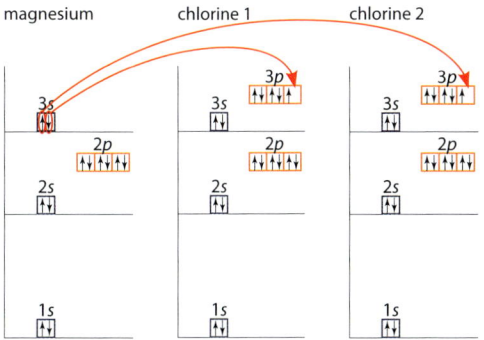

Figure 5.4. Electron transfer between magnesium and chlorine atoms. The magnesium atom becomes a Mg^{2+} ion, and each of the two chlorine atoms becomes a Cl^- ion. All three atoms acquire octets.

Chapter 5

symbol for the anion. Thus, the formula for sodium chloride is NaCl. In the formula, subscripts are used to indicate the whole number ratio of atoms in the compound. If the number of atoms in the ratio is one (as with sodium and chlorine being in a one to one ratio), no subscript is used. In magnesium chloride, two atoms of chlorine are needed for each atom of magnesium, so the formula is $MgCl_2$. The subscripts show the whole-number ratios of atoms in the compound.

▼ Example 5.1

Write the chemical formula for calcium sulfide.

As an alkaline-earth metal, calcium ionizes to +2. Sulfur, a chalcogen, ionizes to −2. Since two electrons are removed from each calcium atom and two electrons are added to each sulfur atom, the ratio of calcium atoms to sulfur atoms is one to one. Thus, the formula is CaS.

▲

▼ Example 5.2

Write the chemical formula for beryllium phosphide.

As an alkaline-earth metal, beryllium ionizes to +2. Phosphorus, a Group-15 nonmetal, ionizes to −3. Two electrons are removed from each beryllium atom, but three electrons are added to each phosphorus atom. The only way for this to work out is for the electrons to be exchanged in quantities based on the least common multiple of two and three, which is six. As shown in Figure 5.5, three beryllium atoms provide a total of six electrons, and these are divided equally among two phosphorus atoms. Since the ratio of beryllium atoms to phosphorus atoms is three to two, the formula is Be_3P_2.

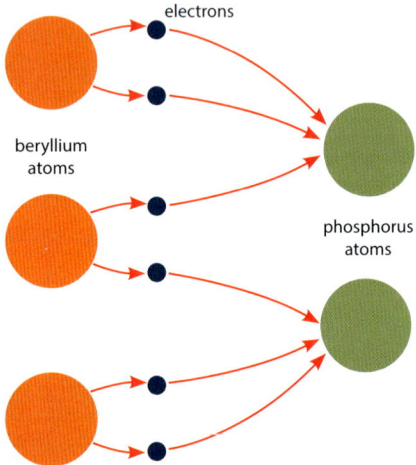

Figure 5.5. Electron transfer between +2 cations and −3 anions requires the ratio of cations to anions to be 3 : 2.

▲

The easiest way to work out formulas for ionic bonds involving metals from Groups 1–4 is to think about the number of valence electrons involved (equal to the group number of the metal) and the numbers of cations and anions that are needed for the electron transfer to work out evenly. But when working out the formulas for ionic compounds with metals in Groups 5 and higher, visualizing numbers of valence electrons from the metal's position in the periodic table isn't so easy. It's easier just to select subscripts for the cation and anion so the product of the subscript and the oxidation state is the same for the cation and anion. This is equivalent to making the amounts of positive and negative charge match, as illustrated in the next example.

▼ Example 5.3

Write the chemical formula for iron oxide, assuming the iron ionizes as Fe^{3+}.

Oxygen ionizes as O^{2-}. To make the positive and negative charges match, we need to use the least common multiple of three (from the Fe^{3+}) and two (from the O^{2-}), which is six. With two +3 cations and three −2 anions the charges are equal. Thus, since the ratio of iron atoms to oxygen atoms is two to three, the formula is Fe_2O_3.

▲

Chemical Bonding

Figure 5.6. Copper(I) oxide, Cu_2O.

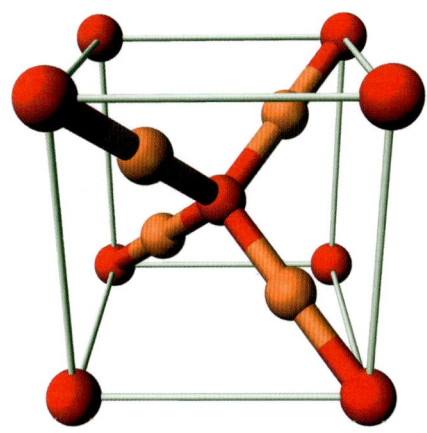

Figure 5.7. A single cell in the lattice structure (one of two possible structures) of copper(I) oxide, Cu_2O. Oxygen atoms are shown in red.

Formulas for ionic compounds are called *binary formulas* because there are two parts in the formula: the symbol for the cation and the symbol for the anion. The formulas for compounds formed with cations and anions from main group elements are easy to figure out because main group metals (Groups 1 and 2) and main group nonmetals (from Groups 14–17) ionize very predictably. All you have to do is look at what group an element is in in the periodic table to see what its oxidation state is when the atom gets its octet. But many transition metals have more than one common oxidation state and can thus make more than one compound with a given nonmetal.

For example, as listed in Figure 4.9, copper commonly forms both Cu^+ and Cu^{2+} ions and as a result can form two different compounds when bonding with oxygen. As Cu^+, copper forms Cu_2O, known as cuprous oxide or copper(I) oxide. (We address these naming schemes next.) This substance is a red powder, shown in Figure 5.6. Figure 5.7 shows one cell in the lattice for one of the two possible bonding structures for Cu_2O.

As Cu^{2+}, copper bonds with oxygen in a one-to-one ratio producing CuO, known as cupric oxide or copper(II) oxide. This substance is a black powder, shown in Figure 5.8. One cell of the lattice structure is shown on the left in Figure 5.9. On the right of the figure is an image showing how the crystal lattice looks when several cells are joined together.

Figure 5.8. Copper(II) oxide, CuO.

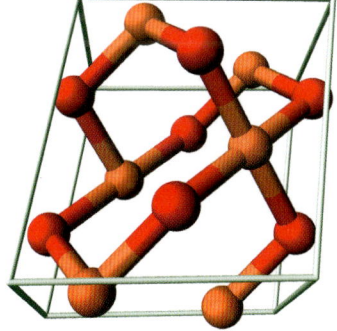
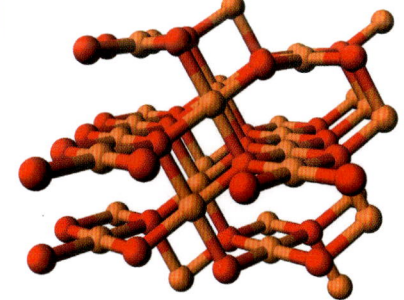

Figure 5.9. One cell of the lattice for copper(II) oxide, CuO (left), and an extended portion of the lattice structure (right).

5.2.2 Naming Ionic Compounds

Names of ionic compounds always have two parts—the name of the cation placed first, followed by the name of the anion.

Chapter 5

The first case to consider is when an ionic compound is formed by a *monovalent* cation and a *monatomic anion*. Monovalent cations have only one oxidation state, and monatomic ions are composed of only one atom, such as all the ions we have discussed so far. The rule for this case is that the name of the cation is simply the name of the metal, and the name of the anion is the name of the nonmetal with the suffix *–ide* attached.

▼ Example 5.4

Write the names and formulas for the compounds formed by the following pairs of ions: Ca^{2+} and F^-; K^+ and O^{2-}; and Al^{3+} and Br^-.

In each case, we state the name of the metal followed by the name of the nonmetal with the *–ide* suffix. The formulas must balance the charges of the ions, just as we have previously seen.

calcium fluoride, CaF_2

potassium oxide, K_2O

aluminum bromide, $AlBr_3$

The second case is for *multivalent* cations, cations with more than one oxidation state, and monatomic anions. There are two different naming systems in use. In both systems, the cation is written first, just as with monovalent cations, and the anion is written with the suffix *–ide*.

The contemporary system is called the *Stock system*, illustrated in Table 5.1. In the Stock system,

Oxidation State	Example Cation	Example Compound
lower	Fe^{2+} iron(II) Cu^+ copper(I)	iron(II) oxide, FeO copper(I) fluoride, CuF
higher	Fe^{3+} iron(III) Cu^{2+} copper(II)	iron(III) sulfide, Fe_2S_3 copper(II) chloride, $CuCl_2$

Table 5.1. Examples of the Stock system for naming ionic compounds.

the cation is the name of the metal followed by parentheses with upper-case Roman numerals indicating its oxidation state. (No space is placed between the name of the metal and the parentheses.) The multivalent cations are the transition metals and some of the metals in Groups 13–15. The most common oxidation states are shown in Figure 4.9. You may assume all Group 1 and Group 2 metals are monovalent, along with Al^{3+}, Ni^{2+}, Zn^{2+}, and Ag^+. (Each of these last four elements has other oxidation states, but they are quite uncommon.) The parentheses are unnecessary in compounds with monovalent cations. The rest of the metals should have the oxidation state written in parentheses as described above.

The older system, illustrated in Table 5.2, is important to know because it still appears often in reference literature and on bottles of chemicals. In this system, the Latin name of the cation is used with the suffix *–ous* to represent the lower oxidation state or the suffix *–ic* to represent the higher oxidation state.

Oxidation state	Cation Ending	Example Cation	Example Compound
lower	*–ous*	Fe^{2+} ferrous Cu^+ cuprous	ferrous oxide, FeO cuprous fluoride, CuF
higher	*–ic*	Fe^{3+} ferric Cu^{2+} cupric	ferric sulfide, Fe_2S_3 cupric chloride, $CuCl_2$

Table 5.2. Latin-based cation naming system.

Chemical Bonding

▼ Example 5.5

Write the names and formulas for the compounds formed by the following pairs of ions: Hg^{2+} and Cl^-; Mn^{4+} and S^{2-}; Pb^{2+} and O^{2-}; and Sn^{4+} and Br^-. Include names in both the older system and the contemporary Stock system.

We note that each of the cations is multivalent and so must have the oxidation state specified.

mercury(II) chloride, mercurous chloride, $HgCl_2$

manganese(IV) sulfide, manganic sulfide, MnS_2

lead(II) oxide, plumbous oxide, PbO

tin(IV) bromide, stannic bromide, $SnBr_4$

Note that for some of the metals with a chemical symbol based on a Latin name (e.g., copper, iron, lead, tin, and gold) the old style name is based on the Latin element name. In other cases (e.g., mercury, cobalt, chromium, and manganese), English names are used.

▲

5.2.3 Energy in Ionic Bonds

Typically, the formation of ionic bonds is an exothermic process—the reaction releases energy in the form of heat and light. We can analyze the process in terms of the ionization energy and electron affinity. Remember, chemistry is all about minimizing energy, so we can understand ionic bonding in these terms.

Let's use sodium chloride as an example. The reaction between sodium and chlorine happens spontaneously and a great deal of energy (light and heat) is released. Since the reaction is spontaneous, any energy required to make the reaction occur comes from the reactants themselves.

The ionization energy—energy that must supplied—for a sodium atom is 5.14 eV. The electron affinity—energy released—for a chlorine atom is 3.61 eV. Now, these values may suggest that we have a problem. We have to put more energy in to ionize the sodium than we get back out when the chlorine atom takes on the electron! This seems to indicate that the reaction between sodium and chlorine couldn't release so much heat, or even occur at all. In fact, our energy problem is even worse. For the sodium and chlorine to react, a piece of sodium metal must be vaporized—the atoms have to be in the gaseous state so they can interact with individual chlorine atoms—and this requires 1.12 eV per atom. Also, the diatomic chlorine molecules have to be separated, requiring an additional 1.25 eV per atom. So all together, 5.14 eV + 1.12 eV = 6.26 eV of energy must be supplied per sodium atom, plus 1.25 eV per chlorine atom, for sodium and chlorine to react, a total of 7.51 eV per pair of NaCl atoms is required. However, only 3.61 eV per atom is released when a chlorine atom accepts an electron. And further, there is energy in the sodium chloride crystal lattice after the reaction is complete. Where does all the extra energy come from? Obviously, there's more to this story.

The ionization energy is the energy required to remove an electron far away from the cation so the electron becomes completely free. Likewise, the electron affinity is the energy released when an electron comes from far away to be united to a neutral atom. This means that in energy terms, we imagine that the electron transfer from the sodium atoms to the chlorine atoms happens with all these atoms far apart. Then after the electron transfer is over, we allow all the ions to collapse together, releasing energy in the process, to form the crystal lattice.

Of course, if we actually combine sodium metal and chlorine gas, the atoms do not spread out, react, and then attract back together from electrical forces. But to perform the energy accounting using the definitions and energy values we have, we imagine they do. Imagining it in energy terms, there is a whopping amount of energy released when all the cations and anions come crashing together. This is where the extra energy comes from, even though it is not apparent simply by looking at the ionization energy and electron affinity values.

Part of the energy released goes to binding the ions together in the crystal lattice. This energy is called the *lattice energy*. Considered this way, the lattice energy is a negative value because the ions go into lower energy states when they come together. Another way to think about lattice energy is as the energy required to separate completely all the ions in the crystal lattice from each other—to pull the crystal apart, ion by ion. Considered this way, the lattice energy is positive. Both definitions are acceptable and common.

Lattice energy values for a few common ionic compounds are given in Table 5.3. To help you get a feel for the quantities of energy involved, let's run through an energy calculation *just for fun*. You may find the results surprising.

▼ **Example 5.6**

Compare the ionization energy of sodium and the electron affinity of chlorine to the lattice energy of sodium chloride.

It makes intuitive sense to specify lattice energies in terms of a bulk quantity such as a mole. So instead of working on an atom-by-atom basis, we work on a per-mole basis. Let's begin by converting the energy values given in the text above from eV for a single particle to kJ/mol:

$$\frac{7.51 \text{ eV}}{\text{particle}} \cdot \frac{1.602 \times 10^{-19} \text{ J}}{1 \text{ eV}} \cdot \frac{6.022 \times 10^{23} \text{ particle}}{\text{mol}} = 725{,}000 \frac{\text{J}}{\text{mol}} = 725 \frac{\text{kJ}}{\text{mol}}$$

So for one mole of sodium and one mole of chlorine, 725 kJ are required to get the Na and Cl atoms apart and ionize the sodium atoms. Similarly, 3.61 eV of energy released when each chlorine atom receives an electron works out to 349 kJ/mol. This leaves an energy deficit of 376 kJ. The energy released as the lattice forms supplies this, and the lattice energy as well. From Table 5.3, the lattice energy of NaCl is 769 kJ/mol. So bringing the crystal together releases enough energy to cover the energy deficit and still have 769 kJ to form the crystal lattice. This is the lattice energy in one mole of NaCl, which is only 58.44 grams (see Figure 2.26).

So how much energy is 769 kJ? Well, one watt of power is an energy rate of one joule per second. Let's pick an electrical device we are familiar with and calculate how long it would run with this much energy to power it. Since we are all familiar with good old fashioned 60-W light bulbs (at least for one more generation or so of students) we will use that:

$$\frac{769{,}000 \text{ J}}{60 \frac{\text{J}}{\text{s}}} = 12{,}817 \text{ s} \cdot \frac{1 \text{ hr}}{3600 \text{ s}} = 3.6 \text{ hr}$$

Have you ever felt the heat from a 60-W light bulb? If you could trap all this heat and light for over three and a half hours you would have an amount of energy equal to the lattice energy holding together the crystals in 1/4 cup of salt!

Compound	Calculated Lattice Energy, kJ/mol
LiF	1030
NaF	910
KF	808
LiCl	834
NaCl	769
KCl	701
NaBr	732
NaI	682
$MgCl_2$	2477
$CaCl_2$	2268
Na_2O	2481
K_2O	2238
MgO	3795
CaO	3414

Table 5.3. Lattice energies for a few ionic compounds.

Chemical Bonding

When 1/4 cup of salt is dissolved in a beaker of water, 769 kJ of energy must be supplied to the salt to pull apart the crystal lattice and allow the ions to go into solution. Now, all the energy needed for this to happen comes from the water. So why doesn't the water's temperature plummet when salt is dissolved in it? The answer relates to the polarity of the water molecule and the ever-present tendency to minimize energy. Since the water molecules are polar, they rush toward and surround the dissolved ions, releasing enough energy in the process to keep the water temperature roughly constant. It is amazing that there is so much energy involved in the process at the atomic level, and that when we stir salt into a glass of water we are completely unaware of it. We return to this topic when we get to the chapter on solutions.

5.2.4 Hydrates

In some ionic compounds, it is common for water molecules to integrate into the crystal lattice if the crystals are grown in an aqueous (water) solution. These compounds are call *hydrates*. The presence of the water molecules can allow compounds to form continuous crystals, when without the water molecules the compound would decompose into a powder. A nice example is the compound copper sulfate pentahydrate, $CuSO_4 \cdot 5H_2O$, shown in Figure 5.10. The coefficient in front of the H_2O (5 in this case) indicates how many water molecules are in the crystal lattice for each formula unit of atoms in the ionic formula.

The complex structures of $CuSO_4$ (without the water molecules) and $CuSO_4 \cdot 5H_2O$ are shown in Figure 5.11. If you study these images closely you can see that in the $CuSO_4$ lattice, the copper-colored copper atoms are attached to the red oxygen atoms in the yellow and red sulfate ions (SO_4^{2-}). But interestingly, the water molecules in the hydrate crystal (shown in red and white) are connected to the copper atoms and the sulfate ions are left floating in the lattice.

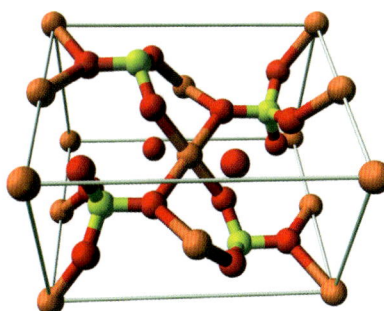

Figure 5.10. Crystals of copper sulfate pentahydrate, $CuSO_4 \cdot 5H_2O$.

5.2.5 Intensive and Extensive Properties

We often have occasion to discuss the physical properties of substances. In that regard, chemists tend to classify the properties of substances into two types: *intensive* and *extensive*. An *intensive property* is a bulk property, meaning the property

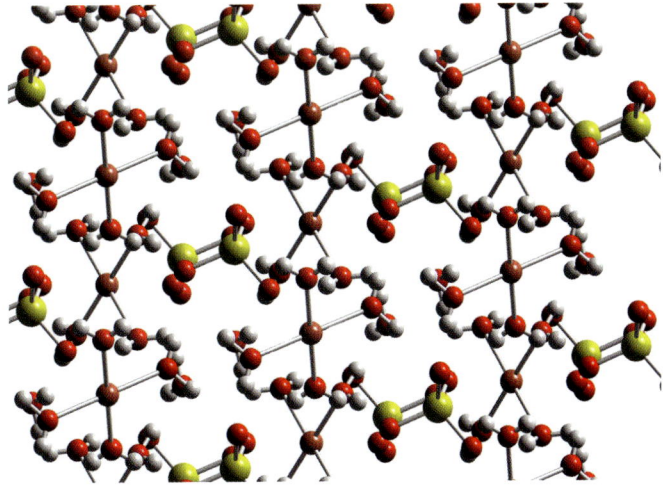

Figure 5.11. One cell of the $CuSO_4$ lattice (left) and an extended image of the $CuSO_4 \cdot 5H_2O$ crystal (right). Colors are red: oxygen, yellow: sulfur, white: hydrogen, and copper: copper.

Chapter 5

does not depend on how much of the material is present. Examples of intensive properties include brittleness, electrical conductivity, boiling point, and density. *Extensive properties* depend on how much material is present. Examples of these include mass and volume.

5.2.6 Physical Properties of Ionically Bonded Substances

The high lattice energies in ionic compounds are due to the strong, symmetrical network of electrical attractions and repulsions in the crystal lattice, illustrated in Figure 5.12. The green arrows in this figure indicate electrical repulsions, and the red arrows indicate electrical attractions. (For clarity, the diagonal repulsions between the blue ions are not shown.) As you see, each ion is bound in place by the symmetrical forces all around it.

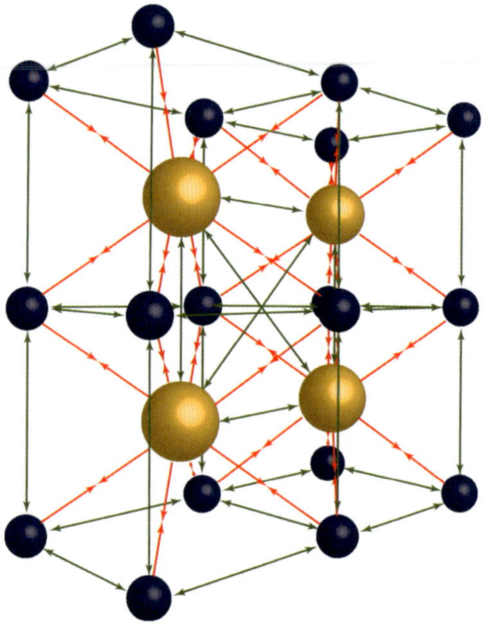

Many of the intensive physical properties of ionic compounds are due to the strong electrical forces in the crystal lattice. One of these is brittleness, the tendency of a substance to shatter instead of bend when stressed. The classic shattering of a crystal glass from the sound waves from an opera singer's voice (a legend confirmed by MythBusters, although a bit of amplification was required) is a great example of this. Ionic crystals can flex to some extent. This is why vibrating pieces of quartz are used now to regulate the speed of just about every cheap battery-powered clock in the world. But when subjected to a high enough stress, a crystal lattice breaks apart instead of flexing. Whack a lump of iron with a hammer and you may put a dent in it. Whack any ionic compound with a hammer (including all the rock minerals) and watch it shatter. Although the iron is crystalline, like all pure metals, it is not held together by ionic bonds. We consider bonding in metals in the next chapter.

Figure 5.12. The symmetry of attractive and repulsive electrical forces in the ionic bond make the crystal lattice very rigid—if stretched too far it will break rather than bend.

High melting point and high density are two other physical properties of ionic compounds. A few of these values are shown in Table 5.4. The contrast with covalently bonded compounds should be obvious. Water and ammonia, for example, melt at 0°C and −78°C, respectively. And water, with a density of 1 g/cm^3, is fairly dense for a covalent compound, but garden variety ionic compounds are two to three times as dense.

An interesting (although somewhat esoteric) property some ionic compounds exhibit is *birefringence*, or double refraction. Figure 5.13 shows this happening as light passes through a piece of calcite, which is calcium carbonate, $CaCO_3$. If you look closely, you can see there

Compound	Melting Point, °C	Density, g/cm^3
LiF	848.2	2.64
NaF	996	2.78
KF	858	2.48
LiCl	610	2.07
NaCl	800.7	2.17
KCl	771	1.99
NaBr	747	3.20
NaI	661	3.67
$MgCl_2$	714	2.33
$CaCl_2$	775	2.15
Na_2O	1134	2.27
K_2O	740	2.35
MgO	2825	3.6
CaO	2613	3.34

Table 5.4. Two physical properties for a few ionic compounds.

Chemical Bonding

is a double image. This is caused by light reflecting from the letters on the paper and refracting twice as it passes upward through the crystal. Birefringence is caused by the polarization of the light waves interacting with the planes of atoms in the crystal structure.

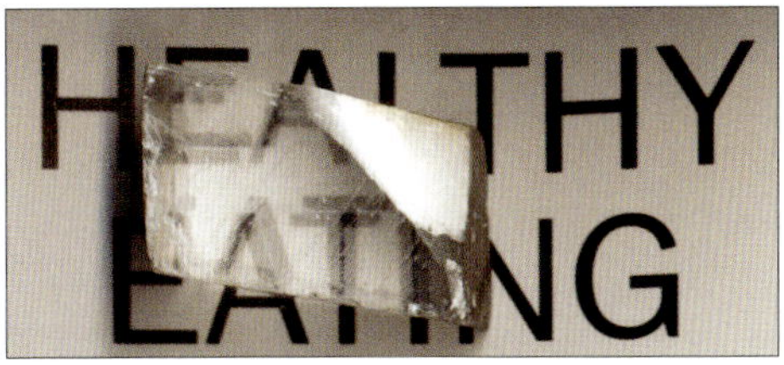

5.3   Covalent Bonding

Figure 5.13. Calcite exhibits the double refraction known as birefringence.

5.3.1 Covalent Bonds and Molecules

Covalent bonds are formed when atoms of nonmetals bond with each other. Instead of transferring electrons to achieve octets, atoms *share electrons in pairs* to accomplish the same thing. Since there are no ions involved, we don't have thousands of electrical forces binding the atoms into a lattice. Instead, it is the actual sharing of electrons in orbitals that binds the atoms together. And when atoms bond this way, they bond in groups or clusters—that is, molecules.

Later, we get into the details of theoretically modeling the orbitals of atoms. For now, we focus on the numbers that make the bonds work. The goal is for atoms to share valence electrons in pairs so that all atoms achieve their octets (or for hydrogen, its complete pair). There are a few important exceptions to the ways atoms follow the octet rule, which we will address in a bit.

To begin, let's consider water, H_2O. To visualize the bonding, we draw the symbol for each of the atoms in the molecule along with dots to represent their valence electrons, a representation called a *Lewis symbol*. Each hydrogen atom has one valence electron and the oxygen atom has six valence electrons. This makes a total of eight valence electrons that are involved in the bonding. For a given atom, the first four valence electrons are drawn with one on each side of the element symbol. After each side of the element symbol has one electron, the remaining electrons are placed to form pairs, with a maximum of two electrons on each side of the element symbol. The Lewis symbols for the atoms involved in the water molecule are:

H·   H·   ·Ö·

Covalent bonding involves atoms sharing electrons in pairs to achieve their octets (or pair for hydrogen). Each atom contributes one electron to make up a shared pair. The reason they share in pairs is that each atomic orbital holds two electrons, and in covalent bonding the orbitals of two atoms meld together into a single *molecular orbital* that holds the shared pair of electrons.

In the case of the water example, each of the hydrogen atoms shares one pair of electrons with the oxygen atom, represented this way:

H:Ö:H

As you see from this representation, we still have eight dots, representing the eight valence electrons. Each hydrogen atom has a pair of electrons next to it, representing its filled $1s$ subshell. The oxygen atom is surrounded by eight electrons, representing its complete octet.

Our next example is carbon dioxide, CO_2. Here are the atoms, each with their valence electrons:

·Ċ·   ·Ö·   ·Ö·

The carbon atom is in the center with an oxygen bonded to either side of it, sharing pairs of electrons. Let's start there.

·Ö:Ċ:Ö·

131

Now we see there are still four unpaired electrons, and none of the atoms has an octet yet. This problem is solved by the carbon atom sharing not one but *two pairs* of electrons with each of the oxygen atoms, symbolized like this:

Ö::C::Ö

The double pairs of electrons symbolize two pairs of electrons being shared. Both pairs are counted in the octet for each atom in the bond. Now, every atom in the molecule has an octet, as illustrated in Figure 5.14. In this figure, the electrons associated with each atom are circled to show the atom's octet of eight electrons.

Figure 5.14. Circles indicate the octets for each of the three atoms in the CO_2 molecule.

Sharing two pairs of electrons is called a *double bond*. There are also triple bonds in which three pairs of electrons are shared. This is illustrated best by the *diatomic molecules* formed by nitrogen atoms. We should pause here for a bit of historical background.

In 1811, Italian scientist Amedeo Avogadro introduced a new molecular theory of gases. Part of Avogadro's theory is the idea that atoms of some elements form molecules this way—by atoms bonding in groups to make molecules, including diatomic molecules. At the time, Avogadro's theory was widely rejected; even John Dalton, the father of modern atomic theory, rejected it. But in 1860, almost 50 years later and just a few years after Avogadro's death, an Italian chemist named Stanislao Cannizzaro gave a presentation in Germany at the first international conference for chemists showing that Avogadro had been correct all along. Acceptance of Avogadro's theory began gaining ground after that.

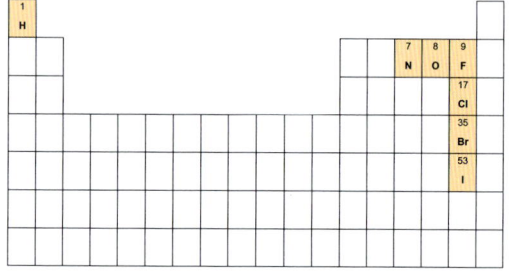

Figure 5.15. The seven elements that form diatomic molecules.

Back in Section 2.4.2, I mentioned the main substance of Avogadro's new theory—that the volume of a gas at a given temperature and pressure was proportional to the number of particles of the gas (atoms or molecules), regardless of the identity of the gas. Because of his work establishing molecular theory, Avogadro is now considered the father of atomic-molecular theory. And as you have already learned, Jean Perrin honored Avogadro by naming the Avogadro constant after him.

There are seven elements that appear in nature as diatomic molecules. These are identified in Figure 5.15. Atoms of hydrogen and all the halogens are lacking only one electron to have a full octet (or pair, for H). Atoms of each of these elements bond together in pairs and share a single pair of electrons, as shown in Table 5.5. As the table shows, the diatomic oxygen molecule is held together by a double bond. Since oxygen atoms have six valence electrons, they need to share two pairs to achieve an octet. Nitrogen atoms have only five valence electrons. Filling up the octet means sharing three pairs of electrons, and thus the *triple bond*. There are many more molecules

Molecule	Lewis Symbol Representation
H_2	H:H
N_2	:N⋮⋮N:
O_2	Ö::Ö
F_2	:F̈:F̈:
Cl_2	:C̈l:C̈l:
Br_2	:B̈r:B̈r:
I_2	:Ï:Ï:

Table 5.5. Lewis symbol representations for the diatomic molecules.

to explore. We take these up shortly with an enhanced symbolism after an excursion into the topic of polyatomic ions.

5.3.2 Polyatomic Ions

For many covalent molecules, when the atoms all get their octets there are more electrons in the molecule than protons, giving the molecule an overall negative charge. These molecules are called *polyatomic ions*, and these ions will form ionic bonds with cations. In two common cases, the ammonium and hydronium ions, the polyatomic ion has an excess of positive charge.

Sulfate, SO_4^{2-}, carbonate, CO_3^{2-}, and nitrate, NO_3^-, are common polyatomic ions. The bonding in the sulfate ion is as follows:

$$\left[\begin{array}{c} :\ddot{O}: \\ :\ddot{O}:S:\ddot{O}: \\ :\ddot{O}: \end{array} \right]^{2-}$$

There are 32 valence electrons shown in this molecule. But if we add up the valence electrons from sulfur (6) and four oxygen atoms (6 each) we get 30. Since there are two extra electrons in the molecule than there would be if only the five neutral atoms bonded together, the molecule has a net charge of −2. When drawing the bonding arrangements for molecules that have a net charge, we place the molecule inside square brackets and show the net charge on the upper right corner.

Similarly, the nitrate ion bonds as follows:

$$\left[\begin{array}{c} :\ddot{O}: \\ :\ddot{O}:N:\ddot{O}: \end{array} \right]^{-}$$

Each atom has an octet, and there are a total of 24 electrons. But adding up the valence electrons for three oxygen atoms and one nitrogen atom gives a total of 23. For this reason, the molecule has a net charge of −1. We discuss molecular shape in much more detail later. But for now you should take note of the general principle that atoms bonded to a central atom distribute themselves evenly around the central atom, so the structures we draw typically simulate this.

Most of the common polyatomic ions have a negative charge, but there are two common positively charged polyatomic ions—ammonium and hydronium. The ammonium molecule is just like the ammonia molecule except with an extra hydrogen atom in the molecule. Ammonia is NH_3, and the three hydrogen atoms make three single bonds with the central nitrogen atom. In ammonia, there is a spare pair of electrons on the central nitrogen atom, and it is easy for a hydrogen ion to attach to this pair to form NH_4^+. In a very similar way, the hydronium ion is just like the water molecule with an extra hydrogen atom attached to the central oxygen atom.

Some of the most common polyatomic ions are listed in Table 5.6. The hydronium ion is included because of its importance later when we discuss acids. But it is highly unlikely that you will find the hydronium ion attached to an anion in an ionic compound.

Ion	Formula
ammonium	NH_4^+
hydronium	H_3O^+
acetate	CH_3COO^-
bicarbonate	HCO_3^-
chlorate	ClO_3^-
hydroxide	OH^-
nitrite	NO_2^-
nitrate	NO_3^-
permanganate	MnO_4^-
carbonate	CO_3^{2-}
chromate	CrO_4^{2-}
sulfite	SO_3^{2-}
sulfate	SO_4^{2-}
phosphate	PO_4^{3-}

Table 5.6. Common polyatomic ions.

5.3.3 Ionic Compounds with Polyatomic Ions

Writing the binary formulas for ionic compounds incorporating polyatomic ions is similar to the process we saw earlier for formulas involving only monatomic ions, ions composed of a single atom. When writing formulas involving polyatomic ions, you should determine the subscripts by making the charges in the formula balance. In other words, the ion subscript times the ion's charge must yield the same product (with opposite signs) for both cation and anion.

▼ Example 5.7

Write the chemical formulas for calcium sulfate, potassium chromate, ammonium sulfite, and magnesium phosphate.

Since calcium ionizes to Ca^{2+}, its charge already balances with sulfate, so the formula is $CaSO_4$.

For K^+ and CrO_4^{2-} to balance, the K^+ needs a subscript of 2, giving the formula K_2CrO_4.

NH_4^+ needs a subscript of 2 to make its charge balance with SO_3^{2-}, giving $(NH_4)_2SO_3$. Note here that placing a subscript on a polyatomic ion requires the ion's formula to be inside parentheses. As a reminder, the subscript indicates a multiple of everything inside the parentheses. Thus, one formula unit of ammonium sulfite contains two ammonium ions and one sulfite ion, for a total of two nitrogen atoms, eight hydrogen atoms, one sulfur atom, and three oxygen atoms.

To balance Mg^{2+} with PO_4^{3-} requires subscripts on both that multiply their charges to the least common multiple (6), resulting in the formula $Mg_3(PO_4)_2$.

5.3.4 Polyatomic Ion Names

Most of the time you will write names or formulas of compounds with polyatomic ions by using the information in Table 5.6 from memory. However, there are quite a few more polyatomic ions than those listed there, and some involve prefixes. Lest you spend the rest of your student days in bewilderment as to the logic behind these names, let's take a moment here to consider the conventions. (Some day you'll thank me.)

First, there are a few polyatomic ions with *–ide* suffixes, such as:

hydroxide, OH^- cyanide, CN^- peroxide, O_2^{2-}

Second, many polyatomic ions contain oxygen and are accordingly called *oxyanions*. In the common oxyanions formed from elements in Groups 14–17, the number of oxygen atoms bonded in the molecule with the other atoms can vary. The convention for these oxyanions is to assign the suffix *–ate* to the most common one, and the suffix *–ite* to the oxyanion with the same charge but with one oxygen atom fewer. Common examples are:

Most Common	sulfate SO_4^{2-}	nitrate NO_3^-
Same Charge, One Less Oxygen	sulfite SO_3^{2-}	nitrite NO_2^-

Third, when a series of oxyanions extends to four members, as with those involving halogens, the prefixes *per–* and *hypo–* are used to form the names of the ions with the most and the least oxygen atoms, respectively. The most common example is the chlorine series of oxyanions:

One More Oxygen than the *-ate* Ion	perchlorate	ClO_4^-
Most Common	chlorate	ClO_3^-
One Less Oxygen	chlorite	ClO_2^-
One Less Oxygen than the *-ite* Ion	hypochlorite	ClO^-

It can definitely be challenging to remember all the different charges and oxygen numbers for the different oxyanions. A section of the periodic table is shown in Figure 5.16 to help you see and remember the pattern. Each oxyanion shown is the most common for the particular element except perchlorate. The ions in Period 2 have three oxygens and those in Period 3 have four. Beginning with the rightmost oxyanion in each period, charges start at –1 and become more negative toward the left. If you remember the oxygen numbers and charges for the oxyanions in this diagram, especially the element in each period that starts the sequence from right to left, you can deduce the charges and oxygens for the others.

The fourth and final rule for naming the polyatomic ions pertains to the addition of one or more hydrogen atoms to the ion. If one hydrogen atom is added, the word *hydrogen* is placed before the name of the ion. If two hydrogen atoms are added, the term *dihydrogen* is placed before the name of the ion. Notice that adding hydrogen atoms reduces the charge on the ion by one unit per hydrogen atom. This indicates that hydrogen atoms join these molecules as ions, that is, as lone protons. Examples:

	14	15	16	17
2	C carbonate, CO_3^{2-}	N nitrate, NO_3^-		
3		P phosphate, PO_4^{3-}	S sulfate, SO_4^{2-}	Cl perchlorate, ClO_4^-

Figure 5.16. Patterns in the oxyanions: oxygens increase going down in the periodic table, charges increase going to the left, and all are the common *-ate* ions except perchlorate.

HCO_3^-	hydrogen carbonate
$H_2PO_4^-$	dihydrogen phosphate

An older convention, still very common, is to use the prefix *bi–* for the single hydrogen oxyanions. Thus, hydrogen carbonate, HCO_3^-, is still commonly called bicarbonate, and hydrogen sulfate, HSO_4^-, is sometimes called bisulfate.

5.3.5 Naming Acids

Our detailed study of acids comes later in the course. But while we are studying names of polyatomic ions, let's also take a moment to discuss naming of acids.

For our present purposes, you may think of acids as compounds that contribute H^+ ions when they dissociate in water (that is, when the acid molecules come apart and form + and – ions in solution). You can recognize acids by the H in the front of the formula, most often followed by a halogen or a polyatomic ion. (Occasionally other anions such as S^{2-} are involved.) Examples are hydrochloric acid, HCl, and nitric acid, HNO_3. Strictly speaking, we say that HCl is hydrogen chloride, named according to the covalent molecule naming rules we address later. But when HCl is dissolved in water—an aqueous solution—we have hydrochloric acid. Here are the conventions for acid naming:

First, acids containing anions with an *–ide* suffix have the *–ide* changed to *–ic* with the prefix *hydro–* added. Examples:

Acid Formula	Anion	Acid Name
HF	fluoride, F^-	hydrofluoric acid
HCl	chloride, Cl^-	hydrochloric acid
HCN	cyanide, CN^-	hydrocyanic acid

Second, acids formed from anions with *–ate* or *–ite* suffixes are named by changing *–ate* to *–ic* and *–ite* to *–ous*. Prefixes on the anions are retained in the acid name. Examples:

Acid Formula	Anion	Acid Name
H_2SO_4	sulfate, SO_4^{2-}	sulfuric acid
HNO_2	nitrite, NO_2^-	nitrous acid
$HClO_4$	perchlorate, ClO_4^-	perchloric acid

An important general term to know is *oxyacid*, which is an acid containing oxygen and at least one other element—usually a nonmetal—bonded to the hydrogen.

▼ Example 5.8

Name these acids: HBr, H_2CO_3, and H_2SO_3.

HBr is formed from the bromide ion, Br^-. Thus, the name is hydrobromic acid.

H_2CO_3 is formed from the carbonate ion, CO_3^{2-}. Thus, the *–ate* is changed to *–ic* and the name is carbonic acid (an oxyacid).

H_2SO_3 is formed from the sulfite ion, SO_3^{2-}. Thus, the *–ite* is changed to *–ous* and the name is sulfurous acid (another oxyacid).

5.3.6 Lewis Structures

In this section, we streamline our Lewis notation a bit and present a formal procedure to enable you to construct diagrams of covalent molecules. We are streamlining the Lewis symbols we have been using in two ways. First, we are replacing shared pairs of electrons involved in bonds, the *bonding pairs*, with a short line segment to indicate the bond. Second, we will now show the other pairs of electrons, the *nonbonding pairs*, only if they are on the central atom in the molecule. This is fine to do because nonbonding pairs only affect the shape of the molecule if they are on the central atom. Otherwise, we know they are there but we can leave them off because they don't affect anything else. The diagrams we produce with this notation are called *Lewis structures*.

Lewis structures were introduced by the great American chemist Gilbert Lewis, frequently referred to as G.N. Lewis. Although he was at the forefront of the theory of chemical bonding in the early 20th century, Lewis never won the Nobel Prize. In a 1916 paper, Lewis first proposed the idea that atoms bonded by sharing electrons in pairs, the phenomenon now known as covalent bonding. He is also the one who in 1926 coined the term *photon* for a quantum of light energy (although this is sometimes erroneously attributed to Albert Einstein).

Table 5.7 shows the Lewis structures for some of the molecules we have discussed so far. In the water molecule the nonbonding pairs are shown on the oxygen atom because it is the central atom in the molecule. The twin pairs of electrons in the double bonds in CO_2 and O_2 have been replaced by double line segments, and the triple bond in N_2 is represented by triple line segments. Each line segment in a Lewis structure represents one pair of electrons.

Our next task is to outline a procedure you can use for constructing Lewis structures on your own for any given simple molecule. Then we will go through several examples showing how to use the procedure. Drawing Lewis structures is one of the most effective ways to learn about how covalent bonding works. To build a Lewis structure, follow the steps described in Table 5.8 on the next page. Read through the steps before reading the examples that follow.

▼ **Example 5.9**

Draw the Lewis structures for methane, CH_4, phosphorus trichloride, PCl_3, sulfur dioxide, SO_2, and boron trifluoride, BF_3.

Beginning with methane, we place the carbon in the center with the hydrogens around it. There are $4 + 4(1) = 8$ valence electrons. These we place around the central atom to give each hydrogen a pair, forming an octet on the carbon in the process. Finally, we replace the electron pairs with line segments. The steps look like this:

Molecule	Lewis Structure
H_2O	H—Ö—H
CO_2	O=C=O
H_2	H—H
N_2	N≡N
O_2	O=O
F_2	F—F
Cl_2	Cl—Cl
Br_2	Br—Br
I_2	I—I

Table 5.7. Lewis structures for some covalent molecules.

```
    H              H            H
              
H  C  H      H : C : H      H—C—H
    H              H            H
```

For PCl_3, the phosphorus is in the center with the chlorines around it. There are $5 + 3(7) = 26$ valence electrons. Place them to make an octet around each chlorine atom. Place the remaining pair on the central phosphorus atom. The nonbonding pair on the phosphorous atom takes its own place around the central atom. Finally, replace the bonding pairs with line segments. Here are the steps:

```
     Cl            :Cl:              Cl
                                      |
Cl   P   Cl    :Cl : P : Cl:     Cl—P—Cl
```

For SO_2, we begin with the sulfur in the center. There are $6 + 2(6) = 18$ valence electrons. Begin by forming octets around the oxygen atoms and placing the remaining pair on the central sulfur atom. The sulfur does not have an octet, so a double bond is required. We remove one pair from one of the oxygen atoms and use it to make a double bond between that oxygen and the central sulfur atom. This gives us an octet on each atom.

```
O   S   O      :Ö : S : Ö:     :Ö::S : Ö:      O=S—O
```

Chapter 5

How to Construct Lewis Structures

1. Write down the chemical symbols for the atoms in the molecule. If there are more than two, then one of them is in the center. If carbon is present, it is the one in the center. If the molecule contains one of one kind of atom and two or more of a different atom, then the single one is in the center. Otherwise, the least electronegative element is in the center. Hydrogen is never in the center. Halogens usually aren't in the center unless there is one of them and more than one of something else.

2. Arrange the atoms around the central atom. If there is more than one carbon, the carbons chain together to make a central spine instead of a single central atom.

3. Add up all the valence electrons. For polyatomic ions, add another electron to account for each negative charge on the molecule. Or, subtract one electron to account for each positive charge on the molecule.

4. Arrange the electrons in pairs to give an octet around each of the atoms bonded to the central atom (except hydrogen, which only gets a pair).

5. Place the remaining electron pairs on the central atom, even if this means the central atom has more than an octet. (There are some exceptions to the octet rule that we address shortly.)

6. If you don't have enough electron pairs to give the central atom an octet, then you need a double bond (or rarely, triple). (The exception is boron, which can bond without making an octet.)

7. If there is a nonbonding pair on the central atom, give it space around the atom as if there were an atom bonded to it.

8. One more tip: almost always, carbon makes four bonds, halogens make one, and hydrogen makes one. Use these guidelines to narrow your options as you try to figure out the bonding arrangement in a molecule.

Table 5.8. Steps for constructing Lewis structures.

Obviously, we could have formed the double bond on the other side, $O-\ddot{S}=O$. The two structures differ only in the way the double bonds are formed, and are called *resonance structures*. We discuss resonance structures in more detail soon. Finally, I drew the structures above as linear molecules—all the atoms are in a straight line. But the nonbonding pair arranges symmetrically around the central atom with the bonding pairs, resulting in the two resonance structures looking like this:

$$\overset{\ddot{S}}{\underset{O}{\diagup}\diagdown}_{O} \qquad \overset{\ddot{S}}{\underset{O}{\diagup}\diagdown}_{O}$$

We discuss the theory of molecular geometry in the next chapter.

Any time you make a molecule with boron, you should be alert to the fact that boron often breaks the octet rule. For BF_3, begin with boron in the center. We have $3 + 3(7) = 24$ valence electrons. Place them around the fluorine atoms to make their octets. This leaves the boron with only six electrons, short of an octet. With any other central atom except boron, we would try forming a double bond to get an octet. But in boron's case, this is not energetically favorable. Now that we know there is no nonbonding pair on the central atom, we need to distribute the three fluorine bonds symmetrically around the central atom. The sequence of steps to work through this structure looks like this:

F B F   :F̈:B:F̈:   F—B—F   F—B(—F)(—F with F on top)

Next we work through some examples of drawing Lewis structures for polyatomic ions. Notice that in each case, the structure is drawn inside square brackets with the charge of the ion shown at the upper right.

▼ Example 5.10

Draw the Lewis structures for carbonate, CO_3^{2-}, chlorite, ClO_2^-, and acetate, CH_3COO^-.

For carbonate, we begin with the carbon atom in the center. We know in advance that carbon needs to make four bonds. There are $4 + 3(6) = 22$ valence electrons. To this, we add two electrons to account for the -2 charge on the molecule, giving a total of 24 electrons. We then place an octet around each oxygen atom. Carbon has now only three bonds, and does not have an octet. So we select one of the oxygen atoms and move one of its nonbonding pairs around to make a double bond with the carbon. With this double bond, carbon now has its four bonds and its octet. Finally, we arrange the three bonded oxygen atoms symmetrically around the central atom. The steps to drawing this structure are as follows:

$$[O\ C\ O\ (O)]^{2-} \quad [\ddot{O}:C:\ddot{O}:(\ddot{O})]^{2-} \quad [\ddot{O}:C::\ddot{O}:(\ddot{O})]^{2-} \quad [O-C(=O)-O]^{2-} \quad [O-C(=O)(\ldots)]^{2-}$$

Once again, the choice of the upper oxygen atom for the double bond was arbitrary. We could have chosen either of these as well:

$$[\text{resonance structure 1}]^{2-} \quad [\text{resonance structure 2}]^{2-}$$

These three structures are identical except for the placement of the double bond, so once again they are resonance structures.

In the chlorite ion, there are $7 + 2(6) = 19$ valence electrons, and we add one more electron to account for the -1 charge on the ion, giving us a total of 20. With the chlorine in the center, we place an octet around each oxygen, and the two remaining pairs are placed around the central atom giving it an octet. In the final drawing, the nonbonding pairs on the central atom are shown because although this molecule looks linear on paper it is not, as we will see later.

$$[O\ Cl\ O]^- \quad [:\ddot{O}:\ddot{Cl}:\ddot{O}:]^- \quad [O-\ddot{Cl}-O]^-$$

The acetate ion is an example of a molecule with a carbon chain in it, although the chain is a short one with only two atoms. Notice that the way the molecular formula is written gives a clue about the structure. The CH_3 indicates that the three hydrogens are all attached to one of the carbon atoms. Knowing this, and that the carbons chain together, and that since the first carbon now has four bonds the two oxygens must attach to the other carbon, we can begin by laying out the general arrangement of atoms. There are 2(4) + 3(1) + 2(6) = 23 valence electrons. We add one more electron to account for the −1 charge on the ion, giving a total of 24. Place pairs at each hydrogen and an octet around each oxygen. The carbon on the left is complete, but the one on the right has only three bonds. So move one pair from one of the oxygens and use it to form a double bond between the carbon atom and that oxygen atom. The steps are shown below.

$$\begin{bmatrix} & H & & O & \\ H & C & C & \\ & H & & O & \end{bmatrix}^- \quad \begin{bmatrix} & H & & \ddot{\ddot{O}} \\ H:C:C & \\ & H & & \ddot{\ddot{O}} \end{bmatrix}^- \quad \begin{bmatrix} & H & & \ddot{O} \\ H:C::C & \\ & H & & \ddot{\ddot{O}} \end{bmatrix}^-$$

The placement of the double bond could have been on either oxygen atom, so these two arrangements are resonance structures.

$$\begin{bmatrix} & H & & O \\ H-C-C & \\ & H & & O \end{bmatrix}^- \quad \begin{bmatrix} & H & & O \\ H-C-C & \\ & H & & O \end{bmatrix}^-$$

5.3.7 Exceptions to the Octet Rule

We have already seen that boron sometimes acts as an exception to the octet rule by bonding without a full octet. Boron is not the only element that does this, but it is the most common example. You know that elements in the same group in the periodic table exhibit similar chemical properties, so it should come as no surprise that aluminum does it too. There are theoretical ways of explaining why a molecule forms without an octet on the central atom rather than forming a double bond somewhere. As usual, minimizing the energy in the molecular structure is the idea.

There are two other types of exceptions. One is the case when there is an odd number of valence electrons, as with nitric oxide, NO. There are 5 + 6 = 11 valence electrons to share between the nitrogen and oxygen atoms, but the odd number is going to leave one of them short one electron. There are two Lewis structures for this molecule, giving us another pair of resonance structures.

$$\ddot{N}=\ddot{O} \qquad \ddot{N}=\ddot{O}$$

The NO molecule is an example of a chemical species known as a *free radical*. A free radical is any atom, molecule, or ion that has a lone, unpaired electron. Because of the unpaired electron, free radicals are highly reactive, and bond with just about anything—including other entities like itself—to obtain the additional electron and complete the octet.

The third exception to the octet rule is atoms that support *more* than an octet of electrons. An interesting example is the compound antimony pentafluoride, SbF_5. At room temperature, this compound is a colorless oily liquid, and it is famous for its property of reacting with nearly everything. Its Lewis structure is simple to work out if you know that the Sb can have five pairs of electrons around it:

The reason for drawing the Lewis structure the way I did instead of arranging the fluorine atoms evenly around the antimony atom is because as shown, the diagram resembles the way the molecule looks in three dimensions. Molecular structure is one of our topics in the next chapter.

The elements that bond this way are *p*-block elements in periods three and higher. The explanation normally given for why the central atom in these compounds accepts more than an octet of electrons is that since the central atom is a *p*-block element, it has a completely unfilled *d* subshell that can take on the extra pair. Antimony, for example, has valence electrons in the 5*s* and 5*p* subshells, but it also has a 5*d* subshell with nothing in it. (It also has a full 4*d* subshell, of course, but recall that for *p*-block elements the electrons in filled, lower *d* subshells do not act as valence electrons.) Some chemists are now of the opinion that there are different factors behind this type of exception to the octet rule, so you see that our efforts at improving our models continue.

5.3.8 Resonance Structures

We have already seen several instances of resonance structures, such as the two for SO_2:

The name for these structures originated because chemists formerly thought that molecules exhibited a sort of resonance where their structures would switch back and forth from one structure to the other. However, experiments have shown that the molecules don't switch structures this way. Instead, the molecular structure is now understood to be a blend of the separate resonance structures. The electrons in such structures are said to be *delocalized*, being shared in orbitals that extend across several atoms. Resonance structure bonds are typically depicted with a dashed line. For SO_2 this depiction is as follows:

Another example is the carbonate ion, CO_3^{2-}. The Lewis structures we saw previously were:

Shown as a single resonance structure this Lewis structure is like this:

Perhaps the most famous of resonance structures is the benzene ring. Benzene, C_6H_6, is a highly flammable molecule found naturally in crude oil. It has a sweet or *aromatic* odor. It is an important industrial compound because it is used as a starting point or *precursor* for the manufacture of many other chemicals. English scientist Michael Faraday (famous for Faraday's Law

Chapter 5

of Magnetic Induction) was the first to isolate and identify this compound in 1825. The Lewis structures for benzene are:

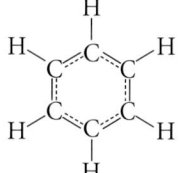

The structure of this molecule was so puzzling that it took another forty years before anyone figured it out. The solution finally came to a German chemist named Friedrich August Kekulé (Figure 5.17). Kekulé had long been working on the nature of carbon bonds and had already figured out the four-bond pattern characteristic of carbon. Kekulé recounted that after working on the bonding structure for carbon for many years, the structure for benzene finally came to him in a daydream or vision in which he saw a snake eating its tail. The discovery of the benzene structure has been of enormous significance for the field of organic chemistry—the study of compounds based on carbon. Because of his ground-breaking work, Kekulé is considered to be the father of chemical structure theory.

Figure 5.17. German chemist August Kekulé (1829–1896).

Shown as a resonance structure, the Lewis structure for benzene is depicted in several different ways, including these:

The benzene ring comes up so often in organic chemistry that for simplicity it is often shown with a circle replacing the dashed lines, as in the second image. And for the utmost in simplicity, the ring is simply shown as a hexagon with a circle inside, the carbon atoms, hydrogen atoms, and resonance structure all being assumed. Figure 5.18 shows computer models of the benzene molecule. The model on the left is called a *ball-and-stick model*. This type of model is good for showing the chemical bonds between specific atoms. The model on the right is called a *space-filling* model. The space-filling model gives a more accurate portrayal of the space the molecule occupies.

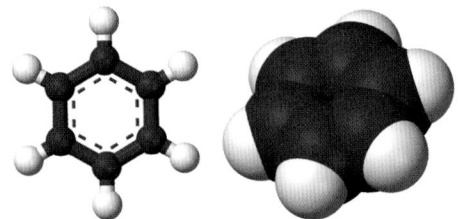

Figure 5.18. Ball-and-stick (left) and space-filling (right) models of the benzene molecule.

5.3.9 Naming Binary Covalent Compounds

When nonmetals form molecules by covalent bonding, the numbers of atoms in the molecules are not predictable the way they are in ionic bonds. Thus, in naming covalent compounds we use prefixes on the atoms involved to denote how many of them are present in the molecule. The Greek prefixes used are listed in Table 5.9.

Many covalent compounds consist of two elements. These are called *binary compounds* and naming them is our topic here.

First, as with ionic compounds, the second element has the suffix *-ide*. Second, both elements in the name have a prefix indicating the number of atoms present in the molecule. However, the *mono-* prefix is typically not used with the first element named. Third, the element farthest to the left in the periodic table is named first unless it is oxygen, which is never written first. If both elements are in the same group in the periodic table, the element with the highest atomic number is written first. Fourth, when a prefix is applied to oxygen, the vowel at the end of the prefix is dropped (except for *di–* and *tri–*).

Examples are as follows:

Formula	Name
SO_2	sulfur dioxide
PCl_3	phosphorus trichloride
N_2O_4	dinitrogen tetroxide

Number of Atoms	Prefix
1	*mono–*
2	*di–*
3	*tri–*
4	*tetra–*
5	*penta–*
6	*hexa–*
7	*hepta–*
8	*octo–*
9	*nona–*
10	*deca–*

Table 5.9. Prefixes used for naming binary covalent compounds.

Note that there are other naming conventions in use in addition to the one described here. The compounds NO and N_2O are commonly called nitric oxide and nitrous oxide, respectively. And some sources use the Stock system, as in the name phosphorus(III) bromide, PBr_3.

5.3.10 Energy in Covalent Bonds

From the perspective of quantum mechanics, the covalent bond and the ionic bond are quite different. The ionic bond is held together by the electrostatic attraction between the ions, whereas the covalent bond is held together because of the energy minimization that occurs within the atomic orbitals when the orbitals share electrons to fill subshells. Nevertheless, we can analyze the *bond energy* in a covalent bond in the same way we did for the lattice energy in ionic bonds because in both cases, the atoms seek the arrangement that minimizes the electrostatic potential energy between them.

Consider two hydrogen nuclei far enough apart from each other so that the potential energy between them is essentially zero. As illustrated on the right side of Figure 5.19, as the nuclei approach each other, the potential energy between them decreases. This is due to the attraction

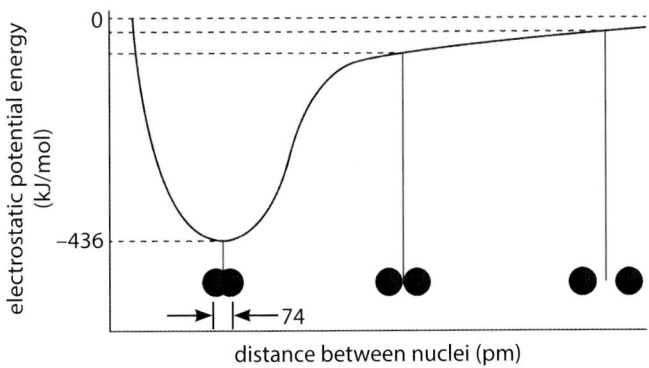

Figure 5.19. Covalent bond energy curve for hydrogen, showing the bond energy and internuclear distance at the stable minimum energy point.

Chapter 5

Figure 5.20. The ball is trapped in an "energy valley" where its potential energy is as low as possible.

between the electron cloud around each atom and the nucleus of the other atom. The minimum energy point is reached when the hydrogen nuclei are 74 pm apart. The molecule is now in an energy valley, quite analogous to the ball in a valley of Figure 5.20. If the nuclei are slightly pushed together or pulled away from each other, they spring back to the minimum-energy distance of 74 pm. Separating the two hydrogen atoms requires an energy input of 436 kJ/mol. The distance between the atomic centers in a molecule is called the *bond length*.

Table 5.10 shows the bond energies and bond lengths for a number of different covalent bonds. There are two key points to notice from these data. The first is that, generally speaking, double bonds are stronger (higher bond energy) than single bonds, and triple bonds are stronger than double bonds. Stronger bonds mean greater amounts of energy are required to break them. This helps explain why nitrogen gas is so unreactive. The triple-bonded N_2 molecule makes up about 79% of our atmosphere. But while double-bonded oxygen molecules react readily (fires, explosions, rust, organic decay, metabolism) the nitrogen molecules don't.

Bond	Bond Energy (kJ/mol)	Bond Length (pm)
H—H	−436	74
H—F	−570	92
H—Cl	−431	127
C—C	−348	154
F—F	−159	141
Cl—Cl	−436	199
C=C	−618	124
S=S	−425	199
O=O	−498	121
N=O	−632	115
C≡C	−839	121
C≡N	−750	117
C≡O	−1076	113
N≡N	−945	110

Table 5.10. Covalent bond energies and lengths.

As you see, HF and S_2 are exceptions to the general trend. The high bond energy of HF is explained by the fact that the fluorine atom is the smallest of the halogens, so the electrons are more tightly bound to the nucleus. The S_2 exception is the opposite: sulfur is below oxygen in the period table, and thus has its valence electrons in a higher shell where they are bound less tightly by the nucleus.

The second point to notice in the table is that as the strength of the bond goes up, the length of the bond shortens. Again, there are exceptions, particularly in molecules incorporating the small hydrogen atom, but in general, tighter bonds pull the atoms closer together.

5.3.11 Physical Properties of Covalently Bonded Substances

The lattice energies in ionic compounds run from several hundred to several thousand kJ/mol. Some of the covalent bond energies shown in Table 5.10 are a bit lower, but in general they are in the same league of several hundred kJ/mol. But there are huge differences between the physical properties of ionic compounds and those of covalent compounds. The basic reason for the differences is that the lattice energy in ionic compounds binds the entire crystal structure together. But the bond energy in covalent compounds only holds together the atoms in the individual molecules. The forces holding one molecule to another, which we study in the next chapter, are much weaker.

One of the results of the weak forces holding molecules together is that many common covalent compounds are gases or liquids at room temperature. Some common covalent compounds

Chemical Bonding

have characteristic colors, such as the bromine and chlorine shown in Figure 5.21, but many other common covalent compounds are colorless. The melting and boiling points (intensive properties, as you recall) for a few common covalent compounds are shown in Table 5.11. Compare these values to the melting points for ionic compounds listed in Table 5.4. The ionic compounds are all several hundred or thousand degrees, whereas all the covalent melting points are below or close to zero.

Compounds that are liquids at room temperature include water, bromine (one of only two elements that are liquids at room temperature), acetic acid, ethanol, benzene, and acetone. The rest of the elements and compounds in the table are gases at room temperature.

Compared to ionic compounds, covalent compounds are generally less soluble in water, less electrically conductive, and less thermally conductive.

5.4 Electronegativity, Polarity, and Bond Character

5.4.1 Polarity and Dipoles

At the end of Chapter 4, we discussed the fact that differences in electronegativity between atoms in a molecule cause the molecule to be polar. Electronegativity is a measure of the relative attraction atoms have for the electrons involved in the bond between them. Electronegativity values are lowest in the lower left of the periodic table (Fr) and highest at the upper right (F), as shown in Figures 4.14 and 4.15.

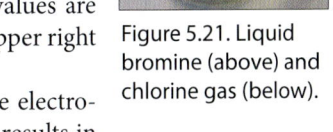

Figure 5.21. Liquid bromine (above) and chlorine gas (below).

In a molecule composed of two identical atoms such as H_2, the electronegativity values of the two atoms are the same (2.20 for H). This results in a nonpolar covalent bond, as illustrated by the image on the left side of Figure 5.22. However, in a molecule such as HCl, the electronegativity values differ: 3.16 for Cl and 2.20 for H. The higher electronegativity of Cl results in the shared electrons being pulled over toward the Cl atom, as illustrated on the right side of the figure. The polar object that results from the electron imbalance is called a *dipole*. (Actually, the term *dipole* applies to any object that has one end more positive than the other, regardless of how that happens.) For polar molecules, we symbolize the positive and negative ends of the dipole with the delta-plus (δ+) and delta-minus (δ−) symbols, respectively, as shown in the HCl molecule in Figure 5.22. In print, the polarity is shown like this: $\overset{+\longrightarrow}{H-Cl}$. In this notation, an arrow is drawn above the element symbols, pointing toward the most electronegative element. A crossbar above the least electronegative element suggests the positive end of the molecule.

The polarity of the water molecule that we discussed previously is so important that we

Compound	Melting Point, °C	Boiling Point, °C
H_2O	0	100
O_2	−219	−183
N_2	−210	−196
Cl_2	−102	−34
Br_2	−7.2	59
CO_2	−79 (sublimes)	
ammonia, NH_3	−78	−33
methane, CH_4	−182	−161
propane, C_3H_8	−188	−42
acetic acid, CH_3COOH	17	118
ethanol, C_2H_5OH	−114	78
benzene, C_6H_6	5.5	80
acetone, $(CH_3)_2CO$	−95	56

Table 5.11. Melting and boiling points for a few covalent compounds.

Chapter 5

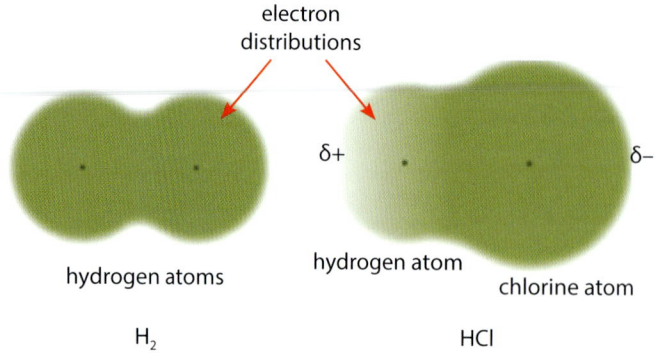

Figure 5.22. Molecules illustrating a nonpolar covalent bond (left) and a polar covalent bond (right).

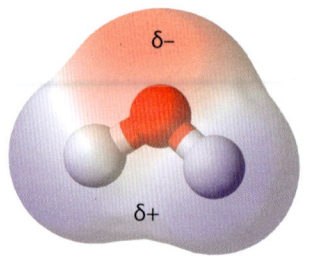

Figure 5.23. The shading shows how the electrical potential around the water molecule changes from positive to negative in the region around the molecule.

consider it again here just for a moment. The shading of the region around the water molecule in Figure 5.23 illustrates how the electrical potential changes from positive on the hydrogen side to negative on the oxygen side of the molecule.

5.4.2 The Nature of the Bond

Let's briefly recall again that chemistry is all about modeling complex entities we cannot see. Our descriptions of the electrostatic forces in ionic bonds and the electron sharing in covalent bonds are simplistic and both fall short of being complete descriptions. In fact, we now understand most bonds to include both ionic and covalent characteristics, depending on the relative electronegativities of the atoms. We can improve our models of atomic bonding by taking this into account.

The scale in Figure 5.24 is used to characterize atomic bonds as nonpolar covalent, polar covalent, or ionic. If the electronegativities are the same or nearly the same, the bond is considered a nonpolar covalent bond. If the difference is significant but is below 1.7, the bond is considered to be polar covalent. When the difference is 1.7, the bond is considered to be 50% ionic, and for differences greater than 1.7, the bond is considered to be essentially ionic in character.

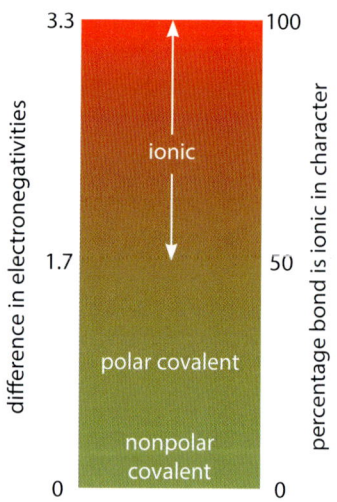

Figure 5.24. Scale for describing the nature of the bond.

The difference between oxygen (3.44) and hydrogen (2.20) is 1.24, indicating a distinctly polar covalent bond in water molecules. The same is the case for the HCl bond pictured above, with a difference of 0.96. By contrast, the electronegativity difference for NaCl is 2.23, indicating a distinctly ionic bond, though not without a little bit of covalent character as well.

There is a close correlation between the percent ionic character of a bond and the color of a compound in aqueous solution. When the percent ionic character is above 20%, compounds in aqueous solution are colorless. As the percent ionic character decreases from 20%

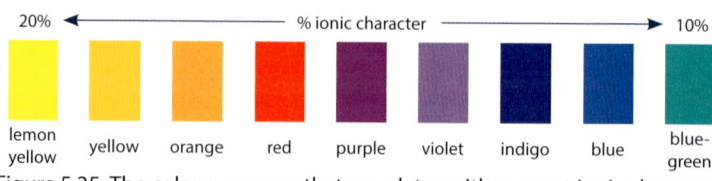

Figure 5.25. The color sequence that correlates with percent ionic character.

Chemical Bonding

down to 10%, solutions take on colors beginning with lemon yellow and going through the sequence of colors shown in Figure 5.25. For example, when bonded to cations of increasing electronegativity, bromine forms solutions that go from yellow to red to brown. Copper forms solutions that are blue and head toward indigo as they become more covalent. Nickel forms compounds that go from yellow to green to blue to violet.

Figure 5.26. Transition metals in aqueous solution. From left to right, the substances in solution are: cobalt(II) nitrate, $Co(NO_3)_2$ (red); potassium dichromate, $K_2Cr_2O_7$ (orange); potassium chromate, K_2CrO_4 (yellow); nickel(II) chloride, $NiCl_2$ (blue-green); copper(II) sulfate, $CuSO_4$ (blue); and potassium permanganate, $KMnO_4$ (purple).

Explaining *why* the percent ionic character correlates to color goes beyond what we can cover here, but it has to do with the way electrons in the orbitals of various compounds absorb light. A more complete explanation will have to wait for future study, but at least now you have some idea of where the beautiful colors in aqueous solutions originate. Figure 5.26 shows six different solutions of transition metals.

Chapter 5 Exercises

SECTION 5.1
1. Explain the octet rule.
2. Explain why hydrogen does not fit in with the octet rule.

SECTION 5.2
3. Explain what lattice energy is and why the lattice energy of ionic compounds is so large.
4. Distinguish between ionic and covalent bonds.
5. Write the formula for the ionic compound formed by each of the following cation-anion pairs:

   a. Sr and Br
   b. Mg and O
   c. Cu(II) and Cl
   d. Li and Se
   e. Ni^{2+} and P
   f. K and N
   g. Be and Cl
   h. Na and I
   i. Cr^{3+} and O
   j. Ag and Se
   k. Mn(IV) and F
   l. Sb^{3+} and Br
   m. Rb and S
   n. Sc and O
   o. Al and F
   p. Mg and F
   q. Zn and Cl
   r. Li and S
   s. Co(III) and O
   t. V^{5+} and I

6. Write the names for each compound in the previous question. For multivalent cations, write the name in both the Latin-based system and the Stock system.
7. Explain what a hydrate is and the conditions under which a hydrate forms.
8. Explain why ionic compounds are brittle and cannot be bent into shapes the way metals can.
9. Explain why ionic compounds generally have high melting points.
10. Since energy is required to separate the ions in a crystal (the lattice energy), why doesn't the temperature of the water drop when salt is dissolved in water?

SECTION 5.3
11. Explain why covalent compounds tend to be gases or liquids at room temperature, while ionic compounds tend to be solids.
12. Describe the three general cases of exceptions to the octet rule (other than hydrogen and helium), and give an example for each. Show the Lewis structures of your examples.

147

Chapter 5

Hmm... Interesting.   The molecular structure of glass and quartz

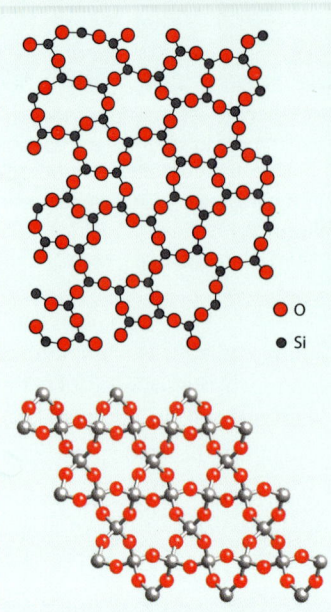

○ O
● Si

In this chapter, we have looked at ionically bonded substances, which have regular crystal lattices. We have also looked at molecular substances, which bond in clusters of atoms. But there is a common substance that bonds in an *irregular* lattice—glass, known in the chemistry world as vitreous silicon dioxide, SiO_2. Because of its completely irregular bonding structure (shown to the left), glass is said to be *amorphous*, a term from Greek that means "no form." Molecular solids, in general, are amorphous, and glass can be thought of as one giant molecule.

There are also forms of SiO_2 with a regular crystal lattice—quartz (shown at the lower left). The varieties of quartz are endless because of different metallic impurities found in the crystal lattice. These impurities cause the quartz crystals to take on different variations in color. The color variations make them desirable as gem stones, especially since the regular crystal lattice means they can be cut into the beautiful symmetrical shapes desired in gems.

13. Draw the Lewis structures for the following compounds and ions. Indicate which of the molecules have resonance structures and show them. (In each of the items shown, when H and O are both present, the H bonds to O.)

   a. C_2HCl
   b. NH_4^+
   c. CCl_4
   d. NO_3^-
   e. PCl_5
   f. NO_2^-
   g. CO
   h. $SiCl_4$
   i. O_3
   j. CrO_4^{2-}
   k. AlF_3
   l. $NOCl$
   m. PO_4^{3-}
   n. H_2S
   o. N_2O_4
   p. C_2H_6
   q. H_3O^+
   r. BCl_3
   s. $HClO_4$
   t. ClO_2^-
   u. H_2CO_3
   v. CCl_2S
   w. C_2H_4ClF
   x. $C_2H_6O_2$

14. State the names for each of the binary compounds listed in the previous exercise. (Several of the compounds are not binary compounds.)

15. Draw the Lewis structures for the following compounds and ions. Identify (i) those that do not obey the octet rule (explain why), (ii) resonance structures (show all), and (iii) free radicals (explain why).

   a. AlH_3
   b. CS_2
   c. $SbCl_5$
   d. N_3
   e. CH_2Cl_2
   f. H_2SO_3
   g. OH
   h. NO_2

16. Explain what resonance structures are.

Chemical Bonding

17. State the names of the following acids.

    a. H_2SO_4
    b. HF
    c. H_2CO_3
    d. $HClO_4$
    e. HNO_2
    f. $HBrO_2$
    g. HBr
    h. H_3PO_4

18. Write the formulas for the ionic compounds formed from each of the ion pairs given below, and state the name of each using the Stock system. (Of course, naming these is a snap.)

    a. magnesium and carbonate
    b. sodium and chromate
    c. calcium and phosphate
    d. ammonium and phosphate
    e. copper(II) and sulfate
    f. sodium and bicarbonate
    g. iron(III) and chlorate
    h. aluminum and hydroxide
    i. chromium(VI) and sulfite
    j. calcium and acetate

SECTION 5.4

19. Explain why water molecules are polar and chlorine molecules are not.

20. For each of the following sets of bonds, compute the difference in electronegativity and arrange the bonds in order of increasing percent ionic character. Identify the most polar and least polar in each set. Identify each bond as nonpolar covalent, polar covalent, or ionic in character.

    a. Be—F, O—F, and C—F
    b. F—F, B—F, and S—O
    c. O—Cl, S—Cl, and C—P

21. When chlorine bonds to the three ions below, it forms compounds that are yellow, orange, or red in aqueous solution. Predict the color each compound displays.

    a. As(III)
    b. Cd(II)
    c. Sb(III)

GENERAL REVIEW EXERCISES

22. Determine the number of oxygen atoms present in 100.00 g limestone, which is composed of calcium carbonate.

23. Determine the percent composition for any sample of sulfur hexafluoride.

24. Write the electron configurations for Cl, Sc, Cd, and Gd.

25. Determine the number of chlorine atoms present in 1.000 cm^3 magnesium chloride. The density of this compound is listed in Table 5.4.

26. Why are the noble gases unreactive?

27. Without referring to any actual radius data, determine the correct order by size, from smallest to largest, for the following ions and atoms: Mg^{2+}, Rb^+, Ca^{2+}, K^+, Rb.

28. Determine the empirical formula for a sample with a percent composition of 12.84% S, 25.45% Cu, 25.63% O, and 36.07% H_2O by mass. Knowing that the empirical formula is the same as the molecular formula, name this compound.

29. In a certain laser, the electron transition generating the laser beam has an energy of 3.73×10^{-19} J. Determine the wavelength of the beam generated by this laser.

Chapter 6
Molecular Theory and Metallic Bonding

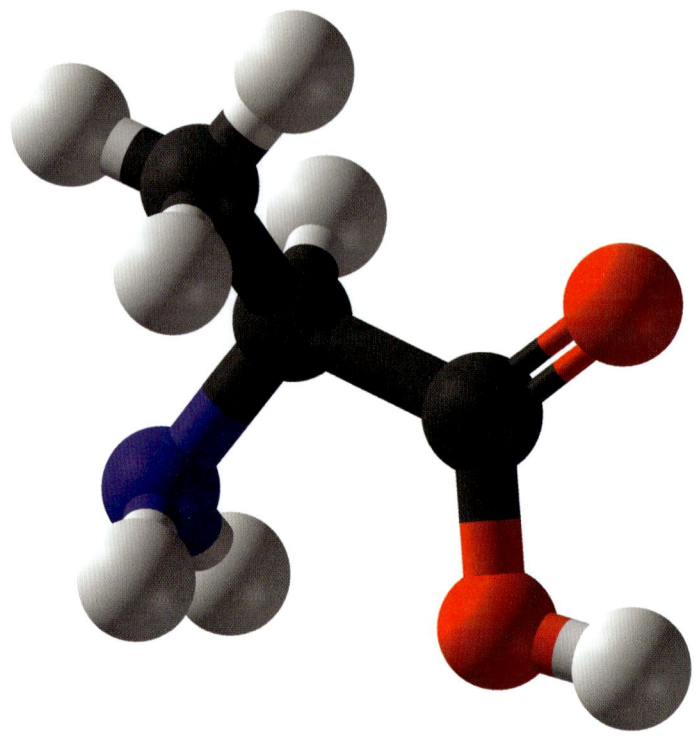

There are 22 "standard" amino acids, of which 21 are found in humans. These molecules are joined together in chains to produce the all-important proteins that are at the center of nearly every cellular process.

The model above is a representation of an amino acid called *L-alanine*. Carbon, hydrogen, oxygen, and nitrogen atoms are shown in black, white, red, and blue. In this chapter, we discuss theoretical models that account for the orientations of the bonds that join atoms together. The examples we consider in the chapter entail simpler molecules than this one, but the same ideas apply to L-alanine, the other amino acids, and much larger molecules as well.

Objectives for Chapter 6

After studying this chapter and completing the exercises, you should be able to do each of the following tasks, using supporting terms and principles as necessary.

SECTION 6.1

1. Describe the central theoretical problem addressed by VSEPR theory.
2. Distinguish between electron domain geometry and molecular geometry.
3. Use VSEPR theory to predict the electron domain geometries and molecular geometries of given molecules.
4. State the names of the five basic electron domain geometries and the seven basic molecular geometries and draw diagrams illustrating each.
5. Explain the conditions under which the bent, trigonal bipyramidal, and octahedral geometries form.
6. Explain how nonbonding electron pairs affect the bond angles in molecules and use this idea to predict when bond angles will depart from the standard angles of pure geometric figures.
7. Apply the VSEPR theory and the principles of electron domain geometry to describe the geometries and estimate bond angles of larger molecules.

SECTION 6.2

8. Give a detailed description of metallic bonding.
9. Compare and contrast covalent bonds and metallic bonding.
10. Explain why the lattice structures of atoms in metals are referred to as "close packing."
11. Explain why metals are good conductors of electricity and heat, exhibit malleability and ductility, have a high luster, and are typically silver in color.

SECTION 6.3

12. Define hydrogen bonding and give several examples of phenomena caused by it.
13. Describe the three different types of Van der Waals forces and give specific examples of each.

Chapter 6

6.1 Molecular Structure

6.1.1 Covalent Bond Theory

In the previous chapter, we used the shared-pair bonding theory developed by G.N. Lewis as a first step in understanding molecular bonding. In the present chapter, we add another layer to our theoretical understanding about molecules and covalent bonds. I like to think of these theoretical components as addressing specific questions. So we begin by articulating the question addressed in what we have seen so far, and then we move on to the second question addressed in this new chapter.

Here is the question addressed by the 1916 covalent bonding theory of Gilbert Lewis:

1. What drives atoms to cluster together in particular ways to form molecules?

Lewis' covalent bonding theory addresses this question. As we saw, atoms seek to fill up their octets and form different clusters depending on how many valence electrons each atom brings to the party. The bonds occur by atoms sharing electrons in pairs, and the atomic clusters formed when the atoms do this are called molecules.

6.1.2 Valence Shell Electron Pair Repulsion (VSEPR) Theory

2. What factors govern the shapes of the molecules that form when atoms are covalently bonded together?

The Lewis structures you learned how to draw in the previous chapter show how many and what kind of atoms bond together in molecules, and whether the bonds involve one, two, or three electron pairs. But Lewis' covalent bond theory doesn't explain the shapes of the resulting molecules. The theory that addresses this question is called *Valence Shell Electron Pair Repulsion (VSEPR) theory*. VSEPR theory is quite successful in explaining molecular geometry.

In VSEPR theory, we consider the regions around atoms in molecules where the electrons are found. These regions are called *electron domains*, and there are two basic types. The first is the *bonding domains* between atoms in the molecule, where shared pairs of electrons reside. These electron domains could be due to single bonds, double bonds, or triple bonds; VSEPR theory treats them equally. The second type of electron domain is the *nonbonding domain*, the region where nonbonding pairs (often called *lone pairs*) on a central atom reside. The bonding domains are located in a region between atoms; the nonbonding domains are located on the side of a central atom where no other atom is attached.

For example, recall the Lewis structure for the water molecule:

H:Ö:H

There are four electron domains around the oxygen atom in this molecule. Two of them are bonding domains—the two regions where the hydrogen atoms are attached by sharing an electron pair with the central oxygen atom. The other two are the nonbonding domains associated with the nonbonding pairs shown in the Lewis structure.

The basic idea behind VSEPR theory is simple. All the electron domains around an atom repel each other, and thus try to push as far away from each other as possible. The result of this is very predictable molecular shapes as the electron domains spread themselves out in two or three dimensions around the central atom in the molecule.

As we look at these geometries, it is important to keep in mind that nonbonding domains on a central atom repel other electron domains just as bonding domains do. In fact, they push a bit harder than the bonding domains do, a detail we will come back to later. We also need

Molecular Theory and Metallic Bonding

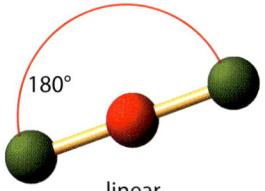
linear

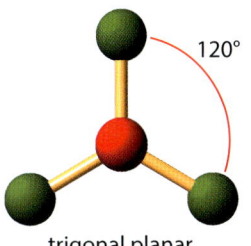
trigonal planar

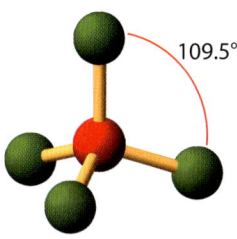
tetrahedral

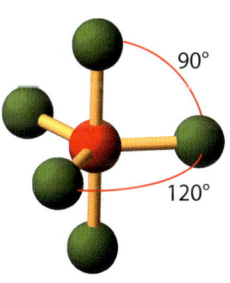
trigonal bipyramidal

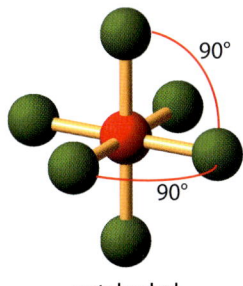
octahedral

Figure 6.1. The five basic electron domain structures.

to distinguish between the electron domain geometry and the molecular geometry. Sometimes these are the same. However, since nonbonding domains on a central atom influence molecular shape just as bonding domains do, the geometry of the molecule itself can be different from the geometry of the electron domains.

There are five basic geometries that can form as the electron domains around a central atom push away from each other, shown in Figure 6.1. Outlines of the three-dimensional geometries are related to standard polyhedra in Figure 6.2. More details concerning the five geometries, as well as example molecules, are shown in Table 6.1. Refer to all these figures and tables as we go through the details.

The first geometry is the *linear* arrangement. (The linear arrangement also occurs, of course, with just two atoms, but that arrangement is trivial and not shown.) A linear arrangement occurs when two atoms are bonded to a central atom and there are no nonbonding domains. Generically, we can designate molecules like this as of the type AB_2 (two B atoms attached to one A atom). A classic example is the CO_2 molecule. As the Lewis structure in Table 6.1 shows, there are two carbon atoms bonded to the oxygen and no nonbonding domains.

In the *trigonal planar* geometry, three electron domains spread around the circumference of the central atom, resulting in a *bond angle*—the angle between bonds—of 120°. There are two ways in which electron domains can take on this geometry. The first is with a molecule of the generic type AB_3 such as boron trifluoride, BF_3. The second way is with a molecule that has two bonding domains and one nonbonding domain on the central atom, such as SO_2. We can represent this structure generically as type AB_2E, where the E stands for the nonbonding domain.

Figure 6.3 illustrates the molecular geometry resulting from molecules of the type AB_2E. In the figure, the black line is in the same position as the third bond in a molecule of the type AB_3. But in this case, there is no atom attached there; instead, there is a nonbonding domain there. This molecular geometry is referred to as *bent*.

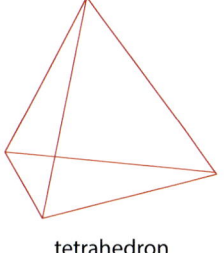
tetrahedron

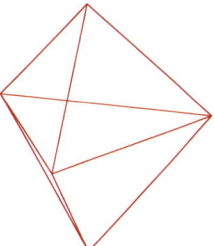
trigonal bipyramid

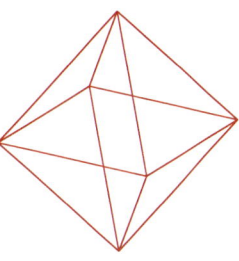
octahedron

Figure 6.2. Outlines of the three 3-D molecular shapes.

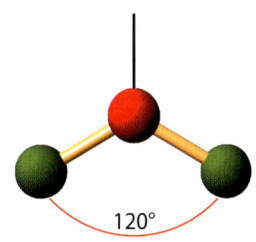

Figure 6.3. The bent geometry resulting (ideally) from molecules of the type AB_2E. (The actual angle is 119°, as discussed in the next section.)

153

Bond Type	Bonding Pairs	Nonbonding Pairs	Electron Domain Geometry	Molecular Geometry	Example Molecule	Example Lewis Structure
AB_2	2	0	linear	linear	CO_2	O=C=O
AB_3	3	0	trigonal planar	trigonal planar	BF_3	
AB_2E	2	1	trigonal planar	bent	SO_2	
AB_4	4	0	tetrahedral	tetrahedral	CH_4	
AB_3E	3	1	tetrahedral	trigonal pyramidal	NH_3	
AB_2E_2	2	2	tetrahedral	bent	H_2O	
AB_5	5	0	trigonal bipyramidal	trigonal bipyramidal	SbF_5	
AB_6	6	0	octahedral	octahedral	SF_6	

Table 6.1. Bond types and their geometries.

So for an AB$_2$E molecule, the electron domain geometry is trigonal planar, but the molecular geometry is bent.

The third basic geometry is the *tetrahedral* geometry, so called because lines drawn connecting the four outer atoms form the outline of a tetrahedron. When there are more than three atoms bonding to the central atom, the electron domains minimize their distance from each other by spreading out in three dimensions.

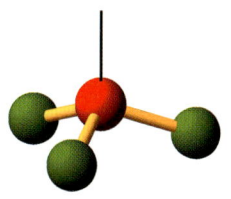

Figure 6.4. The pyramidal molecular geometry resulting from molecules of the type AB$_3$E.

As the table shows, there are three ways this electron domain geometry can form, represented generically as AB$_4$, AB$_3$E, and AB$_2$E$_2$. Methane, CH$_4$, is a perfect example of the AB$_4$ type. This molecule is shaped just like the tetrahedral shape shown in Figure 6.1. If the molecule has a nonbonding domain in place of one of the bonding domains, as with ammonia, NH$_3$, the molecular geometry is as shown in Figure 6.4. This is called a *pyramidal* geometry. As before, the black line in this figure represents a nonbonding domain. There are only three atoms attached to the central atom, but since the nonbonding domain pushes just like the three bonding domains do, we have a tetrahedral electron domain geometry and a pyramidal molecular geometry.

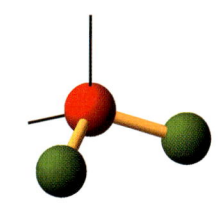

Figure 6.5. The bent geometry resulting from molecules of the type AB$_2$E$_2$, of which of H$_2$O is the most important example.

The water molecule, H$_2$O, is the perfect example of the AB$_2$E$_2$ type. Figure 6.5 shows how a tetrahedral electron domain geometry with only two bonding domains results in the molecular geometry being bent. Note from Figure 6.1 that the bond angle in the tetrahedral geometry is smaller than the bond angle in trigonal planar geometry. As a result, the H$_2$O and SO$_2$ molecules are both bent, but the angle in the H$_2$O molecule is smaller.

As should be clear from our study of covalent bonding in the previous chapter, the last two geometries pertain to molecules that break the octet rule. Antimony pentafluoride, SbF$_5$, has five bonding domains around the central atom. This results in a *trigonal bipyramidal* geometry. Three of the Sb—F bonds lie in the same plane, 120° apart. The other two are on a line perpendicular to this plane, resulting in 90° bond angles between them and the other three.

Sulfur hexafluoride has six bonding domains surrounding the central atom, resulting in the *octahedral* geometry where every bond angle is 90°.

As with the AB$_3$ and AB$_4$ types, there are cases of molecules with AB$_5$ or AB$_6$ electron domain geometries, but with fewer than five or six atoms attached to the central atom. This happens when the total number of bonding domains and nonbonding domains on the central atom is greater than four. We do not explore the details of these geometries further here.

▼ Example 6.1

Use VSEPR theory to predict the molecular shapes of ozone, O$_3$, and the sulfite ion, SO$_3^{2-}$.

The Lewis structures for O$_3$ are

O=Ö—O   and   O—Ö=O

These two resonance structures may be represented as O⋯Ö⋯O . This molecule is type AB$_2$E. Thus, the electron domain geometry is trigonal planar and the molecular geometry is bent.

Chapter 6

As you can verify, the Lewis structure for the sulfite ion is

$$\left[\begin{array}{c} O \\ | \\ O-\underset{..}{S}-O \end{array} \right]^{2-}$$

This molecule is of the form AB₃E. With four electron domains, the electron domain geometry is tetrahedral. But since there are only three bonded atoms, the molecular geometry is pyramidal.

6.1.3 The Effect of Nonbonding Domains on Bond Angle

Figure 6.6 illustrates the relative shapes of the two types of electron domains. Nonbonding domains, like the two lone pairs on the oxygen atom in a water molecule, extend out in a wider region from the atom than the bonding domains do, and this has a slight but definite effect on molecular geometry. You can think of it this way: when the electron domains spread themselves out in 3-D space around an atom, repelling the other domains because of the negative charges inside them, the nonbonding domains push a bit harder than the bonding domains do because they are wider. The result is that the bond angles between atomic bonds in the molecule are less than they would be if all were equal. This effect is seen consistently in covalent molecules. Figure 6.7 illustrates how the effect works in the cases of SO₂, an AB₂E type molecule, and H₂O, an AB₂E₂ type. The electron domain geometry for SO₂ is trigonal planar, with angles of 120°. But with a nonbonding domain in there, the angle between the other two domains is squeezed down to 119°. The same thing happens with H₂O. In this case, the 109.5° bond angle of the ideal tetrahedron is reduced to 104.5°.

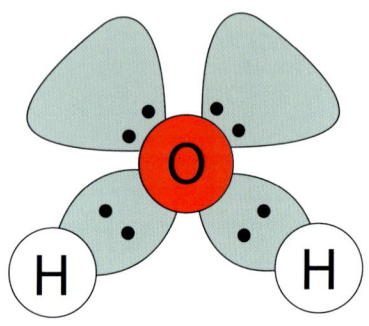

Figure 6.6. The influence of bonding domains does not extend as far to the side as it does for nonbonding domains.

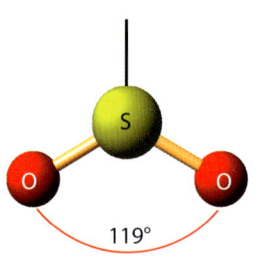

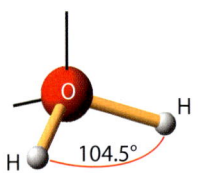

Notice in Figure 6.7 that both the molecular geometries are bent, but the bond angles are different. In the case of SO₂, the ideal angle of the trigonal planar geometry is 120°, and the single nonbonding domain reduces that by 1°. In the case of water, the ideal bond angle of the tetrahedron is 109.5°, but in this case, there are two nonbonding domains involved and together they squeeze down the bond angle by 5°.

Figure 6.7. In a type AB₂E molecule such as SO₂, the nonbonding domain pushes the bonding domains away slightly, reducing the bond angle from the ideal 120° to 119°. In a type AB₂E₂ molecule like H₂O, the bond angle is reduced from the ideal 109.5° to 104.5°.

▼ Example 6.2

Use VSEPR theory to predict the bond angle in the ammonia molecule, NH₃.

The Lewis structure for NH₃ is

Molecular Theory and Metallic Bonding

```
    H
    |
H—N—H
    ..
```

This is a type AB_3E molecule. The electron domain geometry is tetrahedral and there is one nonbonding domain. Since the two nonbonding domains in H_2O (which also has tetrahedral electron domain geometry) reduce the bond angle by 5°, we should expect the bond angle in NH_3 to be reduced by about half as much, putting it at around 107°.

(In fact, the bond angle for NH_3 is 107°.)

6.2   Metallic Bonding

6.2.1 Metallic Lattices

We come at last to metallic bonding. The atoms in a lump of metal arrange themselves in a crystal lattice that minimizes the energy between them, just as the atoms in ionic solids do. However, since all the atoms in an elemental metal are the same (except for isotopic variations, which do not affect metallic bonding), there are just two main lattice structures that occur. These structures are described as "close packing" because the atoms are more densely packed than they would be if the atoms were arranged in a rectangular grid or if the layers were directly above each other.

The first close-packing arrangement is called a *body-centered cubic* (bcc) arrangement, shown down the left side of Figure 6.8. In this arrangement, the third layer is directly above the first layer. The second type of lattice is called a *face-centered cubic* (fcc) arrangement, shown down the right side of Figure 6.8. In the fcc lattice, the positions of the atomic layers repeat in sets of three. To make the geometry of these layers more clear, Figure 6.9 shows outlines of one cell of each.

Some metals take on both these structures depending on the temperature. For example, iron, the primary metal in steel, transitions from the bcc structure, called *ferrite*, to the fcc structure, called *austenite*, above 906°C.

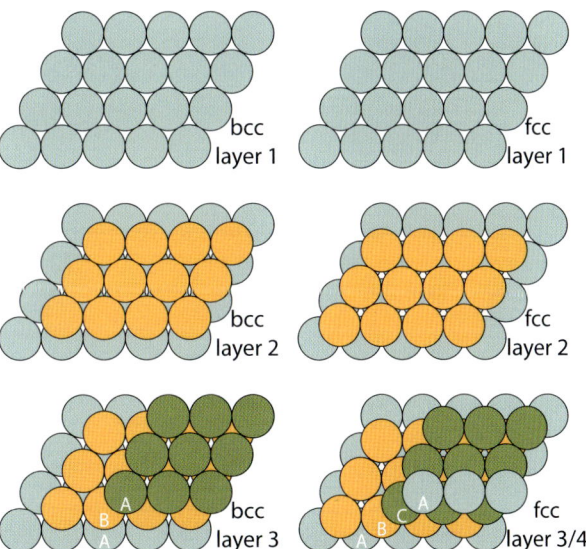

Figure 6.8. Metal atoms can stack in alternating layers (left) of the bcc lattice, or in layers that repeat in sets of three (right), in the fcc lattice.

Stainless steel alloys are made of austenitic iron with 18–20% chromium and 8–10% nickel in the alloy. Once blended in the alloy, the iron preserves its austenitic character when cooled and the result is a family of alloys that possess many superior mechanical qualities. These stainless steel alloys are designated by numbers such as 304 and 316, and are widely used in industry for high-performance machine parts, surgical instruments, and other applications.

The atoms in a metal are held together by the same mechanism that holds together the atoms in a molecule—electrons occupying the orbitals of more than one atom at a time. When an

157

Chapter 6

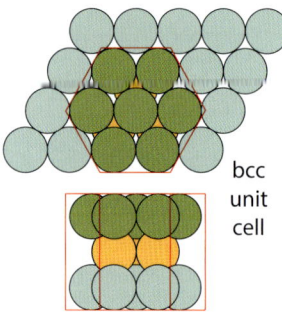

Figure 6.9. One cell of the bcc lattice (top and center), and one cell of the fcc lattice (bottom).

electron simultaneously occupies the orbitals of two or more atoms, each atom has a grip on the electron and the atoms are thus bound to each other.

However, there are major differences between covalent bonds and metallic bonding. This should be obvious, since solid metals possess vastly different properties than the gases and liquids of covalently bonded substances. Metal atoms have a large number of unfilled orbitals and these orbitals overlap one another all over the place in the regions between the metallic nuclei. Consequently, all the valence electrons of the atoms freely move about through the overlapping orbitals in the lattice, and are delocalized—just like the electrons in resonance structures. It is the very nature of metallic bonding that it cannot be thought of as taking place between just two or three atoms at a time. It is the result of the large numbers of orbitals all overlapping one another in a continuous array of atoms.

The diameter of an atomic nucleus is on the order of 100,000 times smaller than the diameter of an atom. This means that the atomic nuclei occupy very little space in the lattice compared to the space occupied by the overlapping orbitals. So the trillions of electrons from the valence shells of all the atoms have plenty of room to move around and almost nothing to hold them in one place or prevent them from moving in the presence of electrical forces.

The free electrons in the metallic lattice are called *conduction electrons*, and there are so many of them that they are often described as an ocean or sea of electrons, or even as a gas of electrons flowing through the lattice. Commonly, the conduction electrons are called the *electron sea*.

6.2.2 Physical Properties of Metals

Let's now review some of the intensive properties we associate with metals, and consider how metallic bonding provides us with explanations for them. The electron sea obviously explains why metals have such *high electrical conductivity*. Electric current is the flow of electric charge, and since the huge numbers of conduction electrons in metals can freely slide from orbital to orbital in the metallic lattice, all one has to do is establish a difference in voltage (also called *electrical potential*) in one part of the metal relative to another part and the electrons, being negatively charged, all go screaming toward the most positive end of the metal.

The conduction electrons are also one of the reasons metals have *high thermal conductivity*. Heating a metal adds kinetic energy to the electrons, causing them to collide with each other, spreading the thermal energy throughout the lattice. The other reason metals conduct heat so well is the ease with which vibrations are passed from atom to atom

Metal	Thermal Conductivity (W/m·K)	Ionic Compound	Thermal Conductivity (W/m·K)
Na	141	NaCl	≈6
Cu	401	Cu_2O	≈5
Mn	7.8	MnO	≈4
Al	237	Al_2O_3	≈30
K	102	KCl	≈7
Ca	200	$CaCO_3$	≈4

Table 6.2. Thermal conductivities of some metals and their ionic compounds. The unusual units W/m·K work like this: To determine the rate of heat flow through a sample, you multiply the thermal conductivity value by the cross-sectional area of the material, and divide by the length. This gives the rate of heat flow in watts per degree of temperature difference at the two ends.

Molecular Theory and Metallic Bonding

through the crystal lattice. But in some metals, the conduction electrons can make a big difference. As examples, Table 6.2 lists the thermal conductivity values for several metals and compares them to the thermal conductivities of ionic compounds made with the same metals. The thermal conductivity of the pure metal is anywhere from two to 80 times as high as that of the ionic compound.

Recall our discussion of malleability and ductility from Section 2.2.3. Malleability indicates the ability of a material to deform rather than shatter when compressed, such as when a blacksmith shapes a metal by pounding it. Ductility indicates the ability of a material to deform under tension, such as when copper is drawn into a wire. The explanation of why metals exhibit these properties lies in the fact that the atoms are held together in the lattice by the conduction electrons sharing orbitals. Planes of atoms can slide past each other without fracturing the lattice, as illustrated in Figure 6.10. This is because movement of one section of atoms still allows the orbitals

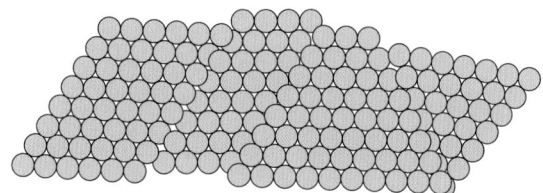

Figure 6.10. Planes of metal atoms can slide past each other without fracturing the metallic lattice.

of the atoms in that section to overlap with the orbitals of an adjacent section of the metal. By contrast, ionic crystals are held together by balanced forces between ions. We saw in the previous chapter that when an ionic material is stressed enough to change the alignment between positive and negative ions, the layers of atoms push apart instead of pulling together and the material shatters.

Metals are also known for their silvery *luster*, typified by the sample of chromium shown in Figure 6.11. The shininess of metals is due to the large numbers of unfilled orbitals in the atoms' crystal lattice. There are so many orbitals available for electrons to occupy when they absorb incoming energy that metals absorb and re-emit wavelengths across the visible spectrum. Since the combination of all colors produces white light, this is what makes them shiny and neutral in color (silver/gray).

There are a few metals that are not silver or gray, and that have natural colors instead. These include copper (red-orange), gold (yellow), and osmium (bluish tint). We can understand this phenomenon in copper's case by looking at copper's electron configuration. Copper's electron configuration does not follow the regular order specified by the Madelung rule. The configuration is

Cu: $[Ar]4s^1 3d^{10}$

Figure 6.11. The silvery-gray color of chromium is typical of most pure metals.

As you see, one of the electrons from the $4s$ subshell is in the $3d$ subshell. The energy difference between the filled $3d$ subshell and the half-filled $4s$ subshell corresponds to orange light. Electrons hopping back and forth between the orbitals in the $4s$ and $3d$ subshells are responsible for copper's unique color. The explanation for gold's color is essentially the same, gold being in the same group as copper in the periodic table.

Hmm... Interesting.   Tin pest

At cold temperatures, tin transitions from a type of bcc structure called white tin or beta-tin to a semiconductor state called gray tin or alpha-tin. In the alpha state, tin completely loses its metallic properties. This transition is called "tin pest" because once it starts it affects the entire metal. If cold enough for long enough, mechanical objects made of tin simply turn to powder. The lore surrounding tin pest goes way back, and some have blamed Napoleon's defeat by the Russians in 1812 on the deterioration of the soldiers' tin buttons in the Russian winter. Tin pest has also been blamed for the deterioration of food and fuel canisters on Robert Scott's 1910 expedition to the South Pole. Roald Amundsen beat him there, but on Scott's return trip in 1912, the canisters failed, causing the loss of food and fuel for Scott and his team. Scott and his two remaining men died 11 miles from a supply depot. The role of tin in both these incidents has been seriously questioned, but without final verdict. Thus, the legends continue.

6.3 Intermolecular Forces

To conclude our study of chemical bonding, we review the forces we have encountered so far that operate at the atomic level. We also add hydrogen bonding to the list, as well as several forces known collectively as *Van der Waals forces*.

6.3.1 Bonding Forces

We have seen two basic types of forces that hold atoms together in chemical bonds. The first is the electrostatic force between ions, between polyatomic ions and ions, and between polyatomic ions and other polyatomic ions. These electrostatic forces hold together the ions in the crystal lattice of an ionic compound.

The second bonding force occurs in covalent bonds and metallic bonding. The force is caused by electrons residing in the orbitals of two or more atoms simultaneously, and since each positive nucleus has the electron trapped in the "energy well" of one of its orbitals, the involved atoms become indirectly attached to one another. The force of this kind of bond can occur on a small scale, as when two atoms bond in a molecule, or on a massive scale, as in the vast array of atoms in a metallic lattice.

6.3.2 Intermolecular Forces

The forces *between* molecules—that is, *intermolecular forces*—include the force of hydrogen bonding and *Van der Waals forces*. Intermolecular forces are far weaker than the bonding forces that hold molecules and crystals together, which means that much less energy is required to break the bonds. The lattice energy in a typical ionic crystal might be 2200 kJ/mol (23 eV/particle), and bond energies in common covalent molecules are in the range of 600 kJ/mol (6 eV/particle). By contrast, the total energy involved in hydrogen bonding in a substance might be 60 kJ/mol (0.6 eV/particle). Energies associated with Van der Waals forces are down in the range of 5 kJ/mol (0.05 eV/particle), or even less.

6.3.3 Hydrogen Bonding

We encountered hydrogen bonding back in the Introduction. Now it is time for a more detailed treatment. The term *hydrogen bonding* refers to the attraction between nonbonding electron pairs on the central atom of a molecule and hydrogen atoms that are attached to oxygen, nitrogen, or fluorine atoms. The high electronegativities of O, N, and F ensure that when bonded to hydrogen, the hydrogen atom has a substantially positive character, symbolized by δ^+. These δ^+ hydrogen atoms are attracted to the nonbonding pairs of electrons on surrounding molecules, regardless of whether those molecules are of the same compound.

The abundance of water on this planet and in our bodies means that hydrogen bonding involving water molecules is going on everywhere around us and inside us. Figure 6.12 illustrates hydrogen bonding at work between a pair of water molecules. The oxygen atom in water has two nonbonding pairs of electrons. Additionally, the greater electronegativity of oxygen causes the shared electrons in the water molecule to crowd over toward the oxygen atom. The result, as we have seen, is a dipole molecule that is attracted to other similar molecules.

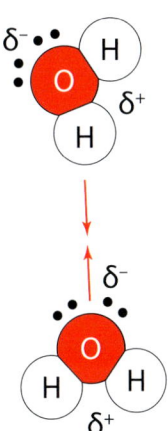

Figure 6.12. The attraction present in hydrogen bonding.

The attraction between water molecules is what causes water molecules to spread out and orient themselves in hexagonal patterns as they cool, causing the density of water to decrease right before it freezes. Thus, ice floats and ponds freeze from the top down, allowing aquatic wildlife to flourish. Now, one can say that this is all perfectly explained by hydrogen bonding, and it is. But this explanation *doesn't account for why the laws of nature are the way they are* so as to create this extremely helpful and very unusual behavior of water. Almost nothing else expands as it freezes, but water does and it just so happens that the planet is covered with it. Life isn't just marginally present on earth; it's *flourishing* on earth. And life flourishes on earth because creation was *designed* for it to flourish. Thanks be to God.

Another excellent example of hydrogen bonding at work is in the absorbance of cotton towels. Cotton is composed of *polymers*, long chain molecules. The polymers in cotton are called *cellulose*, and the smaller

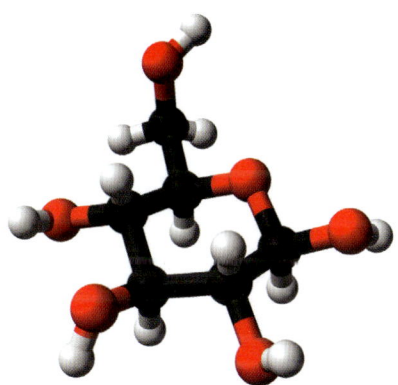

Figure 6.13. The β-glucose molecule.

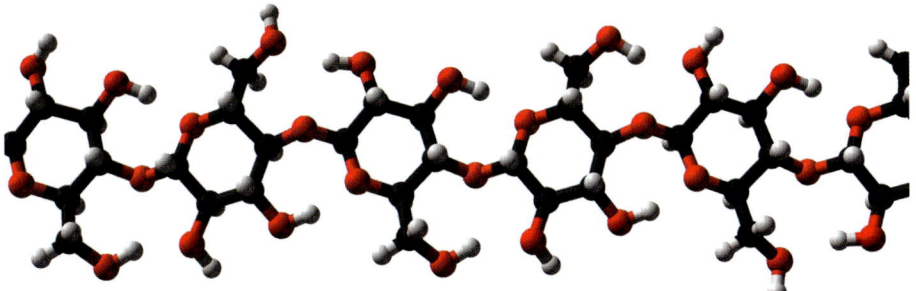

Figure 6.14. A short section of cellulose.

Chapter 6

molecular units—or *monomers*—the polymers are made of are called *β-glucose* (beta-glucose), a carbon ring molecule depicted in Figure 6.13. In this figure, carbon atoms are black, oxygens are red, and hydrogens are white. In cellulose, hundreds or thousands of β-glucose monomers are attached together in long chains. A short section of such a chain is shown in Figure 6.14. As you can see, the outside of this molecule is covered in —OH pairs or *hydroxyl groups*—hydrogen atoms bonded to oxygen atoms that are in turn bonded into the molecule. The hydrogen in every one of

Figure 6.15. High-performance athletic socks use hydrogen bonding to keep feet dry.

these hydroxyl groups is $δ^+$ in character because it is bonded to an oxygen atom. As a result, the nonbonding pairs of any water molecules that may be around are attracted to the cellulose chain at one of the hydrogen sites.

Cotton absorbs moisture in every direction and holds it there, and this explains why it takes your blue jeans a long time to dry if they get wet. But the manufacturers of high-performance athletic wear have figured out how to engineer the hydrogen bonding process so that moisture is pulled away from the skin and sent to the outside of the garment where it evaporates. This action is called "wicking," and those of us who have discovered the power of wicking to keep our feet absolutely dry, like the marathon runners in Figure 6.15, will never run in regular cotton socks again. (Please do not reply that you run without socks. That's just nasty.)

6.3.4 Van der Waals Forces

In addition to hydrogen bonding, there are three other force mechanisms between molecules. These all have to do with attractions between dipoles, and collectively they are referred to as *Van der Waals forces*. Since hydrogen bonding also has to do with forces between dipoles, you should take note that Van der Waals forces are all the forces involving dipoles *other* than hydrogen bonding.

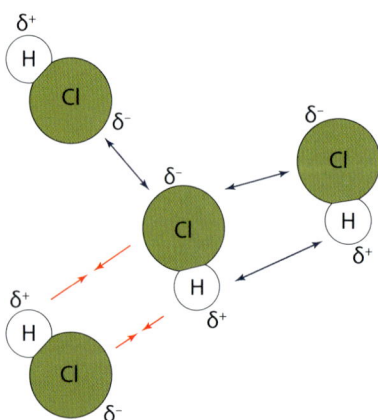

Figure 6.16. Keesom forces are forces between permanent dipoles.

The first is the *Keesom force*, a force between two *permanent dipoles*. The term *permanent dipole* is to be understood as distinguishing between molecules that are dipoles inherently, because of the difference in electronegativities of the atoms in the molecule, and those that possess only momentary and fleeting dipole character, which we get to in a moment. An example of a permanent dipole we have met before is the HCl molecule, illustrated in Figure 6.16. Chlorine is substantially more electronegative than hydrogen, so the molecule is a permanent dipole. The $δ^+$ end of the HCl dipole attracts the $δ^-$ end and repels the $δ^+$ end of other dipoles. As you see in Figure 6.16, Keesom forces can be attractive or repulsive, depending on the orientation of the two dipoles involved.

The second Van der Waals force is the *Debye force*. At this point you need to try to imagine the unimaginable—the micro world of molecules. Molecules are always present in unbelievable numbers and moving at unbelievable speeds. So imagine that a permanent dipole like an HCl molecule has a close encounter with a symmetrical, nonpolar molecule, such as CO_2. As illustrated in Figure 6.17, the charged character of the ends of the HCl dipole can temporarily induce dipole character on another molecule. This happens, for example, because the δ+ hydrogen atom attracts the electrons in the nonpolar molecule. Since the charge distribution in the nonpolar molecule is symmetrical to begin with, when the electrons are attracted toward one end of the molecule, that end becomes negative relative to the other end, and the molecule momentarily takes on a temporary dipole character. Debye forces can be induced on any neutral atom or nonpolar, symmetrical molecule.

Note that since Debye forces are induced, the induced charge is always the opposite of the charge inducing it, leading to attraction between the permanent dipole and the induced dipole. Thus, Debye forces are attractive. Note also that since molecules are constantly in motion, the Debye force is fleeting and temporary, and weaker than the Keesom force.

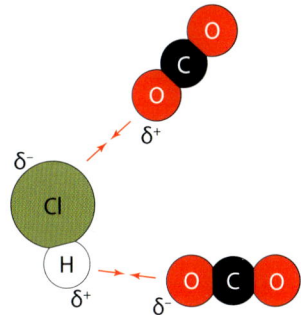

Figure 6.17. The permanent dipole HCl molecule inducing temporary dipoles in CO_2 molecules, resulting in Debye forces between the HCl and CO_2 molecules.

The third of the Van der Waals forces is the *London dispersion force*. London dispersion forces are forces between two temporary dipoles and are the weakest and most fleeting of all intermolecular forces. London forces are a result of the constant motion of the electrons in orbitals. As the electrons move around, there are very brief moments when the electrons surrounding a molecule or an individual atom (as in a noble gas, for example) happen randomly to congregate on one side of the molecule. They don't stay like that, of course, because they repel each other and quickly redistribute themselves. Nevertheless, for an instant, the atom or molecule becomes a dipole: the negative end is where the electrons are all gathered and the positive end is where they came from. The force between two such instantaneously created dipoles is the London dispersion force.

It should be apparent that since London dispersion forces are caused by the random motions of fast moving electrons, they appear constantly in gases and liquids all around us but only last for an instant.

Figure 6.18. The ability of a gecko to walk up a plate of glass is probably due to Van der Waals forces.

Figure 6.18 is an image of a gecko walking up a pane of glass. Researchers have published one study after another analyzing the ability of a gecko's feet to stick to a pane of glass, but the issue does not seem to be settled yet. However, many who have studied the question are convinced that Van der Waals forces are responsible, which means electrical forces. Once again, we can say that chemistry is all about electrical forces!

Table 6.3 summarizes all the forces we have reviewed here.

Chapter 6

Bonding Forces			Intermolecular Forces		
electrostatic	force between ions and/or polyatomic ions		hydrogen bond	force between nonbonding electron pairs and a hydrogen atom bonded to an oxygen, nitrogen, or fluorine atom	
covalent	force holding atoms together due to shared electrons in overlapping orbitals		Van der Waals	Keesom force	force between two permanent dipoles
				Debye force	force between a dipole and an induced dipole caused by it
				London dispersion force	force between two instantaneously formed dipoles

Table 6.3. Summary of the various atomic-level forces involved in chemical bonding and molecular attraction.

Chapter 6 Exercises

SECTION 6.1

1. Describe G.N. Lewis' contribution to molecular theory.

2. Describe the central theoretical problem addressed by VSEPR theory.

3. Use VSEPR theory to predict the shapes of the following molecules. For each one listed, draw the Lewis structure, state the electron domain geometry, the molecular geometry, and sketch a simple ball-and-stick figure of the molecule.

   a. phosphine, PH_3
   b. carbon tetrachloride, CCl_4
   c. sulfur dioxide, SO_2
   d. SCl_2
   e. NH_2Cl
   f. NH_4^+
   g. ClO_2^-
   h. ozone, O_3
   i. acetylene, C_2H_2
   j. CS_2
   k. SO_3
   l. Cl_2O
   m. hydrochloric acid, HCl
   n. hydronium, H_3O^+
   o. hydrogen cyanide (known in the old days as prussic acid), HCN

4. Compare the geometries of sulfur dioxide and sulfur trioxide.

5. Explain what is meant by the term *electron domain*.

6. What is the difference between a bonding domain and a nonbonding domain?

7. Give approximate bond angles for the bonds indicated in the following molecules:

164

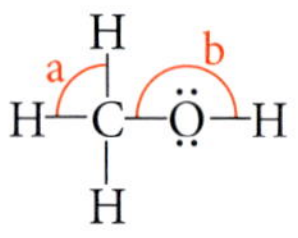

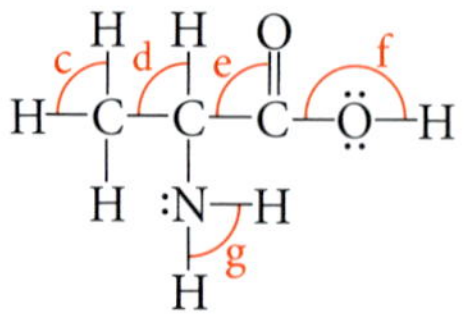

8. Explain why bond angles are sometimes slightly smaller than the angles in pure geometric figures.

9. Consider the molecules described below and state the electron domain geometry and the molecular geometry for each.

   a. two bonding domains and two nonbonding domains.

   b. four bonding domains

   c. three bonding domains and one nonbonding domain

   d. five bonding domains

   e. four bonding domains and two nonbonding domains (Hint: The nonbonding domains are on opposite ends of the molecule, and the resulting molecular shape uses a common geometrical term not introduced in the chapter.)

10. Why is it that the electron domain geometry in a molecule can be different from the molecular geometry?

11. Under what conditions does octahedral molecular geometry occur?

12. Consider a molecule with four bonding domains and one nonbonding domain.

    a. Describe the two options for the molecular geometry of this molecule.

    b. In fact, one of the two options minimizes the energy in the molecule more than the other and is the shape the molecule takes. Identify which one this is and explain why.

13. Look again at the L-alanine molecule on the opening page of this chapter, and shown to the right. Use VSEPR theory to explain the geometry of the molecule at each of the five connection points. These are circled in red in the figure to the right.

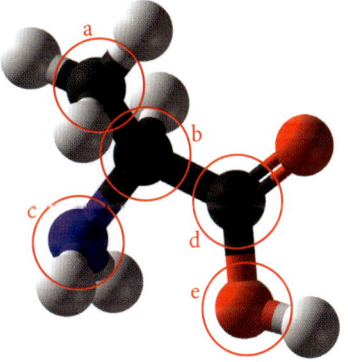

SECTION 6.2

14. Explain what holds the atoms together in metallic bonding.

15. Explain why it is inaccurate to say that metallic bonding is just like covalent bonding.

16. For each of the following typical properties of metals, use what you know about metallic bonding to explain why metals exhibit the property.

    a. high electrical conductivity

    b. high thermal conductivity

    c. high luster

    d. silver or gray color

    e. malleability

17. Explain why copper has a natural orange-red color.

SECTION 6.3

18. Use what you have learned about intermolecular forces to explain why the boiling point of water is so much higher than that of other molecular compounds of comparable molar mass.

19. Explain what a dipole is.

Chapter 6

20. For each of the following molecules, explain why the molecule does or does not engage in hydrogen bonding.

   a. HCN
   b. NH_3
   c. CCl_4
   d. Cl_2
   e. H_2
   f. HCOOH
   g. CH_3COOH
   h. H_2O

21. Use what you know about hydrogen bonding to construct a general explanation of the differences in the boiling points of the compounds listed in Table 5.11, and shown again below.

Compound	Melting Point, °C	Boiling Point, °C
H_2O	0	100
O_2	−219	−183
N_2	−210	−196
Cl_2	−102	−34
Br_2	−7.2	59
CO_2	−79 (sublimes)	
ammonia, NH_3	−78	−33
methane, CH_4	−182	−161
propane, C_3H_8	−188	−42
acetic acid, CH_3COOH	17	118
ethanol, C_2H_5OH	−114	78
benzene, C_6H_6	5.5	80
acetone, $(CH_3)_2CO$	−95	56

22. Distinguish between permanent and temporary dipoles.

23. Describe two ways a temporary dipole can form.

24. Explain why cotton towels can absorb so much moisture.

25. Why are London dispersion forces the weakest of the intermolecular forces?

26. What are Van der Waals forces?

GENERAL REVIEW EXERCISES

27. Determine the molar mass and the percent composition of L-alanine, pictured on the previous page.

28. The bright orange line in the copper emission spectrum has a wavelength of about 601 nm. Determine the energy associated with each photon of light at this wavelength. State your answer in eV.

29. Identify each of the following bonds as covalent, polar covalent, or ionic in character.

   a. H—N
   b. H—F
   c. Br—Br
   d. C—O
   e. Cl—S
   f. P—O

Molecular Theory and Metallic Bonding

30. Use VSEPR theory and electronegativities to determine which of the following molecules are polar.
    a. CF_4
    b. Cl_2O
    c. NH_3
    d. SO_2
    e. N_2
    f. SF_2

31. Use your knowledge of atomic and ionic size trends in the periodic table to place the following atoms and ions in order of increasing size: Cl^-, F^-, Te^{2-}, F, and I^-.

32. Write the condensed electron configurations for Hg, Br, Ba, Mn, and S.

33. Define the unified atomic mass unit.

34. Define the mole.

35. Determine which of the following is likely to have the greatest difference between the second and third ionization energies: Sc, Ar, F, P, and Sr.

36. Distinguish between electronegativity and electron affinity.

37. Determine the number of moles present in each of the following:
    a. 65.5 g water
    b. 1,250 mg vitamin C (ascorbic acid), $C_6H_6O_6$
    c. 400 mg aspirin, $C_9H_8O_4$
    d. 14.0 kg saltpeter, KNO_3
    e. 1,050 g perchloric acid
    f. 953.00 g hydrogen fluoride

38. Determine the number of metal atoms present in each of the following:
    a. 55 g mercury(II) sulfide
    b. 3.00 kg iron(III) oxide
    c. 1.0000 mol calcium carbonate
    d. 45 g strontium nitrite
    e. 2.000 mol sodium chromate
    f. 5.05 kg calcium acetate

39. Use data from Table 2.6 to determine the atomic mass of iron.

40. Explain the Aufbau principle, the Madelung rule, and Hund's Rule.

41. State the Pauli exclusion principle.

42. Describe the different ways that atoms can possess energy.

43. Glucose is a carbohydrate that is the major building block for the polymers cellulose and starch. A certain sample of glucose consists of 4.62 g C, 0.776 g H, and 6.154 g O. The molar mass of glucose is 180.157 g/mol. Determine the percent composition, the empirical formula, and the molecular formula of glucose.

Chapter 7
Chemical Reactions and Stoichiometry

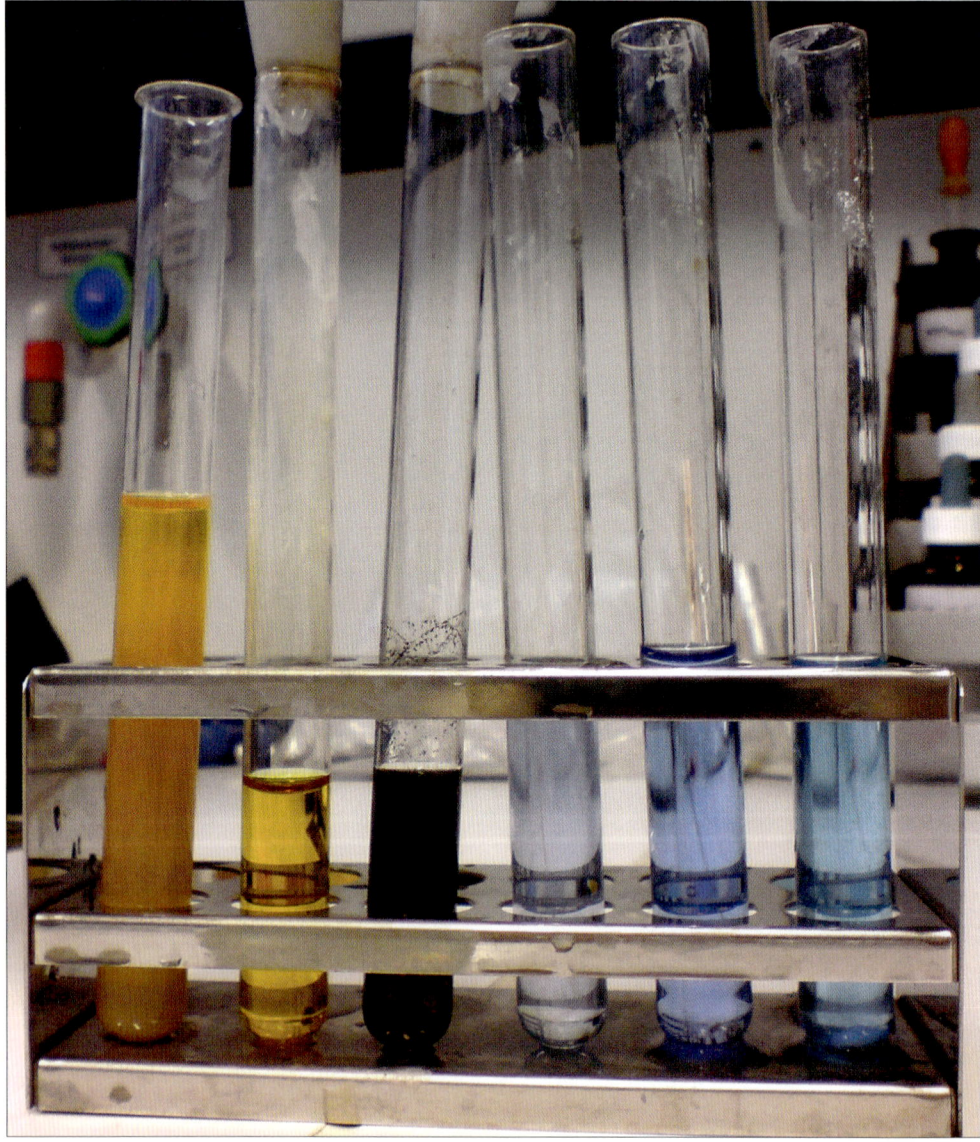

These test tubes contain various solutions and reaction products involving copper(II) nitrate, $Cu(NO_3)_2$. In some cases, precipitates formed; in others, the products remained in solution. We use some of these reactions as examples in this chapter.

Objectives for Chapter 7

After studying this chapter and completing the exercises, you should be able to do each of the following tasks, using supporting terms and principles as necessary.

SECTION 7.1
1. State the law of conservation of mass for chemical reactions.
2. Recognize, interpret, and use the standard symbols involved in chemical equations.
3. Given a verbal description of the reactants and products in a chemical reaction, write the formula equation and balance it.
4. Define the terms *precipitate, oxidation, reduction, oxidizer, oxidation state,* and *oxidation number.*
5. Use the standard oxidation state guidelines from memory to determine the oxidation states of atoms in pure elements, in polyatomic ions, and in compounds.

SECTION 7.2
6. Classify chemical reactions as synthesis, decomposition, single replacement, double replacement, combustion, acid-base neutralization, or oxidation-reduction. Also, identify when a reaction may be classified as two or more of these types simultaneously.
7. Use the activity series of metals to predict whether a given single replacement reaction with metals will occur.
8. Recognize and identify common hydrogen compounds as acids and common hydroxide compounds as bases.
9. Explain what happens in an acid-base neutralization reaction and use this information to form a definition for the term *salt.*
10. Explain the term *redox reaction* and describe what happens in such reactions.

SECTION 7.3
11. Describe the origin of the term *stoichiometry.*
12. Given a mole or mass quantity of one of the reactants or products in a specified chemical reaction, perform stoichiometric calculations to determine the quantities required or produced, in moles or mass, of any of the other reactants or products in the reaction.
13. Define the terms *limiting reactant* and *limiting reagent.*
14. Use a balanced chemical equation and available quantities of reactants to identify the limiting reactant in the reaction.
15. Use the available quantity of a limiting reactant to determine the theoretical yield for the reaction.
16. Use the theoretical yield and actual yield in a reaction to determine the percent yield for a given reaction product.

Chapter 7

7.1 Introduction to Chemical Equations

7.1.1 Fascinating Chemistry

Chemical reactions take place around us—and in us—all the time, but unless a fire or explosion occurs, we often don't notice. You probably already know that fires and explosions are chemical reactions, but so are the drying of paint, the filling of the air bags in a car during a collision, and the digestion of your food. Chemical reactions are of supreme importance in industry, and without them you and I might have electricity and running water, but we would be without nearly all the products we enjoy. Without the chemical research of the 19th and 20th centuries, we would not have computers, plastic products, batteries, synthetic fibers (such as nylon), or just about anything else composed of nonmetallic man-made materials.

There are several tell-tale signs that a chemical reaction is taking place. The first is the release of heat and light. When you are burning logs at a camp fire, you can be assured a chemical reaction is taking place. The second is the production of a gas, such as the carbon dioxide that fizzes out of a soft drink after the can has been opened. A third is a color change. Figure 7.1 shows the dramatic color change that occurs when beautiful blue-green copper carbonate is heated to above 290°C. The copper carbonate decomposes to dark brown copper oxide, releasing carbon dioxide gas. Finally, a *precipitate* may form. A precipitate is any solid substance that forms when two solutions are combined. Figure 7.2 shows a solid copper(II) hydroxide precipitate coming out of solution because of the reaction that occurs when aqueous solutions of $CuSO_4$ and NaOH are combined.

Figure 7.1. Copper carbonate decomposes to copper oxide when heated.

Now, you may be of the opinion that chemistry is really only enjoyed by science nerds, but I suggest it should be otherwise. Because chemistry is so important in contemporary society, and because so many of the processes and products around us involve chemical reactions that are easily understood, I would expect anyone with a normal, healthy level of curiosity about the world to be interested—if not fascinated—to learn about it. In this chapter, we begin diving in to some of these ubiquitous chemical reactions.

Figure 7.2. Solid copper(II) hydroxide precipitates out of solution when aqueous solutions of $CuSO_4$ and NaOH are combined.

7.1.2 The Law of Conservation of Mass in Chemical Reactions

French chemist Antoine Lavoisier (Figure 7.3) is regarded as the father of modern chemistry. He was the first to understand that combustion was a chemical reaction involving oxygen. He was also responsible for helping to turn chemistry from a descriptive science into a quantitative one. This was because of the famous tin experiments Lavoisier conducted in 1774. In his tin experiments, Lavoisier investigated the formation of tin oxide from tin, and in the process he

Chemical Reactions and Stoichiometry

discovered that when tin turns to tin oxide, it does so by reacting with the oxygen in the air. Before the experiments, no one knew what air was composed of. Through the experiments, Lavoisier demonstrated that air is composed of more than one gas, and that oxygen accounts for about 20% of it.

Also in 1774, English chemist Joseph Priestley (Figure 7.4) was the first to isolate oxygen, which he accomplished by heating mercury oxide. Priestley assumed he had produced a particularly pure form of air. Lavoisier identified the gas that Priestley had produced as being the same as the gas that reacted with the tin and named the gas *oxygen*.

Figure 7.3. French chemist Antoine Lavoisier (1743–1794).

In the tin experiments, Lavoisier heated tin in sealed vessels so that he could carefully weigh the contents before and after the reaction. He found the weights before and after to be the same. As a result of his measurements, Lavoisier demonstrated conclusively the principle now known as the *law of conservation of mass in chemical reactions: in any chemical reaction, the total mass of the products equals the total mass of the reactants.* This law is the basis for the methods we use to balance chemical equations, a critical step that undergirds all the calculations involving chemical reactions. We consider the procedure for balancing chemical equations momentarily. First, we consider the notation used in writing chemical equations.

Figure 7.4. English chemist Joseph Priestley (1733–1804).

7.1.3 Reaction Notation

The chemical reaction Joseph Priestley used to isolate oxygen is the following:

$$2HgO(s) \xrightarrow{\Delta} 2Hg(l) + O_2(g)$$

To perform this reaction, solid mercury(II) oxide, shown in Figure 7.5, is heated to produce liquid mercury and gaseous diatomic oxygen. The formulas for each of these compounds and elements are written in the equation with their subscripts. Subscripts in chemical equations are part of the chemical formulas that identify the specific compounds involved in a reaction. The coefficients in front of the compound formulas are determined during the process of balancing the equation.

The symbols used in a chemical equation convey a lot of information about the reaction. The physical states of each of the compounds are shown in parentheses after the chemical symbol or formula. In addition to s (solid), l (liquid), and g (gas) in the equation above, (aq) is used to indicate a substance in aqueous solution. The Δ symbol (*delta*, the Greek upper-case D) over the "yields" arrow indicates heating, and if heating to a particular temperature is required, this temperature is written below the arrow. Other symbols may also be placed over the arrow to indicate

Figure 7.5. Mercury(II) oxide, HgO.

Chapter 7

Symbol	Meaning
→	yields
⇌	reaction is reversible; proceeds in both directions
s	solid
l	liquid
g	gas
aq	in aqueous solution
$\xrightarrow{\Delta}$ or $\xrightarrow{heat}$	heating
$\xrightarrow[290°C]{\Delta}$	heat to 290°C
$\xrightarrow{2.5\ atm}$	reaction occurs at a pressure of 2.5 atm
$\xrightarrow{Fe}$	chemical symbol or formula for a catalyst that must be present

Table 7.1. Symbols used in chemical equations.

7.1.4 Balancing Chemical Equations

As you will see throughout this chapter, the chemical equation is a very important tool for analysis of chemical reactions. To write a chemical equation, first the formulas of the reactants and products are written. This is called the *formula equation*. Using Priestley's mercury(II) oxide reaction again as an example, the formula equation is

$$HgO(s) \xrightarrow{\Delta} Hg(l) + O_2(g) \quad \text{(formula equation)}$$

Once the correct formulas for the reactants and products are present, the equation must be balanced so that it conforms correctly to the law of conservation of mass in chemical reactions. As it stands, the formula above has HgO on the left side, a formula indicating one atom of oxygen. But the O_2 on the right indicates a molecule containing two oxygen atoms.

The *balanced equation* for the mercury(II) oxide reaction is

$$2HgO(s) \xrightarrow{\Delta} 2Hg(l) + O_2(g) \quad \text{(balanced equation)}$$

The coefficients in front of the HgO and Hg apply to the entire compound they precede. Thus, 2HgO means "two units of HgO, and thus two mercury atoms and two oxygen atoms." Note that this is similar in meaning to the mathematical expression $2(a + b)$, but it is completely different in meaning from the expression $2ab$.

While balancing equations, it is most convenient to read a chemical equation as representing numbers of atoms. Read this way, we can simply count atoms of each element on both sides of the equation and manipulate the coefficients until the equation is balanced. But very importantly, the equation can also be read in terms of the bulk quantity moles. And since most reactions involve bulk quantities of atoms, this is the usual way to read an equation once the balancing is complete. Read this way, the balanced mercury(II) oxide equation reads, "two moles of solid mercury(II) oxide yield two moles of liquid mercury and one mole of diatomic oxygen gas."

Just in case you've never seen liquid mercury before, it is shown in Figure 7.6. It may look hot, as if it were molten metal, but it is not. It is one of only two elements that are liquids at room temperature. (The other is bromine.) When I was in high school, we were allowed to play with

Figure 7.6. Pouring liquid mercury.

mercury in our hands. It is almost as dense as lead (look at the periodic table), so a blob of mercury feels unbelievably heavy in your hand. Alas, however, playing with mercury is no longer allowed. Once scientists discovered that it is absorbed through the skin, is quite toxic, can enter the body by breathing the vapor, and that its effects are cumulative, the party was over. With prolonged exposure, mercury destroys brain cells and as a result, people lose brain function. Thus the expression, "mad as a hatter." Hat makers in the old days used mercury to shape the felt hats they made. Not good. Oh, and please make sure you and your family don't toss used fluorescent lamps or compact fluorescent lamps in the trash. They all contain mercury. Take them to a recycle center such as one of those giant hardware stores. Not kidding.

Now, back to balancing equations. There are some key points to attend to when writing down the formula equation. Once those are satisfied, then the equation may be balanced.

1. All formulas for compounds must be correct, with the correct subscripts.

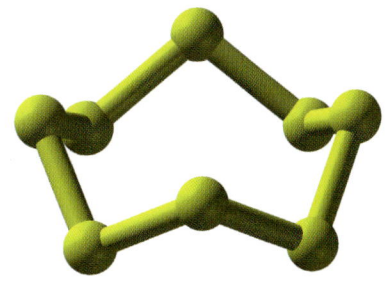

Figure 7.7. A sample of sulfur (above) and a model of cyclooctosulfur, S_8.

2. For pure elements in an equation, there are several cases when a pure element appears in molecular form, and has a subscript. Seven of these are the seven diatomic substances: H_2, N_2, O_2, F_2, Cl_2, Br_2, and I_2. Sulfur and phosphorus form molecules with themselves in the natural state and appear as the molecules S_8 and P_4. The most common form of sulfur, for example, takes the form of the eight-atom ring called *cyclooctosulfur*. A sample of sulfur and the cyclooctosulfur ring are shown in Figure 7.7. And since we are talking about sulfur, I can't resist showing the beautiful images of burning sulfur in Figure 7.8. When it burns, sulfur melts into a blood-red liquid. The flame is bright blue, but cannot be seen in the daytime photo. The second photo, made in the dark, shows it well.

3. The equation must contain all the reactants and products that are part of the reaction that actually occurs.

The process of balancing a formula equation is the process of adding coefficients in front of either reactants, products, or both until the equation displays equal numbers of atoms of each element involved on both sides of the equation. For the relatively simple reactions we consider in this chapter, balancing is performed by inspection and trial and error, although there are some guidelines that help expedite the process. Let's consider the guidelines, and then move to examples. Here they are:

1. If an element appears in an odd number on one side of the equation and an even number on the other, you need to make them both even. Start by placing a coefficient (usually 2) on the formula with the odd number.

Figure 7.8. Burning sulfur in the daylight (above) and in the dark (below).

Chapter 7

2. Usually, treat polyatomic ions in formulas as single units. It is easier to count sulfates than it is to count individual sulfur and oxygen atoms. However, if the polyatomic ion does not appear on both sides of the equation, you must count individual atoms.

3. Tackle the more complicated formulas first. Save placing coefficients on elements that appear alone in the equation (such as the Hg and O_2 terms in the mercury(II) oxide reaction) until last. Placing a coefficient on one element doesn't affect anything else. Placing coefficients on compounds affects two or more elements simultaneously.

4. When you finish balancing, make sure there is no common multiple (other than one) in all the coefficients in an equation. So for example, if all your coefficients are even, divide them all by two.

We now illustrate the process of balancing equations with a number of examples.

▼ Example 7.1

Write the formula equation and balanced equation for the following reaction: zinc metal is combined with hydrochloric acid in aqueous solution. The reaction produces a solution of zinc chloride and hydrogen gas.

We begin by writing the formula equation. Hydrochloric acid is HCl. From Figure 4.9, the oxidation state of zinc is Zn^{2+}, so the formula for zinc chloride is $ZnCl_2$.

$Zn(s) + HCl(aq) \rightarrow ZnCl_2(aq) + H_2(g)$ (formula equation)

Notice that in the reactants there is a single atom of chlorine and a single atom of hydrogen, but in the products there are two of each. We resolve this odd-even problem by placing a coefficient of 2 on the HCl.

$Zn(s) + 2HCl(aq) \rightarrow ZnCl_2(aq) + H_2(g)$ (balanced equation)

This appears to have balanced the equation, but count to make sure. Zn: one on the left, one on the right. H: two on the left, two on the right. Cl: two on the left, two on the right. The equation is balanced.

▼ Example 7.2

One of the forms rust can take is iron(III) oxide (Figure 7.9). Write the formula equation and balanced equation for the combination of iron with oxygen to produce rust.

Oxygen is a diatomic gas, so it appears as O_2 in the reactants.

$Fe(s) + O_2(g) \rightarrow Fe_2O_3(s)$ (formula equation)

Figure 7.9. Rust on iron.

We have the odd-even problem with both Fe and O. Oxygen, with two on the left and three on the right, is more complicated, so we start there. To make the oxygen match, we need the least common multiple of two and three, which is six. To get six on each side, we need a coefficient of 3 on the left and 2 on the right. Note that the coefficient on the right goes in front of the entire compound formula.

Chemical Reactions and Stoichiometry

$$Fe(s) + 3O_2(g) \rightarrow 2Fe_2O_3(s)$$

The oxygens are balanced now, but the iron is not. There are four iron atoms represented in the product. Since iron is by itself in the reactants, a simple coefficient on the iron finishes the balancing to give us

$$4Fe(s) + 3O_2(g) \rightarrow 2Fe_2O_3(s) \qquad \text{(balanced equation)}$$

We now double check the count of each atom to verify that the equation is balanced.

▼ Example 7.3

Aluminum sulfate (Figure 7.10) is used in water treatment plants to remove impurities in the water. When combined in aqueous solution with calcium hydroxide, the reaction produces aluminum hydroxide and calcium sulfate, neither of which are soluble in water. The two solids that are formed attract particulate impurities in the water, and as they settle, the impurities are taken with them. Write the formula equation for this reaction and balance it.

The formula equation is straightforward, given the oxidation states of aluminum and calcium (Figure 4.9) and the polyatomic ion list (Table 5.6).

Figure 7.10. Aluminum sulfate.

$$Al_2(SO_4)_3(aq) + Ca(OH)_2(aq) \rightarrow Al(OH)_3(s) + CaSO_4(s) \qquad \text{(formula equation)}$$

There are two polyatomic ions that appear on both sides of this equation. We treat these as single objects during the balancing. The aluminum sulfate is as good a place to start as any. There are three sulfates on the left and one on the right, so we place a coefficient of 3 on the calcium sulfate.

$$Al_2(SO_4)_3(aq) + Ca(OH)_2(aq) \rightarrow Al(OH)_3(s) + 3CaSO_4(s)$$

Calcium is now out of balance, so we address that with a coefficient of 3 on the calcium hydroxide.

$$Al_2(SO_4)_3(aq) + 3Ca(OH)_2(aq) \rightarrow Al(OH)_3(s) + 3CaSO_4(s)$$

Now we look at the hydroxides. We have six on the left, so a coefficient of 2 on the aluminum hydroxide gives us six on the right.

$$Al_2(SO_4)_3(aq) + 3Ca(OH)_2(aq) \rightarrow 2Al(OH)_3(s) + 3CaSO_4(s) \qquad \text{(balanced equation)}$$

By adding the coefficient of 2 on the aluminum hydroxide we see that the aluminum becomes balanced at the same time as the hydroxide. We now double check the count of each atom and polyatomic ion to verify that the equation is balanced.

Chapter 7

▼ Example 7.4

On the opening page of this chapter, the first test tube contains a solution of white copper(I) iodide precipitate in an aqueous solution of iodine (shown again in Figure 7.11). These products are actually formed in a second reaction that is part of a two-step reaction sequence. The sequence begins with combining aqueous solutions of copper(II) nitrate and potassium iodide to produce an aqueous solution of potassium nitrate and solid copper(II) iodide. The copper(II) iodide immediately decomposes to form white copper(I) iodide precipitate and iodine in aqueous solution. Write the formula equations for each of these reactions and balance them.

The formula equation for the first reaction is as follows:

$$Cu(NO_3)_2(aq) + KI(aq) \rightarrow KNO_3(aq) + CuI_2(s) \qquad \text{(formula equation)}$$

We start with the nitrate. There are two on the left and one on the right, so we add a coefficient of 2 on the potassium nitrate.

$$Cu(NO_3)_2(aq) + KI(aq) \rightarrow 2KNO_3(aq) + CuI_2(s)$$

Now we see that the potassium and iodine are both single on the left and double on the right. Adding a coefficient of 2 on the potassium iodide takes care of both of them and gives us the balanced equation.

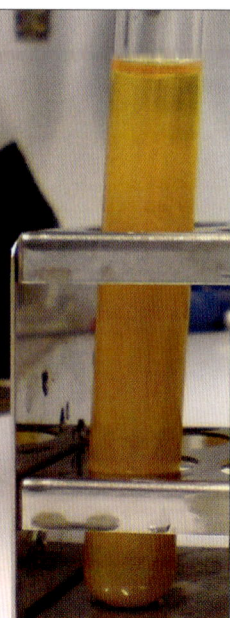

Figure 7.11. White copper(I) iodide in a yellow iodine solution.

$$Cu(NO_3)_2(aq) + 2KI(aq) \rightarrow 2KNO_3(aq) + CuI_2(s) \qquad \text{(balanced equation)}$$

Be sure to double check the count of each atom and polyatomic ion to verify that the equation is balanced.

The formula equation for the second reaction is

$$CuI_2(s) \rightarrow CuI(s) + I_2(aq) \qquad \text{(formula equation)}$$

With three iodines on the right, the thing to do is double the single one so we have an even number on the right. This gives

$$CuI_2(s) \rightarrow 2CuI(s) + I_2(aq)$$

We now have four iodines on the right and two on the left. Placing a coefficient of 2 on the CuI_2 balances the iodine and the copper at the same time.

$$2CuI_2(s) \rightarrow 2CuI(s) + I_2(aq) \qquad \text{(balanced equation)}$$

Finally, double check the count of each atom and polyatomic ion to verify that the equation is balanced.

▲

7.1.5 Oxidation States

We introduced the concept of oxidation states back in Chapter 4 in the context of ionization. It is now time to develop this further so that you are able to determine the oxidation state of any element in any compound. When an atom, such as a metal, loses one or more electrons, we say it has been *oxidized*. When an atom gains one or more electrons we say it has been *reduced*.

The terms *oxidation* and *reduction* go back to the days of Antoine Lavoisier, just a few years before the French Revolution. At the time, it was thought that oxygen was the only element that caused oxidation, hence the name of the process. Lavoisier coined the term reduction, referring to the weight loss of metals during smelting. In this process, ore is converted to pure metal by the removal of oxides and other substances (giving the metal atoms their electrons back). The resulting pure metal weighs less than the ore, so it has been "reduced."

We will use sodium and chlorine to illustrate oxidation and reduction. If an atom of sodium is ionized and loses its one valence electron to a chlorine atom, it is *oxidized* to become Na^+ and its oxidation state is +1. Chlorine is the *oxidizer* or *oxidizing agent* in the process. When the chlorine atom receives that electron, the chlorine atom is reduced to become Cl^- and its oxidation state is –1. Sodium is the *reducer* or *reducing agent*. These two ions join together by electrostatic attraction to become part of the crystal lattice of NaCl.

If this chemistry is driven in the opposite direction (which can be done by running an electric current through molten sodium chloride), the Cl^- ion loses an electron to become Cl; the chlorine ion is now the one that is oxidized. The sodium ion receives that electron to become Na, so the sodium ion is reduced. This is what happens during smelting: the metal atoms in the ionic compounds in the rock get their electrons back, cease to be ions, and become elemental metal. A tried and true mnemonic to help you remember oxidation and reduction is shown in Figure 7.12.

The oxidation state of an element is indicated by an oxidation number such as +2, +1, 0, –2, –1, and so on. The oxidation state of any neutral unbonded atom is 0. As illustrated just above, in ionic compounds the oxidation state of an element is directly related to the number of electrons the element has gained or lost, and the oxidation number is simply the charge on the ion. This charge is usually obvious because of the element's position in the periodic table, or because it is indicated in the name of the compound. In magnesium chloride, $MgCl_2$, the oxidation number of magnesium is +2 because as an alkaline-earth element, magnesium ionizes by losing two electrons. The oxidation number of chlorine is –1 because chlorine, like all the halogens, ionizes by gaining one electron. In copper(I) iodide, CuI, the Roman numeral in the name indicates the oxidation state of the metal (+1).

However, in covalent bonds where electrons are being shared, determining the oxidation state of an element is not as simple as keeping track of gaining or losing electrons. Instead, we use electronegativities and a list of rules of thumb to determine oxidation states. In such compounds, we think of the oxidation state as indicating *partial* gain or loss of electrons. Because of this, it is probably more helpful to you if you think of oxidation and reduction this way: *oxidation occurs any time the oxidation state of an element increases; reduction occurs any time the oxidation state of an element decreases*. This definition also aids your memory—you can associate the term *reduction* with a *decrease* in the oxidation state (and the oxidation number).

Here are the rules we follow to determine the oxidation state of elements in covalent compounds:

Figure 7.12. A classic memory aid for oxidation and reduction. Say it over and over to yourself. Say it now.

Chapter 7

1. The operating rule is this: *in a molecular compound, the oxidation states of the elements involved must add up to the charge on the molecule.* In a neutral molecule, the oxidation numbers must add up to zero. In a polyatomic ion, they must add up to the charge on the ion.

2. In pure elements in their natural state, the oxidation number is 0. This also applies to diatomic molecules such as H_2 and the other elements that naturally make molecules with themselves, S_8 and P_4.

3. The most electronegative element in the molecule has the oxidation state it would have if it were an anion. Since fluorine has the highest electronegativity, its oxidation state is always −1. Oxygen has the second-greatest electronegativity, so it is usually assigned an oxidation state of −2. One exception is when oxygen forms peroxides as the molecule O_2^{2-}, in which case its oxidation state is −1. A common peroxide is hydrogen peroxide, H_2O_2. Another exception is when oxygen bonds with fluorine to form OF_2, in which case its oxidation state is +2 (since F is always −1).

4. The oxidation number for halogens is −1. The exception is when they bond with oxygen to form polyatomic ions with negative charge (oxyanions). In that case, they have positive oxidation states (except for fluorine).

5. When hydrogen forms a compound with an element that is more electronegative than itself (a nonmetal), its oxidation state is +1. In compounds with metals, the oxidation state of H is −1.

▼ Example 7.5

Determine the oxidation states for each atom in the following compounds or ions: $MgCl_2$, SF_6, BaO_2, $HClO_3$, PO_4^{3-}.

$MgCl_2$ is an ionic compound. The oxidation numbers for Mg and Cl are thus +2 and −1, respectively. These add up to zero: $2 + 2(-1) = 0$.

In SF_6, fluorine has an oxidation number of −1. Since there are six fluorine atoms in the molecule, the contribution to the overall charge of the molecule is $6(-1) = -6$. Since the overall charge on the molecule is zero, the oxidation numbers must add up to zero. Thus, the oxidation number of sulfur is +6.

Barium peroxide, BaO_2, is a nontypical ionic compound—it is a peroxide. We would normally expect Ba and O to form BaO since their typical ionizations are Ba^{2+} and O^{2-}. But in the compound BaO_2, the oxidation state of the metal is the one that runs the show. The oxidation number for Ba is +2 so the oxidation number for O must be −1. Oxygen is not very happy in peroxides, so the compounds are unstable and like to go *boom*. See the accompanying box for more about that exciting subject.

Chlorine and oxygen are both more electronegative than hydrogen, so in $HClO_3$ the oxidation state of H is +1. The chlorate ion, ClO_3^-, is an oxyanion, so we expect the chlorine to have a positive oxidation state. The oxidation state of oxygen is −2 and there are three of them, so the charge contribution from oxygen is $3(-2) = -6$. Adding the hydrogen gives $-6 + 1 = -5$. For the oxidation numbers in the molecule to add to zero, the oxidation state of chlorine must be +5.

Hmm... Interesting.   Why nitrates and nitros blow up

To a chemist, any substance that causes an element to oxidize is an oxidizing agent, or oxidizer. The oxygen in air causes iron to oxidize and form rust, Fe_2O_3, and is an oxidizing agent in this general chemical sense.

But a more narrow definition of an oxidizing agent is used in classifying substances as dangerous materials. In such classifications, an oxidizing agent is a substance that causes other substances to combust rapidly. In barium peroxide, BaO_2, the oxidation state of oxygen is -1, whereas oxygen's preferred oxidation state, the state in which its valence shell is full, is -2. This means the compound is unstable and the oxygen in BaO_2 aggressively reacts with other substances. BaO_2 is used in fireworks to produce an intense green color. Hydrogen peroxide is also an aggressive oxidizer and is sometimes used as a rocket fuel. It was also the oxidizer used in the 2005 London bombings that killed 52 people. But don't worry about the hydrogen peroxide solution you may have in the medicine chest at home. That's only a 3% aqueous solution of H_2O_2. (Researchers now say that while hydrogen peroxide solution is useful for disinfecting inanimate objects and removing blood stains, it is not the best for disinfecting wounds because it causes tissue damage and thus slows healing.)

Potassium nitrate, KNO_3, is the oxidizer used in gunpowder. Nitrate compounds in general are aggressive oxidizers. Ammonium nitrate, NH_4NO_3, a common fertilizer, was used as an explosive in the 1995 Oklahoma City bombing (which killed 168 people) and the 1993 World Trade Center bombing (which killed 6 people but injured several thousand). (The WTC bombing occurred before the WTC was destroyed by terrorists on September 11, 2001.) Ammonium nitrate is also the substance that exploded and destroyed a fertilizer plant in West, Texas in 2013, killing 15 people. In the 1947 Texas City disaster, a ship containing 2,300 tons of ammonium nitrate detonated, killing 581 people—the deadliest industrial accident in U.S. history.

When it comes to explosive compounds, so-called *nitro* compounds are the most explosive of all. Nitro compounds include TNT, nitroglycerine (which can explode just by being jostled), and the stabilized form of nitroglycerine called dynamite. Nitro compounds contain carbon rings with nitronium ions (NO_2^+) attached, as illustrated by the Lewis structure for TNT, trinitrotoluene, shown above. Nitro compounds are not oxidizers, but are explosive for the same reason the nitrates are: because of the nitrogen and oxygen atoms they contain. When the nitrogen atoms are freed from the molecule, they form diatomic molecules of nitrogen gas, N_2. The oxygen atoms also form gases during reactions, typically CO_2, but also CO. The N_2 (and CO) molecules contain triple bonds, which means they are very strong and release a lot of energy while forming (see Table 5.10). The presence of a lot of heat while gases are forming means the gases expand *very* rapidly. Put all this together and you have huge explosions. The reason the nitro compounds explode more vigorously than the nitrate oxidizers is that the nitro compounds have multiple carbons (which get oxidized in the explosion, releasing additional heat), nitrogens, and oxygens *in the same molecule*. You can't get any closer together—thus faster—than that.

One more interesting tidbit: dynamite was invented by Alfred Nobel, and the fortune he made from it was used to create the Nobel Foundation, which awards the annual Nobel Prizes.

Oxygen is more electronegative than phosphorus, so its oxidation state in PO_4^{3-} is -2. The four oxygens make a charge contribution of $4(-2) = -8$. The oxidation numbers must add up to the charge on the ion, which is -3. Thus, the oxidation state of phosphorus is $+5$.

7.2 General Types of Chemical Reactions

7.2.1 Synthesis Reactions

In *synthesis reactions*, separate elements or compounds are combined to form a single, new compound. Synthesis reactions are described by the general equation form

A + B → AB

The classic example of a synthesis reaction is the formation of water by the combustion of hydrogen:

$2H_2(g) + O_2(g) \rightarrow 2H_2O(g)$

Another example is the production of calcium hydroxide from calcium oxide:

$CaO(s) + H_2O(l) \rightarrow Ca(OH)_2(s)$

Calcium hydroxide, shown in Figure 7.13, is known as "slaked lime" for the same reason people use the phrase "slake your thirst": both are accomplished by adding water. The production of $Ca(OH)_2$ from CaO is an important step in the curing of concrete.

Figure 7.13. Calcium hydroxide.

Because of its environmental importance, an important synthesis reaction to know about involves the production of acids in the atmosphere, which leads to *acid rain*. Sulfur dioxide, SO_2, is a common product of the combustion of coal in electric power stations. In the atmosphere, SO_2 combines with moisture to produce sulfur trioxide, SO_3. This compound reacts easily with moisture in the air to produce sulfuric acid:

$SO_3(g) + H_2O(l) \rightarrow H_2SO_4(aq)$

Acidic rainfall was first detected back in the 17th century, but after the Industrial Revolution, atmospheric levels began to skyrocket. By the late 1960s, forests like the one in Figure 7.14 were being destroyed, as well as outdoor stone sculptures and architectural works made of marble or stone. New legislation to control emissions of SO_2 and other pollutants from power stations went into effect in the 1980s and 1990s, and as a result acid rain levels have dropped by 65% since 1976 (for a total cost much lower than originally predicted).

Figure 7.14. A forest destroyed by acid rain.

7.2.2 Decomposition Reactions

In synthesis reactions, compounds or elements are brought together; in *decomposition reactions*, compounds are taken apart. The general equation form is as follows:

$AB \rightarrow A + B$

The decomposition reaction you see happening most often is probably the decomposition of carbonic acid, H_2CO_3:

$H_2CO_3(aq) \rightarrow H_2O(l) + CO_2(g)$

Carbonic acid, an oxyacid, is added to soft drinks to make them fizzy when the can is opened. At atmospheric pressure, carbonic acid spontaneously decomposes to water and carbon dioxide. The CO_2 that *evolves* (i.e., is given off) during the reaction is the gas that makes the drink fizzy.

The heating of mercury oxide by Joseph Priestley, mentioned at the beginning of the chapter, is a decomposition reaction in which oxygen evolves:

$2HgO(s) \xrightarrow{\Delta} 2Hg(l) + O_2(g)$

A third example goes back to the calcium oxide synthesis reaction mentioned above. To obtain the CaO in the first place, limestone, which is calcium carbonate, $CaCO_3$, is roasted (that's what they call it) at high temperature, producing the following decomposition reaction:

$CaCO_3(s) \xrightarrow{\Delta} CaO(s) + CO_2(g)$

This is the starting point for the manufacture of "Ordinary Portland Cement," the most common cement in the world.

7.2.3 The Activity Series of Metals

To understand single replacement reactions (coming up next), you need to know about the *activity series of metals*, a list of metals in order of their *chemical activity*, that is, how aggressively they react relative to one another. The activity series is shown in Table 7.2. The metals in this list are listed from top to bottom in order of their chemical activity, with the most chemically aggressive metals at the top and the metals that react least at the bottom. In aqueous solutions of compounds of these metals, any metal on the list replaces any metal below it in a solution. Figure 7.15 illustrates this with a solution of silver nitrate, $AgNO_3$, which is colorless. When a coil of copper wire is placed in the solution, the copper begins going into solution, which turns the solution blue-green. The silver comes out of solution, precipitating in fluffy tufts onto the copper wire.

The chemical activity of the metals was determined empirically, that is, through experiments. Using the information on the list, you can determine if certain reactions occur. Table 7.2 also indicates the reactivity of the metals with water, oxygen, and acids. Active metals like lithium and sodium react very vigorously in water. All metals except the most unreactive react with oxygen to

Figure 7.15. In a solution of silver nitrate, the copper replaces the silver in solution.

Activity of Metals	Activity with Water	Activity with Oxygen	Activity with Acids
Li			
Rb			
K		reaction (to form oxides)	reaction (replacing hydrogen)
Ba	reaction		
Sr			
Ca			
Na			
Mg			
Al			
Mn	reaction with steam, but not cold water	reaction	reaction
Zn			
Cr			
Fe			
Cd			
Co			
Ni	no reaction	reaction	reaction
Sn			
Pb			
Sb			
Bi			
Cu	no reaction	reaction	no reaction
Ag			
Hg			
Pt	no reaction	no reaction	no reaction
Au			

Table 7.2. The activity series of metals and other activity information.

form oxides. Most metals also react with acids.

▼ Example 7.6

What do you expect to happen if a steel nail is placed in a container of copper(II) chloride solution?

Steel is mostly iron and iron is above copper in the activity series of metals. When placed in the solution, ions of iron begin going into solution and copper ions begin precipitating out of solution by attaching to the nail.

The phenomenon described in the previous example is shown in the sequence of images in Figure 7.16. After about 30 minutes, the solution becomes colorless, indicating that all the copper ions have been reduced to copper atoms and have precipitated out of solution. The reaction is a single replacement reaction, our next topic.

7.2.4 Single Replacement Reactions

In *single replacement reactions*, a lone element replaces another in a

Figure 7.16. Iron displaces the copper from solution, and the nail is buried in copper.

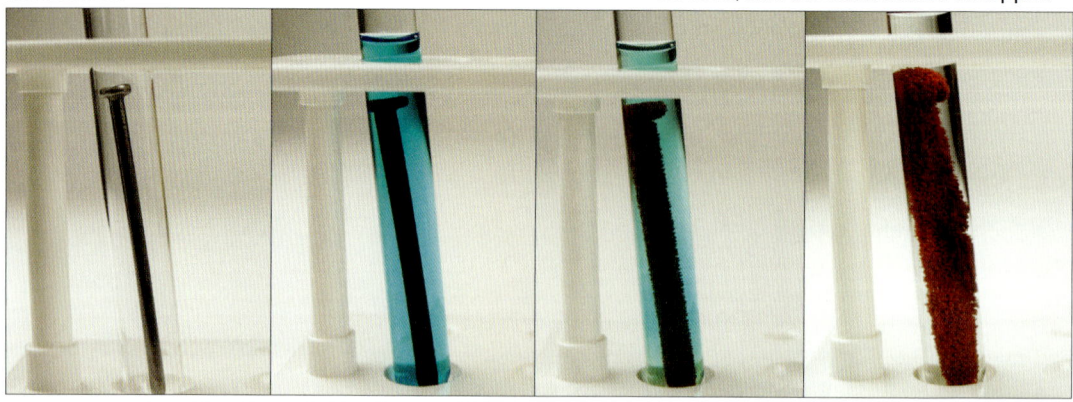

empty test tube with nail   CuCl₂ solution added   Cu precipitation underway   Cu precipitation complete

compound. Typically, this happens in aqueous solutions, and may be generically represented by the equation

A + BC → B + AC

Three common single replacement reactions are: one metal replacing another in solution, as in the copper(II) chloride example above; a metal replacing some of the hydrogen atoms in water to form a hydroxide solution; and a metal replacing the hydrogen atoms in an acid to form a salt.

Iron replacing copper in solution in Figure 7.16 is a typical example of the first type. The equation for the reaction is as follows:

$$Fe(s) + CuCl_2(aq) \rightarrow Cu(s) + FeCl_2(aq)$$

In this single replacement reaction, the copper(II) chloride is in aqueous solution. As we explore further in Chapter 10, this means the copper and chlorine ions have *dissociated* in the water—the crystal comes apart and the individual ions are separated from each other. Iron is a more active metal than copper, so when the steel nail is placed in the solution, the Fe atoms are oxidized to become Fe^{2+} ions in solution. The Cu^{2+} ions in the solution are reduced to become Cu atoms, copper metal. The steel nail provides a landing place where atoms of solid Cu can attach to a crystal lattice as they precipitate.

The reaction of sodium in water is a well-known example of the second type of single replacement reaction. The equation is:

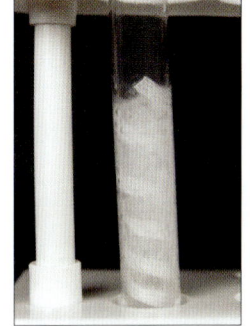

$$2Na(s) + 2H_2O(l) \rightarrow 2NaOH(aq) + H_2(g)$$

Figure 7.17. Magnesium replaces the hydrogen in the hydrochloric acid molecules, forming a salt and evolving hydrogen gas. The photo on the left is at the beginning of the reaction, when H_2 is rapidly evolving. The photo on the right was taken after about 5 minutes.

Sodium is high in the activity series, and reacts very vigorously in water. The hydrogen gas evolves quickly and a lot of heat is released—so much, in fact, that the hydrogen easily combusts, causing an explosion.

The third common single replacement reaction is when a metal replaces the hydrogen in an acid. A typical example is the formation of the metal salt magnesium chloride when magnesium is placed in hydrochloric acid:

$$Mg(s) + 2HCl(aq) \rightarrow MgCl_2(aq) + H_2(g)$$

This reaction is shown in Figure 7.17. The bubbles you see are the evolving hydrogen gas. As you see from Table 7.2, every metal in the activity series from lithium down through lead reacts this way with acids. The last seven metals listed generally do not. However, both platinum and gold dissolve in a mixture of acids known from the medieval period as *aqua regia*, Latin for "royal

Figure 7.18. *Aqua regia.*

Chapter 7

> ### Hmm... Interesting.   A story about aqua regia
>
> After 1935, Adolph Hitler prohibited Germans from accepting Nobel Prizes. German physicists Max von Laue and James Franck sent their Nobel Prize gold medals to Niels Bohr at the Bohr Institute in Denmark for safe keeping. In 1940, Germany invaded Denmark. Fearing that the Nobel Prize medals might be confiscated by the Nazis, Bohr directed Hungarian chemist George de Hevesy to dissolve the medals in *aqua regia*, and place the flask of red liquid with the gold in it on a shelf in the lab. The Nazis never knew what was in that flask and after the war it was still sitting there. The Nobel Foundation took back the gold and recast it into two Nobel Prize medals.
>
>

water." This extremely corrosive substance, shown in Figure 7.18, is composed of one part nitric acid and two or three parts hydrochloric acid.

Here is one final type of single replacement reaction: the replacement of one halogen by another. The chemical activity of the halogens is highest with fluorine and decreases down Group 17 in the periodic table. Thus, any halogen replaces any of the ones below it. This is illustrated by fluorine gas replacing the chlorine atoms in sodium chloride:

$$F_2(g) + 2NaCl(aq) \rightarrow 2NaF(aq) + Cl_2(g)$$

When the fluorine replaces the chlorine in solution, chlorine gas evolves.

7.2.5 Double Replacement Reactions

Double replacement reactions may be represented by the following generic equation:

$$AB + CD \rightarrow AD + CB$$

In this type of reaction, the four ions in two compounds all "switch partners." Notice that in the product ionic compounds in this generic equation, the cations A and C are still written first.

Double replacement reactions often result in the formation of a precipitate or the evolution of a gas when product compounds are insoluble in water. One example is the reaction discussed in Example 7.4. Another, also involving copper(II) nitrate, is the following reaction:

$$Cu(NO_3)_2(aq) + H_2S(aq) \rightarrow CuS(s) + 2HNO_3(aq)$$

The copper monosulfide[1] product is a black powder that is insoluble in water, so it forms a precipitate that turns the solution black, as shown in the third test tube from the left in the image on the opening page of this chapter, and in Figure 7.19.

The evolution of a gas in a double replacement reaction is illustrated by the reaction of iron(II) sulfide and hydrochloric acid:

$$FeS(s) + 2HCl(aq) \rightarrow H_2S(g) + FeCl_2(aq)$$

---

1 There are quite a few different copper sulfide compounds. To distinguish CuS from the others, the prefix *mono-* is added to the sulfide in the naming of the compound.

Notice that here, just as in the Mg and HCl single replacement reaction, a metal replaces hydrogen in an acid. In the single replacement reaction, a sample of pure metal is placed in the acid solution. Here, an ionic compound of the metal is placed in the acid solution.

7.2.6 Combustion Reactions

Combustion reactions involve the reaction of a fuel with oxygen to produce carbon dioxide and water. Combustion reactions are not actually a separate class of reactions; they are oxidation-reduction reactions (discussed below), and some are also synthesis reactions. But they are so common and important that they warrant their own section here.

The natural gas commonly burned in homes for heating and cooking is primarily composed of methane, CH_4. The methane combustion reaction is as follows:

$$CH_4(g) + 2O_2(g) \rightarrow CO_2(g) + 2H_2O(g)$$

The combustion of most fuels, such as methane, propane, and all other hydrocarbon fuels, produces carbon dioxide and water. One important exception is the combustion of hydrogen, a synthesis reaction:

$$2H_2(g) + O_2(g) \rightarrow 2H_2O(g)$$

Figure 7.19. The black precipitate copper monosulfide forms when copper nitrate combines with hydrogen sulfide.

Burning hydrogen produces nothing at all but pure water vapor, and this makes hydrogen the ultimate clean fuel. If we could develop efficient and safe ways to prepare, transport, and burn hydrogen, pollution and CO_2 emissions from cars could be eliminated (although the air in cities would be somewhat more humid).

7.2.7 Acid-Base Neutralization Reactions

In Chapter 11, we explore what acids and bases are and look with more detail into acid-base chemistry. However, *acid-base neutralization reactions* are so important that they warrant a brief mention here. Careful definitions for the terms *acid* and *base* must wait until Chapter 11. However, we can say roughly that acids are recognizable as compounds with hydrogen and a halogen, or hydrogen and one of the polyatomic ions. Examples are HCl and H_2SO_4. Equally roughly, bases are recognizable as compounds composed of a metal and hydroxide such as NaOH and $Ca(OH)_2$.

Though it may be tricky to define acids and bases, it is not tricky to define what you get when an acid reacts with a base: you get a *salt*, and often nothing else but water. In fact, that is the definition for a salt—the product of an acid-base reaction. Acid-base chemistry typically takes place in aqueous solution, and is so indicated in the chemical equations. The reaction between an acid and a base is referred to as *neutralization*. Both acids and bases are corrosive. But when an acid and a base are combined in chemically equivalent ratios, their corrosive properties are neutralized as they are converted into a salt and water.

A good example of an acid-base reaction involves the acid in the human stomach, gastric acid, HCl. Gastric acid plays a key role in food digestion by causing the large protein molecules in the foods we eat to unravel so that the digestive enzymes in the stomach can break down the molecules in the food. However, an unbalanced diet sometimes leads to excess gastric acid, a condition known as *acid indigestion* or *heartburn*. For minor symptoms, over-the-counter antacids can be taken to neutralize excess gastric acid. Rolaids, a popular antacid, contains magnesium hydroxide. The gastric acid neutralization reaction from taking Rolaids is as follows:

$$2HCl(aq) + Mg(OH)_2(aq) \rightarrow MgCl_2(aq) + 2H_2O(l)$$

The magnesium chloride salt formed from this reaction is composed of the cation from the base and the anion from the acid. This is always the case in acid-base neutralization reactions.

Tums, another popular antacid, contains calcium carbonate, which, you may recall, is the same compound that composes chalk and limestone. Even though $CaCO_3$ does not contain the hydroxide ion, it still acts as a base to neutralize the HCl in the stomach. The gastric acid neutralization reaction from taking Tums is as follows:

$$2HCl(aq) + CaCO_3(aq) \rightarrow CaCl_2(aq) + H_2O(l) + CO_2(g)$$

The carbonate base in this reaction results in the production of CO_2.

7.2.8 Oxidation-Reduction Reactions

As with acid-base reactions, oxidation-reduction reactions, or *redox* reactions as they are fondly known, are so important that we treat them separately in a later chapter. But since we are talking about different types of chemical reactions here, we need to include a brief look at redox reactions.

Any time an element's oxidation state increases, oxidation occurs. And any time oxidation occurs, reduction occurs as well. These two sentences apply in reverse as well. If an element's oxidation state decreases, reduction occurs, and oxidation also occurs at the same time.

Combustion reactions are always redox reactions because oxidation always occurs. (The oxidation is caused by the oxidizer, or oxidizing agent, which we discussed a few pages back.) For example, here is the reaction for the combustion of propane, C_3H_8:

$$C_3H_8(g) + 5O_2(g) \rightarrow 3CO_2(g) + 4H_2O(g)$$

Let's look at the oxidation states of the carbon in propane and carbon dioxide. The electronegativity of carbon is higher than that of hydrogen, so hydrogen's oxidation state is +1 in propane. There are eight hydrogen atoms in propane, giving a total charge contribution from hydrogen of 8(+1) = +8. The oxidation numbers must all add up to zero, which means carbon's oxidation state in propane, x, must solve the equation: $3x + 8 = 0$, giving an oxidation state of –8/3.

In carbon dioxide, the oxidation state of oxygen is –2, so the oxidation state for carbon must be +4. Thus, the oxidation state of carbon increases from –8/3 to +4. The carbon is oxidized.

The oxidizing agent in this reaction is O_2. In redox reactions, the oxidizing agent is the compound that causes the oxidation of another compound in the reaction. The result for the oxidizer itself is that it is reduced. We can easily see this in the chemical equation because as O_2, the oxidation state of oxygen is 0, but its oxidation state in both CO_2 and H_2O is –2. The oxygen is reduced.

Note that although the carbon is the element oxidized in this reaction, it is common parlance to say that the *propane* is oxidized. I know this sounds confusing, but it is just the way everyone talks, so students must simply get used to it. At first, this kind of language will drive you crazy because you will be wondering about specifically which element is the one that is oxidized in the reaction. To preserve your sanity, just sit down and figure it out like we did above with the carbon in propane. It takes a lot of practice for one's intuitions about oxidation and reduction to become automatic. Just stay with it.

Chemical Reactions and Stoichiometry

Tossing a metal into an acid solution also produces a redox reaction. Let's look again at the Mg and HCl reaction from a few pages back:

$$Mg(s) + 2HCl(aq) \rightarrow MgCl_2(aq) + H_2(g)$$

Here the pure metal Mg is oxidized to Mg^{2+} to form the ionic compound $MgCl_2$. The oxidizing agent is the hydrochloric acid, but here we should note that more specifically it is the hydrogen in the HCl that is reduced. In HCl, hydrogen's oxidation state is +1; in H_2 it is 0. Thus, the hydrogen is reduced.

The number of different redox reactions is beyond telling. We look at a few more of them in the redox chapter.

7.3 Stoichiometry

Everyone who has ever taken a chemistry class remembers performing stoichiometric calculations. The term *stoichiometry* comes from the Greek words *stoicheion*, meaning "element," and *metron*, meaning "measure." When we do stoichiometry, we are measuring the elements. There are many different basic types of calculations we perform in chemistry, but stoichiometric calculations might just be the most basic of the basic. And here's the good news: even though this section has a long, strange foreign name, the calculations themselves are *easy*.

You really have to know only four things to do stoichiometric calculations. First, you need to be able to set up and solve a proportion, a basic math skill you no doubt learned how to do years ago in Prealgebra. Second, you need to be able to compute the molar mass of a compound. This topic is covered in Chapter 2 and should not be difficult at this point. Third, you need to be able to convert from moles to grams and vice versa. This, too, is covered in Chapter 2 and is not difficult. Finally, you need to be able to set up and balance a chemical equation, which we have just covered and which you have (no doubt) just mastered.

So now, with your confidence meter reading "high," let us proceed.

7.3.1 Stoichiometric Calculations

Stoichiometry is all about calculating the quantities of compounds involved in chemical reactions. For example, just look back at previous page at the reaction of C_3H_8 and O_2. If we have, say, 10.0 kg of propane, how much oxygen is required to burn it all? How much CO_2 and water are formed in the process? These calculations are standard stoichiometry.

Let's begin with RULE NUMBER 1.

RULE NUMBER 1
Perform stoichiometric calculations in moles.

If the problem asks for masses in grams, or supplies you with masses in grams, that's fine. If your given quantities are masses in grams, you simply convert given masses into number of moles, as you already know how to do. If you are required to state your answer as a mass in grams, you just convert to grams at the end. But the stoichiometric calculation is performed in moles.

Now for RULE NUMBER 2.

RULE NUMBER 2
The mole ratios for performing the stoichiometric calculations come from the coefficients in the balanced chemical equation.

That's right. That's why this is easy. Remember that while we are balancing a chemical equation, we think of the coefficients as helping us to figure *numbers of atoms*. But remember also that earlier in this chapter, in Section 7.1.4 on balancing chemical equations, I write that chemical reactions happen with bulk quantities of atoms, and in chemistry our favorite bulk quantity is moles. So think of the coefficients in the balanced equation as telling us the *numbers of moles of compounds* or elements that participate in the reaction.

Now, thinking in moles, look again at that propane equation on the previous page. The coefficients in the equation say this: *one* mole of propane reacts with *five* moles of oxygen to produce *three* moles of carbon dioxide and *four* moles of water. That's how you do it. The rest is just the details, which I demonstrate below in a few examples.

Before we start in on the examples, let's briefly consider how to use *mole ratios* to perform calculations when the quantities you are given do not match the coefficients. (They never do.) Looking again at the propane equation, the equation says that for every five moles of oxygen involved, four moles of water form. If you are given a quantity of oxygen and asked to determine how much water forms, you simply use the 5 : 4 ratio as a conversion factor to figure it out. We can write this conversion factor two ways:

$$\frac{5 \text{ mol } O_2}{4 \text{ mol } H_2O} \quad \text{or} \quad \frac{4 \text{ mol } H_2O}{5 \text{ mol } O_2}$$

Now here's how you use one of these ratios to solve a problem. Suppose the problem is to determine how many moles of water are produced by the reaction of 13.55 mol O_2 with propane, assuming an unlimited supply of propane. Just take your given quantity and multiply it by the mole ratio that appropriately cancels with the given quantity to give you the quantity you need:

$$13.55 \text{ mol } O_2 \cdot \frac{4 \text{ mol } H_2O}{5 \text{ mol } O_2} = \frac{13.55 \cdot 4}{5} \text{ mol } H_2O = 10.84 \text{ mol } H_2O$$

Like I said, easy. Now two more comments before we proceed. First, it should be pretty obvious to you that I write the mole ratio in that computation as I do because I am given moles of oxygen and I want that to cancel to give me moles of water. That's why the oxygen is on the bottom in the conversion factor. If I am given moles of water and asked to find moles of oxygen, I flip the conversion factor the other way. Second, the coefficients in the chemical equation are *exact*; they are not approximations. When the equation says 5 mol O_2 react to produce 4 mol H_2O, these figures are exact. Accordingly, they play no part in determining the significant digits to use in your results. In the computation above, the given quantity of oxygen has four significant digits. That's why the result is written with four significant digits.

Now we are ready to dive into some examples. The following examples should illustrate everything you need to know. Keep Rule Number 1 in mind: perform the computations in moles. If you are given a quantity as a mass, use the compound's molar mass to convert it to number of moles for the calculation. If you are asked for a result to be stated as a mass, use the molar mass to convert it at the end after you determine the result in moles.

▼ Example 7.7

In submarines and spacecraft, the breathing air has to be "scrubbed" of excess carbon dioxide to remove the carbon dioxide continuously being exhaled by those on board. One way to purify the breathing air of excess CO_2 is with the compound lithium hydroxide, shown in Figure 7.20. Solid lithium hydroxide reacts with carbon dioxide to produce solid lithium carbonate and water. If the average amount of carbon dioxide exhaled each day is 880 g/person, determine the number of moles of lithium hydroxide consumed per person per day in the reaction.

First, the given quantity is in grams, so we convert this to number of moles. Referring to the periodic table, we determine the molar mass of CO_2, which is 44.010 g/mol. Converting the given quantity to moles, we have

$$880 \text{ g } CO_2 \cdot \frac{1 \text{ mol}}{44.010 \text{ g}} = 20.0 \text{ mol } CO_2$$

Figure 7.20. Lithium hydroxide.

We know our final answer must be stated with two significant digits because of the given quantity. I have kept an extra significant digit in the gram/mole conversion because it is an intermediate result. Final rounding is performed at the end.

Next, write the formula equation:

$$LiOH(s) + CO_2(g) \rightarrow Li_2CO_3(s) + H_2O(l)$$

Then, balance the equation.

$$2LiOH(s) + CO_2(g) \rightarrow Li_2CO_3(s) + H_2O(l)$$

Now we are ready for the stoichiometry. In the problem, we are given a quantity of CO_2 and asked to find the corresponding quantity of LiOH in the reaction. From the coefficients in the balanced equation, we see that the mole ratio of LiOH to CO_2 in the reaction is 2 : 1. So we use this mole ratio as a conversion factor to convert the mole quantity of CO_2 into a corresponding mole quantity of LiOH.

$$20.0 \text{ mol } CO_2 \cdot \frac{2 \text{ mol LiOH}}{1 \text{ mol } CO_2} = 40.0 \text{ mol LiOH}$$

This is our final result, but it is stated with three significant digits. However, we require two significant digits in our result, and the only way to round 40.0 down to two significant digits is to write it in scientific notation. Thus, our answer is

4.0×10^1 mol LiOH

▲

▼ Example 7.8

One of the body's metabolic processes—from which we get our fuel—is the oxidation of solid glucose by reaction with the oxygen we get from breathing.[2] Glucose, $C_6H_{12}O_6$, is one of the fundamental carbohydrates we take in from eating starch foods such as potatoes and rice. The products of the oxidation reaction are carbon dioxide, which we exhale, and water. (We eliminate more water than we take in; the extra water comes from the oxidation of glucose.) Determine the mass of water produced by the oxidation of 2.00 g glucose.

---

2 Here is another instance of using the term *oxidation* in reference to an entire molecule. If you look at the oxidation states, you see that it is the carbon in the glucose that is being oxidized.

Again, we are given a quantity in grams, so we convert this to moles right at the beginning. Using data from the periodic table, the molar mass of glucose is found to be 180.157 g/mol. Converting the given mass of glucose into moles, we have

$$2.00 \text{ g } C_6H_{12}O_6 \cdot \frac{1 \text{ mol } C_6H_{12}O_6}{180.157 \text{ g } C_6H_{12}O_6} = 0.01110 \text{ mol } C_6H_{12}O_6$$

Again, I have an extra significant digit in this intermediate calculation to avoid the build up of rounding error. I round to the required three significant digits at the end. Next, we write the formula equation for the oxidation of glucose:

$$C_6H_{12}O_6(s) + O_2(g) \rightarrow CO_2(g) + H_2O(l)$$

Now we balance the equation:

$$C_6H_{12}O_6(s) + 6O_2(g) \rightarrow 6CO_2(g) + 6H_2O(l)$$

This problem gives us a quantity of glucose and asks about a quantity of water, so these are the two compounds to look at in the equation. From the coefficients in the balanced equation, we see that the mole ratio of glucose to water in this reaction is 1 : 6. Now we write down the given quantity of glucose (in moles) and use the mole ratio to convert this into a mole quantity of water.

$$0.01110 \text{ mol } C_6H_{12}O_6 \cdot \frac{6 \text{ mol } H_2O}{1 \text{ mol } C_6H_{12}O_6} = 0.06660 \text{ mol } H_2O$$

Now that we have the amount of water produced, we simply need to use the molar mass of water to convert this into grams as the problem requires. From the periodic table, we determine that the molar mass of water is 18.02 g/mol. Using this value as a conversion factor, we determine the mass of water produced by the reaction:

$$0.06660 \text{ mol } H_2O \cdot \frac{18.02 \text{ g } H_2O}{1 \text{ mol } H_2O} = 0.06660 \cdot 18.02 \text{ g } H_2O = 1.20 \text{ g } H_2O$$

Notice, finally, that this result is rounded to three significant digits as the problem requires.

It should be apparent from these two examples that a stoichiometric calculation can be performed using any or all of the products or reactants in a reaction. It doesn't matter which ones are asked about in the problem; you simply set up mole ratios using two compounds (or elements) at a time. We use this principle in our final example.

▼ Example 7.9

The photosynthesis reaction in plants is identical to the oxidation of glucose running in reverse. In the plant, CO_2 from the air and water from the soil are converted (using sunlight as an energy source, of course) into glucose, which becomes fuel and building material for the plant, and oxygen, which is released into the atmosphere. For each 5.00 g H_2O consumed by a plant, determine the masses of the CO_2 consumed, the O_2 released, and the glucose produced.

Since this reaction is identical to the one in the previous problem, only running in reverse, we can write down the balanced equation immediately:

$$6CO_2(g) + 6H_2O(l) \rightarrow C_6H_{12}O_6(s) + 6O_2(g)$$

In this example, we are required to calculate everything, so we may as well calculate all four molar masses right now. Since the mass given in the problem has three significant digits, we write down each of the molar masses with four significant digits.

CO_2: 44.01 g/mol

H_2O: 18.02 g/mol

$C_6H_{12}O_6$: 180.2 g/mol

O_2: 32.00 g/mol

Next, let's convert the given quantity from grams to moles.

$$5.00 \text{ g } H_2O \cdot \frac{1 \text{ mol } H_2O}{18.02 \text{ g } H_2O} = 0.2775 \text{ mol } H_2O$$

(This value has the extra significant digit required for an intermediate calculation.) We are given an amount of H_2O in the problem, and from this quantity we must calculate quantities of three other substances. So we need mole ratios for water and the three other substances. From the coefficients in the balanced equation, these ratios are

$H_2O : CO_2$ 6 : 6, which is equal to 1 : 1

$H_2O : C_6H_{12}O_6$ 6 : 1

$H_2O : O_2$ 6 : 6, which is equal to 1 : 1

Now we just use the given amount of water, in moles, with one of these ratios to compute the quantities of the other three substances.

$$0.2775 \text{ mol } H_2O \cdot \frac{1 \text{ mol } CO_2}{1 \text{ mol } H_2O} = 0.2775 \text{ mol } CO_2$$

$$0.2775 \text{ mol } H_2O \cdot \frac{1 \text{ mol } C_6H_{12}O_6}{6 \text{ mol } H_2O} = 0.04625 \text{ mol } C_6H_{12}O_6$$

$$0.2775 \text{ mol } H_2O \cdot \frac{1 \text{ mol } O_2}{1 \text{ mol } H_2O} = 0.2775 \text{ mol } O_2$$

The last step is to use the molar masses to convert each of these mole quantities into grams.

Chapter 7

$$0.2775 \text{ mol CO}_2 \cdot \frac{44.01 \text{ g CO}_2}{1 \text{ mol CO}_2} = 12.2 \text{ g CO}_2$$

$$0.04625 \text{ mol C}_6\text{H}_{12}\text{O}_6 \cdot \frac{180.2 \text{ g C}_6\text{H}_{12}\text{O}_6}{1 \text{ mol C}_6\text{H}_{12}\text{O}_6} = 8.33 \text{ g C}_6\text{H}_{12}\text{O}_6$$

$$0.2775 \text{ mol O}_2 \cdot \frac{32.00 \text{ g O}_2}{1 \text{ mol O}_2} = 8.88 \text{ g O}_2$$

These quantities are all rounded to three significant digits as required. We have completed the problem, and there are a lot of calculations involved. A great way to perform an overall error check on the calculations is to verify that the law of conservation of mass in chemical reactions is satisfied. From the equation, the masses of the water and carbon dioxide must add up to equal the masses of the glucose and oxygen.

mass CO_2 + mass H_2O = 12.2 g + 5.00 g = 17.2 g

mass $C_6H_{12}O_6$ + mass O_2 = 8.33 g + 8.88 g = 17.21 g

Rounding that second value to three significant digits, we see that the reaction consumes 17.2 g of reactants and produces 17.2 g of products. Mass conservation is confirmed, so it is highly likely that our calculations are correct.

7.3.2 Limiting Reactant

Let's say you have a portable propane burner and a small canister of propane to fuel it. We know that the combustion of propane is a chemical reaction with oxygen that produces carbon dioxide and water. In the combustion of the propane in this canister, what is the limiting factor that determines how much water and carbon dioxide are produced? By *limiting factor*, what I really mean with this question is *limiting reactant*—the reactant that runs out first, once the reaction starts.

Clearly, the oxygen available for this reaction, which is in the atmosphere, is available without limit. Thus, the limiting reactant is the propane, and when the propane is completely consumed, the reaction ceases. Every chemical reaction has a limiting reactant. One of the reactants is in shorter supply than the others and runs out first. When it does, the reaction ceases. The quantities of products produced by the reaction are determined by the quantity of the limiting reactant available for the reaction.

The limiting reactant is also often called the *limiting reagent* (pronounced ree-A-gent), a reagent simply being one of the compounds consumed in the reaction.

If you have known quantities available for a chemical reaction, you can determine which reactant is the limiting reactant simply by looking at the mole quantities available and the mole ratios from the balanced chemical equation. Once you know which compound is the limiting reactant, you can calculate the quantities of the products produced by the reaction based on the quantity of the limiting reactant supplied. As with the stoichiometric calculations we just reviewed, all the calculations associated with determining the limiting reactant are performed in moles. When you are given mass quantities, begin your solution by converting masses into numbers of moles.

Chemical Reactions and Stoichiometry

▼ Example 7.10

One of the most important industrial chemical processes in the world is the Haber-Bosch process for producing ammonia from nitrogen in the air. Grains need nitrogen as a nutrient, but cannot get it from the air because the nitrogen molecules are held together by the strong nitrogen triple bond. Bacteria in the soil produce the "fixed" nitrogen that plants use, but bacteria cannot supply the large quantities of nitrogen needed for industrial scale agriculture. The nitrogen-based fertilizers used are made from ammonia. Some have estimated that one third of the world's population is sustained by food grown using fertilizer produced by ammonia made with the Haber-Bosch process.

The process occurs at extremely high pressures—around 200 atmospheres, or 3,000 psi—in reactors like the historical 1921 reactor shown in Figure 7.21. For years chemists thought the reaction was impossible, but German chemist Fritz Haber solved the problem in 1909, and with engineering assistance from Carl Bosch, had the commercial production of ammonia up and running by the end of 1910. The reaction is as follows:

$$N_2(g) + 3H_2(g) \rightarrow 2NH_3(g)$$

Consider a test run of this process, with 652 kg N_2 and 175 kg H_2 available for the reaction. Determine whether the N_2 or the H_2 is the limiting reactant. Then assuming that the limiting reactant is completely consumed in the reaction, calculate the mass of ammonia produced by the time the reaction ceases.

We begin by converting our mass quantities from kilograms to grams, and from grams to moles. We use the periodic table to determine the molar masses of H_2 and N_2, and then we use these as conversion factors to get mole quantities:

Figure 7.21. This high-pressure reactor from 1921 is on display at the Karlsruhe Institute of Technology in Germany.

$$175{,}000 \text{ g } H_2 \cdot \frac{1 \text{ mol } H_2}{2.016 \text{ g } H_2} = 86{,}810 \text{ mol } H_2$$

$$652{,}000 \text{ g } N_2 \cdot \frac{1 \text{ mol } N_2}{28.01 \text{ g } N_2} = 23{,}280 \text{ mol } N_2$$

These mole quantities are each written with four significant digits, one more than required in the final result. Looking now at the balanced equation for the reaction, we see that the ratio of hydrogen to nitrogen required is 3 : 1. In order to consume the 23,280 moles of nitrogen we have available, we need 3(23,280) = 69,840 moles of hydrogen. We have a lot more hydrogen than this available, so nitrogen is the limiting reagent.

Now we perform a standard stoichiometric calculation to determine the amount of ammonia produced from 23,280 mol N_2. The mole ratio of N_2 to NH_3 is 1 : 2, so we use this ratio as a conversion factor to determine the amount of NH_3 produced:

$$23{,}280 \text{ mol } N_2 \cdot \frac{2 \text{ mol } NH_3}{1 \text{ mol } N_2} = 46{,}560 \text{ mol } NH_3$$

Finally, we convert this mole quantity to mass as the problem requires. This requires calculating the molar mass of NH$_3$ to use as a conversion factor.

$$46{,}560 \text{ mol NH}_3 \cdot \frac{17.03 \text{ g NH}_3}{1 \text{ mol NH}_3} = 792{,}900 \text{ g NH}_3 \cdot \frac{1 \text{ kg}}{1000 \text{ g}} = 793 \text{ kg NH}_3$$

This final result is rounded to three significant digits as the given information requires.

Example 7.11

A 5.00-g strip of magnesium metal is placed in an aqueous solution containing 15.0 g ZnCl$_2$, causing the following reaction:

$$Mg(s) + ZnCl_2(aq) \rightarrow Zn(s) + MgCl_2(aq)$$

Determine (a) the limiting reactant, (b) the mass of MgCl$_2$ produced, and (c) the mass of the excess reactant that remains.

Converting the given quantities to moles, we have:

$$5.00 \text{ g Mg} \cdot \frac{1 \text{ mol Mg}}{24.305 \text{ g Mg}} = 0.2057 \text{ mol Mg}$$

$$15.0 \text{ g ZnCl}_2 \cdot \frac{1 \text{ mol ZnCl}_2}{136.3 \text{ g ZnCl}_2} = 0.1101 \text{ mol ZnCl}_2$$

The mole ratio for Mg and ZnCl$_2$ in the balanced equation is 1 : 1, so the ZnCl$_2$ is the limiting reactant.

The mole ratio of ZnCl$_2$ and MgCl$_2$ is also 1 : 1, so 0.1101 mol ZnCl$_2$ produces 0.1101 mol MgCl$_2$, giving a mass of

$$0.1101 \text{ mol MgCl}_2 \cdot \frac{95.21 \text{ g MgCl}_2}{1 \text{ mol MgCl}_2} = 10.5 \text{ g MgCl}_2$$

Since the mole ratio for Mg and ZnCl$_2$ is 1 : 1, 0.1101 mol ZnCl$_2$ consume 0.1101 mol Mg. The reaction began with 0.2057 mol Mg, so 0.2057 − 0.1101 mol Mg = 0.0956 mol Mg are left over. This converts to a mass of

$$0.0956 \text{ mol Mg} \cdot \frac{24.305 \text{ g Mg}}{1 \text{ mol Mg}} = 2.32 \text{ g Mg}$$

7.3.3 Theoretical Yield and Percent Yield

In the example of the Haber-Bosch process to produce ammonia, the 793 kg of ammonia we calculated that are produced by the reaction is called the *theoretical yield*. Let's now assume we perform the reaction using the quantities of reactants listed in that example and find that

Chemical Reactions and Stoichiometry

we actually end up with 742 kg NH_3 when all the N_2 is consumed and the reaction ceases. This quantity is called the *actual yield*. We use this quantity to calculate the *percent yield* as follows:

$$\text{percent yield} = \frac{\text{actual yield}}{\text{theoretical yield}} \times 100\%$$

Accordingly, the percent yield associated with an actual yield of 742 kg NH_3 is:

$$\text{percent yield} = \frac{742 \text{ kg } NH_3}{793 \text{ kg } NH_3} \times 100\% = 93.6\%$$

When chemists develop procedures to synthesize compounds, the percent yield is a measure of the effectiveness of the procedure. Yields above 80% are considered very good.

Chapter 7 Exercises

SECTION 7.1

1. Butane, C_4H_{10}, is the fuel used in disposable lighters. Butane burns in oxygen to produce carbon dioxide and water. Write the formula equation for this reaction and balance it.

2. Iron(II) sulfide reacts with hydrochloric acid to produce hydrogen sulfide and iron(II) chloride. Write the formula equation for this reaction and balance it.

3. Acetylene, C_2H_2, is a gas that produces an extremely hot flame when combusting with oxygen, so hot that cutting torches use this gas to cut through steel. The products of the combustion are carbon dioxide and water. Write the formula equation for this reaction and balance it.

4. With the aid of a catalyst, liquid methanol, CH_3OH, is produced from a reaction of carbon monoxide gas and hydrogen gas. Write the formula equation for this reaction and balance it.

5. Potassium carbonate is used to neutralize the sulfuric acid in well water caused by acid rain. The products of the neutralization reaction are potassium sulfate, carbon dioxide, and water. Write the formula equation for this reaction and balance it.

6. The tungsten metal used as filaments in light bulbs is produced by reacting tungsten oxide with hydrogen gas to produce tungsten metal and water. Write the formula equation for this reaction and balance it.

7. Sodium azide, NaN_3, is the original compound formerly used in automobile air bags. Other compounds were used to initiate the reaction, but the bags were filled by the conversion of solid sodium azide into sodium metal and nitrogen gas. Write the formula equation for this reaction and balance it.

8. Identify the oxidation state for each element in the following compounds, elements, and ions:

   a. PBr_3
   b. As_2O_5
   c. ClO_4^-
   d. F_2
   e. UF_6
   f. CO_3^{2-}
   g. H_3PO_4
   h. $Zn(NO_3)_2$
   i. C_4H_{10}
   j. $Cr_2O_7^{2-}$
   k. KH
   l. Fe

9. Why must chemical equations be balanced?

Chapter 7

SECTION 7.2

10. For the reactions in exercises 1–7 above, identify the reactions that can be classified as synthesis, decomposition, single replacement, or double replacement.

11. In each of the following reactions, determine which element is oxidized and which is reduced. Also, where possible, identify the reaction as synthesis, decomposition, single replacement, or double replacement.

   a. $3Fe(NO_3)_2(aq) + 2Al(s) \rightarrow 3Fe(s) + 2Al(NO_3)_3(aq)$

   b. $PbS(s) + 4H_2O_2(aq) \rightarrow PbSO_4(s) + 4H_2O(l)$

   c. $N_2(g) + 3H_2(g) \rightarrow 2NH_3(g)$

   d. $Cl_2(aq) + 2NaI(aq) \rightarrow I_2(aq) + 2NaCl(aq)$

12. Explain what a redox reaction is.

13. For each of the following reactions, write the formula equation, balance it, and identify the reaction by type.

   a. magnesium hydroxide yields magnesium oxide and water
   b. sodium chloride and sulfuric acid yield sodium sulfate and hydrogen chloride gas
   c. calcium oxide and water yield calcium hydroxide
   d. ammonium bicarbonate and sodium chloride yield sodium bicarbonate and ammonium chloride

14. Use the activity series of metals to determine which of the following reactions occur. For those that do, complete the equation and balance it.

   a. $Ca(s) + O_2(g) \rightarrow$
   b. $Fe(s) + ZnCl_2(aq) \rightarrow$
   c. $Pt(s) + CuSO_4(aq) \rightarrow$
   d. $Al(s) + H_2SO_4(aq) \rightarrow$
   e. $Ni(s) + CuCl_2(aq) \rightarrow$
   f. $Ba(s) + H_2O(l) \rightarrow$
   g. $Cu(s) + FeSO_4(aq) \rightarrow$
   h. $Al(s) + Pb(NO_3)_2(aq) \rightarrow$
   i. $Li(s) + KI(aq) \rightarrow$
   j. $Au(s) + O_2(g) \rightarrow$

15. Assuming the following compounds are the result of decomposition reactions, identify the reactant, write the formula equation, and balance it.

   a. lithium chloride and oxygen (Hint: The chlorine and oxygen are in an oxyanion.)

b. copper(II) oxide and water (Hint: The hydrogen and oxygen are in an oxyanion.)

c. carbon dioxide and water (Hint: Think about soft drinks.)

SECTION 7.3

16. Silver bromide is used on film for producing holograms. If 3.5 mol silver nitrate react in a double replacement reaction with excess sodium bromide to produce silver bromide, what mass of silver bromide is formed?

17. Hydrogen, the perfect clean fuel, combusts with oxygen to produce water vapor. If 1,575 mol H_2 react, determine the mole quantities of oxygen required and water produced.

18. Aluminum sulfide reacts with water to produce dihydrogen sulfide (usually just called hydrogen sulfide) and aluminum hydroxide. If 2.290 kg aluminum sulfide react completely with excess water, what mass of aluminum hydroxide is produced?

19. Aluminum hydroxide is a popular compound for use in antacid tablets. This compound reacts with stomach acid, HCl, to produce the salt aluminum chloride and water. If 750 mg aluminum hydroxide are consumed, determine:

   a. the number of moles of hydrochloric acid neutralized.

   b. the mass of water produced.

20. Iron ore contains iron(III) oxide along with several other substances. The iron is reduced by heating it to about 1,250°C in the presence of carbon monoxide (a so-called *reducing atmosphere*) in a blast furnace. This process, which has been in use since around 1491, is shown to the right. The reaction produces iron metal and carbon dioxide. What is the theoretical yield in kilograms of iron from this reaction if 2.50×10^4 kg iron(III) oxide react with excess carbon monoxide?

21. A common acid-base reaction is between sulfuric acid and sodium hydroxide to produce sodium sulfate and water. If 29.55 g sodium hydroxide react with 44.11 g sulfuric acid, determine:

   a. the limiting reactant.

   b. the mass of sodium sulfate produced.

   c. the mass of water produced.

22. Octane, C_8H_{18}, is the major component of gasoline. The combustion reaction between octane and oxygen is:

$$2C_8H_{18}(l) + 25O_2(g) \rightarrow 16CO_2(g) + 18H_2O(g)$$

   a. How many moles of oxygen gas are required to burn 350.0 moles of octane?

   b. How many kilograms of water are produced if the combustion consumes 3.5 kg O_2?

c. The density of octane is 0.692 g/mL. Determine the mass of oxygen required to burn 11.5 gal octane. (Recall, conversion factors are in Appendix A, Table A.3.)

23. Nitric acid is used in the process of *nitration*, in which organic molecules are converted into explosive compounds. The first step in the industrial production of nitric acid involves the following reaction between ammonia and oxygen to produce nitric oxide and water:

$$4NH_3(g) + 5O_2(g) \xrightarrow{cat} 4NO(g) + 6H_2O(g)$$

Assume that 855 g NH_3 are combined with 1,750 g O_2 in this reaction. Determine:

a. the limiting reagent.

b. the theoretical yield of nitric oxide, assuming all the limiting reagent is consumed.

c. the percentage yield of NO, assuming 1,272 g NO are actually produced.

24. Aluminum hydroxide undergoes a double replacement reaction with sulfuric acid. Assume that 31.8 g sulfuric acid are combined with 25.4 g aluminum hydroxide with 100% yield. Determine the following:

a. the limiting reactant.

b. the mass of excess reactant remaining.

c. the mass of each product formed.

25. Dinitrogen tetroxide, N_2O_4, is an important rocket fuel. Its structure is pictured to the right. On one of the Apollo lunar missions, N_2O_4 powered the lunar landing module as it ascended from the surface of the moon. During the ascent, 1.200×10^3 kg N_2H_4, 1.000×10^3 kg $(CH_3)_2N_2H_2$, and 4.500×10^3 kg N_2O_4 were available for the following reaction:

$$2N_2H_4 + (CH_3)_2N_2H_2 + 3N_2O_4 \rightarrow 6N_2 + 2CO_2 + 8H_2O$$

a. Which of the compounds was consumed first in this reaction?

b. How many kilograms of waste water were spewed into space by the reaction?

c. How many nitrogen atoms were left in space by the reaction?

26. For reasons we explore in a later chapter, hydrofluoric acid is classified as a weak acid. Nevertheless, HF is so corrosive it eats right through glass, silicon dioxide, SiO_2, and is used for glass etching. (Needless to say, HF is not stored in glass bottles.) HF has also been used for etching silicon wafers in the process of making semiconductors, although the use of HF has waned in recent years. The reaction between HF and SiO_2 is

$$SiO_2(s) + 6HF(aq) \rightarrow H_2SiF_6(aq) + 2H_2O(l)$$

a. How many moles of HF are needed to react with 542 g SiO_2?

b. How many grams of H_2SiF_6 form when 4.25 mol HF react completely?

c. How many moles of HF are required to produce 2.0 gallons of water? (Remember, the density of water is 0.998 g/mL at room temperature.)

d. If 2.50 kg HF are available to react with 1.205 kg SiO_2, what is the limiting reactant?

e. What is the theoretical yield of H_2SiF_6 if the limiting reagent is completely consumed, assuming the reactant quantities listed in part (d)?

f. Using the theoretical yield from the previous question, what is the percentage yield of H_2SiF_6 if 2.657 kg H_2SiF_6 are produced?

27. When benzene, C_6H_6, reacts with bromine (a type of reaction called *halogenation*), the reaction produces bromobenzene, C_6H_5Br, and hydrogen bromide.

    a. If 45.0 g of benzene reacts with 97.5 g of bromine, what is the theoretical yield of bromobenzene?

    b. If the actual yield of bromobenzene is 63.25 g, what is the percent yield of bromobenzene?

28. To monitor levels of ozone, O_3, in the air, an air sample is processed by an instrument that causes ozone to react with sodium iodide in the following reaction:

$$O_3(g) + 2NaI(aq) + H_2O(l) \rightarrow O_2(g) + I_2(s) + 2NaOH(aq)$$

   a. Determine the number of moles of sodium iodide required to process 2.85×10^{-6} moles of ozone.

   b. Determine how many grams of sodium iodide are required to process 3.00 grams of ozone.

   c. If 1,455 g NaI are available to process a sample of 250.0 g O_3 in a reaction with excess water, determine the limiting reactant and the theoretical yield of I_2.

   d. From part (c), determine the number of iodine atoms present in the theoretical yield of I_2.

GENERAL REVIEW EXERCISES

29. Melatonin is a compound secreted naturally in the brain when it is time to go to sleep. The molar mass of melatonin is 232.281 g/mol. A sample of 25.0 mg of melatonin is found to contain 16.81 mg C, 1.736 mg H, 3.015 mg N, and 3.445 mg O.

    a. Determine the percent composition of melatonin.

    b. Determine the molecular formula for melatonin.

    c. Melatonin is sold as an over-the-counter sleeping aid. Each tablet contains 3.0 mg of melatonin. Determine the number of carbon atoms in each tablet of melatonin.

30. Mercury-vapor lamps are widely used for general lighting in gymnasiums, exhibit halls, and other large spaces. One of the wavelengths in the emission spectrum of mercury is a bright green line with an energy of 2.271 eV. Determine the wavelength of this line.

31. Write the condensed electron configurations for yttrium, tin, titanium, and iodine.

32. Write explanations in one or two sentences describing the Aufbau principle, the Madelung rule, and Hund's Rule.

33. Use the information from Table 2.6 to calculate the atomic mass of silicon.

34. Based on your knowledge of general trends in the periodic table, place the following in order of increasing electronegativity: Fe, F, Se, Fr, Ba, and Cl.

35. Draw the Lewis structures for the following molecules and polyatomic ions:

    a. CCl_2S b. PO_4^{3-} c. NO_2^+ d. N_2H_2
    e. O_3 f. SF_6 g. CO h. H_3O^+

36. Identify the electron domain geometry and molecular geometry for each molecule listed in the previous exercise.

Chapter 8
Kinetic Theory and States of Matter

This researcher is exploring a technology known as *chemical vapor deposition* (CVD). This process is used in the manufacturing of semiconductors (computer chips) for use in computers, mobile devices, digital signal processing, and other digital technologies. In CVD processes, a thin layer of metallic (conducting) or nonmetallic (nonconducting) material is deposited directly from the gas state onto a silicon substrate. Often the gaseous material undergoes a reaction at the surface to form a coating of pure metal (such as tungsten), pure glass, or other material. Sometimes plasmas are used to assist the process, in which case the process is called *plasma-enhanced chemical vapor deposition* (PECVD).

Kinetic Theory and States of Matter

Objectives for Chapter 8

After studying this chapter and completing the exercises, you should be able to do each of the following tasks, using supporting terms and principles as necessary.

SECTION 8.1

1. Describe the relationship between temperature and molecular energy.
2. Describe the Maxwell-Boltzmann velocity distribution and relate it to the particle velocities in gases.
3. Describe the basic principles of the kinetic-molecular theory of gases.
4. Explain the cause of gas pressure.
5. State the name, symbol, and definition for the official SI system unit of pressure.
6. Write a paragraph describing how a mercury barometer works.

SECTION 8.2

7. State the four basic states of matter and give a description of each at the molecular level.
8. Describe the forces that hold substances together in the solid and liquid states, distinguishing between forces acting in ionic crystals, metals, and molecular substances.
9. Explain the causes for the following properties of solids: high melting point (crystals and metals), low melting point (molecular substances), high density, incompressibility, and definite shape.
10. Define and explain the cause of *surface tension* and use surface tension to explain capillary action and wicking.
11. Define and give examples of *diffusion* in liquids and gases.
12. State common examples of plasmas.
13. State and describe the six basic phase transitions that occur between solids, liquids, and gases.
14. On a P-T phase diagram, locate the phase transition points for given pressures or temperatures.
15. Define *triple point*, and locate the triple point on a P-T phase diagram.
16. Define *critical point*, *critical temperature*, and *critical pressure*, and locate the critical point on a P-T phase diagram.
17. Describe the unique feature of the P-T phase diagram for water.
18. Given appropriate thermal constants, calculate the heat that must be added or removed to raise or lower the temperature in a given mass of solid, liquid, or gas substance.
19. Given appropriate thermal constants, calculate the heat that must be added or removed to effect a phase transition in a given mass of material.
20. Define and explain the causes of *evaporation* and *vapor pressure*.
21. Explain why water boils at lower temperatures in high-altitude locations.

Chapter 8

8.1 Temperature, Kinetic-Molecular Theory, and Pressure

8.1.1 Temperature and Molecular Energy

So far we have encountered several different physical mechanisms by which substances can possess energy. In Chapter 3, we saw that the energy atoms possess can be modeled as kinetic energy in the motion of the atoms, or as the energy of electrons when excitation pushes electrons into excited states. In more recent chapters, we have seen that the energy in bulk substances resides in the lattice energy of crystals or bond energy of molecules or metals. In this chapter, we look more closely at the first of these—the energy of molecular motion.

According to *kinetic-molecular theory*, we understand the temperature of a substance to be directly related to the kinetic energy in the motion of atoms and molecules. In solids, atoms are not free to move around, so atomic motion is manifest as atomic vibrations. When an object vibrates back and forth its energy continuously changes from kinetic energy to potential energy, but we don't need to get into that much detail here. We can just think of vibrating atoms as each having an average kinetic energy defined as

$$E = \tfrac{1}{2}mv^2$$

where v represents some kind of average speed. In fluids—liquids and gases—particles are free to move around, so they possess translational kinetic energy. In molecular fluids, molecules also possess rotational kinetic energy because they tumble as they move. In these cases, molecules are continuously colliding and ricocheting all over the place.

In a given substance, the sum of all the kinetic energies of all the particles in the substance due to all their various motions is called the *internal energy*. The variable we call *temperature* is directly related to the internal energy of the substance. A temperature measurement is not an energy measurement; nevertheless, the temperature of a substance is strictly due to the internal energy of the substance. In fact, the internal energy of a substance can be expressed as a function of temperature, and vice versa.

8.1.2 Velocity Distribution of Gases

A very important application of our theory of molecular motion is in the motion of the molecules in a gas. (I use the term *molecules* because at room temperature, all gases are composed of molecules except for the noble gases. The noble gases do not form chemical bonds, and thus consist of individual atoms. So when I write "molecules" of gas, I mean also to include the individual atoms in noble gases.) Gas molecules are in constant motion and are constantly colliding with each other. In room temperature air for example, each molecule of nitrogen and oxygen collides with another molecule about five billion times per second. In these collisions, energy is transferred from one molecule to another. Just as with pool balls on a pool table, sometimes a molecule speeds up in a collision, a gain in kinetic energy, and sometimes a molecule slows down, a loss of kinetic energy.

Figure 8.1. Scottish physicist James Clerk Maxwell (1831–1879).

At a given temperature, the molecules of a particular gas possess a predictable average speed, but the speeds of individual molecules are spread out in a well-known distribution called the *Maxwell-Boltzmann distribution*. This speed distribution was derived from mathematical theory in 1860 by James Clerk Maxwell (Figure 8.1), the same Scottish physicist who developed Maxwell's equations explaining electricity and magnetism. Ludwig Boltzmann (Figure 8.2) was an Austrian physicist whose name is now synonymous with theory in thermodynamics and statistical mechanics. In the late 19th century, Boltzmann developed much of the explanatory theory supporting Maxwell's speed distribution curves.

The Maxwell-Boltzmann speed distributions for the noble gases at 25°C are shown in Figure 8.3. Notice that the peak of the xenon curve is at around 200 m/s and the peak of the helium curve is at around 1,100 m/s. As a point of reference, the speed of sound in air at this temperature is around 342 m/s.

The first important thing to get from the distribution curves is the fact that the speeds lie in a *distribution*. Corresponding to a given gas temperature is a specific average particle speed, but any particular atom can have a speed lying anywhere in the distribution at any given time. For example, an atom of neon at 25°C can have a speed ranging anywhere from near zero to about 1,300 m/s, with most neon atoms having speeds close to 500 m/s.

Figure 8.2. Austrian physicist Ludwig Boltzmann (1844–1906).

The second important thing to notice from the distributions is that *the heavier the atom, the lower the average speed*. Connect this to what we have just been considering about molecular energy. At a given temperature, the particles of any gas have the same average energy because the temperature of the gas depends only on the energy, and vice versa. The kinetic energy of an atom, $\frac{1}{2}mv^2$, is proportional to the mass and speed of the atom. So on average, heavier atoms move at slower speeds.

8.1.3 The Kinetic-Molecular Theory of Gases

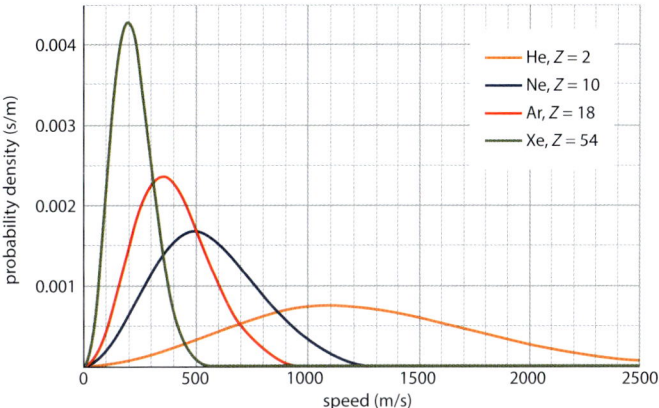

Figure 8.3. The Maxwell-Boltzmann speed distribution for noble gases at 25°C. Interpret the units of measure on the vertical axis on this graph this way: A section of area under the curve represents the proportion of atoms moving with speeds in that speed range. That area is calculated by multiplying the horizontal width of the section by its height. But a proportion is a unit-less number, thus the units of measure on the vertical axis must be the reciprocal of those on the horizontal axis so that they cancel out when multiplied.

The *kinetic-molecular theory of gases* is the major theory supporting our present view of gases, pressure, and the gas laws. The theory as we state it today comes directly from the work of Maxwell and Boltzmann. (Unlike many other scientists at that time, Maxwell and Boltzmann both accepted the actual existence of atoms.) It is important here to remind ourselves what theories are and what role they play in scientific inquiry. As discussed briefly in the Introduction, a scientific theory is a model describing and accounting for some aspect of the natural world. The kinetic-molecular theory accounts very well for the behavior of gases, with some limitations described below.

The following is an outline of the major principles included in the kinetic-molecular theory of gases.

1. A gas consists of a very large number of molecules (or atoms, in the case of noble gases).

2. The total volume occupied by all the molecules of gas in a container is negligibly small compared to the volume of the container. This means we can model the molecules as point particles that occupy no space.

3. When molecules collide, they experience *elastic collisions*. An elastic collision is one in which the total kinetic energy of the objects after the collision is the same as the total kinetic energy prior to the collision. In other words, no energy is converted to other forms of energy in the collision. A molecule might slow down or speed up in an interaction with another molecule, but the total kinetic energy of the colliding molecules is unaffected by the collision and any transfer of energy that occurs.

4. Electrical forces of attraction and repulsion between molecules are negligible. This is due to a couple of factors. One is that neutral molecules have no net charge, so there are no ionic attractions. Another is that on average gas molecules are relatively far apart—something like a thousand times farther apart than the molecules in a liquid are—and thus intermolecular forces are so weak as to be negligible.

5. The average kinetic energy of the molecules in a gas is proportional to the temperature. At any given temperature, the gas molecules all have the same average energy.

One of the immediate results of the kinetic-molecular theory is an explanation for gas pressure (our next topic). Gases that behave according to the assumptions listed here are called *ideal gases*. It turns out that under certain conditions, many gases do exhibit ideal gas behavior. For example, at low pressure there are fewer gas particles in a given amount of space, so the second assumption listed above becomes an accurate description of the conditions of the gas. Also, the more energy the gas particles have (i.e., higher temperature) the less any intermolecular forces affect the particle motion, which means the third and fourth assumptions listed above are accurate descriptions.

In light of the fourth statement above, gases whose molecules are nonpolar exhibit ideal gas behavior to a larger degree than gases whose particles are polar because polar molecules are attracted to one another by Keesom forces, the strongest of the Van der Waals forces. Intermolecular forces between nonpolar molecules are limited to the much weaker London dispersion forces (see Section 6.3.4).

8.1.4 Gas Pressure

According to the kinetic-molecular theory, pressure in gases is caused by the collisions of molecules with each other and with the walls of the container. Higher internal energy means faster moving molecules, resulting in harder collisions, and consequently, higher pressure. This is why heating a gas confined in a container causes the pressure to increase. The heat increases the internal energy and thus also increases the average velocity of the gas particles.

The official SI System unit for pressure is the *pascal* (Pa), named after French scientist, mathematician, and philosopher Blaise Pascal (Figure 8.4). In the 17th century, Pascal developed many of the general principles pertaining to pressure in both liquids and gases. To help demonstrate how pressure is defined and how much pressure a pascal represents, Figure 8.5 depicts a chamber with a freely sliding side wall. The sliding wall is one meter square, so its area is 1 m². Imagine that the chamber contains ordinary air, and that a force of one newton (1 N, which is about 1/4 lb) is applied to the sliding wall. The wall slides in until the force of the air pushing back on the

Figure 8.4. French scientist Blaise Pascal (1623–1662).

wall also equals 1 N. The forces on both sides of the wall are then equal at 1 N each. But the force on the inside of the wall is due to the gas pressure inside (all those molecules of gas colliding with the inside of the wall). The force of 1 N applied by the gas molecules inside the chamber to the area of 1 m² is defined as a pressure of 1 pascal (1 Pa). This is how the pascal is defined. In other words,

$$1\,\text{Pa} = 1\,\frac{\text{N}}{\text{m}^2}$$

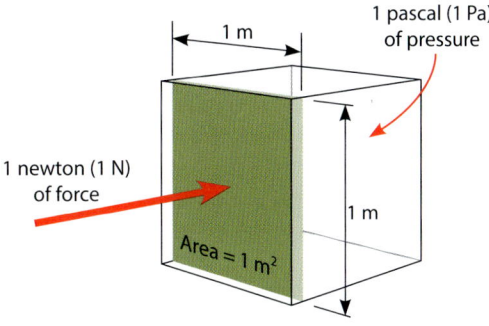

Figure 8.5. One pascal of pressure is one newton of force per square meter.

Scientists have been experimenting with pressure for a long time, and as a result there are many different units for pressure in use, more than for just about any other variable you can think of. Some of the most common units are listed in Table 8.1 along with their unit symbols, definitions, and the average atmospheric pressure in each unit. In the U.S., pounds per square inch (psi) is the most common pressure unit for general use. Weather reporters like to use a scale based on "inches of mercury." The liquid pressure 29.9 inches beneath the surface in a column of mercury with a vacuum on top is equivalent to atmospheric pressure, as explained in the box on the next two pages. Because of the historical significance of barometers and their continued use for measuring atmospheric pressure, I've included some of the history in the box.

Pressure Unit Name (unit symbol)	Definition	Average Atmospheric Pressure at Sea Level in these Units
pascal (Pa)	1 Pa = 1 N/m²	101,325 Pa
kilopascal (kPa)	1 kPa = 1,000 Pa	101.325 kPa
atmosphere (atm)	1 atm = 101,325 Pa	1 atm
bar (bar)	1 bar = 100,000 Pa	1.01325 bar
pounds per square inch (psi)	1 psi = 6894.757 Pa*	14.696 psi
torr (Torr)	1 Torr = 1/760 atm	760 Torr
inches of mercury (in Hg)	height, in inches, of mercury column in a barometer	29.9 in Hg
millimeters of mercury (mm Hg)	height, in millimeters, of mercury column in a barometer (1 mm Hg = 1 Torr)	760 mm Hg

Table 8.1. Common pressure units and their definitions. In the definitions column, the one labeled with an asterisk (*) is approximate; all others are exact.

Hmm... Interesting.   *How barometers work*

The air of the earth's atmosphere is obviously not in a container, but nevertheless air pressure is still caused by the collisions of gas molecules. A helpful way to think about atmospheric pressure is in terms of the earth's gravity pulling the gas molecules toward the earth—the molecules are held against the earth by the force of gravity. The pressure at a given altitude is due to the molecules at that altitude colliding with the molecules just above. The pressure can be modeled in terms of the weight of the gas above. This weight in turn depends on the density of the molecules and the frequency with which they collide with each other. At higher elevations, the earth's gravitational attraction is weaker, so the molecules are more spread out. This lowers the density, reduces the frequency of molecular collisions, and thus reduces the pressure.

Atmospheric pressure varies with the weather from about 2% above normal to about 5% below normal. These variations are the cause of wind, as pressure differences between one region and another cause air to flow toward regions of lower pressure. Low pressures occur when the wind is moving rapidly, so when atmospheric pressure drops, it means stormy weather is headed your way. Atmospheric pressure is measured with a *barometer*, so weather reporters often refer to the value of atmospheric pressure as the *barometric pressure*.

Italian physicist and mathematician Evangelista Torricelli (1608–1647).

The mercury barometer for measuring air pressure was invented in 1643 by Italian physicist and mathematician Evangelista Torricelli. The diagram below shows how we might make one of these devices today, using a vacuum pump to remove the air from a glass tube set into a bowl of mercury. (Torricelli didn't have a vacuum pump, so he accomplished this a different way, but the result is the same.) As the air is removed from the glass tube, the mercury rises into the tube up to a height of approximately 760 mm. The tube is then capped off and the device registers changes in atmospheric pressure

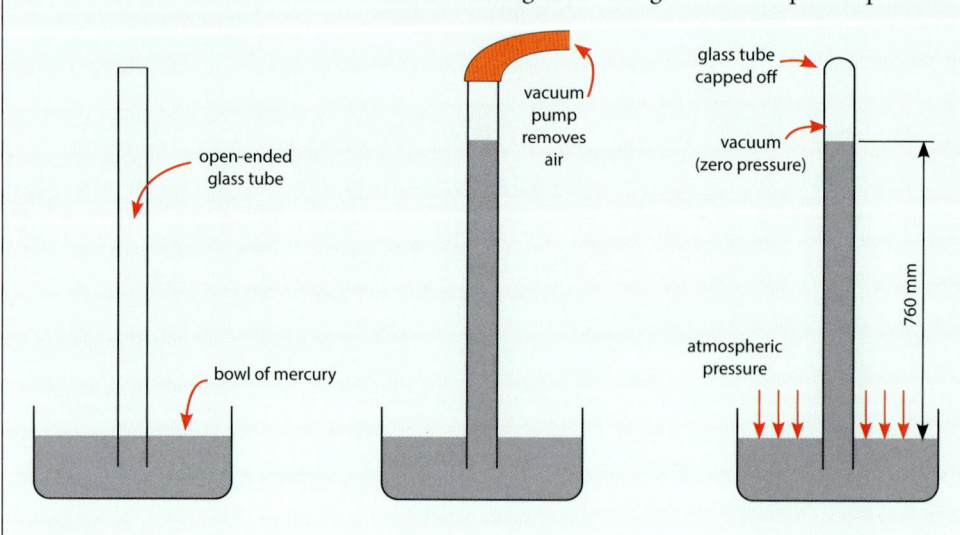

caused by the weather. The pressure of the atmosphere pushes the mercury up into the tube, and because of the vacuum, there is no gas pressure on top of the mercury column to push it back down.

As atmospheric pressure changes, the height of the mercury column in the glass tube rises and falls. At normal atmospheric pressure at sea level, the height of the mercury is 760 mm, which is about 29.9 inches, and sometimes weather reporters refer to atmospheric pressure in terms such as "thirty inches of mercury." Normal barometric pressure variation in the U.S. is from about 725 mm to 775 mm of mercury, which corresponds to a range of 28.5 to 30.5 inches. The higher the atmospheric pressure is, the farther it pushes the mercury up into the barometer tube. So if the local barometric pressure is 30.4 inches of mercury, that is a high-pressure day and the weather is likely to be clear, sunny, and calm. When the barometric pressure falls, it means a change of weather as high-pressure regions push wind and moisture toward the now low-pressure region. You might be interested to know that the world record low barometric pressure (other than the pressure inside a tornado, which is extremely low) is 25.7 inches, measured in 1979 during a typhoon in the Pacific Ocean. The pressure unit *torr* (symbol: Torr) is named after Evangelista Torricelli. Normal atmospheric pressure is equal to 760 torr.

A vacuum is a complete void in space in which there is no matter of any kind. In 1643, scientists did not believe that a vacuum could exist (because of things Aristotle had written long before), so at the time the barometer was invented, no one could formulate a correct explanation for how it worked. The big question was about the mysterious force that holds the mercury up in the tube. Blaise Pascal explained how the barometer works in 1647, theorizing that a vacuum does exist in the glass tube above the mercury. On the basis of his own experiments, Pascal explained that there is no air pressure above the mercury column in the tube, and that the pressure of the atmosphere acting on the surface of the mercury in the bowl pushes the mercury up into the tube into the vacuum region. This explanation is correct, but since at the time no one accepted the possibility of a vacuum, scientists did not want to accept Pascal's bold theory. But as with all scientific theories, a theory gains strength when it repeatedly leads to successful hypotheses, and this one did.

Pascal's explanation is easy to understand if we calculate the pressure in the tube 760 mm below the surface of the mercury. As with the pressure in the atmosphere, the pressure under a stationary liquid can be modeled in terms of the density and depth of the liquid, and the acceleration of gravity, g. If you work out the mathematics of the weight of liquid applied to a certain amount of area you get the standard equation for the pressure at a specific depth under the surface of a liquid:

$P = \rho g h$

where P is the pressure in pascals, ρ is the liquid density in kg/m^3, and h is the depth below the surface in meters. The density of mercury is 13,600 kg/m^3. Using the pressure equation above, the pressure at a depth of 760 mm or 0.76 m comes out to 101,300 Pa, which agrees with the value for atmospheric pressure in Table 6.1 to four significant digits. Thus, the atmospheric pressure outside the tube is equal to the pressure due to the weight of the liquid inside the tube. This calculation demonstrates that the pressure at the same height in the mercury inside and outside the tube is the same, a principle discovered by Pascal.

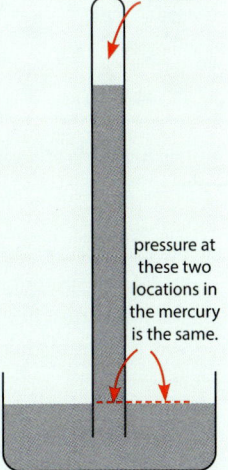

vacuum

pressure at these two locations in the mercury is the same.

Chapter 8

8.2 States of Matter

8.2.1 The Four Basic States of Matter

No doubt you are already familiar with the basic states (or phases[1]) of matter, or at least three of them. Our objective here is for you to be able to distinguish the four states of matter and their characteristics at the atomic or molecular level.

The four basic states of substances are *solid*, *liquid*, *gas* (or *vapor*), and *plasma*. The gas state may be called gas or vapor, depending on whether a substance normally exists in nature as a gas or a liquid. If a substance exists at ordinary temperature and pressure as a gas, we refer to a substance in this state as a gas. But if a substance is a liquid or a solid at ordinary temperature and pressure, we conventionally refer to the substance in the vapor state as a vapor. Steam, for example, is water *vapor* because at room temperature water is a liquid.

The internal energy of a substance is what distinguishes the states from one another. Beginning with a pure substance in the solid state, if the substance is continuously heated, the substance transitions from one state to the next as the kinetic energy of its molecules gets higher and higher. Some substances can exist as solids or liquids, some as liquids or gases, and some as all three. Plasmas are a completely unique state of matter, which we briefly address later.

8.2.2 Solids

If a substance has a low internal energy, this means its temperature is low. If the energy is low enough, the substance will be in the solid state. For a substance normally in the liquid state, we say it is frozen when it is in the solid state. Since the atoms in the solid state have a low internal energy, the atoms' kinetic energies are low and they are vibrating relatively slowly. This low energy state for the atoms means they cling strongly to one another by electrical attraction (in the case of ions in crystals) or metallic bonding (in the case of metals) forming a solid. In this state the atoms are not free to move around; they are stuck in place and can only vibrate where they are. The atoms *do* have kinetic energy, so they are not at rest. But they do not have enough energy to break free of the electrical attractions or metallic bonding holding them all together, so their kinetic energy forces them to vibrate in place. This is what is going on with the atoms in any solid. Figure 8.6 depicts the atoms in an ionic crystal solid.

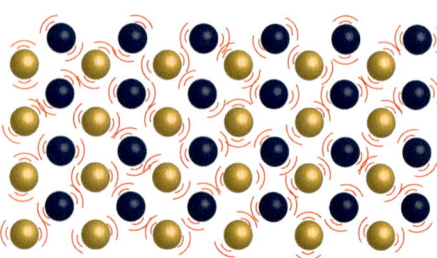

Figure 8.6. Atoms in a solid crystal vibrating in place.

The close packing of atoms in metallic and ionic solids explains their high density relative to gases. Forces of electrical attraction and repulsion in the crystal lattice, or metallic bonding in the metallic lattice, account for the rigidity and *incompressibility* (tendency to withstand being compressed into a smaller volume) of these solids.

Of course, molecules can form solids as well—ice, for example. Many molecular solids such as glass and plastics are amorphous because the molecules are not organized in an orderly fashion the way the atoms are in crystals (ice being a notable exception). Because of their relatively low energy, the molecules in a molecular solid are more closely packed than in molecular liquids. As a result, intermolecular forces (hydrogen bonding and Van der Waals forces) in molecular solids

---

1 Physics texts often refer to "phases" of matter, while chemistry texts tend to use the phrase "states" of matter. These phrases are equivalent. When discussing transitions between states, the phrase *phase transition* is commonly used by everyone.

are much greater than in liquids or gases, and these intermolecular forces hold the molecules together in a solid.

The intermolecular forces in molecular solids are much weaker than the covalent bonds holding the individual molecules together. For this reason, molecular substances have relatively low melting points compared to the much higher melting points of metals and ionic crystals.

We think of solids as having a definite, permanent shape. For the most part, this is correct. However, amorphous solids such as glass can "flow" at an extremely slow rate over an extended period of time. Windows in homes built around 100 years ago are noticeably thicker at the bottom than they are at the top because of the very slow downward flow of the glass due to gravity.

Table 8.2 summarizes a few of the physical properties we have touched on in this section.

Property	Cause
high melting point in metals and ionic crystals	high bonding forces between ions or atoms
relatively low melting point of molecular solids	intermolecular forces in molecular solids are much lower than ionic forces in crystals or metallic bonding forces in metals
high density	close packing of particles due to low kinetic energies
incompressibility (definite volume)	close packing and high forces between particles (relative to gases) resist compression
definite shape	close packing and high forces between particles cause particles to hold their position, maintaining the shape of the substance

Table 8.2. Some physical properties of solids.

8.2.3 Liquids

Liquids are common here on the earth, but liquid is the least common state of matter in the universe. The range of temperatures and pressures where liquids can exist is very narrow, and is generally only found on relatively temperate planets.

At higher internal energies—higher temperatures—the kinetic energies in atoms or molecules of a substance can be high enough to overcome the attractions that bind them together in a solid. Every pure substance has a particular temperature at which the higher kinetic energy in the atoms overcomes the electrical attraction between the atoms, allowing the atoms or molecules to come apart. At this temperature—the *melting point*—a solid transitions to the liquid state. The atoms in a liquid move quickly, and there are a lot of them. The result is that the particles in a liquid, depicted in Figure 8.7, are continuously colliding with each other.

In the liquid state, the atoms or molecules of a substance are still held together loosely by ionic attractions or intermolecular forces, causing liquids to stay together in an open container, puddles, or drops. As mentioned back in the box on page 6 of the Introduction, small liquid drops are nearly spherical because this shape minimizes the surface area of the drop, and thus minimizes the potential energy between molecules or atoms at the surface. This is yet another instance of energy minimization.

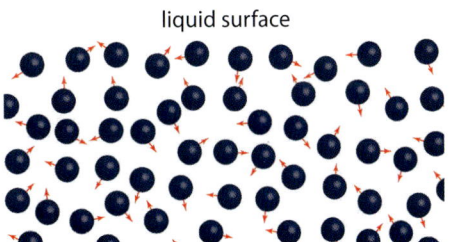

Figure 8.7. Particles in the liquid state.

Chapter 8

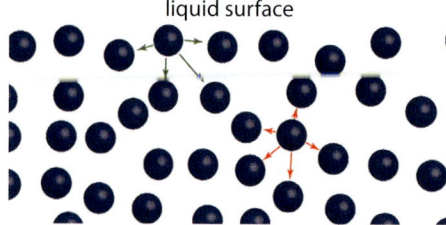

Figure 8.8. Surface tension is due to the fact that the combined effect of the attractive forces on a molecule at the surface of a liquid tends to pull the molecules toward the inside of the liquid. Molecules in the liquid interior tend to be pulled equally in all directions.

Figure 8.8 depicts the unbalanced forces on a molecule at the liquid surface compared to the more balanced forces present on a molecule in the liquid interior. At the surface, the forces tend to hold the surface molecules together with each other and with the liquid interior. The unbalanced forces at the surface give rise to a liquid characteristic called *surface tension*. The surface tension holding surface molecules together is what allows insects to walk on the surface of water. The combination of surface tension and forces between surface molecules and a polar surface such as glass gives rise to several effects, including the ability of small water droplets to cling to a vertical glass surface, the meniscus that forms on the surface of water in a glass container (Figure 8.9), and *capillary action*.

Capillary action occurs in small spaces such as the inside of thin glass tubes or the open spaces in porous rocks or sponges. In a small space, the forces between the surface molecules in the liquid and the polar surface of the solid can pull the liquid up into the space. Examples are seen in the way water flows upward against gravity in a thin glass tube and in porous bricks (Figure 8.10). Back in Section 6.3.3, we discussed the fact that hydrogen bonding between water molecules and the cellulose molecules in cotton is the cause for the high absorbency of cotton. When that absorbency is combined with tiny channels in a fabric that allow liquid surface molecules to be pulled by polar molecules in the fabric, wicking occurs. Wicking is seen in the way that liquid candle wax or lamp oil travels up a wick to be burned, and in the fabrics used in athletic clothing that wick moisture away from the skin.

Figure 8.9. A liquid meniscus, the bowl-shaped curve on a liquid surface, is a result of surface tension and attraction between surface molecules and the polar surface of the glass.

Diffusion, the tendency for molecules spontaneously to mix and mingle, is a phenomenon exhibited by both liquids and gases.[2] In these states, the particles are energetic and free to move around. Their random motions result in complete mixing of a soluble liquid in another liquid, forming a homogeneous mixture (a solution). Diffusion is illustrated by the food coloring added to water in Figure 8.11. With no shaking or stirring at all, the molecules of the dye thoroughly mix with the water molecules after short time.

Figure 8.10. Capillary action draws water up into the pores in a brick.

Intermolecular forces hold the molecules together in a liquid, so the particles stay together in a container. But high particle energies (due to the temperature) mean that a liquid has no fixed shape, so it takes the shape of its container.

---

2 Interestingly, diffusion occurs in solids, too. For example, foreign atoms in a crystal lattice, such as the carbon atoms in the iron lattice of steel alloys, migrate through the lattice. (Alloys are discussed in Section 10.2.5).

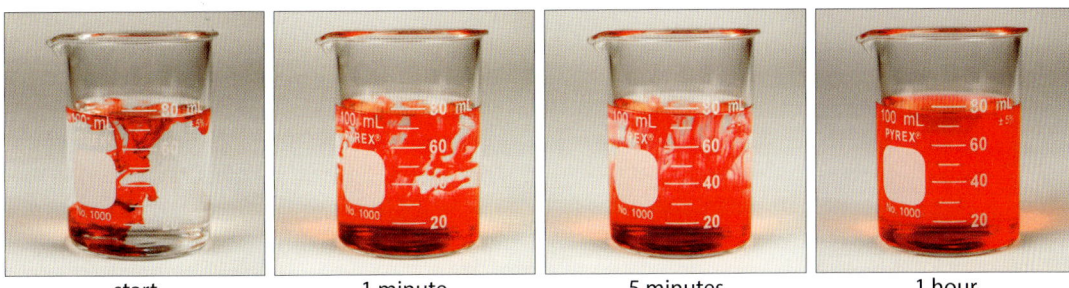

Figure 8.11. Diffusion leads to 100% mixing of a soluble liquid solute, resulting in a homogeneous mixture.

In the liquid state, the molecules of a substance are not much farther apart from each other than they are in the solid state. Thus, liquid densities are very close to solid densities. Further, liquids are essentially incompressible. Even enormous pressure does not decrease the volume of a liquid more than a few percent.

8.2.4 Gases

In gases, the internal energy of the atoms or molecules is so high that it completely overcomes the forces of attraction that hold the particles together in solids and liquids. Instead of staying together, if the particles are not contained they have so much energy they fly away at great speed. Because of their higher speeds, the atoms or molecules spread apart, still colliding with one another, but spreading out as much as they are able. If they are contained in a container of some kind, they still fly around at great speed, but instead of flying off into the atmosphere, they furiously collide with each other and with the walls of the container, creating pressure inside the container.

The average distance between particles in gases is on the order of 1,000 times that of particles in liquids. In Figure 8.12, in order for the gas particles to be visible in the figure, the gas particles are shown only about 10 times as far apart as the liquid particles in Figure 8.8. In fact, to make the picture more to scale, the tiny spheres representing the gas particles would have to be 100 times smaller than they are in the figure.

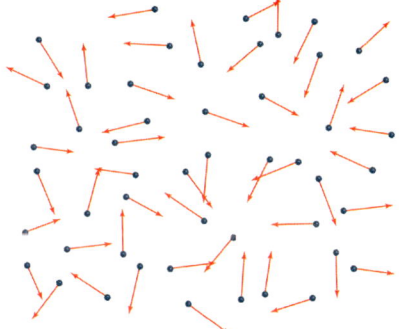

Figure 8.12. Particles in gases are far apart (on average) and move at very high speeds compared to particles in liquids and solids.

We have already seen (Figure 8.3) that even at ordinary temperatures, the average velocity of gas particles is in the same ballpark as the speed of sound in air. At room temperature, molecules in air collide with each other around five billion times per second! Because of such high energies, the molecules in a gas diffuse much faster than liquid molecules do. Recall from Figure 8.3 that the lighter the molecules are, the higher their average velocities. One can conduct a simple diffusion experiment with vinegar and ammonia to test this, as described in the box on the next page.

Since the particles in a gas are spread so far apart, the density of gases is orders of magnitude lower than liquid and solid densities. As we will see, the volume of a gas is a function of both the temperature and pressure. As a result, gases are quite compressible: doubling the pressure on a gas reduces its volume by half. We return to study gases in more detail in the next section and in the next chapter.

Chapter 8

> ### Hmm... Interesting.   Gas diffusion
>
> Here is a simple experiment you can perform to observe gas diffusion and compare rates of diffusion for molecules with different masses. We using two household substances, white (distilled) vinegar and ammonia, both of which are available down at the grocery store.
>
> The molecular formula for acetic acid (vinegar) is CH_3COOH, which means its molecular mass is 60.2 u. The molecular mass of ammonia, NH_3, is 17.0 u. Since vinegar molecules are 3.5 times as massive as ammonia molecules, we expect them to diffuse through air more slowly. The lighter ammonia molecules move at higher speeds, bounce more rapidly between molecules of oxygen and nitrogen in the air, and thus we expect them to diffuse through the air more rapidly.
>
> To make the contest fair, you need to make sure the concentrations of your ammonia and vinegar are about the same. Both products are dilute aqueous solutions. When I tried this, the label on the vinegar said it was "5% acidity," which I took to mean 5% by volume acetic acid. The ammonia label did not specify the concentration. A few minutes of online research led to the Safety Data Sheet (SDS) for the ammonia brand (which is also sold under many other brand names). The SDS states that the ammonia concentration was "<3%." From this I knew to mix some vinegar about 50-50 with water to bring its concentration down to be approximately equal to the ammonia concentration. (Experimental methods such as *titration* can be used to determine the concentrations of both the acetic acid and the ammonia more precisely. We discuss titration when we address acids and bases in Chapter 11.)
>
>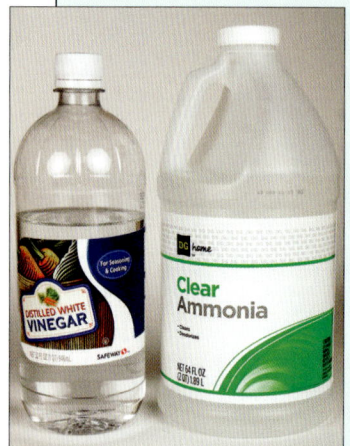
>
> For the experiment, use two small bowls with tight-fitting plastic lids. To prevent odors from contaminating the room where your experiment is conducted, take the bowls to a separate room, fill each one with one of the solutions, and put on the lids to seal the bowls.
>
> Perform the experiment in a room with still air (no fans, windows closed, etc.) You need two people to conduct the experiment. One person can be the odor sniffer and operate the stopwatch. The other person can be the one to open the bowls. Do each solution separately. Place one bowl on a table about 24 inches in front of the sniffer. As one person removes the lid, the sniffer starts the timer then sits quietly, breathing deeply, waiting for the fumes to arrive. As soon as the sniffer detects the odor, he or she stops the timer. Wait a few minutes for the air to clear and repeat the test with the other solution. When I tried this, my sniffer smelled the vinegar in 34 seconds and the ammonia in 11 seconds, just as we expected.

8.2.5 Plasmas

A *plasma* is an ionized gas. At the very high energies where ionization occurs, molecules break apart, so it is not really meaningful to think of molecules in the context of plasmas. Instead, think in terms of single-atom gases, such as the noble gases. Neon signs, for example, are illuminated by the high-energy neon plasma inside the glass tube. As you recall from Chapter 3, each atomic species has its own emission spectrum, and the colors of light emitted by a plasma depend on the atomic species involved. The neon atoms ionized in a neon sign produce an intense orange light, shown in the lettering of the left image in Figure 8.13. Nitrogen ions emit a

Kinetic Theory and States of Matter

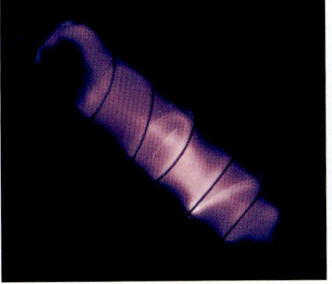

Figure 8.13. Plasmas produced by a neon sign (left), a tube of energized nitrogen gas (middle), and lightning (right).

beautiful violet light, shown in the nitrogen gas tube in the center image of the figure. During a lightning strike, O_2 and N_2 molecules in the air are ripped apart and atoms of nitrogen and oxygen are ionized to produce a plasma. Since air is 78% nitrogen, the bright violet glow of a lightning strike in the right-hand of the figure is similar in color to the light emitted by the nitrogen gas tube.

The energy necessary to ionize a gas can be supplied by intense heat, as in the colored flames of a fire or welder's cutting torch—also plasmas—or by high-voltage electricity. Heat, as in a fire, excites the electrons in the atoms so much that they are broken free from the atoms; high voltage, as in neon signs and lightning, attracts the electrons so forcefully that they are torn away from the atoms. Either way, the result is a gas of ions, electrons, protons, and neutrons. Since there are so many charged particles in a plasma, plasmas conduct electricity and are affected by magnetic fields. The night-time sky colors of the *aurora borealis* or "northern lights" (Figure 8.14) are the result of a plasma. Ions from the solar wind emitted by the sun shower the earth constantly. The magnetic field of the earth, called the *magnetosphere*, shields life of the earth from the harmful effects of all this so-called *ionizing radiation*. But the ions interact with the magnetic field in the earth's upper atmosphere, resulting in the beautiful light displays visible at night in far northern and southern latitudes.

Figure 8.14. The sky light produced by the plasma in one of the polar auroras.

Plasmas are exciting to see (pardon the pun) and fascinating to study. Unfortunately, they play no significant role in the phenomena we study at this level in chemistry, so I leave them for you to study more about later and move on to more detailed analysis of the other three states of matter.

8.2.6 Phase Transitions and Phase Diagrams

As described in the previous few pages, for a given pure substance, there is a distinct temperature at which a solid substance melts, depending on the energy necessary for the atoms or molecules to break free of the forces of the bonds holding them together. The change of state or phase we call *melting* is an example of a *phase transition*. When studying phase transitions, a use-

Chapter 8

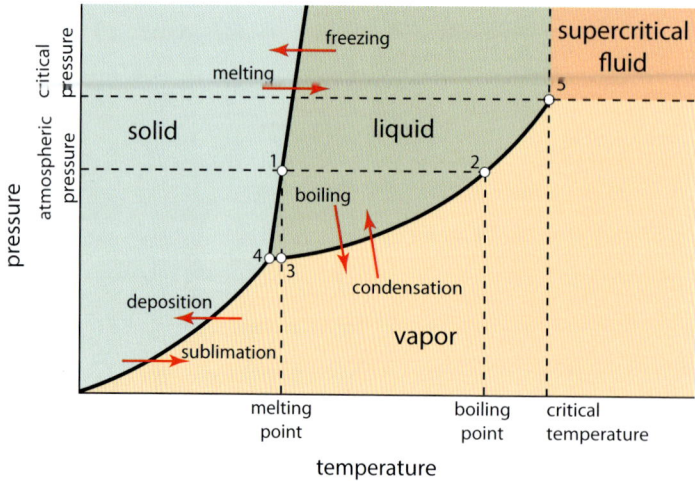

Figure 8.15. Generic P-T phase diagram.

ful graphical aid is the *phase diagram*. The phase diagram for each substance is unique, but a typical, generic phase diagram is shown in Figure 8.15. This graph indicates the states of a substance for various combinations of pressure and temperature, so to distinguish this diagram from another type of phase diagram we will see shortly, we will call this a "P-T phase diagram."

The axes on the P-T phase diagram are pressure and temperature, and the large regions on the diagram represent the conditions of temperature and pressure that result in each of the different phases or states. Notice the dashed horizontal line at atmospheric pressure. The temperature where that line crosses the boundary from solid to liquid (point 1) represents the melting point (also the freezing point) of the substance at atmospheric pressure, and the temperature where that line crosses the boundary from liquid to vapor (point 2) represents the boiling point at atmospheric pressure. From this diagram, it is clear that at lower pressures, the melting and boiling points are lower. One consequence of this is that water boils at a much lower temperature at high altitude because the pressure at high altitude is substantially lower than at sea level. When preparing foods such as rice or vegetables that require boiling water, cooking times at high altitudes are longer than usual because the temperature of the boiling water is lower than the 100°C people are accustomed to at low altitudes. At high altitudes, some food preparations require water to be at temperatures higher than the boiling point, which is not possible to achieve without a pressure cooker. A pressure cooker has a lid that clamps onto the cooking pot and seals it. As water boils and produces steam inside the pot, the pressure increases, which increases the boiling point so that the food can cook properly.

Looking again at Figure 8.15, notice that at the temperature and pressure of point 1, the substance can exist as a solid. If you have a solid at these conditions and hold the temperature constant while reducing the pressure, the solid first melts and then when the pressure gets low enough, the substance boils as it transitions to the vapor state (point 3).

There are six phase transitions labeled on the diagram. You are probably familiar with most of these. You have probably seen dry ice (frozen CO_2) exhibiting *sublimation*. Dry ice does this because on the phase diagram for carbon dioxide, the curves are higher up the graph and the horizontal atmospheric pressure line passes from the solid region to the vapor region. So when the temperature of dry ice is increased at atmospheric pressure, the ice undergoes the phase transition from solid to vapor—sublimation.

Deposition occurs when a vapor transitions directly to the solid state, the reverse of sublimation. The formation of frost is an example of deposition, as water vapor in the air deposits onto a cold surface and freezes. This can result in some beautiful patterns, as illustrated in Figure 8.16. Semiconductor manufacturers use deposition to apply a thin layer of a metallic compound such as titanium nitride or an insulating compound such as SiO_2 to a silicon substrate. This process is called *chemical vapor deposition* (CVD). (See the photo on the opening page of this chapter.)

Back again to Figure 8.15, point 4 in the diagram is called the *triple point*. This is the temperature and pressure where all three states of a substance can coexist in equilibrium. Point 5 in the diagram is called the *critical point*, located by the *critical temperature* and *critical pressure*. The critical temperature is the highest temperature where the liquid state of the substance can exist. Above this temperature, regardless of the pressure, the substance does not liquefy. The critical pressure is the lowest pressure at which the liquid state can exist at the critical temperature. At conditions above the critical temperature and critical pressure, the substance is in a state referred to as a *supercritical fluid*. In this region, distinct liquid and vapor states do not exist. At pressures close to the critical pressure, the fluid is more vapor-like. At very high pressures and close to the critical temperature, the fluid is more liquid-like. One of the most common applications of a supercritical fluid is in producing decaffeinated coffee. This process uses supercritical CO_2, and the caffeine extracted from the coffee is sold to soft drink manufacturers for use in carbonated beverages.

Figure 8.16. Deposition of water vapor can create beautiful crystal shapes.

Critical constants for several substances are shown in Table 8.3. The critical constants are very important for engineering applications with fluids. It would be pointless, for example, to try to liquefy propane at temperatures above 96.7°C. Liquid propane cannot exist at such temperatures. Ammonia is easy to liquefy—room temperature is well below its critical temperature. Oxygen can be liquefied if it is first chilled to below −118.6°C.

The P-T phase diagram for water is shown in Figure 8.17. The dashed horizontal line at 1 atm represents atmospheric pressure at sea level. The two points shown on that line are the melting point and boiling point of water, at 0°C and 100°C, respectively.

The phase diagram for water is unique. Notice that the line separating the solid and liquid regions has a negative slope. In almost every other substance, this boundary has a positive slope, as shown in the generic diagram of Figure 8.15. In the case of water, the negative slope of this line is closely related to the fact that the density of liquid water decreases as the water is cooled from 4°C to 0°C. The density of other substances increases as the substance freezes. But as we have discussed before, the hydrogen bonding between water molecules causes the molecules to spread out and form hexagonal patterns just before freezing.

The phase diagrams help explain this unique behavior. In the generic diagram of Figure 8.15, if a substance is at the triple point and the pressure is increased while the temperature is held constant (a line straight up from the triple point), you see that the substance must become entirely solid. However, for water, the same change results in liquid water. Increasing the pressure packs the molecules closer together, disrupting the hexagonal hy-

Substance	Critical Temperature (°C)	Critical Temperature (K)	Critical Pressure (atm)
ammonia, NH_3	132.4	405.6	112.1
carbon dioxide, CO_2	30.9	304.1	72.8
nitrogen, N_2	−147.0	126.2	33.5
oxygen, O_2	−118.6	154.6	49.8
propane, C_3H_8	96.7	369.9	41.9
water, H_2O	373.9	647.1	217.8

Table 8.3. Critical constants for a few substances.

Chapter 8

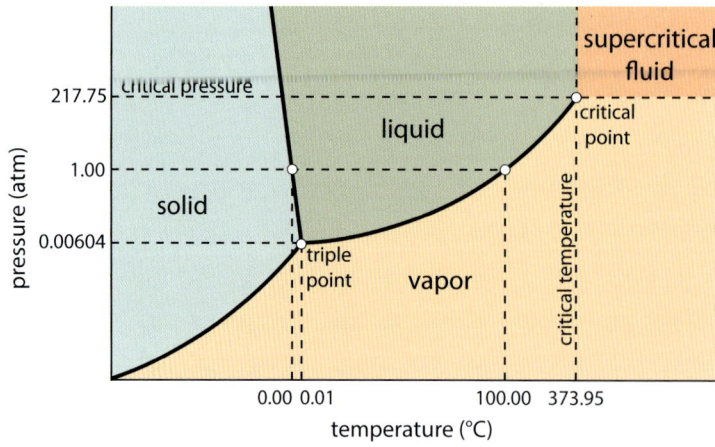

Figure 8.17. P-T phase diagram for water. (In order to make small features visible, the scales are not drawn linearly and are not in proportion.)

drogen bonding in the ice and increasing the density. Notice that the same thing happens at atmospheric pressure: increasing the pressure at 0°C turns ice into water. Even if the temperature is below 0°C, if the pressure is increased enough, the ice melts. This is what makes ice skating possible. The weight of the skater is supported only by the small area of contact between the thin skate blade and the ice (and basically on only one blade at a time while the skater is moving). The small area results in a very high pressure under the skate blade. This high pressure under the blade causes a layer of water to form, which lubricates the skate so it slides easily over the ice.

8.2.7 Heat Capacity, Heat of Fusion, and Heat of Vaporization

We have seen that the difference between solids, liquids, and gases is in the internal energy of a substance. If you take a solid substance at a temperature below its melting point and heat it continuously, the added heat causes the substance to go through five different changes, illustrated in the case of H_2O in Figure 8.18. In the figure, the horizontal axis represents the amount of heat being added to (or removed from) the H_2O. The vertical axis is the temperature of the H_2O. As heat is added to the H_2O, we can imagine the H_2O moving along the line on the graph. In region 1 on the graph, the H_2O is a solid. If we begin at –20°C and begin adding heat to this ice, the following five changes occur (the numbers here match the numbers in the graph):

1. The temperature of the solid increases to the melting point, 0°C.

2. The substance melts, maintaining the temperature at the melting point as it does.

3. As a liquid now, the temperature of the liquid begins at the melting point and increases steadily to the boiling point, 100°C.

4. The substance vaporizes, maintaining the temperature at the boiling point as it does.

5. As a vapor now, the temperature of the substance begins at the boiling point and increases.

Each of these five situations requires a specific amount of heat to accomplish, depending on the substance involved. If we begin with a vapor at a temperature above the boiling point and continuously cool the substance, it passes through the same five changes in reverse, requiring the same quantities of heat, the only difference being that heat is being removed from the substance by cooling instead of added by heating.

Figure 8.18 is a second type of phase diagram. In this diagram, the axes are labeled temperature and heat, so we will call this a "T-H phase diagram." Similar diagrams can be drawn for other substances, but we will limit ourselves here to discussing H_2O.

The purpose of the T-H phase diagram is to show the quantities of heat that must be added to or removed from a quantity of ice, water, or steam in order to change its temperature or effect a phase transition. On the three diagonal sections of the graph, the H_2O is in the solid, liquid, or

Kinetic Theory and States of Matter

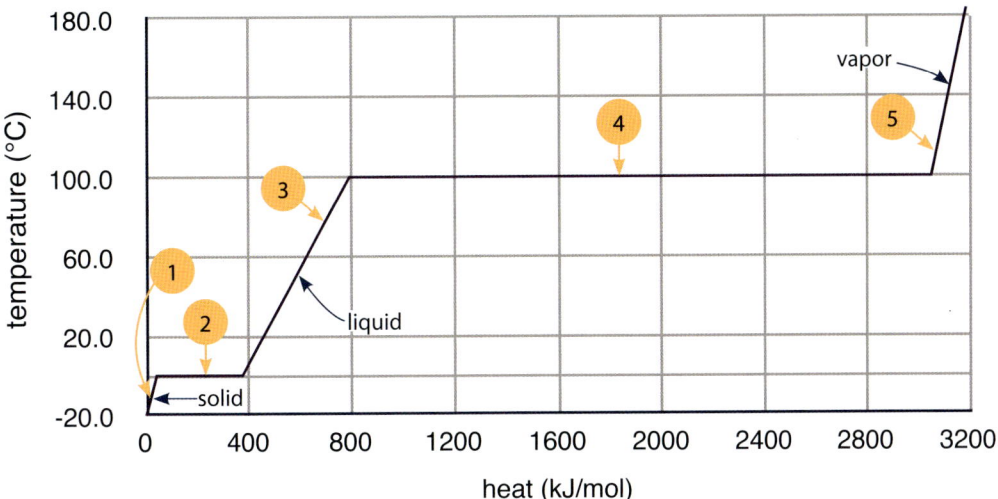

Figure 8.18. T-H phase diagram for H_2O showing five places the H_2O can be as the temperature is increased from below 0°C to above 100°C.

vapor state (from lower left to upper right). The horizontal sections represent the phase transitions between ice and water (the one at 0°C) and between water and steam (the one at 100°C). From our discussion in previous pages, you know that the melting and boiling points of a substance depend on the pressure, so it should be clear that a T-H phase diagram is only valid for a specific pressure. We will focus our discussion on water at atmospheric pressure.

To further illustrate how to read the T-H phase diagram, Figure 8.19 shows H_2O in two different situations. At point A, liquid water is at 22°C and warming up, as indicated by the upward pointing arrow. At point B, the H_2O is a mixture of water and steam at 100°C because it is in the middle of a phase transition. The arrow indicates that the mixture is on its way to becoming completely vapor.

To move the H_2O along the curve in the graph, we have to add or remove heat. In the SI system, the unit of measure for quantities of heat (which is energy) is the joule (J). We symbolize the amount of heat added or removed as Q. We can calculate the amount of heat required to move between any two points on the graph using two simple equations and a few constants. One of the equations is used to calculate the amount of heat required to increase or decrease the temperature. The other is used to determine the amount of heat required to effect a phase transition. We will look at both equations and an example of each calculation, beginning with the calculation of the heat required for a temperature change.

On the three diagonal lines where temperature changes occur, the amount of heat needed to change the temperature is calculated as

$$Q = Cn\Delta T \qquad (8.1)$$

In this equation, C is a constant called the *molar heat capacity* of the substance, n represents the number of moles present, and ΔT represents the change in temperature. The molar heat capacity of a substance is defined as the quantity of heat required to raise or lower the temperature of one mole of the substance

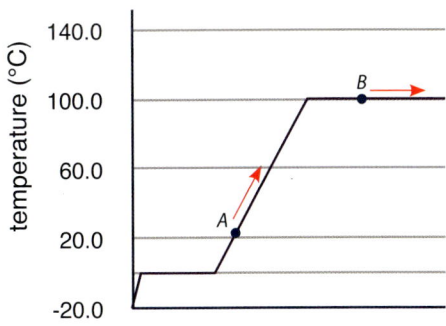

Figure 8.19. At point A, liquid water is at 22°C and warming up. At point B, a mixture of water and vapor is at 100°C and transitioning to vapor.

Chapter 8

by 1 K (which is the same as an increase of 1°C). The higher C is for a given substance, the more heat is required to raise the temperature. The value of the molar heat capacity actually depends on the state the water is in. The three values for water as solid, liquid, and vapor are listed in Table 8.4.[3] (The reference information for water is also in Table A.5 in Appendix A.)

In Equation (8.1), the change in temperature is always defined as $\Delta T = T_f - T_i$, where T_f and T_i are the final and initial temperatures, respectively. (The greek letter delta, Δ, is often used this way to mean *difference* or *change*.) Used this way, a positive value for Q indicates that heat is being *added* to the substance, because $T_f > T_i$. When a substance is being cooled down, $T_f < T_i$, so the value of Q is negative. This indicates that heat is being *removed* from the substance. Note that since Celsius degrees and Kelvin "degrees" are the same size, you can calculate the quantity ΔT with either one and get the same result.

Constant	Value
C (ice, −10°C)	0.0364 $\frac{kJ}{mol \cdot K}$
C (water)	0.0752 $\frac{kJ}{mol \cdot K}$
C (steam, 110°C)	0.0369 $\frac{kJ}{mol \cdot K}$
H_f	6.01 $\frac{kJ}{mol}$
H_v	40.7 $\frac{kJ}{mol}$

Table 8.4. Thermal constants for water.

▼ Example 8.1

Determine the quantity of heat that must be added to 75.0 g of water to heat it from 22.0°C to 35.0°C.

We use Equation (8.1) to solve this problem. Let's begin by listing the given information, including the value of the molar heat capacity from Table 8.4:

$m = 75.0$ g

$C = 0.0752 \; \dfrac{kJ}{mol \cdot K}$

$T_i = 22.0°C$

$T_f = 37.0°C$

$Q = ?$

Before using Equation (8.1), we need to compute the number of moles of water we have. We use the periodic table to compute the molar mass of water, which is 18.02 g/mol. Using this value as a conversion factor, we calculate the number of moles:

$$75.0 \text{ g } H_2O \cdot \frac{1 \text{ mol } H_2O}{18.02 \text{ g } H_2O} = 4.162 \text{ mol } H_2O$$

Now we are ready to compute the heat required using Equation (8.1):

$$Q = Cn\Delta T = 0.0752 \; \frac{kJ}{mol \cdot K} \cdot 4.162 \text{ mol} \cdot (37.0°C - 22.0°C) = 4.69 \text{ kJ}$$

---

[3] Heat capacity values for ice and steam vary quite a bit with temperature, but within about 20°C of the melting or boiling point, the values given are reasonable approximations.

Kinetic Theory and States of Matter

Again, the temperature units cancel because the size of one degree in the Celsius and Kelvin scales is the same, so a temperature change of so many degrees is the same on both scales. Note that the molar heat capacity is given in kilojoules (kJ), so the units of our result are also kJ. Finally, I use four significant digits for the intermediate calculation for the number of moles, but the final result is rounded to three significant digits as the problem requires.

Table 8.5 lists all the variables and constants used for these calculations, along with those used in the second type of calculation, which we address next.

The amount of heat that must be added or removed from a substance to effect a phase transition is calculated using one of these two equations:

$$Q = nH_f \qquad (8.2a)$$
$$Q = nH_v \qquad (8.2b)$$

The only difference between these two equations is the constant involved. As before, n represents the number of moles that are present. The constant H_f is called the *molar heat of fusion*, defined as the quantity of heat required to melt or freeze one mole of the substance. Thus, the first equation listed is the one to use for calculating the heat required to melt or freeze a substance. The constant H_v is called the *molar heat of vaporization*, defined as the quantity of heat required to boil or condense one mole of the substance. Thus, the second equation listed is the one to use for calculating the heat required to boil or condense a substance. You simply use the appropriate constant for the phase transition involved. To calculate the heat associated with melting or freezing, use the molar heat of fusion, H_f; to calculate the heat associated with boiling or condensing, use the molar heat of vaporization, H_v.

▼ Example 8.2

Determine the quantity of heat that must be added to 5.75 moles of water at 100.0°C to vaporize it.

This problem is about a phase transition, boiling, so we need to use Equation (8.2b) with the molar heat of vaporization, H_v, from Table 8.4. We write the equation, insert the values, and compute the result.

Quantity	Units	Definition
Q	J	variable representing the quantity of heat that must be added to or removed from the substance
C	J/(mol·K)	*molar heat capacity* of the substance, defined as the quantity of heat required to raise or lower the temperature of one mole of the substance by 1 K
n	mol	variable representing the number of moles of the substance
ΔT	K	variable representing the change in temperature of the substance
H_f	J/mol	*molar heat of fusion*, defined as the quantity of heat required to melt or freeze one mole of the substance
H_v	J/mol	*molar heat of vaporization*, defined as the quantity of heat required to boil or condense one mole of the substance

Table 8.5. Thermal variables and constants for calculating quantities of heat in phase transitions and temperature changes.

$$Q = nH_v = 5.75 \text{ mol} \cdot 40.7 \frac{\text{kJ}}{\text{mol}} = 234 \text{ kJ}$$

▲

As you see, this is a much larger quantity of heat than we calculated in the previous problem, even though the amounts of H$_2$O involved were similar. If you look back at Figure 8.18, you see that the line segment labeled "4" spans across a greater energy difference than any other line in the diagram, and this means the quantities of heat involved in vaporizing water are larger than the quantities required for any of the other processes we are considering here. The reason it takes so much energy to vaporize water goes back to the hydrogen bonding we have discussed. The water molecules are clinging to each other because of hydrogen bonding, and they must have enough kinetic energy to separate in order to enter the vapor state.

8.2.8 Evaporation

The kinetic-molecular theory tells us that in a liquid at a given temperature, there is a distribution of speeds among the liquid molecules, just as there is with the molecules in a gas. At ordinary temperatures, there are many molecules in a liquid that have enough energy to transition into the vapor state if they can only break free of the other molecules blocking their paths to the open air. Near the surface of the liquid, these higher energy molecules aren't blocked by other liquid molecules. Consequently, they transition to the vapor state, flying away from the liquid and joining the molecules in the air. The phase transition from liquid to vapor that occurs this way is called *evaporation*, illustrated in Figure 8.20. The warmer the liquid is, the higher the average energy of the molecules in the liquid, and the faster evaporation occurs.

In the process of evaporation, it is the highest energy molecules in the liquid that transition to the vapor state. As the molecules with higher energy depart the liquid, the average energy of the molecules left behind decreases. (This is analogous to the fact that every time a graduating class of seniors leaves a school, the average age of the students left behind drops because all the oldest students have left.) Since temperature varies with internal energy, the temperature of the liquid decreases during evaporation, an effect known as *evaporative cooling*. In warm-weather locations, evaporative cooling is used at outdoor seating areas at restaurants to keep patrons cool. Misting systems spray a fine mist of water into the air. As the molecules of water in the tiny water droplets transition to the vapor state (evaporate), the temperature of the water, and thus of the surrounding air, decreases. The amount of cooling can be calculated using H_v.

Evaporative cooling is used by the human body to keep the body cool. The human body is a thermodynamic engine, and like every other engine it constantly produces waste heat, just as boilers emit waste heat through their smoke stacks and cars through their tailpipes. In a room at 72°F, the temperature difference between the body (98.6°F) and the air is high enough that waste heat escapes and keeps the body cool. This happens by conduction between the body and air molecules near the surface of the skin, and to some extent through infrared electromagnetic radiation. But in a hot environment, where heat

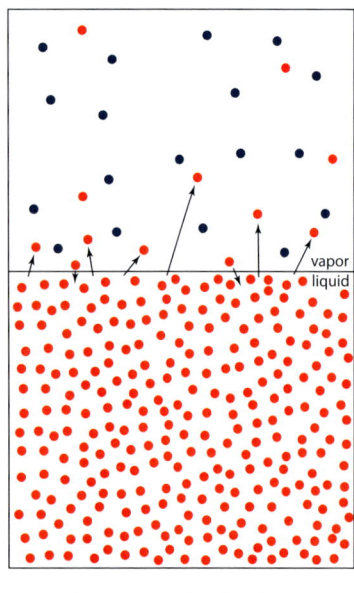

- N$_2$ and O$_2$ molecules of air
- liquid and evaporated liquid molecules

Figure 8.20. Liquid particles transitioning to the vapor state during evaporation.

cannot escape from the body fast enough, the body begins secreting water onto the surface of the skin through the sweat glands, shown in Figure 8.21. The drops of moisture on the skin then evaporate, lowering the temperature of the remaining moisture and cooling the body.

I am compelled to pause here to note *how stupendous this is*. Sweat glands (called *sudoriferous* glands in the world of anatomy) are found in our skin on almost the entire surface of our bodies. In a warm environment, an exquisite and complex control process called *thermoregulation*, controlled by the *hypothalamus* in the brain, calls upon the sweat glands to begin channeling water to the surface of the skin where evaporative cooling assists in maintaining the body's temperature. Thermoregulation, like any control process, requires a system of input sensors, a decision-making algorithm that compares current conditions to desired conditions, and an output mechanism that controls some kind of variable that has the power to change current conditions and bring them toward the desired conditions.

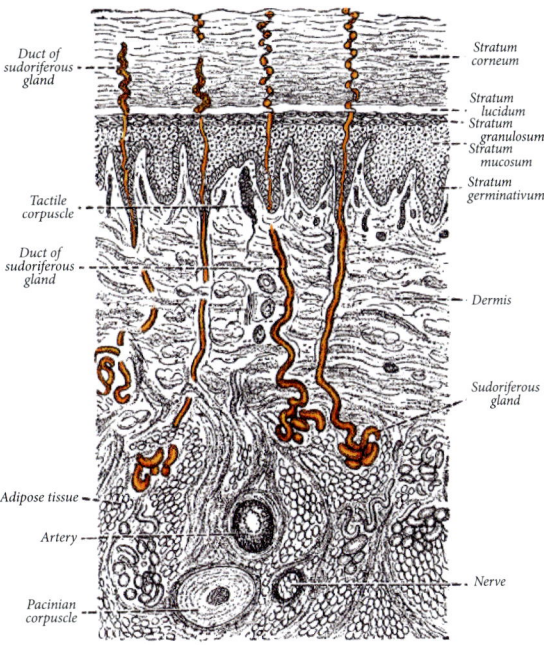

Figure 8.21. Cut-away view of the skin showing the sweat glands (orange). This drawing is from Henry Gray's classic 1858 text, *Gray's Anatomy*.

A diagram of the thermoregulation control system in the human body looks exactly like a sophisticated control system in a manufacturing plant. The system even includes special information channels to cause the system response to kick in more rapidly in the case of a sudden alarm—the so-called *fight or flight response*. In anticipation of a fast-approaching temperature increase due to increased exertion while contending with danger—running, rescuing people, and so on—the cooling system kicks in before the body temperature even increases. (This is why you can start sweating almost instantly when faced with a sudden danger, such as almost having a car crash.) In fact, many other systems in our bodies increase production prior to actual need in the same way, including increased heart rate in anticipation of increased oxygen demand by the cells, increased respiratory rate to bring in more oxygen, dilation of blood vessels to allow faster blood flow, and release of energy sources such as fat and glycogen in anticipation of immediate muscle use. By itself, the thermoregulation system in our bodies is stunning. Add in all these other systems, each of which is equally amazing, and the wonder of our bodies is breathtaking.

The more you contemplate the sophistication of the way our bodies are designed, the more obvious it becomes that they were in fact *designed*—they did not acquire their present sophistication by accident. No doubt you are well aware that different people have different views about how our bodies originated. Among Christians, too, there are different views about how God brought us into existence. But regardless of how God brought forth these amazingly-designed bodies, the sophistication of processes like thermoregulation is a testament to the wisdom of our Heavenly Designer. Next time you break out in a sweat, you may just pause for a moment to praise God for the fact that we are "fearfully and wonderfully made" (Psalm 139:14).

Returning briefly to Figure 8.20, notice in the figure that a few of the liquid molecules that have already evaporated are recondensing into the liquid state. This might happen, for example, if a high-energy liquid molecule transitions to the vapor state, immediately collides with a molecule in the air near the liquid surface, and having lost a lot of its energy in the collision, condenses back into the liquid state.

8.2.9 Vapor Pressure

If a liquid is placed in a sealed container, molecules of the liquid begin to evaporate into the vapor space in the container above the liquid, causing the pressure in the vapor to begin increasing. This continues until an equilibrium is reached. In this equilibrium, the density of the molecules in the vapor state is high enough that the rate of molecules transitioning into the vapor state equals the rate of molecules transitioning back into the liquid state, as illustrated in Figure 8.22. The pressure at which this equilibrium is reached is called the *vapor pressure*. The vapor pressure is a measure of the rate of evaporation of a liquid and its *volatility*—its tendency to become a vapor. More volatile substances have higher vapor pressures.

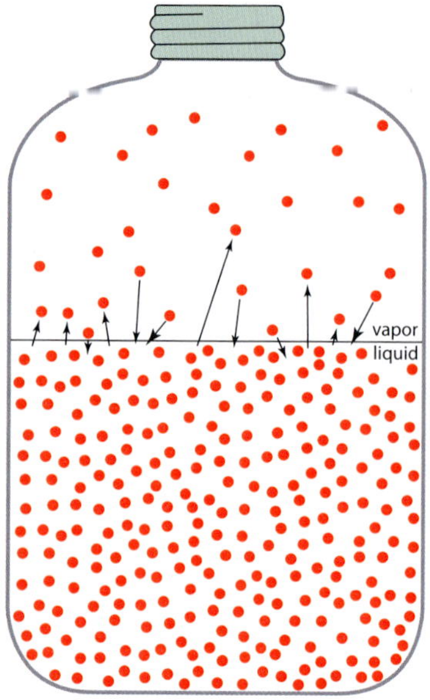

Figure 8.22. The vapor pressure of a substance in a sealed container causes an equilibrium in which vaporization and condensation are happening at equal rates.

You have probably noticed that if you open a container of a liquid with a high vapor pressure, such as paint thinner or gasoline, you briefly hear the sound of gas escaping from the container when you break the seal of the container's lid. This effect is due to the vapor pressure of the liquid inside the container.

The vapor pressure of liquids is strongly dependent on temperature, as indicated by the curve for the vapor pressure of water in Figure 8.23. Looking carefully at the curve of the graph, we see that at 100°C, the boiling point of water, the vapor pressure appears to be—and is—760 Torr, which is normal atmospheric pressure. This is because *the vapor pressure curve also indicates the pressure at which a liquid boils at a given temperature.* At 100°C, water boils at a pressure of 760 Torr. At 40°C (104°F) water boils at a pressure of about 55 Torr. We never see water boiling at 40°C because atmospheric pressure is never as low as 55 Torr. But when water at 40°C is placed in a vacuum chamber and the pressure is reduced to about 55 Torr, the water begins boiling, *and it is not that hot—40°C is the temperature of hot bath water*. When water is at atmospheric pressure and 100°C, it boils because atmospheric pressure is the vapor pressure of water at that temperature. At that temperature and pressure, all the water begins vaporizing at once. And by

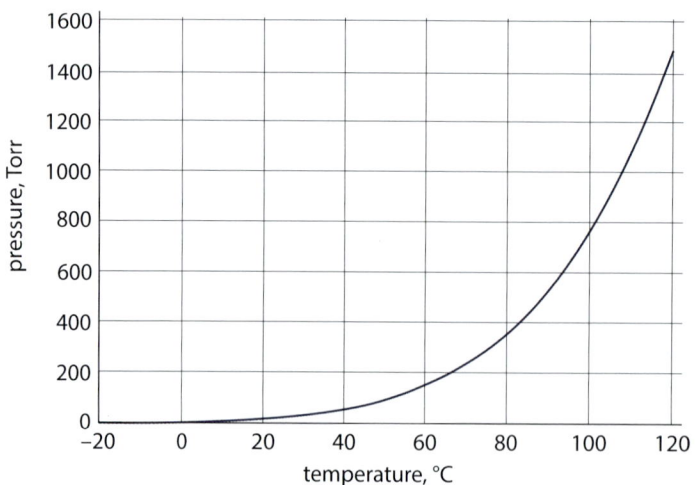

Figure 8.23. Vapor pressure curve for water.

the way—the bubbles erupting from a boiling liquid are *not* air bubbles. They are bubbles of *vapor*—liquid that has transitioned to the vapor state and is on its way out.

Recall the discussion earlier about longer cooking times due to water boiling at lower temperatures at higher altitudes. In Leadville, Colorado, atmospheric pressure is about 502 Torr. From the graph in Figure 8.23, you see that water boils at that pressure at about 89°C (192°F). At the top of Mt. Elbert, the highest mountain in Colorado, atmospheric pressure is down to 418 Torr, resulting in a boiling point of 84°C.

Chapter 8 Exercises

Note: Constants and conversion factors are in Tables A.2 and A.3 in Appendix A.

SECTION 8.1

1. Give a physical description of how to interpret a temperature measurement. In other words, what does temperature *mean*?
2. What is the Maxwell-Boltzmann distribution?
3. State the approximate range of velocities and the mean velocity found in Argon atoms at 25°C (see Figure 8.3).
4. Explain what holds the mercury up in the tube of a mercury barometer.
5. As indicated in the box on page 207, the pressure under the surface of a liquid is given by $P = \rho g h$. Use this equation to calculate the height of the fluid column in a barometer that uses water instead of mercury. (Again, see Appendix A for reference data.)
6. The pressure inside an industrial cylinder of oxygen gas is 1,850 psi. State this pressure in Pa, kPa, Torr, atm, and bar.
7. Consider a typical ice hockey skate. The contact surface between the blade and the ice when the skater is gliding is 2.9 mm wide and 57 mm long. Determine the pressure on the ice under a 200.0-lb skater with one skate in contact with the ice. State your answer in atm.
8. Of the five statements given in this section outlining the kinetic-molecular theory of gases, some of them may be taken as fully accurate descriptions and some of them are based on obvious approximations. State and describe the approximations inherent in the kinetic-molecular theory of gases.
9. Explain why containers of gas, such as cans of spray paint or mosquito spray, include warning labels telling people not to leave the cans in the direct sunlight. Address this question in detail at the molecular level.
10. Describe conditions under which a real gas might exhibit ideal gas behavior.
11. Describe which of the following gases you would expect to exhibit ideal gas behavior to the largest degree, and which to the smallest degree: Ar, HF, H_2O, H_2, O_2, He, and NH_3.

SECTION 8.2

12. Write paragraphs describing each of the four states of matter.
13. Explain why solids and liquids are essentially incompressible, while gases are compressible.
14. Explain surface tension, then use surface tension to explain why capillary action and wicking occur.
15. Explain diffusion and its cause.
16. Which would you expect to diffuse faster in air: molecules of ethanol, C_2H_5OH, or butane, C_4H_{10}? Explain your response.

17. Explain evaporation and its cause.

18. Explain why evaporative cooling helps remove heat from the body in a hot environment.

19. Calculate each of the following. (Again, see Table A.5 in Appendix A for reference data.)

    a. the amount of heat required to warm 3.47 moles of water from 25°C to 55°C.

    b. the mass of steam required to release 3.65×10^5 kJ on condensation.

    c. the amount of heat required to melt 12.00 kg of ice at 273.15 K.

    d. the volume of water that can be heated from 20.0°C to 30.0°C with 7.32 kJ of heat. (Use the density at 25°C—the approximate average density of water over this temperature range.)

    e. the total amount of heat required to warm 5.00 kg of ice from −25.0°C to 0.0°C and then melt it.

20. The P-T phase diagram for carbon dioxide is shown below. Use this diagram to answer the following questions.

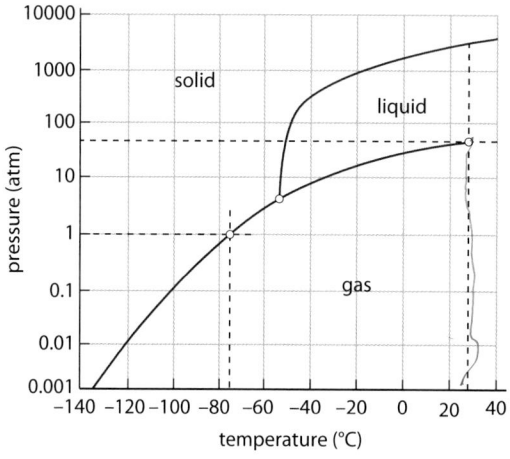
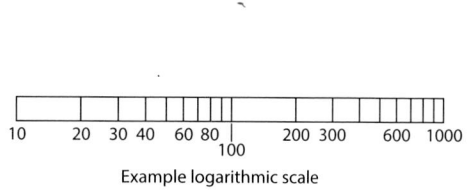

   a. At atmospheric pressure, estimate the temperature at which CO_2 sublimates.

   b. Estimate the pressure and temperature at the triple point, and state your results in both atm and Torr, and both °C and K. (Note that the pressure scale is logarithmic, not linear. See the reference scale in the figure.)

   c. What are the melting and boiling points at a pressure of 7,600 Torr?

   d. What is the lowest pressure at which liquid CO_2 can exist at 28°C?

   e. At a temperature of −20°C, what are the pressures at which CO_2 boils and freezes?

   f. At a temperature of 0°C, what is the vapor pressure?

   g. At 40°C, what happens to gaseous CO_2 if the pressure is increased to 10,000 atm?

21. When you squeeze an ice cube between your teeth—but not hard enough to crush it—the ice begins to melt away. Use the P-T phase diagram for water to account for this phenomenon.

22. If you want to store liquid water in an environment that is at 120°C, what pressure does the water have to be under? (See Figure 8.23.)

23. On a P-T phase diagram, why does the curve separating the liquid and gas regions end at the critical point?

Kinetic Theory and States of Matter

24. At atmospheric pressure, the melting and boiling points of O_2 are −218°C and −183°C, respectively. The triple point for O_2 is at −219°C and 0.0015 atm, and the critical point is at −119°C and 49.8 atm.

   a. Sketch the phase diagram for O_2, showing the four given points and the regions of the three states.

   b. Determine whether solid O_2 floats on liquid O_2.

   c. At atmospheric pressure, when solid O_2 is heated, does it sublime or melt?

25. Cabin pressure in a commercial jet is around 594 Torr. When the flight attendants make coffee, at what temperature does the water boil? (See Figure 8.23.)

GENERAL REVIEW EXERCISES

26. Write the electron configurations for bromine and yttrium.

27. The main component of gasoline is octane, C_8H_{18}. The combustion reaction when octane burns is as follows:

$$2C_8H_{18}(l) + 25O_2(g) \rightarrow 16CO_2(g) + 18H_2O(g)$$

   a. How many moles of O_2 are required to completely burn 5.00 mol C_8H_{18}?

   b. At 20.0°C, the density of octane is 0.692 g/mL. How many grams of O_2 are required to burn 1.00 gal of octane?

28. Solid ammonium nitrate decomposes to yield nitrogen gas, oxygen gas, and water vapor. Write the formula equation for this reaction and balance it.

29. A 2.50-g strip of zinc metal is placed in an aqueous solution of silver nitrate containing 3.00 g $AgNO_3$. The following single replacement reaction occurs:

$$Zn(s) + 2AgNO_3(aq) \rightarrow 2Ag(s) + Zn(NO_3)_2(aq)$$

   a. Which reactant is the limiting reagent?

   b. In a complete reaction, how many grams of $Zn(NO_3)_2$ form?

   c. How many grams of the excess reactant remain after the reaction?

30. Ibuprofen, the pain reliever in Advil and Motrin, contains 75.69% C, 8.80% H, and 15.51% O by mass. The molar mass is 206.3 g/mol. Determine the empirical and molecular formulas for ibuprofen.

31. Determine the oxidation state of S in H_2SO_4 and C in CO_3^{2-}.

32. Identify the following bonds as covalent, polar covalent, or ionic in character.

   a. C—N   b. I—I   c. Al—Cl

33. Explain why metals tend to be malleable.

34. For each of the following molecules, draw the Lewis structures and predict the molecular geometry:

   a. SO_4^{2-} b. H_2S c. ClO_4^-

Chapter 9
The Gas Laws

The burners in hot air balloons use liquid propane fuel to heat the air inside the balloon. According to the gas laws, if volume and pressure are held constant as the temperature of the air is raised, the number of molecules of air inside the balloon decreases. Fewer molecules inside the balloon means less dense air, so the balloon floats. The volume of a typical hot air balloon these days is in the range of 2,800 m^3. Heated to 99°C, such a balloon provides a lift force (simply called *lift*) of nearly 1,600 lb. This is about where hot air balloons operate once they have reached altitude. To get up there requires acceleration, and thus more force. To achieve this force, balloonists heat the air up to around 120°C, acquiring a lift of around 1,900 lb.

Balloons achieving lift strictly by hot air are called *Montgolfier* balloons. The Montgolfier brothers were French, and invented the hot air balloon in 1783. In October of that year, one of them went up in their balloon. Others went up that same day and over the next few weeks. The new phenomenon of manned flight caused a sensation. In recognition of their achievement, the Montgolfier brothers' *father* was elevated to the nobility by King Louis XVI of France. (Not sure if the brothers reaped any reward for themselves, but at least Dad enjoyed himself.)

Though lots of fanfare surrounded the Montgolfier's achievement, the honor of the first manned balloon flight is not theirs. They flew the first hot air balloon, but as we will see in this chapter, Jacques Charles' flight in a hydrogen balloon just two months before was the first manned balloon flight.

Objectives for Chapter 9

After studying this chapter and completing the exercises, you should be able to do each of the following tasks, using supporting terms and principles as necessary.

SECTION 9.1
1. Describe and explain Boyle's law, Charles' law, and Avogadro's law.
2. Given initial conditions of a gas, use Boyle's law and Charles' law to compute gas temperature, volume, or pressure.
3. Explain how Charles' law and Avogadro's law demonstrate some of the assumptions in the kinetic-molecular theory of gases.

SECTION 9.2
4. State the conditions associated with standard temperature and pressure (STP).
5. Compare ideal gases to real gases and explain how these concepts apply to practical chemistry.
6. Use the ideal gas law and the ideal gas constant to solve gas problems involving temperature, volume, pressure, and number of moles.
7. Use the ideal gas law, density equation, and molar mass to solve gas problems involving densities and masses of gases.

SECTION 9.3
8. Use Dalton's law of partial pressures to compute partial pressures, mole fractions, and the total pressure in a gas mixture.
9. Use the law of partial pressures and water vapor pressure data to determine the mass of a gas collected over water.

SECTION 9.4
10. Solve stoichiometric problems involving gas volumes.
11. Distinguish between diffusion and effusion.

Chapter 9

9.1 Early Formulations of the Gas Laws

9.1.1 Boyle's Law

The names of the gas laws are a sort of hall of fame of the great scientists whose work laid the foundations of modern science. In 1660, the Irish-English genius, inventor, and scientist Robert Boyle (Figure 9.1) first referred to the gas law that now bears his name. *Boyle's law states that at a given temperature, the volume of a fixed amount of gas varies in inverse proportion to the pressure.* In mathematical form, the law may be written as

$$V = \frac{k}{P} \tag{9.1}$$

Figure 9.1. Irish-English scientist Robert Boyle (1627–1691).

where k is a constant that depends on the units of measure. Graphically, this relationship between volume and pressure is as shown in Figure 9.2. This law can be rewritten as

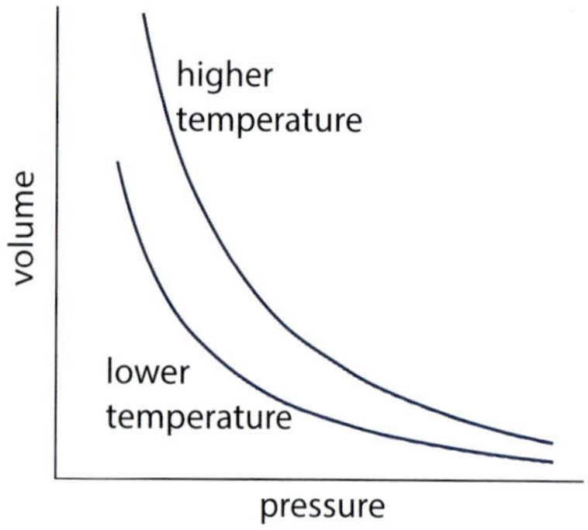

$$VP = k \tag{9.2}$$

In this form, the equation shows that the product of the volume and pressure of a gas is a constant. In particular, this also means that the product of volume and pressure at one set of conditions must equal the product at a different set of conditions, or

$$V_1 P_1 = V_2 P_2 \tag{9.3}$$ **Boyles' Law**

Figure 9.2. Graphical representation of the relationship between volume and pressure for a gas at constant temperature, Boyle's law.

Equation (9.3) represents the most common form Boyle's law is written in today. Note that Boyle's law is simply a pair of ratios. This is clear when the law is written in the form

$$\frac{V_1}{V_2} = \frac{P_2}{P_1} \tag{9.4}$$

Since there are only ratios in this equation, we can solve problems with Boyle's law using any convenient units of measure. If we let V_1 and P_1 represent certain known initial conditions, we can use Boyle's law to determine the new volume, V_2, of the gas at a new pressure, P_2, or vice versa. (Keep in mind, though, that Boyle's law only applies when the temperature is constant.) This is illustrated by the following example.

▼ **Example 9.1**

The volume of a certain cylinder can be varied by means of an adjustable piston, as shown in Figure 9.3. The cylinder contains 0.75 L of a gas at an absolute pressure of

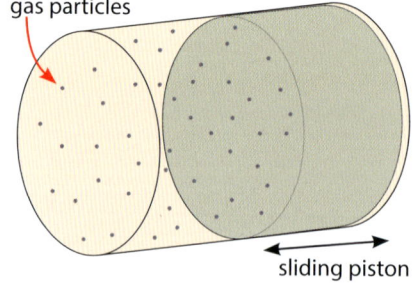

Figure 9.3. Cylinder-piston system.

850 Torr. If the volume is reduced to 0.50 L at a constant temperature, determine the new gas pressure.

We let V_1 and P_1 represent the initial conditions of the gas. The new volume is V_2, and the new pressure—our unknown—is P_2. We first solve the Boyle's law equation for the unknown, P_2.

$$P_2 = P_1 \frac{V_1}{V_2}$$

Next, insert the given values and compute the new pressure. The volume units cancel, giving the new pressure value in the same units as the initial pressure.

$$P_2 = P_1 \frac{V_1}{V_2} = 850 \text{ Torr} \cdot \frac{0.75 \text{ L}}{0.50 \text{ L}} = 1275 \text{ Torr}$$

Rounding this value to the correct number of significant digits, we have

$$P_2 = 1300 \text{ Torr}$$

It is important to note here that in ratio equations like Boyle's law, all values must be in absolute units. Typical pressure gauges, such as those used by mechanics to measure tire pressure, do not read absolute pressure because the "zero" of the pressure gauge is at atmospheric pressure. On an absolute pressure scale the zero of the scale is at a complete vacuum. We consider this in more detail in Section 9.2.

9.1.2 Charles' Law

By the late 18th century, many scientists in England and France were investigating the properties of gases. This work led to the rise of ballooning as a popular sport for scientifically minded (and fearless) individuals. In August of 1783, Jacques Charles, a French scientist, inventor, and balloonist, flew to an altitude of about 1,800 feet in the world's first manned balloon flight, an event captured in the illustration shown in Figure 9.4. The gas in the balloon was hydrogen. Hydrogen is the lightest gas there is (four times lighter than helium), but is, as you know, highly flammable. Use of hydrogen as a gas to lift manned balloons came to an end with the disastrous burning of the *Hindenburg* in 1937.

In the 1780s, Jacques Charles formulated the law now known as *Charles' law*, but his work went unpublished. In 1802, over 140 years after Boyle's law had been published, French scientist Joseph Louis Gay-Lussac (Figure 9.5) published this gas law and credited the law to Jacques Charles. Charles' law

Figure 9.4. Jacques Charles (1746–1823) in the first manned balloon flight in 1783.

Figure 9.5. French scientist Joseph Gay-Lussac (1778–1850).

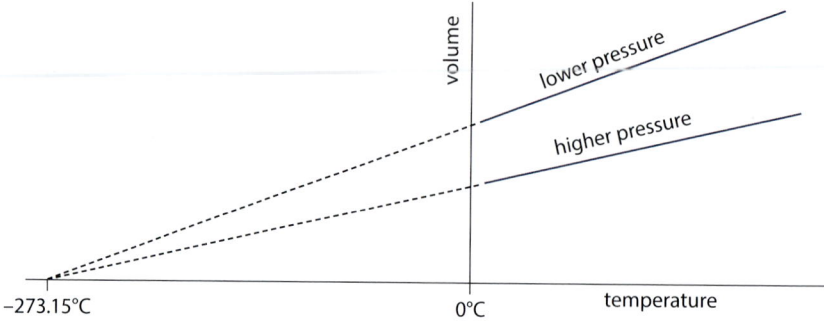

Figure 9.6. Graphical representation of the relationship between volume and temperature for a gas at constant pressure, Charles' law.

states that *at a constant pressure, the volume of a fixed amount of gas is directly proportional to its temperature*, or

$$V = kT \qquad (9.5)$$

As before, k is a constant. Graphically, the relationship between volume and temperature is as shown in Figure 9.6. Charles' law states that volume is a linear function of temperature. An important and interesting feature of this law is that the lines on the graph representing different pressures all converge to the same point on the temperature axis, −273.15°C (0 K), the temperature known as *absolute zero*. This obviously suggests that the volume of a gas goes to zero at absolute zero, which may sound puzzling at first. The resolution of this puzzle lies in the kinetic-molecular theory of gases discussed in the previous chapter. There we saw that the kinetic-molecular theory of gases includes the assumption that the atoms or molecules in a gas are so far apart that the total volume of the particles is *negligible*. Obviously, the total volume of the particles in a gas is not literally zero; atoms do take up space. According to current scientific theory, at absolute zero, molecular motion ceases. So the particles in the gas do not need any room to move around, but they still take up the space of their own total volume.

The interesting thing about Charles' law is that it relates directly to the kinetic-molecular theory. Both laws include the same assumption about the total volume of the particles in a gas. This assumption is also embedded in the *ideal gas law*, which we come to soon. Under many conditions, gases behave as if the total volume of the particles in the gas really is zero, and this means that Charles' law may be used in many cases. However, under extremes of pressure or temperature, all gases depart from the simple curves of Charles' law (and Boyle's law, too).

Charles' law may also be written as

$$\frac{V}{T} = k \qquad (9.6)$$

Written this way, we can see that the ratio of volume to temperature for a gas at constant pressure is always the same. This means that, as before with Boyle's law, we can equate the volume-temperature ratio at one set of conditions to the ratio at another set of conditions:

$$\frac{V_1}{T_1} = \frac{V_2}{T_2} \qquad (9.7) \quad \textbf{Charles' Law}$$

This equation allows us to use V_1 and T_1 as initial conditions to compute either the volume or the temperature at a different set of conditions. The principles for using Boyle's law also apply to problems using Charles' law. First, the equation consists of two ratios, so it doesn't matter what the units of measure are. Second, the temperatures must be in absolute units, which means kelvins. (There *is* an absolute scale related to the Fahrenheit scale called the Rankine scale. If you simply must use U.S. Customary Units, you can use temperatures in °R. However, the sensible thing to do in most circumstances is to convert temperatures into kelvins, an absolute scale, and the SI system base unit for temperature.) We touch on this again in Section 9.2. Finally, keep in mind that Charles' law only applies when the pressure is constant.

▼ Example 9.2

A special cylinder and piston arrangement is designed so that the piston is weighted and slides freely in the cylinder. This cylinder is oriented vertically, as shown in Figure 9.7. The weight of the piston provides a constant pressure on the gas, and the volume of the gas increases and decreases with changes in temperature.

A chemist observes that when 1,275 cm³ of gas are heated to 166.0°F, the gas expands to 1,565 cm³. Determine the original gas temperature and state the result in degrees Celsius.

We can leave the volumes in cm³, but the temperature must be in kelvins, so we begin by executing this conversion.

$$T_C = \frac{5}{9}(T_F - 32°) = \frac{5}{9}(166.0° - 32°) = 74.4°C$$

$$T_K = T_C + 273.15 = 74.4 + 273.15 = 347.6 \text{ K}$$

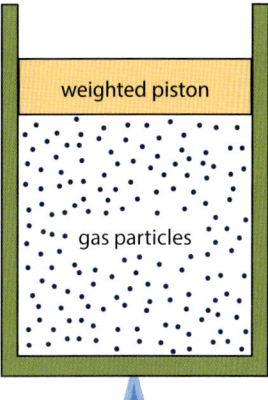

gas particles

weighted piston

We seek to determine the original temperature of the gas, which is T_1. So we must solve the Charles' law equation for this variable. Begin by cross multiplying and dividing by V_2:

$$\frac{V_1}{T_1} = \frac{V_2}{T_2}$$

$$V_1 T_2 = V_2 T_1$$

$$T_1 = T_2 \frac{V_1}{V_2}$$

Figure 9.7. Apparatus for visualizing Charles' Law.

Next we insert the values and calculate the original temperature.

$$T_1 = T_2 \frac{V_1}{V_2} = 347.6 \text{ K} \cdot \frac{1275 \text{ cm}^3}{1565 \text{ cm}^3} = 283.2 \text{ K}$$

Finally, we convert this to °C as the problem requires, and round to three significant digits.

$T_K = T_C + 273.15$

$T_C = T_K - 273.15 = 283.2\,\text{K} - 273.15 = 10.0°\text{C}$

9.1.3 Avogadro's Law

There are several more important historic gas laws from the 19th century for us to examine. We review one of them here, and save the others for later in the chapter. Amedeo Avogadro (Figure 9.8)[1] was an Italian scientist whose work in molecular theory was ahead of its time. In 1811, a time when relatively few scientists accepted the existence of atoms, Avogadro was already hypothesizing the existence of molecules—clusters of atoms—an idea that no one else accepted at the time. Avogadro correctly predicted that some gases, such as O_2, exist as diatomic molecules. General acceptance of Avogadro's molecular hypothesis did not occur until demonstration of it was given by Stanislao Cannizzaro in 1860, four years after Avogadro's death.

Part of Avogadro's 1811 molecular hypothesis is now known as *Avogadro's law*. This law states that *at equal temperatures and pressures, equal volumes of gas contain equal numbers of particles.* It is difficult to convey the immense importance of this discovery. The fact that equal volumes of gas contain the same number of particles means that gases can be weighed to determine relative molecular masses and atomic masses. This is how early 19th-century measurements of atomic mass and molecular mass were achieved.

One of the consequences of Avogadro's law is that the volume of a gas is directly proportional to the number of particles of gas present. Or, using the number of moles of gas, n, as a measure of particles,

Figure 9.8. Italian physicist and chemist Amedeo Avogadro (1776–1856).

$$V = kn \tag{9.8}$$

As we will see below, one mole of any gas at atmospheric pressure and 0°C has a volume of 22.4 L. This is a well-known consequence of Avogadro's law. Notice that Avogadro's law, like Charles' law, is consistent with the kinetic-molecular theory of gases. Equal quantities of all gases at the same temperature and volume exert the same pressure, regardless of the size of the gas particles. This is consistent with the assumption that the total volume of the gas particles is negligible. It is also consistent with our understanding of the internal energy of the gas as the sum of the energies of gas particles, each being $E = \frac{1}{2}mv^2$, since larger molecules move with lower velocities (Figure 8.3).

---

[1] I have often observed that when first seeing this image, students show disrespect by laughing at Avogadro's personal appearance. I remind them that it is wrong to laugh at someone for something they have no control over. Avogadro was not only a brilliant scientist and professor, he was a devoted Christian believer and faithful family man with six children. He was a man made in God's image—someone to admire, not to despise.

9.2 The Ideal Gas Law

9.2.1 Standard Temperature and Pressure

Since the volume of a given amount of gas is completely dependent on the pressure (Boyle's law) and temperature (Charles' law), gases are often compared by citing volumes at *standard temperature and pressure*, or *STP*. In chemistry, STP conditions are defined as an absolute pressure of 100 kPa (0.987 atm, 1 bar) and a temperature of 0°C (273.15 K).

In defining STP, the pressure is designated as "absolute pressure" because of the convention in industry of specifying pressures using atmospheric pressure as a zero reference rather than using a vacuum as the zero reference. For example, we would say that a flat tire on a car has no pressure (pressure = 0 psi). After the tire is repaired and pressured up, the tire might be pressurized to 35 psi. It is common to refer to pressures this way, using atmospheric pressure as a zero reference, and stating pressure values as the pressure above atmospheric pressure. A pressure value referenced this way is called a *gauge pressure*, or P_{gauge}.

A mechanic's pressure gauge for measuring tire pressure reads gauge pressure. A gauge pressure of 0 psi means atmospheric pressure, which means that gas molecules (such as air) are still present at a pressure of 14.7 psi. The term *absolute pressure*, designated as P_{abs}, is used to denote a pressure reading based on a complete vacuum (no gas particles) being the zero reference. The absolute pressure in a tire inflated to 35 psi is 35 psi + 14.7 psi = 49.7 psi. In summary, atmospheric pressure, designated P_{atm}, relates to gauge pressure and absolute pressure as:

$$P_{abs} = P_{atm} + P_{gauge} \tag{9.9}$$

Gauge pressure values are commonly used in industry and in specifications involving consumer products such as bicycle tires. However, gas calculations in chemistry and physics require all pressure and temperature values to be in absolute units. For this reason, pressure values in chemistry are usually stated as absolute pressure values.

By contrast, temperature values are often measured or stated in Celsius degrees, which is not an absolute temperature scale. Before performing any gas calculations, temperatures in Celsius degrees must be converted to kelvins.

9.2.2 The Ideal Gas Law

If we combine together the three gas laws we have considered so far, we have one of the most important and well-known laws in chemistry—the *ideal gas law*. We examine the ideal gas law in detail shortly, but before we do so, some words about its applicability are in order.

Chemists often refer to "ideal gases" and "real gases." The ideal gas law is a good model and useful in many circumstances when routine calculations are adequate. However, there is actually no such thing as an "ideal gas," a gas that perfectly obeys the gas laws under all conditions. As we saw previously, many gases do exhibit ideal gas behavior over a wide range of conditions, which is why the ideal gas law is so useful to chemists and physicists. But at extreme pressures—or even moderate pressures in some cases—gas particles are compressed together densely enough that their volumes are no longer negligible, and departures from ideal gas behavior are observed. Similar factors apply to temperature extremes. In this course, our calculations involve only ideal gases. But you should be aware that much more sophisticated models of gas behavior now exist. You can read a bit more about how scientists have attempted to model the behavior of "real gases" in the accompanying box.

It is important to understand how the ideal gas law emerges from the three laws we have considered so far. According to Boyle's law, expressed in a basic form in Equation (9.1), the volume of a gas is inversely proportional to the absolute pressure, or

> **Hmm... Interesting.**   **The gas laws as models**
>
> After the ideal gas law had been worked out, the first effort at modeling the behavior of "real gases" was introduced in 1873, when Dutch physicist Johannes Diderik Van der Waals (1837–1923, shown in the photo) derived the equation that now bears his name. Van der Waals is the same scientist for whom are named the Van der Waals forces. For his work in these areas, Van der Waals was awarded the 1910 Nobel Prize in Physics. The Van der Waals equation, which applies to both liquids and gases, takes into account the particle volumes and the intermolecular forces between them. The Van der Waals equation is an improvement over the ideal gas law, but is still not adequate for highly rigorous calculations. For chemists at that level, much more complex gas models must be used.
>
>
>
> As we have seen repeatedly, science is not about making truth claims; science is about modeling—developing theories that model nature with increasing accuracy. This may be as good a place as any to quote Gilbert N. Lewis on this subject. We met Lewis when we studied his ideas about the electron pair in covalent bonding, and his development of Lewis structures as a way of representing molecules. In 1925, Lewis was invited to give the Silliman Memorial Lectures, a series of lectures given annually by a prominent scientist. At that time, Lewis stated:
>
>> The scientist is a practical man and his are practical aims. He does not seek the *ultimate* but the *proximate*. The theory that there is an ultimate truth, although very generally held by mankind, does not seem to be useful to science except in the sense of a horizon toward which we may proceed.

$$V = \frac{k}{P}$$

In this expression, k is a constant that depends on the units of measure involved. According to Charles' law, the volume is also directly proportional to the absolute temperature, as Equation (9.5) indicates:

$$V = kT$$

If the gas volume is inversely proportional to pressure and directly proportional to temperature, then we can combine the equations together, lumping the two constants together to be a single new constant. Doing so we have

$$V = \frac{kT}{P} \tag{9.11}$$

We have also seen from Avogadro's law that the volume of a gas is proportional to the number of particles or moles of gas present. From Equation (9.8) we have

$$V = kn$$

Combining this expression with Equation (9.11), and lumping the constants together again, we have

$$V = \frac{nkT}{P}$$

Rearranging this expression, and replacing the letter k with an R to represent the constant, we have the *ideal gas law*, one of the most famous equations in science.

$$PV = nRT \qquad (9.12)$$ **Ideal Gas Law**

The constant R in the ideal gas law goes by various names, including the *ideal gas constant*, the *molar gas constant*, and the *universal gas constant*. The value of the ideal gas constant depends on the units of measure associated with it, and because of all the different units in use for pressure and volume there are a lot of different values for R. A few of the most common of these are listed in Table 9.1.[2] All you have to do to solve ideal gas problems is use a value for R that incorporates the units you have for your pressure and volume values. Your values for n and T are always in moles and kelvins, and all the constants listed incorporate these units.

The fourth value listed in the table uses MKS units, units of measure based on the meter-kilogram-second subsystem of the SI System of units. For this reason, this value is commonly used in physics problems, where the use of MKS units is standard practice. The numerator unit, joules (J), is derived as follows, using the fact that the unit N·m is defined as the joule, a quantity of energy:

$$m^3 \cdot Pa = m^3 \cdot \frac{N}{m^2} = N \cdot m = J$$

As long as you keep your units straight, ideal gas law problems are not difficult. Just always make sure you use absolute pressure and absolute temperature values.

▼ **Example 9.3**

Demonstrate that at atmospheric pressure, 101.325 kPa, and 0.00°C the volume of 1.00 mole of any gas is 22.4 L.

The given information is as follows:

$P = 101.325$ kPa

$n = 1.00$ mol

$T_C = 0.00°C$

$T_K = T_C + 273.15 = 0.00 + 273.15 = 273.15$ K

Now we write the ideal gas law and solve it for the unknown, the volume:

---

2 Since 2019, R has an exact value. See Appendix A.

R Value, with Units	Units for P, V
$8.314 \; \frac{L \cdot kPa}{mol \cdot K}$	kPa, L
$0.08206 \; \frac{L \cdot atm}{mol \cdot K}$	atm, L
$8314 \; \frac{L \cdot Pa}{mol \cdot K}$	Pa, L
$8.314 \; \frac{J}{mol \cdot K}$ *	Pa, m^3
$62.36 \; \frac{L \cdot Torr}{mol \cdot K}$	Torr, L
$62.36 \; \frac{L \cdot mm\,Hg}{mol \cdot K}$	mm Hg, L
$8.314 \times 10^{-5} \; \frac{m^3 \cdot bar}{mol \cdot K}$	bar, m^3
$8.206 \times 10^{-5} \; \frac{m^3 \cdot atm}{mol \cdot K}$	atm, m^3

*These are the MKS units.

Table 9.1. Approximate values for the ideal gas constant.

$$PV = nRT$$

$$V = \frac{nRT}{P}$$

With the pressure in kPa and the required volume in L, we select a value for R from Table 9.1 of

$$R = 8.314 \; \frac{\text{L} \cdot \text{kPa}}{\text{mol} \cdot \text{K}}$$

Inserting this value along with the other information into the equation for volume, we have:

$$V = \frac{nRT}{P} = \frac{1.00 \text{ mol} \cdot 8.314 \; \frac{\text{L} \cdot \text{kPa}}{\text{mol} \cdot \text{K}} \cdot 273.15 \text{ K}}{101.325 \text{ kPa}} = 22.4 \text{ L}$$

▼ Example 9.4

A certain 20.00-L cylinder of oxygen gas contains 5.00 mol O$_2$ at 25.00°C. Determine the pressure in atmospheres in the cylinder, the density of the gas, and the number of oxygen molecules in the cylinder.

For determining the pressure, we use the ideal gas law. The other two computations involve methods addressed in prior chapters.

Our given information is as follows:

$V = 20.00$ L

$n = 5.00$ mol

$T_C = 25.00°C$

$T_K = T_C + 273.15 = 298.15$ K

Now we write the ideal gas law and solve it for the unknown, the pressure:

$$PV = nRT$$

$$P = \frac{nRT}{V}$$

With the volume in L and the required pressure in atm, we select a value for R from Table 9.1 of

$$R = 0.08206 \; \frac{\text{L} \cdot \text{atm}}{\text{mol} \cdot \text{K}}$$

Inserting this value and the other information into the equation for P, we get

$$P = \frac{nRT}{V} = \frac{5.00 \text{ mol} \cdot 0.08206 \; \frac{\text{L} \cdot \text{atm}}{\text{mol} \cdot \text{K}} \cdot 298.15 \text{ K}}{20.00 \text{ L}} = 6.11 \text{ atm}$$

This value is rounded to the required three significant digits. To determine the density, we need the mass and volume of the gas. We already have this volume, and the mass we can obtain from the number of moles and the molar mass of O_2. Looking up the atomic mass for oxygen in the periodic table, we find that the molar mass of O_2 is 32.00 g/mol. Thus the mass is

$$m = 5.00 \text{ mol } O_2 \cdot 32.00 \frac{\text{g } O_2}{\text{mol } O_2} = 160.0 \text{ g } O_2$$

Since this is an intermediate calculation, I keep one extra significant digit. Using this value and the volume, we use the density equation to determine the gas density:

$$\rho = \frac{m}{V} = \frac{160.0 \text{ g}}{20.00 \text{ L}} = 8.00 \frac{\text{g}}{\text{L}}$$

Finally, the number of oxygen molecules is determined from the number of moles and the Avogadro constant.

$$5.00 \text{ mol } O_2 \cdot \frac{6.022 \times 10^{23} \text{ molecules } O_2}{\text{mol } O_2} = 3.01 \times 10^{24} \text{ molecules } O_2$$

The last two values are stated with three significant digits, as the problem requires.

▼ Example 9.5

Figure 9.9 illustrates a flask with a device called a *manometer* attached. A manometer is a U-shaped glass tube with a liquid inside that is used to measure small pressure differences. As shown in the figure, if the pressure on both sides of the manometer is the same, the two liquid columns are the same height (indicated by the arrows on the left figure). If the pressure inside the flask increases, the higher pressure pushes the liquid toward the low-pressure side of the manometer, and the pressure inside the flask is calculated from the difference in the heights of the columns.

A sealed flask contains 1.000 L of nitrogen gas at STP. The gas is then heated until a manometer on the flask indicates the pressure has increased by 10.19 Torr. How many moles of gas are present in the flask, and what is the temperature of the gas at the higher pressure?

We are given a complete list of the initial conditions of the gas, so we use those to determine the number of moles of N_2 present in the flask. Once we know the quantity of gas present, we use the ideal gas law to find the new temperature. The given initial conditions are:

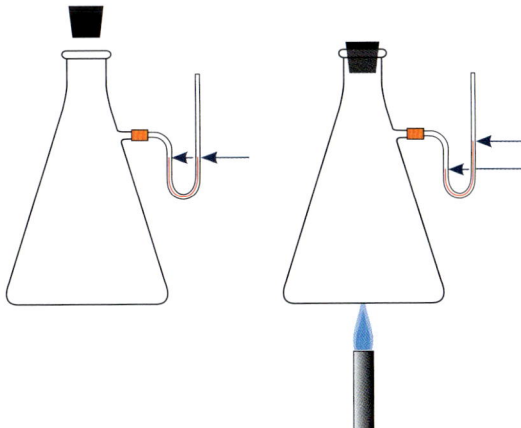

Figure 9.9. Using a manometer to measure small pressure changes at constant volume.

Chapter 9

$V = 1.000$ L

$P = 100.0$ kPa

$T_C = 0.00°C$

$T_K = T_C + 273.15 = 0.00 + 273.15 = 273.15$ K

From Table 9.1, we select the most convenient form of the ideal gas constant:

$$R = 8.314 \ \frac{L \cdot kPa}{mol \cdot K}$$

Now we write the ideal gas law equation, solve for the number of moles, insert the values, and compute the result.

$PV = nRT$

$$n = \frac{PV}{RT} = \frac{100.0 \text{ kPa} \cdot 1.000 \text{ L}}{8.314 \ \frac{L \cdot kPa}{mol \cdot K} \cdot 273.15 \text{ K}} = 0.044034 \text{ mol}$$

I keep an extra significant digit because we use this result in the next calculation. But we report this value as 0.04403 mol.

In determining the new temperature, we are working with pressures in torr, so we select this ideal gas constant from Table 9.1:

$$R = 62.36 \ \frac{L \cdot Torr}{mol \cdot K}$$

To determine the new pressure in torr, we convert the original pressure to torr, then add the pressure increase to obtain the new pressure:

$$100.0 \text{ kPa} \cdot \frac{1 \text{ atm}}{101.325 \text{ kPa}} \cdot \frac{760 \text{ Torr}}{1 \text{ atm}} = 750.06 \text{ Torr}$$

$P = 750.06 \text{ Torr} + 10.19 \text{ Torr} = 760.25 \text{ Torr}$

The information we have now for the temperature calculation is as follows:

$V = 1.000$ L

$P = 760.25$ Torr

$n = 0.044034$ mol

We write the ideal gas equation, solve for the temperature, insert the values, and compute the result.

$$PV = nRT$$

$$T = \frac{PV}{nR} = \frac{760.25 \text{ Torr} \cdot 1.000 \text{ L}}{0.044034 \text{ mol} \cdot 62.36 \frac{\text{L} \cdot \text{Torr}}{\text{mol} \cdot \text{K}}} = 276.9 \text{ K}$$

▲

▼ Example 9.6

An engineer fills a small steel sample cylinder with a sample of process gas. The sample is taken at an absolute pressure of 766 Torr and 28°C. That night, the engineer leaves the sample in her truck outside, where the temperature drops to −4°C. Determine the pressure in the sample cylinder at the new temperature. Assume the volume of the cylinder does not change.

In this case, we do not have complete information about the gas, but we do know that one of the variables T, P, or V is a constant (V in this case). We use the ideal gas law to construct a set of proportions, like those of Boyle's law and Charles' law. (In fact, you don't really need to remember those two laws explicitly because you can work them out from the ideal gas law using the technique shown here.) Starting with the ideal gas law equation, let's get all the constants for this problem on one side and everything else on the other side:

$$PV = nRT$$

$$\frac{P}{T} = \frac{nR}{V}$$

In this problem, everything on the right side of this expression is a constant, so any ratio of temperature and pressure is equal to the same value. This means the ratios of T and P are all equal to each other:

$$\frac{P_1}{T_1} = \frac{P_2}{T_2}$$

Now, we let T_1 and P_1 represent the initial conditions, and T_2 and P_2 represent the new conditions at the colder temperature. This means that P_2 is the unknown we seek. Solving for that variable, we have

$$P_2 = P_1 \frac{T_2}{T_1}$$

Now we are ready to solve for the new pressure. As always with these problems, we must make sure our temperatures and pressures are in absolute units. Here are the data and unit conversions:

$T_1 = 28°C$

$T_{1,K} = 28 + 273.15 = 301.2$ K

$T_2 = -4°C$

$T_{2,K} = -4 + 273.15 = 269.2$ K

$P_1 = 766$ Torr

According to the addition rule for significant digits, the temperature values should have no decimal places. However, I am keeping one extra digit of precision for these intermediate calculations. Putting these values into the expression for P_2, we find the new pressure:

$$P_2 = P_1 \frac{T_2}{T_1} = 766 \text{ Torr} \cdot \frac{269.2 \text{ K}}{301.2 \text{ K}} = 685 \text{ Torr}$$

9.2.3 Using the Ideal Gas Law to Find Molar Mass and Density

Suppose that you use the ideal gas law to determine the number of moles, n, present in a sample of unknown gas. If you also determine the mass, m, of the gas sample, then you can use the number of moles and the mass to compute the molar mass, M, because

$$n = \frac{m}{M} \tag{9.13}$$

Thus, the molar mass, M, can be calculated as

$$M = \frac{m}{n} \tag{9.14}$$

If we insert m/M from Equation (9.13) into the ideal gas law equation in place of n, we have

$$PV = nRT$$

$$PV = \frac{mRT}{M}$$

Since density is mass over volume, we can solve this equation for the density

$$PV = \frac{mRT}{M}$$

$$\frac{m}{V} = \frac{MP}{RT}$$

or

$$\rho = \frac{MP}{RT} \tag{9.15}$$

To save yourself time deriving equations, you may want to go ahead and commit Equation (9.15) to memory. This form of the ideal gas law is very useful with problems involving density or molar mass. Note that the density of a gas is directly proportional to the molar mass and to the pressure.

▼ Example 9.7

A gas sample with a volume of 1.50 L has a mass of 7.74 g at 28.0°C and 0.974 atm. Determine the molar mass of this gas.

Since the mass of the gas is given, we can calculate the molar mass from Equation (9.14) if we have the number of moles. So we use the ideal gas law to solve for the number of moles first, and then use Equation (9.14) to calculate the molar mass.

The given information, with the necessary temperature unit conversion, is as follows:

$m = 7.74$ g

$V = 1.50$ L

$P = 0.974$ atm

$T_C = 28.0°C$

$T_K = 28.0°C + 273.15 = 301.2$ K

Selecting the appropriate ideal gas constant from Table 9.1, we write the ideal gas law equation, solve for n, insert values, and compute the number of moles.

$PV = nRT$

$$n = \frac{PV}{RT} = \frac{0.974 \text{ atm} \cdot 1.50 \text{ L}}{0.08206 \frac{\text{L} \cdot \text{atm}}{\text{mol} \cdot \text{K}} \cdot 301.2 \text{ K}} = 0.05911 \text{ mol}$$

Now we solve for the molar mass:

$$M = \frac{m}{n} = \frac{7.74 \text{ g}}{0.05911 \text{ mol}} = 131 \frac{\text{g}}{\text{mol}}$$

▲

In the next example, we pile it on. There are a lot of data, and the solution involves calculating the density as an intermediate step so that we can use Equation (9.15) to obtain the molar mass. But don't get lost in the details. The procedure simply uses the given information to determine the mass of an unknown gas, then its volume, then its density. Then we use the density form of the ideal gas law, Equation (9.15), to calculate the required molar mass. There is nothing new here, but the problem is a nice example of how one actually goes about making the necessary measurements to determine the molar mass of an unknown gas.

▼ Example 9.8

A chemist uses a sequence of measurements as a pathway to determining the molar mass of an unknown gas. First, a flask is evacuated (that is, there is a vacuum inside) and its mass is found to be 133.785 g. The flask is then filled with the gas to a pressure of 735 Torr at 31.0°C and weighed again. This time the mass is 137.982 g. Finally, the flask is filled completely with water at 31.0°C and weighed again. This time the mass is 1,066.8 g. Use these data to determine the molar mass of the gas. At 31°C, the density of water is 0.997 g/mL.

The difference between the first two mass measurements is the mass of the gas itself:

$$m = 137.982 \text{ g} - 133.785 \text{ g} = 4.197 \text{ g}$$

The first and third mass measurements allow us to determine the mass of water that fills the flask, and from this we determine the volume. The mass of the water is

$$m_{water} = 1066.8 \text{ g} - 133.785 \text{ g} = 933.0 \text{ g}$$

With the mass of the water, we use the density equation to determine the volume of the flask. The ideal gas constant we use below is in liters, so we know we want the volume/density to be in terms of liters. Accordingly, we may as well convert the volume to liters while we are at it.

$$\rho = \frac{m}{V}$$

$$V = \frac{m}{\rho} = \frac{933.0 \text{ g}}{0.997 \frac{\text{g}}{\text{mL}}} = 938.5 \text{ mL} \cdot \frac{1 \text{ L}}{1000 \text{ mL}} = 0.9385 \text{ L}$$

Now we have both the mass and volume of the unknown gas, so we compute its density.

$$\rho = \frac{m}{V} = \frac{4.197 \text{ g}}{0.9385 \text{ L}} = 4.472 \frac{\text{g}}{\text{L}}$$

Finally, we are ready to use Equation (9.15). First, let's list our data:

$$\rho = 4.472 \frac{\text{g}}{\text{L}}$$

$$P = 735 \text{ Torr}$$

$$T_C = 31.0°\text{C} + 273.15 = 304.15 \text{ K}$$

Now we write the equation, solve it for the molar mass, insert values, and compute the result.

$$\rho = \frac{MP}{RT}$$

$$M = \frac{\rho RT}{P} = \frac{4.472 \frac{\text{g}}{\text{L}} \cdot 62.36 \frac{\text{L} \cdot \text{Torr}}{\text{mol} \cdot \text{K}} \cdot 304.15 \text{ K}}{735 \text{ Torr}} = 115 \frac{\text{g}}{\text{mol}}$$

▲

9.3 The Law of Partial Pressures

9.3.1 Dalton's Law of Partial Pressures

As you recall from Chapter 2, John Dalton (Figure 9.10), was the English scientist who formulated the first detailed atomic model in 1803. Also in 1803, just one year after Gay-Lussac published Charles' law, Dalton formulated the law now known as *Dalton's law of partial pressures*. This law states that *for a mixture of non-reacting gases, the total pressure of the mixture, P_T, is equal to the sum of the pressures each gas would exert if it were present alone*, or

$$P_T = P_1 + P_2 + P_3 \ldots \qquad (9.16) \quad \boxed{\textit{Dalton's Law}}$$

where P_1, P_2, P_3, and so on are the so-called *partial pressures* for gases 1, 2, 3, etc.

Essentially, Dalton's law of partial pressures says that at a given set of conditions (T and V), the only factor that determines the total pressure, P_T, of a mixture of gases is the total number of moles of gas that are present, and that each component gas in the mixture contributes its portion of moles to the total number of moles and an identical portion to the total pressure. The portion contributed by each gas is called the *partial pressure*, and the total pressure of the mixture is the sum of the partial pressures.

Figure 9.10. English scientist John Dalton (1766–1844).

Notice that Dalton's law of partial pressures is another application of the kinetic-molecular theory of gases. If the total pressure of combined gases is simply the sum of the individual partial pressures, then we are assuming that the gases do not interact with each other (no intermolecular forces), and that the gas particles to do not take up any space.

In a gas mixture, each component gas contributes a portion of the total moles of gas present. The portion one gas contributes is called the *mole fraction*, symbolized by X. Let's let the number of moles of each component gas be represented by n_1, n_2, n_3, etc., and the total number of moles present by n_T. Then the mole fraction of gas 1 is

$$X_1 = \frac{n_1}{n_T} \qquad (9.17)$$

The partial pressure of gas 1 is the same fraction of the total pressure as the mole fraction is of the total number of moles, or

$$\frac{P_1}{P_T} = X_1 = \frac{n_1}{n_T}$$

Thus, we can write

$$P_1 = \frac{n_1}{n_T} P_T = X_1 P_T \qquad (9.18)$$

Chapter 9

For example, oxygen represents 20.95% of the earth's atmosphere, or a mole fraction of $X = 0.2095$. The pressure of the atmosphere at sea level is 101,325 Pa. Using Equation (9.18), we can compute the partial pressure of oxygen in the atmosphere.

$$P_{O_2} = X_{O_2} \cdot P_T = 0.2095 \cdot 101{,}325 \text{ Pa} = 21{,}230 \text{ Pa}$$

There are two ways to find the partial pressure of a gas in a mixture. First, as in the oxygen illustration above, if you know the total pressure and the mole fraction of the gas, you can just multiply these to get the partial pressure. Second, if you know the number of moles present for each of several gases in a mixture, you can calculate the partial pressure for each component gas using the ideal gas law, and then add the partial pressures to get the total pressure. The following example illustrates the second type of calculation.

▼ Example 9.9

A mixture of gases is made from 6.00 g O_2 and 10.0 g CH_4. The gas mixture is placed in an 18.0-L container at a temperature of 22.0°C. Determine the total pressure in the container and the partial pressure for each gas. State the pressures in torr.

We begin by computing the number of moles present for each gas. To do this, we calculate the molar masses for O_2 and CH_4, which are 32.00 g/mol and 16.04 g/mol, respectively. The numbers of moles are thus

$$n_{O_2} = 6.00 \text{ g } O_2 \cdot \frac{1 \text{ mol } O_2}{32.00 \text{ g } O_2} = 0.1875 \text{ mol } O_2$$

$$n_{CH_4} = 10.0 \text{ g } CH_4 \cdot \frac{1 \text{ mol } CH_4}{16.04 \text{ g } CH_4} = 0.6234 \text{ mol } CH_4$$

Now we use the ideal gas law to calculate each partial pressure. Solving for the pressure from the gas law, we have

$$PV = nRT$$

$$P = \frac{nRT}{V}$$

For the oxygen, we now have these data:

$n = 0.1875$ mol

$V = 18.0$ L

$T_C = 22.0°C$

$T_K = T_C + 273.15 = 22.0 + 273.15 = 295.2$ K

Selecting the appropriate gas constant for liters and torr from Table 9.1, the partial pressure for oxygen is

$$P_{O_2} = \frac{nRT}{V} = \frac{0.1875 \text{ mol} \cdot 62.36 \frac{\text{L} \cdot \text{Torr}}{\text{mol} \cdot \text{K}} \cdot 295.2 \text{ K}}{18.0 \text{ L}} = 191.8 \text{ Torr}$$

Rounding to three significant digits,

$P_{O_2} = 192$ Torr

For the methane, all the values are the same except the number of moles. Using the number of moles for methane, the partial pressure is

$$P_{CH_4} = \frac{nRT}{V} = \frac{0.6234 \text{ mol} \cdot 62.36 \frac{\text{L} \cdot \text{Torr}}{\text{mol} \cdot \text{K}} \cdot 295.2 \text{ K}}{18.0 \text{ L}} = 637.6 \text{ Torr}$$

Rounded to three significant digits,

$P_{CH_4} = 638$ Torr

Finally, the total pressure in the container is the sum of the partial pressures:

$P_T = 191.8 \text{ Torr} + 637.6 \text{ Torr} = 829.4 \text{ Torr}$

Rounding this value to three significant digits, we have

$P_T = 829$ Torr

Notice from the preceding example that the mole fractions and total pressure give us the same partial pressures. The total number of moles present is 0.8109 mol, so the mole fractions are

$$X_{O_2} = \frac{n_{O_2}}{n_T} = \frac{0.1875 \text{ mol}}{0.8109 \text{ mol}} = 0.2312$$

$$X_{CH_4} = \frac{n_{CH_4}}{n_T} = \frac{0.6234 \text{ mol}}{0.8109 \text{ mol}} = 0.7688$$

Using these values with the total pressure of 829.4 Torr and Equation (9.18), the partial pressures are

$P_{O_2} = X_{O_2} P_T = 0.2312 \cdot 829.4 \text{ Torr} = 192 \text{ Torr}$

$P_{CH_4} = X_{CH_4} P_T = 0.7688 \cdot 829.4 \text{ Torr} = 638 \text{ Torr}$

9.3.2 Collecting a Gas Over Water

In chemistry labs, experiments are often performed in which a gas is collected in a glass vessel sealed from the atmosphere in a pan of water. As illustrated in Figure 9.11, the collecting vessel is first filled with water and inverted into a pan of water so that there is no air in the top of the collecting vessel. Then the gas from a reaction is directed into the collection vessel and collected there, pushing the water level in the collection vessel down. After gas collection, the collection vessel is adjusted up or down until the water levels inside and outside the collecting vessel are the same. In this position, the pressure inside the collecting vessel is equal to atmospheric pressure, and the gas volume in the collecting vessel at atmospheric pressure may be read from the scale on the collecting vessel.

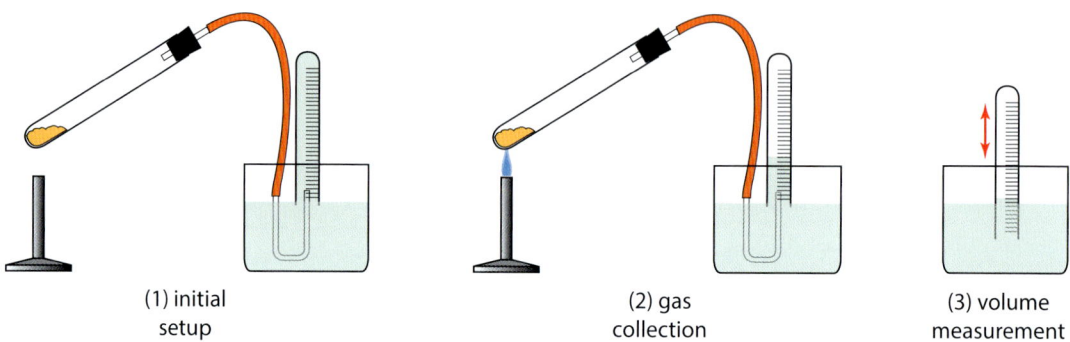

(1) initial setup   (2) gas collection   (3) volume measurement

Figure 9.11. Collecting gas over water.

The gas collected in the collecting vessel is a mixture of the gas from the reaction and water vapor, which evaporates up into the collection vessel along with the desired gas from the reaction. To determine the quantity of gas collected, the experimenter first waits for the temperature and pressure in the collecting vessel to reach equilibrium with the environment in the lab. Then the experimenter measures the volume of the gas mixture in the collecting vessel, along with the barometric pressure and temperature in the lab. The partial pressure of the collected gas is then determined from a partial pressure calculation, using the vapor pressure of water at the measured temperature. (Water vapor pressures at different temperatures are listed in Appendix A, Table A.4.) Finally, the quantity of gas collected is calculated using the ideal gas law with the volume, temperature, and partial pressure for the collected gas. The following example illustrates this technique.

▼ Example 9.10

When potassium chlorate is heated, it decomposes into potassium chloride and oxygen gas (a standard technique for producing laboratory oxygen). A sample of $KClO_3$ is decomposed to produce 0.22 L of gas, which is collected over water. Data from measurement instruments in the lab indicate a temperature of 26°C and barometric pressure of 754 Torr. Determine the number of moles of oxygen gas collected and the mass in grams of the $KClO_3$ that is decomposed.

We begin by using the law of partial pressures and the ideal gas law to determine the amount of O_2 produced. After we have that, then we use stoichiometry to work out the mass of $KClO_3$.

We must determine the partial pressure of the oxygen in the collecting vessel. From Table A.4 in Appendix A, at 26°C the vapor pressure of water—equal to its partial pressure—is 25.35 Torr. The total pressure inside the collection vessel is equal to atmospheric pressure. Dalton's law of partial pressures applied to the gas mixture in the collection vessel gives us

$$P_{atm} = P_{O_2} + P_{H_2O}$$

$$P_{O_2} = P_{atm} - P_{H_2O}$$

From this, we calculate the partial pressure of the O_2 (applying the significant digits rule for addition as we do):

$$P_{O_2} = P_{atm} - P_{H_2O} = 754 \text{ Torr} - 25.35 \text{ Torr} = 729 \text{ Torr}$$

Now we have the following data for oxygen:

$P = 729$ Torr

$V = 0.22$ L

$T_C = 26°C$

$T_K = T_C + 273.15 = 26 + 273.15 = 299$ K

With this information, we solve the ideal gas law equation for n and compute the number of moles of O_2 in the collection vessel:

$$PV = nRT$$

$$n_{O_2} = \frac{PV}{RT} = \frac{729 \text{ Torr} \cdot 0.22 \text{ L}}{62.36 \frac{\text{L} \cdot \text{Torr}}{\text{mol} \cdot \text{K}} \cdot 299 \text{ K}} = 8.60 \times 10^{-3} \text{ mol } O_2$$

We report this result with two significant digits as

$$n_{O_2} = 8.6 \times 10^{-3} \text{ mol } O_2$$

Next, to determine the amount of $KClO_3$ decomposed we need to do a stoichiometric calculation. We need to work out the formula equation and balance it. Doing so gives us

$$2KClO_3 \xrightarrow{\Delta} 2KCl + 3O_2$$

This equation tells us that the mole ratio of O_2 to $KClO_3$ is 3 : 2. Using this ratio as a conversion factor, we have

$$8.60 \times 10^{-3} \text{ mol } O_2 \cdot \frac{2 \text{ mol } KClO_3}{3 \text{ mol } O_2} = 5.73 \times 10^{-3} \text{ mol } KClO_3$$

Finally, we convert this to grams using the molar mass for $KClO_3$:

Chapter 9

$$5.73 \times 10^{-3} \text{ mol KClO}_3 \cdot \frac{122.5 \text{ g KClO}_3}{\text{mol KClO}_3} = 0.70 \text{ g KClO}_3$$

This final result is rounded to two significant digits as the given data require.

9.4 Stoichiometry of Gases and Effusion

9.4.1 Stoichiometry of Gases

We have seen that the coefficients in a chemical equation give us the ratios of molecules and the ratios of moles of the reactants and products. If we make all our measurements at the same temperature and pressure, such as STP, the same ratios of coefficients can be used to give us *volumetric ratios* of the gases involved in the reaction. Since only ratios are involved, the volume units may be any units that are convenient. The following example illustrates.

▼ Example 9.11

Assume that 3.00×10^4 m^3 of propane gas are to be completely combusted with oxygen. Determine the volumes of O_2 required and CO_2 produced. Assume all measurements are made at STP.

Since all measurements are made at the same conditions (STP, which are 0°C and 100 kPa), the coefficients from the balanced chemical equation give us the volumetric ratios we need. The equation is

$$C_3H_8 + 5O_2 \rightarrow 4H_2O + 3CO_2$$

From this equation, we see that the volumetric ratios we need are

$C_3H_8 : O_2$ $1 : 5$

$C_3H_8 : CO_2$ $1 : 3$

You can probably work the math out in your head, but here it is:

$$3.00 \times 10^4 \text{ m}^3 \text{ C}_3\text{H}_8 \cdot \frac{5 \text{ m}^3 \text{ O}_2}{1 \text{ m}^3 \text{ C}_3\text{H}_8} = 1.50 \times 10^5 \text{ m}^3 \text{ O}_2$$

$$3.00 \times 10^4 \text{ m}^3 \text{ C}_3\text{H}_8 \cdot \frac{3 \text{ m}^3 \text{ CO}_2}{1 \text{ m}^3 \text{ C}_3\text{H}_8} = 9.00 \times 10^4 \text{ m}^3 \text{ CO}_2$$

The next example illustrates the use of the ideal gas law in conjunction with a routine stoichiometric calculation.

▼ Example 9.12

Tungsten is the metal used for the filaments inside incandescent light bulbs. Tungsten is found in nature as tungsten oxide, WO_3, and is refined with hydrogen to produce tungsten metal and water. Determine how many liters of hydrogen gas at STP are required to react completely with 1,475 g of tungsten oxide.

248

We first have to work out the balanced chemical equation for this reaction, which is

$$WO_3(s) + 3H_2(g) \rightarrow W(s) + 3H_2O(l)$$

We are given a mass of WO_3. The strategy is that we convert this mass to moles and use standard stoichiometry to determine the number of moles of hydrogen required. Then we use the ideal gas equation to determine the volume of this quantity in liters.

Using the molar mass of WO_3, we find that the given quantity of WO_3 is equal to

$$1475 \text{ g } WO_3 \cdot \frac{1 \text{ mol } WO_3}{231.85 \text{ g } WO_3} = 6.3619 \text{ mol } WO_3$$

From the balanced equation, the ratio of moles WO_3 to moles H_2 is 1 : 3. Using this ratio as a conversion factor, we have

$$6.3619 \text{ mol } WO_3 \cdot \frac{3 \text{ mol } H_2}{1 \text{ mol } WO_3} = 19.086 \text{ mol } H_2$$

Now that we know the quantity of H_2 involved, we use the ideal gas law equation to compute its volume at STP. First let's write down the givens for this calculation, remembering that the values for STP are exact by definition:

$n = 19.086$ mol

$P = 100.0$ kPa

$T_C = 0.00°C$

$T_K = T_C + 273.15 \text{ K} = 273.15 \text{ K}$

Using Table 9.1 to obtain an appropriate value for R, we now use the ideal gas law equation to compute the volume of hydrogen:

$PV = nRT$

$$V = \frac{nRT}{P} = \frac{19.086 \text{ mol} \cdot 8.314 \, \frac{L \cdot kPa}{mol \cdot K} \cdot 273.15 \text{ K}}{100.0 \text{ kPa}} = 433.4 \text{ L}$$

9.4.2 Gas Diffusion and Effusion

We have already discussed the process of diffusion in liquids and gases, in which the random motion of fluid molecules causes them to mix, and over time leads to the uniform mixing of a homogeneous mixture (solution). In the process of gas diffusion, there are always two or more gases present at the same time. For example, when a person wearing a fragrance (perfume or cologne) enters a room, molecules of the fragrance diffuse throughout the molecules of oxygen, nitrogen, and other gases in the room.

Chapter 9

> ### Hmm... Interesting.   Uranium enrichment
>
> Back in the early 1940s, scientists involved in the Manhattan Project were developing the atomic bombs eventually deployed against Japan in WWII. One of the technical challenges was the process known as *uranium enrichment*. The two main isotopes of uranium are U-238, which constitutes about 99.3% of all uranium atoms, and U-235, which makes up about 0.7% of uranium. Unfortunately for those seeking to use uranium as a nuclear fuel, only the U-235 sustains the nuclear chain reaction needed for a nuclear reactor or atomic bomb. Several different methods have been developed for increasing the percentage of U-235 by separating the two isotopes; all of them are painfully slow. Three different methods were used in succession to separate the U-235 used in the manufacture of the two atomic bombs. The first process increased the percentage of U-235 to 2%. The middle process is called *gaseous diffusion*, which uses multiple stages of effusion to increase the percentage of U-235. At the time, this process was performed at the Oak Ridge facility in Oak Ridge, Tennessee.
>
> In the gaseous diffusion process, the uranium is first reacted with fluorine to produce uranium hexafluoride gas, UF_6, an extremely corrosive acid gas. The U-238 isotope is three neutrons heavier than the U-235 isotope (see Table 2.6). This isn't much of a difference, but it is enough that molecules of UF_6 containing U-235 atoms effuse through a tiny opening 0.4% faster than do UF_6 molecules containing U-238 atoms. By effusing the UF_6 through thousands of repeated stages one after the other, the concentration of U-235 was increased from 2% to 23%. Once separated to this level of purity, the uranium was sent through a third process that brought the concentration of U-235 up to its final level of 84%.
>
> The name gaseous diffusion can be a bit confusing. Despite its name, the process works by effusion. The UF_6 molecules pass through openings only 10–25 nanometers in diameter, which is only about 10% of the mean free path of UF_6 molecules (see the footnote below). To make the confusion worse, the material with the tiny holes in it is called a *diffuser*, and effusion theory is used to describe effusion rates.

A similar gas movement phenomenon is *effusion*, in which molecules of a gas pass through a very tiny opening,[3] as illustrated in Figure 9.12. Effusion occurs when molecules of gas pass singly through the opening and begin spreading into the region on the other side. A common instance of effusion occurs when the helium atoms inside a helium balloon escape through microscopic pores in the balloon into the air. The fact that this happens explains why helium balloons eventually loose their pressure and go flat.

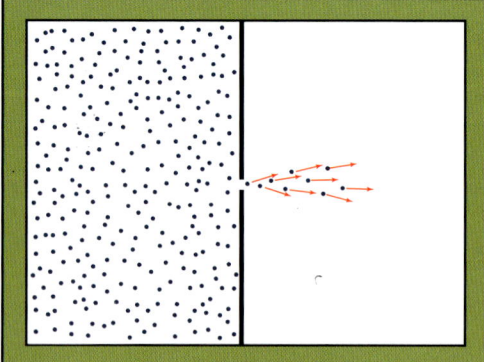

Figure 9.12. Gas effusion through a tiny opening.

---

[3] For effusion, the diameter of the opening is less than the so-called *mean free path* of the particles, the average distance they travel between collisions with each other. With an opening this small, only one particle at a time passes through it. As an example of how small this is, for helium at STP the mean free path is approximately 193 nm—far smaller than the wavelengths of visible light. If the opening is large enough so that several particles can pass through simultaneously, diffusion takes place.

Chapter 9 Exercises

SECTION 9.1

1. A fixed quantity of gas is in a 2.00-L cylinder at 22.0°C and 0.974 atm.

   a. Use Boyle's law to determine the volume the gas occupies when the pressure is reduced to 0.772 atm while the temperature is held constant.

   b. Using the original starting conditions, use Charles' law to determine the new gas temperature if the pressure is held constant and the volume increases to 3.14 L during heating.

2. A gas with a volume of 3.5 m^3 is at 0.275 bar. The volume is reduced to 1.25 m^3 while the temperature is held constant. Determine the new gas pressure.

3. A gas with a volume of 2,225 cm^3 is at a pressure of 655 Torr and a temperature of 37.00°C. In a two-stage process, the gas is first cooled until the temperature is −15°C, while the pressure is held constant. Then the pressure is decreased to 603 Torr, while the temperature is held constant. Determine the final volume of the gas.

4. Explain how the graphical depiction of Charles' law in Figure 9.6 includes assumptions that are also in the kinetic-molecular theory of gases.

5. Explain how Avogadro's law implies two of the assumptions in the kinetic-molecular theory of gases.

6. According to Boyle's law, if the volume of a gas is reduced to 1/5 of its original value at a constant temperature, what happens to the pressure?

7. According to Charles' law, if the absolute temperature of a gas is reduced by 25%, what happens to the volume?

8. According to Charles' law, the volume of a gas at 0 K is zero. Does Boyle's law predict a volume of zero at any pressure? Explain your response.

SECTION 9.2

9. What volume is required to contain 100.0 mol O$_2$ at STP?

10. Determine the temperature of 0.105 moles of a gas in a volume of 1,750 mL at 15.0 bar. State your answer in degrees Fahrenheit.

11. A mass of 2.10 g of methane is in a container at 145 Torr and 276 K. Determine the volume of the container holding this methane.

12. How many moles of gas are present in a cylinder with a volume of 1.75 L if the pressure is 0.922 bar and the temperature is 5°C?

13. A tank of methane holds 1,550 m^3 and is held at STP.

    a. How many moles of methane are in the tank?

    b. How many kilograms of methane are in the tank?

    c. What is the density of the methane?

    d. How many carbon atoms are in the tank?

14. You have probably seen the Goodyear blimps as they fly over sporting events. The blimps hold approximately 4,960 m^3 of helium. Assume the gas is at 17°C and 1.00 atm.

    a. What mass of helium is contained in the blimp?

    b. If the blimp were full of hydrogen instead of helium (it's not and won't be, but just suppose), what would be the mass of hydrogen in the blimp?

Chapter 9

15. A 2.000-L tank is found containing gas at 27.0°C and 23.11 atm.

    a. Determine the number of moles of gas in the tank.

    b. If the mass of the gas is found to be 133.0 g, confirm which halogen is in the tank.

16. A weather balloon is filled with 275 L He at 29°C and 765 mm Hg. The balloon's volume can vary, since both pressure and temperature vary with altitude.

    a. How many moles of helium are in the balloon?

    b. At an altitude of 10,000 m (a common cruising altitude for commercial aircraft), the temperature is 221 K and the pressure is 25.0 kPa. Determine the volume of the balloon at that altitude.

17. A driver fills his tires to 35 psi gauge on a day when the air is 36°C. Over the course of the next few weeks the temperature drops to –5°C. Determine the new pressure in the tires. Assume the tire volume is constant and that there are no air leaks. State your answer in psi gauge. (Note the pressure is given as gauge pressure.)

18. A SCUBA diver tank contains 12.00 L of air at a gauge pressure of 119.0 bar and a temperature of 22.00°C. (Note the pressure is given as gauge pressure.)

    a. Determine the number of moles of air in the tank.

    b. Determine the volume this much air occupies at STP.

19. Find the molar masses of the following gases, measured at the stated conditions:

    a. 0.942 g occupying 0.975 L at 26°C and 868 Torr

    b. 0.651 g occupying 888 mL at –32°C and 2.14 bar

20. Distinguish between "ideal gases" and "real gases."

21. Explain why the existence of liquid and solid states of matter imply that gases, in fact, do not act like ideal gases under all conditions.

22. Is a gas more likely to act as an ideal gas at high temperature and low pressure, or vice versa? Explain why.

SECTION 9.3

23. The *Voyager 1* space probe, which is presently leaving our solar system forever, sent back data that has allowed scientists to determine the composition of the atmosphere of Titan, Saturn's largest moon. The atmosphere consists of 82.0 mol percent N_2,[4] 12.0 mol percent Ar, and 6.0 mol percent CH_4. The total atmospheric pressure at Titan's surface is 1.61 atm. Determine the partial pressure of each of these gases in Titan's atmosphere.

24. Scientists are studying the effect of a synthetic atmosphere on plant growth. The components of the synthetic atmosphere are 80.5 mol percent Ar, 18.0 mol percent O_2, and 1.5 mol percent CO_2.

    a. If the total pressure of the gas mixture is to be 748 Torr, determine the partial pressure for the argon in the mixture.

    b. If this atmosphere is to be contained in a vessel with a volume of 165 L at a temperature of 295 K, how many moles of Ar are needed?

25. A sample of 2.55 g of propane, C_3H_8, is collected in a container at a pressure of 2.55 bar. In a separate container 1.01 g of methane, CH_4, are collected. The two gases are then combined

---

[4] The specification "82.0 mol percent" means the mole fraction is 0.820.

into the container the propane is in originally. Determine the mole fraction and partial pressure for each gas in the mixture, and the pressure in the container after the two gases are combined.

26. In the figure to the right, two flasks containing gases at the conditions shown are connected together with a stopcock[5] in between. The valve is then opened. (The stopcock is shown in the open position.)

    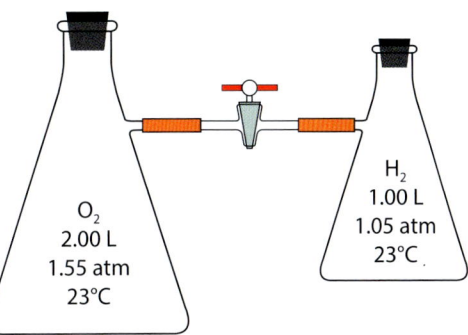

    a. Determine the number of moles present for each gas.

    b. By considering the new volume available to each gas, determine the partial pressure of each gas after the stopcock is opened and the gases have mixed. (Assume the volume of the tubing is included in the volumes given, and the temperature remains unchanged.)

    c. Determine the total pressure in the system after the gases mix.

27. A sample of nitrogen is collected over water. At a total pressure of 1.00 atm and 22°C, the container volume is 285 mL. Determine the following: (Vapor pressure data for water are in Appendix A, Table A.4.)

    a. the number of moles, partial pressure, and mole fraction of the nitrogen in the sample.

    b. the number of moles, partial pressure, and mole fraction of the water vapor in the sample.

28. When zinc reacts with sulfuric acid, the reaction products are zinc sulfate and hydrogen gas according to this equation:

    $$Zn(s) + H_2SO_4(aq) \rightarrow ZnSO_4(aq) + H_2(g)$$

    Assume that 184 mL of H_2 is collected over water at a barometric pressure of 757 mm Hg and a temperature of 28°C. Determine how many grams of zinc are consumed in the reaction.

29. A sample of mercury oxide, HgO, is decomposed by heating to form pure mercury metal and oxygen gas. A volume of oxygen equaling 355 mL is collected over water at 18°C and 725 Torr total pressure. Determine how many grams of mercury oxide are decomposed.

SECTION 9.4

30. Carbon monoxide, CO, reacts with oxygen to produce carbon dioxide. How many liters of oxygen are required to react with 5.0 L of CO, and how many liters of CO_2 are produced by the reaction?

31. Propane, C_3H_8, combusts with O_2 to produce H_2O vapor and CO_2. If 30.00 m³ C_3H_8 at certain conditions are consumed, determine the volumes at the same conditions for the other three compounds in the reaction.

32. Solid iron(III) hydroxide decomposes to produce iron(III) oxide and water vapor. Assume a reaction occurs in which 2.00 L of water vapor are produced at atmospheric pressure and 390 K. Determine how many grams of iron(III) hydroxide are consumed and how many grams of iron(III) oxide are produced.

---

5 A stopcock is a valve in a glass tube that can be opened or closed to connect or isolate the two sections of tubing.

Chapter 9

33. When Jacques Charles made his original 1783 balloon flight, the balloon contained 31,150 L H_2 at atmospheric pressure. He collected the hydrogen gas from the single replacement reaction between iron and hydrochloric acid:

$$Fe(s) + 2HCl(aq) \rightarrow FeCl_2(aq) + H_2(g)$$

Determine the mass in kilograms of the iron needed to produce this much hydrogen if the temperature is 24°C.

34. Suppose Monsieur Charles had used sulfuric acid instead of hydrochloric acid in his iron reaction to produce hydrogen. In this case, the other reaction product is iron(II) sulfate. Determine the amount of iron needed in this case, assuming the same conditions.

35. The reaction between magnesium and hydrochloric acid produces hydrogen gas. If such a reaction produces 0.015 L H_2 at STP, what mass of magnesium is used?

36. A reaction between calcium carbide and water can be used to prepare acetylene gas, C_2H_2:

$$CaC_2(s) + 2H_2O(l) \rightarrow Ca(OH)_2(s) + C_2H_2(g)$$

Calculate the volume of acetylene that is collected over water at 20.0°C by a reaction of 0.883 g CaC_2 if the total pressure of the gas is 735 Torr.

37. In a reaction between an aqueous solution of potassium iodide and chlorine gas at STP, potassium chloride in aqueous solution and 7.75 L of gaseous I_2 are produced. Determine:

   a. the number of moles of each of the reactants and each of the products.

   b. the masses in grams of each of the reactants and each of the products.

38. Solid aluminum reacts with an aqueous solution of HCl to produce an aqueous solution of $AlCl_3$ and hydrogen gas. If 110.0 g HCl are in the solution and 25.00 g Al are added, determine the following:

   a. the limiting reactant.

   b. the volume of H_2 produced at atmospheric pressure and 32.0°C.

39. Methanol, CH_3OH, is also known as wood alcohol. It is manufactured by reacting carbon monoxide and hydrogen gases at high temperature and extreme pressure. Assume 660.0 m³ CO and 1,210.0 m³ H_2 go into the reaction.

   a. Which reactant is present in excess?

   b. How much of the excess reactant remains after the reaction?

   c. What volume of methanol vapor is produced, assuming the same conditions?

40. Nitroglycerine, $C_3H_5N_3O_9$ or $C_3H_5(NO_3)_3$, is an oily liquid at room temperature that explodes just by being jostled too much. (It's packed with nitrates and produces lots of gas—see the box in Chapter 7 on the subject.) When it explodes, the following reaction occurs:

$$4C_3H_5N_3O_9(l) \rightarrow 10H_2O(g) + 12CO_2(g) + 6N_2(g) + O_2(g)$$

Assume that in such an explosion 1.00 g H_2O is produced at 350°C and 50.0 bar. Determine:
   a. the total volume of all the reaction products.

   b. the mass of nitroglycerine consumed in the reaction.

GENERAL REVIEW EXERCISES

41. Determine the wavelength of a photon with an energy of 3.20×10^{-18} J, and state what part of the electromagnetic spectrum it is in.

42. Determine the percent composition of nitroglycerin, $C_3H_5(NO_3)_3$.
43. In Jacques Charles' hydrogen production process (exercise 33 above), if 53.5 kg Fe are used and 19,625 L H_2 are produced,
    a. what is the percent yield of the reaction?
    b. how many molecules of hydrogen are produced?
    c. what is the mass of iron(II) chloride produced (assuming 100% efficiency)?
44. Use only the periodic table to answer the following questions about sodium, bromine, strontium, plutonium, and bismuth.
    a. Which atom has the largest radius?
    b. Which element is a halogen?
    c. Which element has electrons in the $5f$ subshell?
    d. Which element is an alkaline-earth metal?
    e. Which element has the highest electronegativity?
    f. Which element has the highest electron affinity?
45. How much heat is required to melt 22.0 kg of ice at 0°C, and then warm the water to 20.0°C?
46. Carbon dioxide at –60°C and atmospheric pressure is held at a constant temperature while the pressure is increased to 10 atm. According to the phase diagram for carbon dioxide (see Chapter 8, exercise 20), what are the states the CO_2 is in at the beginning and end of this process?

Chapter 10
Solution Chemistry

A solution is formed when a solute dissolves in a solvent. In the two aqueous solutions above, the solute on the left is sodium chloride; the solute on the right is copper sulfate. Both the solutions are saturated—there is as much solute dissolved in the water as the water can hold. You can see the excess solute on the bottom of the beaker in both solutions. This means that each of the solutions is in equilibrium with the undissolved solute—solute is continuously dissolving into solution and recrystallizing out of solution at the same rate.

Both solutions are transparent, and the sodium chloride solution is colorless as well.

Objectives for Chapter 10

After studying this chapter and completing the exercises, you should be able to do each of the following tasks, using supporting terms and principles as necessary.

SECTION 10.1

1. Describe the process of dissolution of an ionic solid, acid, or base in a polar solvent, such as water.

2. Use energy concepts to explain how the heat of solution might be positive, indicating an endothermic dissolution process, or negative, indicating an exothermic dissolution process.

3. Describe entropy, and use the concepts of entropy and energy to explain why an endothermic dissolution process can occur.

4. Define *electrolyte* and explain why electrolytes conduct electricity.

5. Distinguish between strong and weak electrolytes, and account for the difference.

SECTION 10.2

6. Use the solubility guidelines to classify given compounds as soluble or insoluble in water.

7. Explain what is meant by the phrase "like dissolves like" and what factors are involved.

8. Explain why solutes of ionic solids and polar molecules dissolve in polar solvents but not in nonpolar solvents, and explain why solutes of nonpolar molecules do not dissolve in polar solvents such as water.

9. Give examples and descriptions of several solid solutions.

10. Describe the effect of temperature on liquid solutions of solids and gases.

SECTION 10.3

11. Define and calculate *molarity* and *molality*.

SECTION 10.4

12. Use the solubility guidelines to determine if a precipitate forms when given aqueous solutions of ionic solids are combined.

13. Write ionic equations and net ionic equations describing precipitation reactions.

14. Define *spectator ion*.

SECTION 10.5

15. Define *colligative property* and list three colligative properties.

16. Describe, in detail, vapor pressure lowering, freezing point depression, and boiling point elevation.

17. Give examples of freezing point depression from everyday life.

18. Use molal freezing-point and molal boiling-point data to compute freezing points, boiling points, and molal concentrations for solutions.

19. Describe the effect of electrolytes on calculations of freezing and boiling points of solutions.

Chapter 10

10.1 Dissolution

10.1.1 The Process of Dissolving

Today it is difficult for us to imagine the intellectual climate among chemists in the 1880s. Many prominent scientists at that time were still opposed to the theory of atoms (including Dmitri Mendeleev), despite the success of John Dalton's 1803 atomic theory. At this time, a key problem was to explain why some substances dissolve in water while others do not, and why after dissolving in water, some solutions conduct electricity while others do not.

In 1887, Swedish chemist Svante Arrhenius (Figure 10.1) published his theory of ionic dissociation. Despite the combative atmosphere of the time, with scientists ridiculing each other for their belief—or their refusal to believe—in atoms, Arrhenius' theory encountered little opposition and was widely accepted by chemists within just a few years. Arrhenius and three other prominent chemists of the day became known as the "wild army of the Ionists," and they fought tooth and nail for acceptance of dissociation theory. They won the debate, and today dissociation theory is the most fundamental component of our theoretical understanding of solutions.

The process of dissolving—or *dissolution*—involves a *solvent*, often water, into which is dissolved another substance, called the *solute*. If a particular solute dissolves in a given solvent, it is said to be *soluble* in that solvent. So many things are soluble in water that water is often called the "universal solvent," although this term is quite hyperbolic. A lot of stuff *doesn't* dissolve in water, as you know if you have ever tried to rinse oil or grease off your hands with water but no soap.

Figure 10.1. Swedish chemist Svante Arrhenius (1859–1927).

The solubility of different substances in water varies over an extremely wide range. Some substances are essentially insoluble, some are soluble to a degree, and some liquids are infinitely soluble. Substances that can dissolve in a solvent in any proportions are referred to as *miscible*. Ethanol and acetic acid, two compounds we discuss further below, are both miscible in water.

The term *dissociation* refers to the separation or coming apart of an ionic substance in a solvent composed of polar molecules. Common substances that dissociate are many ionic compounds, acids, and bases.

Water, as you know, is composed of polar molecules. Figure 10.2 illustrates what happens when an ionic compound such as sodium chloride dissolves in water. The hydrogen or δ^+ ends of the water molecules exert an electrical attraction on the negative ions in the crystal, the chloride ions in the case of sodium chloride. At the same time, the oxygen or δ^- ends of the water molecules exert an electrical attraction on the positive ions in the crystal, sodium ions in the case of sodium chloride. These attrac-

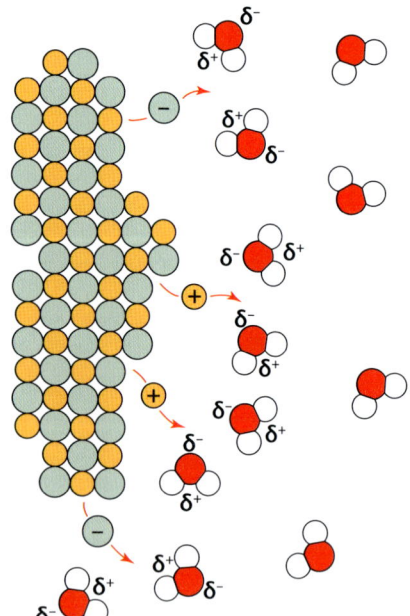

Figure 10.2. Dissociation of an ionic crystal in water. (Note that for clarity, the H$_2$O molecules are drawn much farther apart than they actually are.)

tions are strong enough to pull the ions out of the crystal lattice. This is amazing, when you think about it, because ionic bonds are quite strong. Note that because of hydrogen bonding, the H_2O molecules in water are actually much closer together than indicated by the figure. The molecules are shown spread out in the drawing for clarity.

Once again, this chemistry is all about electrical attractions. But understanding the process of dissociation also requires us to consider the energies involved in these electrical attractions. The lattice energy of a typical ionic crystal is so high that if the energy to pull the crystal apart were to come entirely from the internal energy of the water molecules (that is, the kinetic energy associated with their motion at a certain temperature), the temperature of the water molecules would have to plummet. Not only is energy required to dissociate the crystal lattice, additional energy is required to overcome the hydrogen bonding and separate the water molecules from each other to make room among the H_2O molecules for the ions.

For example, just considering the energy needed to overcome the lattice energy of 10 g NaCl, if the energy were to come from the thermal energy in the water, the temperature of 400 mL H_2O would drop by 141°F! (This is a simple calculation based on the equation from Chapter 8, $Q = Cn\Delta T$.) Obviously, when you dissolve a bit of salt in a glass of water, this doesn't happen. So where is the energy coming from that keeps the water temperature more or less constant while the salt dissolves?

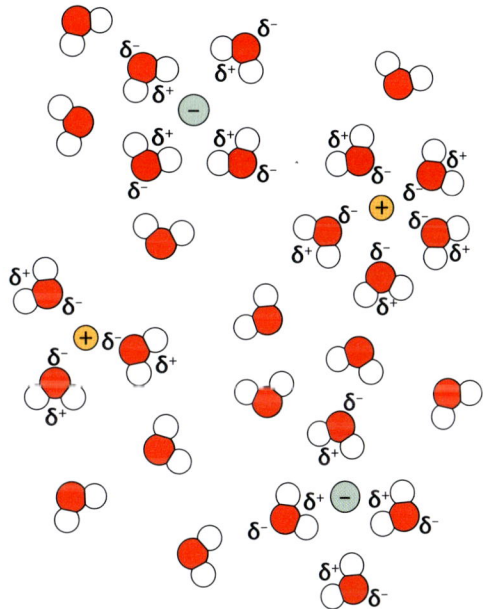

The answer is given by the process of *hydration* that occurs in parallel with the dissociation. Recall that when entities of opposite charge are allowed to fly toward each other, energy is released. Figure 10.3 illustrates what happens as ions from a crystal dissociate in water. The polar water molecules rush in to form cages of molecules around the ions, with either the δ^+ ends or the δ^- ends of the molecules pointing toward the ions, depending on each ion's charge. The decrease in electrical potential energy between the molecules and ions as they come together means energy is released. In the case of salt in water, this release of energy balances (more or less) the energy absorbed by the dissociation of the crystal lattice.

Before we move on, note that the term hydration refers specifically to the ion-surrounding process that occurs in water. Other polar solvents act the same way, and the more general term for solvent molecules surrounding ions in solution is *solvation*. Besides water, other polar solvents include ammonia, ethanol, and methanol.

Figure 10.3. Hydration of ions in aqueous solution. (Again, H_2O molecules are drawn farther apart than they actually are.)

Figure 10.4 shows the forces and energy flow involved during the process of an ionic compound dissolving in water. To dissociate a crystal and push apart the solvent molecules, energy is required, so heat flows into areas where these processes are occurring. Where hydration is occurring, heat is released. The combined effect of these energy absorptions and releases determines whether the dissolving of a solute is an endothermic process (net flow of energy is in) or an exothermic process (net flow of energy is out). If the energy required for dissociation and solvent molecule separation is greater than the energy released during hydration, then more energy is absorbed than released and the temperature of the solution drops—the complete process

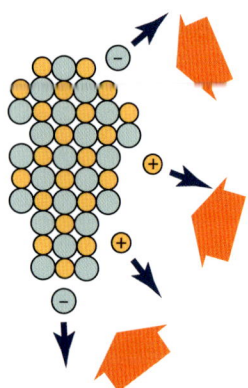 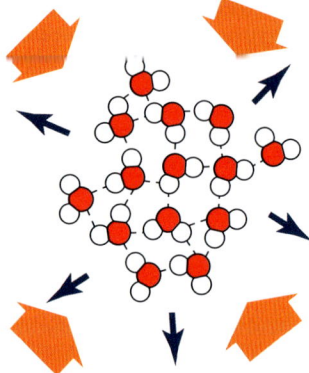 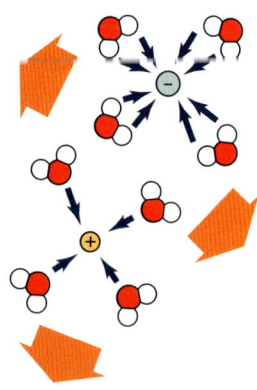

Energy is required to pull apart the crystal lattice.	Energy is required to spread apart the H₂O molecules held together by hydrogen bonds.	Energy is released during hydration as H₂O molecules are attracted to the ions.
Effect: tends to lower temperature of water	Effect: tends to lower temperature of water	Effect: tends to raise temperature of water

Figure 10.4. Forces and energy during dissociation and hydration. Blue arrows show directions of forces involved, orange arrows show direction of energy flow. Thermal energy from the water flows into areas where forces are pulling particles apart, and out of areas where particles are rushing together.

is endothermic. The drop in solution temperature occurs because the energy difference has to come from the thermal energy in the solution, and when some of that energy is removed, the solution temperature decreases. However, if the energy released by hydration exceeds the energy requirements of dissociation and solvent separation, then additional energy is released into the solution, and its temperature increases—so the complete process is exothermic.

When sodium chloride dissolves in water, the temperature does not change noticeably. But temperature changes can be substantial when other ionic compounds dissolve. The instant cold packs popular among sports trainers these days can be made from ammonium nitrate. This compound is separated from water inside the cold pack by a thin barrier. Twisting the cold pack ruptures the barrier, allowing the ammonium nitrate to dissolve into the water. This turns out to be a very endothermic process, so the pack gets cold and can be used to treat injuries such as sprains.

Our discussion so far has concentrated on ionic compounds. However, when acids and bases dissolve in water, the same logic applies. One significant difference, though, is that when acids or bases dissolve in water, the process is generally highly exothermic. Adding even a small amount of many different acids and bases to water generates a lot of heat. For this reason, it is standard procedure in every lab always to add a concentrated acid or base to water, and never the other way around. By adding the acid to water, the active ingredients in the process get diluted. The dissolving processes and the heat generated by this exothermic process are distributed through the water in the container. Conversely, when water is added to a concentrated acid or base, the acid or base remains concentrated, as does the heat released, and in some cases creates explosive fumes, such as hydrogen. Concentrated heat and explosive fumes can cause dangerous boiling, hot containers, and even explosions.

10.1.2 Heat of Solution

The net amount of heat energy absorbed during dissolution is called the *heat of solution*. Table 10.1 lists heat of solution values for a few common *electrolytes*, substances that ionize in aqueous solution. Positive values for the heat of solution indicate a net absorption of heat during the dissolution process. Negative values indicate a net release of heat during dissolution. The large negative values of the acids and bases are due to the large amounts of heat released, which we discussed above. Ammonium nitrate, the compound in the instant cold packs, has a fairly large positive value indicating a substantial absorption of heat. But make sure you have your mittens on if you ever have to mix up a beaker of rubidium perchlorate solution!

10.1.3 Entropy and Free Energy

Before we move on, let's pause here to address a question some readers may have at this point. We have seen repeatedly that the tendency of systems toward energy minimization is a rule of nature with a lot of explanatory power. Charged particles minimize the energy between them by coming together and releasing energy into the environment. But when charges are pulled apart, the energy between them *increases*, and this is what happens to the ions during dissolution. As we have seen, there is a corresponding decrease in energy during hydration. But in the case of an endothermic dissolution, there is a net gain in the energy of the particles in the solution.

Electrolyte	Heat of Solution (kJ/mol)
$HClO_4$	−88.76
HCl	−74.84
KOH	−57.61
$NaOH$	−44.51
$LiCl$	−37.03
HNO_3	−33.28
CH_3COOH	−1.51
$NaCl$	3.88
$AgNO_3$	22.59
NH_4NO_3	25.69
KNO_3	34.89
$KClO_3$	41.38
$RbClO_4$	56.74

Table 10.1. Heat of solution (at 25°C and 1 atm) for some common electrolytes.

How can this occur? In the case of an endothermic dissolution process, the net flow of energy is into the solution, and the energy required for dissociation exceeds the energy released during hydration. The key to understanding this puzzle is in the concept of *entropy*. We first encountered entropy back in the Introduction. There I defined entropy as a measure of the disorder present in a system. Just as minimizing energy is one of the universal principles according to which inanimate objects behave, maximizing entropy is as well. In fact, the second law of thermodynamics is basically a statement saying that in any process, the entropy of the universe increases. When entropy decreases in one location or system, it always increases even more somewhere else.

To bring home the concept of entropy, let's illustrate disorder with some examples. Figure 10.5 depicts four scenarios in which ordered and disordered states are represented side by side. In each case, the disordered state is a condition with higher entropy, entailing mixing, spreading out, randomness, and the erasing of concentrations, distinctions, or boundaries. For example, notice in the hot-cold object scenario (B), when the objects are first brought together, there is a sort of boundary or transition region between the hot object and the cold one. If heat is allowed to flow across this boundary, it moves from the hot object to the cold one. The boundary gradually disappears as thermal equilibrium is reached; at that point entropy is maximized.

The four scenarios represent three different cases with respect to the energy involved. In scenario (A), the house of cards represents a higher energy state than cards laying flat on the floor. In this scenario, nature's tendency to minimize energy and maximize entropy pull in the same direction—to bring down the house. Scenarios (B) and (C) are energy neutral. There is no gain or loss of energy in the overall system, assuming these objects are contained in ways that keep the total thermal energy or total number of gas particles isolated from the outside world.

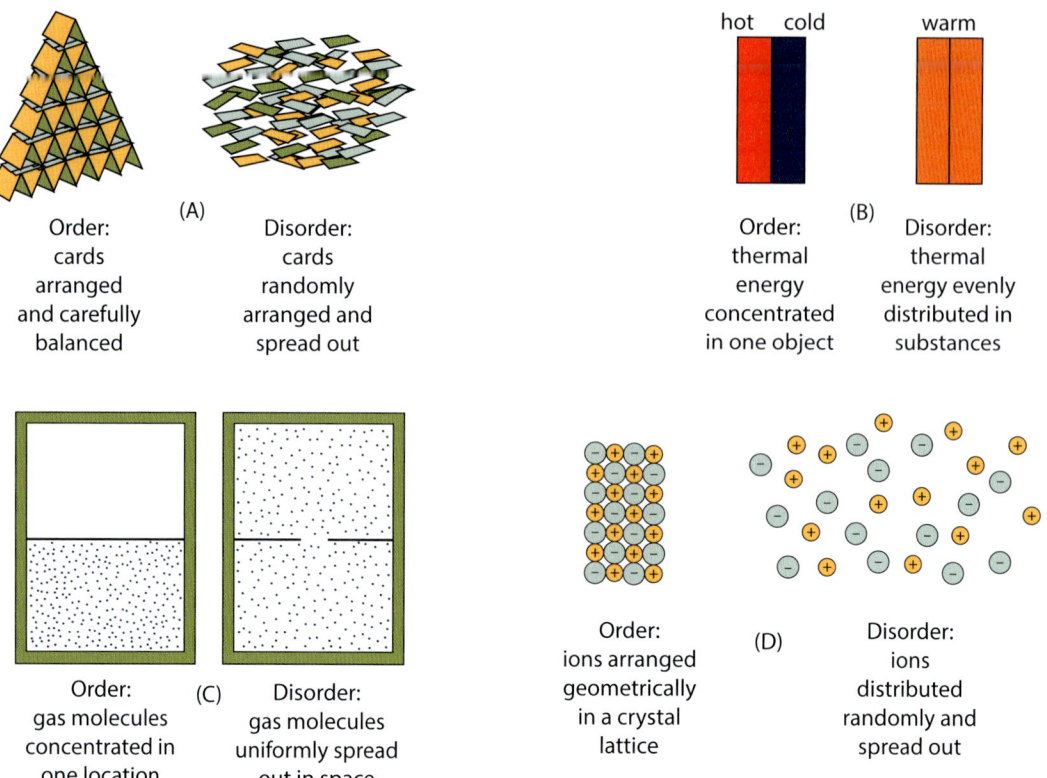

Figure 10.5. Four examples of entropy increases due to an increase in disorder. The following observations hold regarding the energy in these cases: In (A) energy is minimized; (B) and (C) are energy neutral; (D) represents an increase in system energy.

Scenario (D) represents an increase of energy. The circles represent ions, which are electrically attracted to each other and thus must be pulled apart by the input of energy. In this case, the tendencies toward energy minimization and entropy maximization are in opposite directions.

In many physical and chemical processes, the relationship between energy and entropy is like the ions in scenario (D): the tendencies to minimize energy and maximize entropy pull the system in opposite directions. In a case like this, the system can move toward a condition of minimum energy or toward a condition of maximum entropy but not both. Endothermic dissolution of an ionic crystal is one example of this situation. In such cases, the system goes in the direction in which it gains the most (lower energy or higher entropy), and it does so by borrowing, so to speak, from the other side. In the endothermic dissolution of a crystal, energy is borrowed from the thermal energy in the solution. But the borrowing is worth it because of the gain in entropy, so the crystal dissolves.

In any process, there is (a) an increase or decrease in the internal energy in the system, and (b) an increase or decrease in the entropy of the system. These two are related together in a system energy variable called the *Gibbs free energy*. If you will hang with me for just another paragraph or two you will see how this works—and then you will understand why dissolution can occur, even if the system energy increases in the process.

Natural processes proceed in the direction that reduces the Gibbs free energy in the system. There are two terms involved in calculating the Gibbs free energy. For a process such as dissolution, the first term is the heat of solution. The second is the *product* of entropy and temperature.

If we represent the change in the Gibbs free energy as ΔG, the heat of solution as ΔH, and the change in system entropy as ΔS, the Gibbs free energy calculation looks like this:

$$\Delta G = \Delta H - T\Delta S$$

Since the change in entropy is multiplied by the temperature, the determining factor as to whether the system goes in the direction that minimizes energy or the direction that maximizes entropy is the temperature. The higher the temperature is, the greater the subtracted entropy term is compared to the heat of solution term. At a low temperature, ΔH is greater than $T\Delta S$, and ΔG is positive. Such processes do not occur by themselves because, as I wrote above, natural processes go in the direction that reduces the Gibbs free energy. That is, natural processes occur only if ΔG is negative. At a high enough temperature, the $T\Delta S$ term is larger than the ΔH term. This makes ΔG negative—such a process happens spontaneously. When you dissolve salt in water at room temperature, the temperature is high enough that ΔG is negative, so the salt dissolves. But at a low enough temperature, ΔG is positive, and at that temperature the dissolution doesn't happen.

10.1.4 Electrolytes

As I mentioned in Section 10.1.2, electrolytes are compounds that ionize in aqueous solutions. Ions in solution conduct electricity. The compounds that do this are the same as those we have just been considering, namely, compounds that ionize in water. As we have seen, these include soluble ionic compounds, acids, and bases.

The fact that these compounds ionize in aqueous solution explains why they conduct electricity. Ions have a net electrical charge, and are thus affected by electric fields. When two electrodes from the terminals of a battery are placed in the solution, as illustrated in Figure 10.6, positive ions travel toward the negative electrode and negative ions travel toward the positive electrode. "Conducting electricity" means allowing electric current to flow, and electric current is, by definition, the movement or flow of electric charge.

However, as we examine in more detail next chapter, acids and bases ionize in water to differing degrees. Hydrochloric acid dissociates almost completely in water, as represented by this equation:

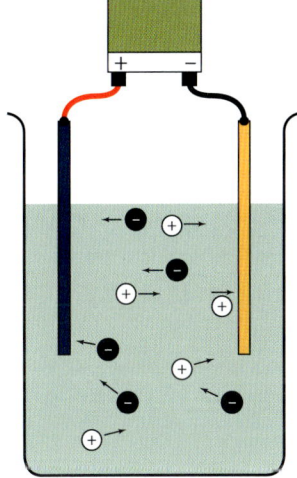

Figure 10.6. Electrolytes in solution conduct electricity.

$$\text{HCl}(aq) \rightarrow \text{H}^+(aq) + \text{Cl}^-(aq)$$

Since HCl dissociates so thoroughly, it is called a *strong electrolyte*. Strong electrolytes ionize readily, and thus in aqueous solution conduct electricity at any concentration. By contrast, acetic acid, CH_3COOH, (the main component of vinegar other than water), is a *weak electrolyte*. When acetic acid dissolves in water, most of the molecules stay intact and do not dissociate to form ions. The few molecules that do dissociate may be represented by this equation:

$$\text{CH}_3\text{COOH}(aq) \rightleftharpoons \text{H}^+(aq) + \text{CH}_3\text{COO}^-(aq)$$

In this equation, the normal reaction "yields" arrow is replaced with a double arrow, indicating that this reaction proceeds in both directions simultaneously. When acetic acid dissolves in water, the solution quickly comes to an equilibrium state in which the rate of dissociation of CH_3COOH molecules equals the rate at which H^+ and CH_3COO^- ions are joining back together to become CH_3COOH molecules. The double arrows indicate this two-directional process.

Just because a substance is soluble in water does not mean it is an electrolyte. Sucrose, or ordinary table sugar, is quite soluble in water. A molecule of sucrose has eight different —OH groups that participate in hydrogen bonding with water molecules, so it readily dissolves. (Recall from Section 6.3.3 that an —OH group joined to a molecule is called a *hydroxyl group*.) However, molecules of sucrose are not ions; they have no net charge. Accordingly, sucrose is not an electrolyte.

States of equilibrium like acetic acid in water are very important in chemical processes. We encountered one example of equilibrium in Chapter 8—when a substance is at its triple point, all three states exist in equilibrium. Molecules are continuously freezing, melting, boiling, and condensing at the same rate. This is also happening in an insulated container of ice water. The ice water is always at 0°C, and molecules of H_2O in the water are freezing and melting at the same rate. (Of course, in a warm room, the ice all eventually melts because there is no perfect insulation, and heat from outside slowly conducts into the container, causing slightly more molecules to melt than freeze.) We encounter more examples of equilibrium in this chapter.

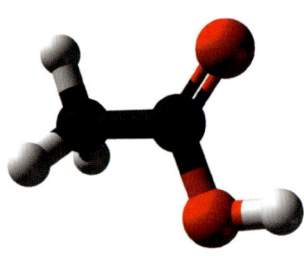

Figure 10.7. The acetic acid molecule, CH_3COOH. As usual, black = C, red = O, and white = H.

By the way, in case you are wondering about the formula for acetic acid, sometimes the formula is written as HCH_3COO so that its formula resembles those of other acids, with the ionizing H atom written first. However, these days the formula is most often written CH_3COOH because this way the formula resembles the sequence of atoms in the molecule, shown in Figure 10.7. The hydrogen atom that dissociates from the molecule is in the hydroxyl group shown at the lower right of the molecule in the figure.

10.2 Solubility

10.2.1 Ionic Solids in Water

Solids that are soluble in water are only so up to a point. There is a maximum amount of a solid solute that dissolves in a given quantity of water. A solution that contains this maximum amount of dissolved solute is said to be *saturated*. Saturation is another example of chemical equilibrium. At saturation, ions are dissociating from the crystal lattice and returning to recrystallize on the lattice at equal rates, as illustrated in Figure 10.8.

For many solutions, the saturation point depends on the temperature, and for many solutes, the solubility increases with temperature. An important exception to this common trend is NaCl. The solubility of NaCl increases only very slightly with increasing temperature.

Just about any ionic solid dissolves in water to at least a minute degree. However, as a working definition for solubility we can say that if at least 0.05 mole of a substance dissolves in one liter of water, then the substance is soluble in water; otherwise the substance is insoluble.

We don't really have a simple method for predicting whether a given ionic solid is soluble in water, but we do have lots of empirical data regarding the solubility of various compounds. These data have been generalized into a set of rules of thumb—guidelines we use to describe the solubility of certain broad classes of compounds.

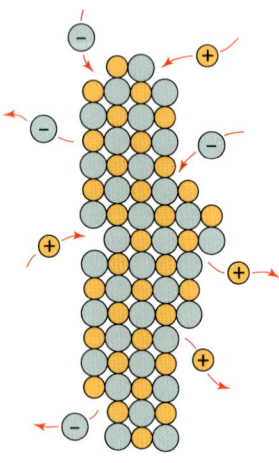

Figure 10.8. At saturation, ions dissociate and recrystallize at the same rate.

Solution Chemistry

These guidelines are listed in Table 10.2. It is important to notice that all ionic solids with an alkali metal or ammonium ion as the cation are soluble. (This is because they are always listed as exceptions to the rules denoting insoluble compounds.) Also notice that all the nitrates and acetates are soluble (because no exceptions are listed).

▼ Example 10.1

Classify each of these compounds as soluble or insoluble in water: (a) barium sulfate, (b) sodium carbonate.

According to Table 10.2, sulfates are generally soluble. However, sulfates with barium as the cation are listed as an exception. Thus, barium sulfate is insoluble.

Carbonates are generally insoluble. However, all ionic compounds with an alkali metal as the cation are soluble, so sodium carbonate is soluble.

▲

10.2.2 Ionic Solids in Nonpolar Solvents

Seeing is believing. If you have never done so, go into your kitchen, pour a bit of cooking oil into a glass, and sprinkle in some salt. You can stir the salt in the oil all you like, but it won't dissolve. Oils like cooking oil, lubricating oil, and motor oil are composed of nonpolar molecules. There are no forces in a solvent of nonpolar molecules strong enough to pull ions out of a crystal lattice, so ionic solids do not dissociate in nonpolar solvents. We return to nonpolar solvents shortly.

Soluble Compounds	
Compounds Containing These Anions	Cations That Are Significant Exceptions
NO_3^-	None
CH_3COO^-	None
Cl^-	Compounds of Ag^+, Hg_2^{2+}, Pb^{2+}
Br^-	Compounds of Ag^+, Hg_2^{2+}, Pb^{2+}
I^-	Compounds of Ag^+, Hg_2^{2+}, Pb^{2+}
SO_4^{2-}	Compounds of Sr^{2+}, Ba^{2+}, Hg_2^{2+}, Pb^{2+}

Insoluble Compounds	
Compounds Containing These Anions	Cations That Are Significant Exceptions
S^{2-}	Compounds of NH_4^+, alkali metal ions, Ca^{2+}, Sr^{2+}, Ba^{2+}
O^{2-}	Compounds of NH_4^+, alkali metal ions, Sr^{2+}, Ba^{2+}
CO_3^{2-}	Compounds of NH_4^+, alkali metal ions
PO_4^{3-}	Compounds of NH_4^+, alkali metal ions
OH^-	Compounds of NH_4^+, alkali metal ions, Ca^{2+}, Sr^{2+}, Ba^{2+}

Table 10.2. Solubility guidelines for ionic compounds in water.

10.2.3 Polar Liquids

You may have heard the rule that when it comes to solvents, "like dissolves like." The like-dissolves-like rule refers to the types of molecular forces present in a liquid solvent and potential solute. Polar solutes dissolve in polar solvents. Nonpolar solutes dissolve in nonpolar solvents. But nonpolar substances do not dissolve in polar substances, and vice versa.

You are already familiar with the polarity of water and ammonia molecules. Each of the substances shown in Figure 10.9 also consists of polar molecules. These substances are all soluble in water, and each can be used as a solvent to dissolve other polar substances. The first three substances are all members of a class of substances called *alcohols*. These molecules are not only polar, but because they each have a hydroxyl group attached, they are able to engage in hydrogen bonding in water. Looking back at the acetic acid molecule in Figure 10.7, you see the same thing. Thus, as a molecule, acetic acid is polar and participates in hydrogen bonding in water.

The fourth molecule shown in Figure 10.9, acetone, does not participate in hydrogen bonding. The molecule is covered with hydrogen atoms, but those hydrogen atoms are not attached to oxygen, nitrogen, or fluorine atoms as required for hydrogen bonding. Nevertheless, because of the strongly electronegative oxygen atom on one side of the molecule, the acetone molecule is polar. The polarity allows the molecules of a polar solvent such as water to form cages around the acetone molecules, just as they do around the ions of acetic acid, and thus acetone is soluble in water and alcohols.

Note that since acetone does not participate in hydrogen bonding, it is quite volatile. Recall from our discussion of vapor pressure in Chapter 8 that volatility is a measure of how readily a liquid evaporates to the vapor state. The more volatile a liquid is, the higher its vapor pressure. Acetone molecules are attracted to one another by Keesom forces, forces between permanent dipoles. But Keesom forces, as with all Van der Waals forces, are much weaker than the forces in hydrogen bonding, and with only weak forces attracting molecules toward one another, the molecules easily break lose and transition to the vapor state.

Alcohols are about twice as volatile as water; acetone is about 10 times as volatile as water. Liquids more volatile than water are also famous for often being flammable—the more volatile, the more flammable. (This explains the use of the term volatile as a metaphor to describe a person's personality.)

As we saw before with the sucrose molecule, all these polar molecules are soluble but none of them is an electrolyte. Again, these molecules have no net charge—they are not ions—and they do not ionize in water. Alcohols and other polar liquids that do not ionize are not electrolytes.

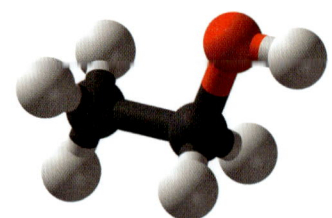

Ethanol, CH_3CH_2OH or C_2H_6O. Also called grain alcohol. This is the type of alcohol found in alcoholic beverages.

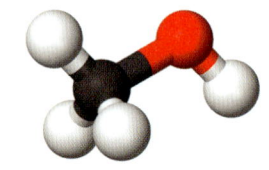

Methanol, CH_3OH, the simplest alcohol. Also called wood alcohol. This substance is toxic.

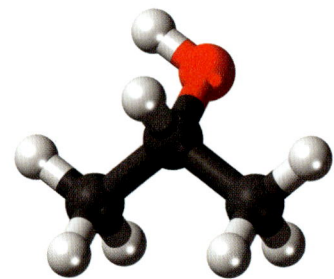

Isopropyl alcohol, written as C_3H_8O, C_3H_7OH, or $(CH_3)_2CHOH$. This is the alcohol used in rubbing alcohol.

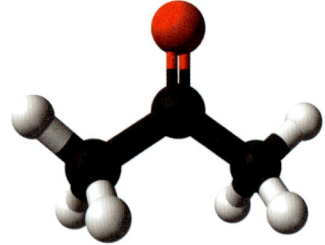

Acetone, $(CH_3)_2CO$, the solvent used in fingernail polish remover.

Figure 10.9. Examples of polar solvent molecules. These substances all dissolve in water.

10.2.4 Nonpolar Liquids

Nonpolar liquids such as cooking oil do not dissolve in water. If you try to mix a nonpolar liquid into water, the hydrogen bonding between the polar water molecules attracts them toward each other and squeezes the nonpolar molecules out. This is why oil and vinegar salad dressing separates after shaking. Vinegar is a solution of about 5% acetic acid and 95% water—all polar molecules that do not mix with nonpolar oil molecules.

Molecules of greases and oils are all nonpolar, including oils from plants like olive oil, peanut oil, and cottonseed oil, as well as fats from animals. Waxes are another example of nonpolar molecules. Nonpolar solvents include gasoline, mineral spirits, benzene, toluene, and carbon tetrachloride. These are insoluble in water, just as oils are.

In nonpolar liquids, the main forces holding the molecules together are London dispersion forces—the random forces between temporary dipoles. Even though nonpolar substances are not soluble in water, they are soluble in each other—another application of the like-dissolves-like rule. Nonpolar liquids dissolve in each other because mixing of the molecules is favored by the increase in entropy that occurs when the molecules mix.

An impressive demonstration of this is that cooking oil is very effective at removing oil-based stain or grease from your hands. After treating my leather boots with waxy mink oil, my hands are usually covered with it and it is hard to remove with soap and water. So I first rub my hands with cooking oil, which dissolves the mink oil. Then I wash off the cooking oil with soap and water. Wood stain is even harder to remove, but the cooking oil takes it right off. Try it—you'll be amazed.

10.2.5 Solutions of Solids

Most of our attention has been focused on liquid solutions. That is as it should be, and we return to them shortly. We also hit on solutions of gases very briefly in the discussion of colloidal dispersions in Section 2.2.2. There we saw that gases are miscible. Gas molecules mix together in any proportion to make a homogeneous mixture of gases—a solution in the gas state. In this section, we look at solutions of solids—*metal alloys*.

Metal alloys are a very important class of solutions. The techniques for making alloys have been widely known and continuously developed technologies for thousands of years. Here we look briefly at several common metal alloys.

There are many steel alloys; basic diagrams of the two most common are shown in Figure 10.10. *Carbon steel* contains 2.1% or less carbon by weight. The carbon atoms are much smaller than the iron atoms, and are located in the spaces or *interstices* between iron atoms. If the carbon content is greater than 2.1%, the alloy is known as *cast iron*. Cast iron is extremely strong, but also very inflexible. A lower carbon content gives the steel more flexibility, enabling it to be drawn into wire (ductility) or hammered into shapes (malleability). Cast iron is so inflexible (brittle) that it must be cast into shapes. It is not malleable at all.

On the right in Figure 10.10 is shown a typical atomic arrangement in *stainless steel*. Stainless steel is important because it resists rust. In stainless steel, atoms of chromium and nickel, which are about the same size as atoms of iron, have taken

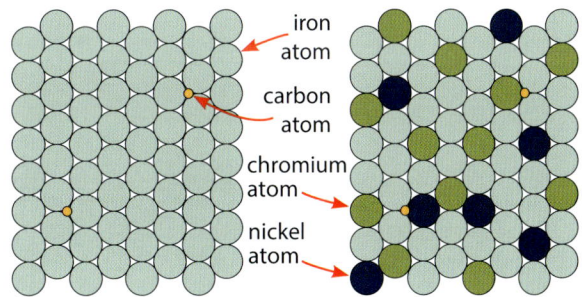

Figure 10.10. Carbon steel has 2.1% or less by weight of carbon atoms in the interstices between iron atoms (left.) Stainless steel (right) has this carbon content, plus typically about 18% chromium atoms and 8% nickel atoms in place of iron atoms in the crystal lattice.

Hmm... Interesting.   How soap works

Be thankful you don't live back in the days before soap was discovered. The ancient Greeks used to rub themselves with sand mixed with olive oil. The olive oil removed the dirt, and then the whole mixture, along with the dead skin cells broken lose by the sand, was scraped from the body with a special tool called a *strigil*. Back then, getting clean was a pain, and the squeeky-clean feeling we are accustomed to today after a shower did not exist.

Soap is obviously soluble in water. But how can it simultaneously bond to a nonpolar substance such as oil? How does soap work? You may have been involved in a soap-making project when you were a kid; what we seek here is to understand how soap does its job.

First, let's look at how it is made. Legend has it that soap was discovered when rain runoff from animal sacrifices made at a temple on Mount Sapo flowed into the Tiber River near Rome. Women washing clothes downstream of the temple found that while it was raining the water cleaned the clothes more effectively. The reason was that the runoff from the temple contained a crude soap formed in a reaction between the fats from the burnt offerings, the wood ash from the fire, and the heat of the fire. The reaction used in making soap is called *saponification*, and the words soap and saponification both derive from the name of Mount Sapo.

The two key ingredients that led to those clean clothes are potash, which is potassium hydroxide, from the ashes of the wood fire, and triglycerides, which are long, three-stranded fat molecules (shown in the first figure) from the animal sacrifices.

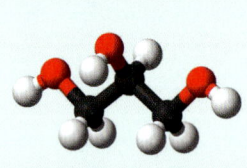

The backbone of each triglyceride strand is a carbon chain with two hydrogen atoms attached at each carbon bond. The red atoms are oxygens. With heat and the potassium hydroxide, the three long strands separate from the triglyceride molecule, leaving behind the center section of the former triglyceride, with hydroxyl groups attached in their places. This center molecule, which is *not* the soap, is called *glycerol* (second figure). The soap is the long strands, which go by various names, including glycerides, stearate ions, or, if a hydrogen atom is attached to the oxygen on the end (third figure), fatty acids. In the saponification reaction, instead of a hydrogen atom attaching to the glyceride molecule, the potassium atom from the KOH attaches there. (The red arrow in

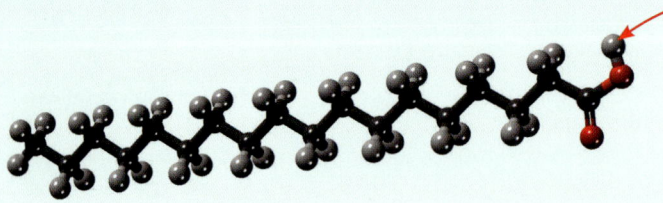

the figure points to the potassium atom). That is a soap molecule. Or, at least it is in the case of soft soap.

The same saponification reaction occurs with sodium hydroxide, NaOH or *lye*, instead of the KOH. (In the old days, sodium hydroxide was called soda ash. It was formed by burning seaweed instead of wood.) In a reaction with NaOH, the atom at the red arrow is a sodium atom. Making soap from NaOH is the most common industrial process for making soap, and results in bars of soap that are hard. Soft soaps are made with KOH; hard soaps are made with NaOH.

Now we come to how soap does its job. In water, the alkali metal atom ionizes (K^+ or Na^+), leaving the long stearate anion. (Actually, the stearate ion is just one possibility. There are others depending on the source of the triglycerides.) With the glyceride ionized to form a stearate ion, the oxygen where the K or Na was attached is now an O^- ion, and is attracted to the δ^+ regions of the water molecules. But the long carbon chains of these anions, being nonpolar, are not part of this action and are constantly being squeezed out of the way by  the hydrogen bonding between water molecules. So the stearate carbon chains are looking for a safe place to get out of the way, and in the presence of particles of nonpolar oil or grease, the carbon chains collect together by inserting themselves into the grease particles, forming what are called *micelles* (last figure). The white spheres in the figure represent the O^- ions at the ends of the stearate ions, and the yellow strands represent the carbon chains. Once the grease particle gets completely covered with the inserted stearate strands, it is no longer being squeezed out of the way. The ball is now just a big negative ion, which becomes surrounded by a cage of water molecules (the δ^+ ends) and flushes away with water, carried along by water's own polar molecules.

the places of iron atoms in the crystal lattice of the metal. In general, the chromium content of stainless steel is 10.5% by weight or higher. We encountered the popular 304 and 316 stainless steel alloys in Chapter 6 in our discussion of metallic bonding. In these alloys, the chromium and nickel contents are about 18% and 8%, respectively.

Another alloy we all encounter is gold. The indication of purity for gold jewelry uses the *carat* system, which specifies the amount of gold out of 24 parts. With 18-carat gold, the alloy is 18 parts gold (75%) and 6 parts (25%) another metal, often silver. At a purity of 24 carat, the gold must be at least 99.9% pure. The atoms of silver in the alloy take the place of gold atoms in the metal crystal lattice, just as the chromium and nickel atoms do in stainless steel. Pure gold is a fairly soft metal. Adding some silver makes jewelry much stronger, while preserving its gold color.

As with gold, pure silver is very soft. A popular silver alloy is called *sterling silver*, which is 92.5% silver and 7.5% other metals, usually copper. Adding the copper increases the strength, while preserving the color and luster. Sterling silver has been popular for centuries as an alloy for making dishes, candlesticks, vases, jewelry, eating utensils, and many other items.

A few copper alloys need mentioning, both because of their historical significance and their continued use today. You may already be familiar with these. *Brass* is an alloy of copper and zinc; the proportions of each vary widely for different purposes. Brass resists corrosion, which makes it an excellent material for use outdoors. Copper has the excellent property of being lethal to bacteria. This makes brass an excellent material for doorknobs, which get touched by many dirty hands.

Copper's antibacterial properties also make it highly appropriate for the alloys used for coins. (Pure copper is a good choice for water supply plumbing for the same reason.) Nowadays, pennies have the least copper of all U.S. coins. Since 1982, pennies have been 98.5% zinc, with just a thin copper plating on the outside. All other currently minted U.S. coins are alloys with at least 75% copper content.

Figure 10.11. An interesting 15th-century bronze door knocker.

Finally, *bronze* is a copper alloy usually containing tin. Bronze has been popular for centuries as a material for tools, machines, weapons, and statuary. As with brass, the copper content varies widely, but bronze is roughly 90% copper and 10% tin. Other metals, including zinc, are sometimes included in the alloy as well. Artisans have long preferred the rich color of bronze, as shown in the 15th-century bronze door knocker from Germany shown in Figure 10.11.

10.2.6 Gases in Liquid Solutions

One more important class of solutions are those in which a gas is dissolved in a liquid. Oxygen molecules dissolve in water, and those molecules are extracted by the gills of fish. Another familiar example is the CO_2 dissolved in soft drinks. Let's use this example to make a few important points about how gases dissolve.

Unlike other solutions we have discussed, the solubility of gases in liquids is strongly affected by pressure. Recall from our discussion of vapor pressure in Chapter 8 that at a given temperature, the gas and liquid states reach an equilibrium in which molecules are evaporating and condensing at the same rate. The same holds for a solution of gas in a liquid such as the sugar solution of a soft drink and a dissolved gas like CO_2. We represent this situation with an equilibrium equation this way:

gas + solvent ⇌ solution

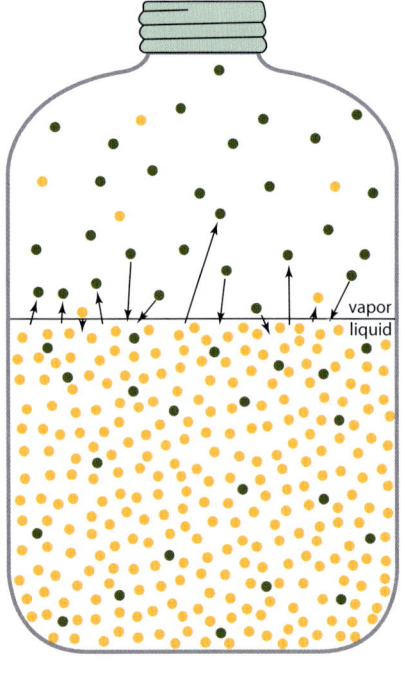

- molecules of (mostly) liquid
- gas molecule

Figure 10.12. When a gas dissolves in a liquid an equilibrium is reached, and the solubility of the gas is directly proportional to the partial pressure of the gas.

Figure 10.12 illustrates this solution equilibrium. Molecules of gas are shown dissolved throughout the liquid and in the vapor space above the liquid. An equilibrium exists in which molecules of the liquid are vaporizing and condensing at the same rate, and molecules of the gas are entering the liquid (dissolving) and leaving the liquid at the same rate.

Increasing the gas pressure results in the gas dissolving into the liquid at a faster rate. With more gas molecules in solution, they are also exiting the solution at a higher rate. A new equilibrium position is reached in which there is a higher concentration of dissolved gas molecules.

When soft drinks are manufactured, CO_2 is added to water at a pressure of 5–10 atm to form carbonated water.[1] This high pressure forces a lot of CO_2 molecules into solution. When the CO_2 pressure is increased, the amount of dissolved CO_2 increases in proportion.

When you open a container of your favorite soft drink, the pressure in the container drops to atmospheric pressure. This reduces the solubility of CO_2 by a factor of five to ten, depending on what the pressure was in the container. This causes the CO_2 to come out of solution until its concentration reaches its solubility at atmospheric pressure. This is why carbonated beverages begin *effervescing* (fizzing) when the container is opened. Over time,

---

1 The CO_2 is the source of the carbonic acid in soft drinks. When CO_2 is dissolved in water, this chemical equilibrium is established: $CO_2(g) + H_2O(l) \rightleftharpoons H_2CO_3(aq)$

if the container is left open, the amount of dissolved CO_2 drops off dramatically. We say that the beverage becomes "flat."

Although we have focused our attention on the CO_2 dissolved in beverages, these same principles apply to all solutions of gases in liquids.

10.2.7 The Effect of Temperature on Solubility

The temperature of a solution can have a pronounced effect on the solubility of both ionic solids and gases in aqueous solution. The graphs shown in Figures 10.13 and 10.14 show solubility curves for assorted ionic solids and gases in water. These curves show the amount of dissolved solute at saturation at different temperatures. In Figure 10.13, the vertical axis indicates the percentage of the total mass of the solution that is due to the mass of the solute. A few particulars in the graph are interesting to note. First, the solubility of NaCl is almost flat; it does increase with temperature, but only increases by about 2% over the temperature range of liquid water. The solubility of Li_2SO_4 actually decreases substantially with increasing temperature, an uncommon trend. By contrast, note the strong upward trend for the solubility of potassium nitrate, KNO_3.

In Figure 10.14 are four curves representing the solubilities of some common gases. The vertical axes for these two curves represent the mole fraction of the solution attributable to the gas, and all the data have been multiplied by 1,000.

These curves indicate solubility at atmospheric pressure, and in every case but one, the solubility decreases with temperature. The exception is oxygen, whose solubility is virtually constant over the temperature range shown. The solubility of oxygen is very low compared to the other gases, but apparently it is high enough for the fish to be happy. What's really interesting is that its solubility is relatively *constant*. It does vary, and at 0°C (not shown) it is about twice as high as it is at 20°C. Still, the concentration of oxygen in ponds does not vary with much temperature, which is good for the fish who live there. This matters most in shallow ponds and in rivers that are alternatively fed by rain water or snow melt. In lakes and oceans, a few meters below the surface the temperature does not vary much.

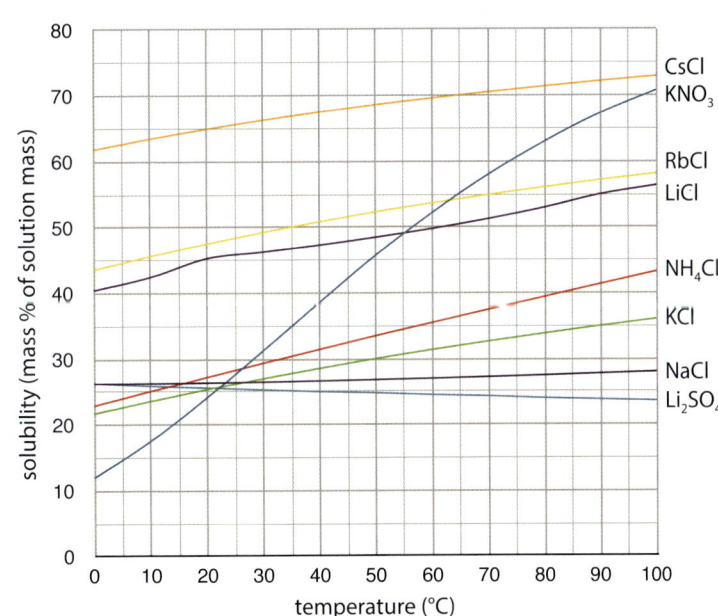

Figure 10.13. Temperature dependence of solubility for a few ionic solids in water.

Chapter 10

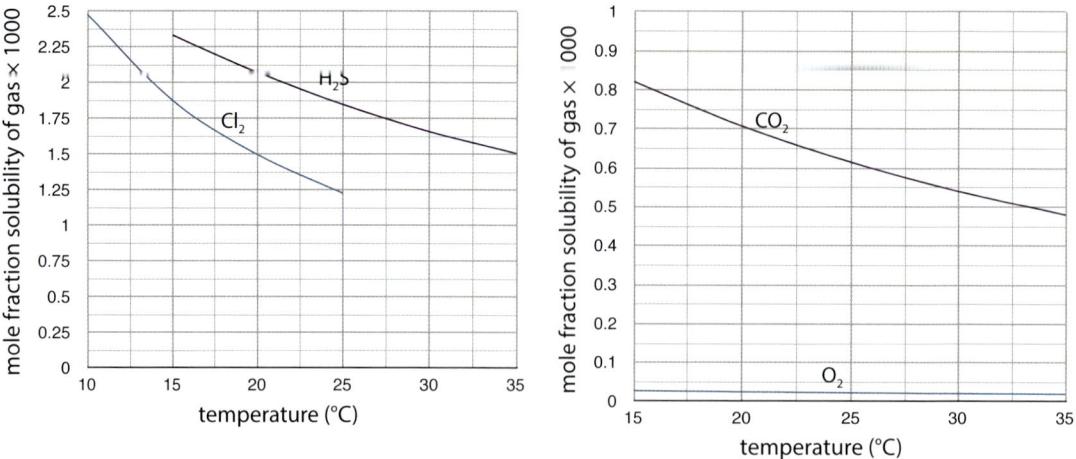

Figure 10.14. Temperature dependence of a few gases in water. The graph on the left shows H_2S and Cl_2. The graph on the right shows CO_2 and O_2. Mole fraction percentages in these figures have been multiplied by 1,000.

10.3 Quantifying Solution Concentration

10.3.1 Molarity

As you might expect, there are standard ways of referring to the concentrations of solutions. The most common is by a solution's *molarity*. Molarity, signified by M, is defined as the number of moles of solute per liter of solution:

$$\text{molarity } (M) = \frac{\text{moles solute}}{\text{solution volume (L)}} \qquad (10.1)$$

A solution with 1.00 mole of solute in 1.00 liter of total solution is referred to as a "1.00 molar solution." This concentration is written as 1.00 M. Note that the denominator in Equation (10.1) is the volume of the *solution*, not the volume of the solvent by itself.

▼ Example 10.2

A solution of sodium nitrate is prepared by adding 45.55 g $NaNO_3$ to enough water to bring the solution volume to 0.100 L. Determine the molarity of the solution.

The molar mass of sodium nitrate is 84.99 g/mol. Converting the given quantity to moles we have

$$45.55 \text{ g NaNO}_3 \cdot \frac{1 \text{ mol NaNO}_3}{84.99 \text{ g NaNO}_3} = 0.5359 \text{ mol NaNO}_3$$

With this amount in moles and a total solution volume of 0.100 L, we now calculate the concentration.

$$\text{molarity } (M) = \frac{\text{moles solute}}{\text{solution volume (L)}} = \frac{0.5359 \text{ mol}}{0.100 \text{ L}} = 5.36 \, M$$

This result is rounded to three significant figures.

Since molarity is defined in terms of total solution volume, to prepare a solution with a certain molarity one must first measure out the required amount of solute, dissolve it in water, and then add water to bring the total volume up to what is needed for the specified molarity. This is usually done using a *volumetric flask*, as shown in Figure 10.15. A volumetric flask has a line etched into the neck of the flask indicating a volume of 0.1 L, 0.25 L, 0.5 L, or 1 L. To make an aqueous solution of a particular molarity, the correct mass of solute is first calculated, based on its molar mass and the volume of the flask. The solute is then weighed out and added to the flask. The flask is filled partially with water to enable the solute to dissolve and then filled to the line with water. The following example illustrates this procedure using the same solute and quantities used for the photos in Figure 10.15.

▼ Example 10.3

A 0.25-L volumetric flask is to be used to prepare a solution of copper(II) chloride with a concentration of 0.100 M. Determine the mass of copper(II) chloride needed for this preparation.

The calculation involves solving the molarity equation for the moles of solute required, and then converting the number of moles to grams using the molar mass of $CuCl_2$. The molar mass calculation indicates the molar mass of $CuCl_2$ to be 134.45 g/mol.

$$\text{molarity } (M) = \frac{\text{moles solute}}{\text{solution volume (L)}}$$

We solve this for the moles of solute. However, instead of writing M for the molarity units, it is helpful to write the units for the solution concentration as mol/L (with a horizontal fraction bar) to make the unit cancellation clear.

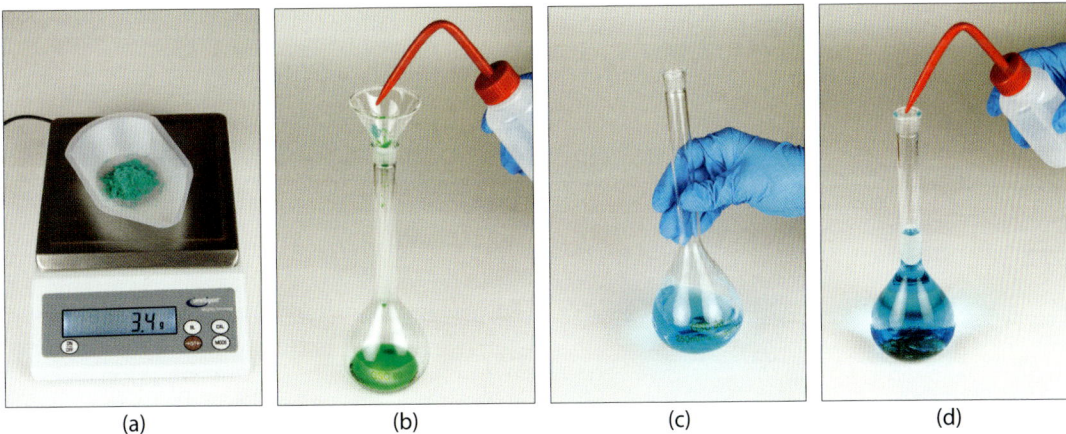

(a) (b) (c) (d)

Figure 10.15. Using a volumetric flask to prepare a standard quantity of solution with a specific concentration. First (a), the calculated mass of solute is measured out. Next (b), the solute is added to the volumetric flask and the solute particles rinsed out of the funnel and into the flask with water. Third (c), the flask is partially filled with water and the flask is swirled to dissolve the solute. Finally (d), the flask is filled to the line with water. As a final step, the flask is stoppered and inverted back and forth several times to ensure complete mixing.

moles solute = molarity · solution volume (L) = 0.100 $\frac{\text{mol}}{\text{L}}$ · 0.250 L = 0.0250 mol

Finally, we convert this mole quantity into grams.

$$0.0250 \text{ mol CuCl}_2 \cdot \frac{134.45 \text{ g CuCl}_2}{1 \text{ mol CuCl}_2} = 3.36 \text{ g CuCl}_2$$

▼ Example 10.4

A chemist needs 50.0 g potassium chromate for a reaction. She finds that the only source of potassium chromate she has on hand is 4.5 L of 8.0 M K_2CrO_4. What volume of this solution should the chemist take in order to have the required 50.0 g K_2CrO_4?

As usual, we begin by calculating the molar mass of K_2CrO_4, which we find is 194.19 g/mol. Now we convert the required quantity from grams to moles:

$$50.0 \text{ g K}_2\text{CrO}_4 \cdot \frac{1 \text{ mol K}_2\text{CrO}_4}{194.19 \text{ g K}_2\text{CrO}_4} = 0.2575 \text{ mol K}_2\text{CrO}_4$$

We use the molarity equation to determine the quantity of 8.0 M solution needed to provide 0.2575 mol K_2CrO_4. Start with the equation and solve it for the solution volume.

$$\text{molarity } (M) = \frac{\text{moles solute}}{\text{solution volume (L)}}$$

$$\text{solution volume (L)} = \frac{\text{moles solute}}{\text{molarity}} = \frac{0.2575 \text{ mol K}_2\text{CrO}_4}{8.0 \frac{\text{mol K}_2\text{CrO}_4}{\text{L of solution}}} = 0.032 \text{ L solution}$$

This result is rounded to two significant digits.

10.3.2 Molality

Another measure of solution concentration is *molality*, defined as follows:

$$\text{molality } (m) = \frac{\text{moles solute}}{\text{solvent mass (kg)}} \qquad (10.2)$$

A solution with 1.00 moles solute in 1.00 kg of solvent is a "1.00 molal" solution, written as 1.00 *m*.

Take careful note of the difference between the ways molarity and molality are calculated. Both ways of specifying solution concentration entail moles of solute. But molarity uses the total volume of the solution, whereas molality uses the mass of only the solvent. While molarity is a more convenient concentration to use for stoichiometric calculations, molality is more convenient in applications of varying temperature, such as vapor pressure. The overall volume of a solution may change at different temperatures, causing a change in the molar concentration. But

Solution Chemistry

the molal concentration does not change because it is based on the mole or mass quantities of the substances used to prepare the solution, not on the volume of the solution.

▼ Example 10.5

A 3.0 m solution of table sugar (sucrose, $C_{12}H_{22}O_{11}$) and water is prepared from 0.75 L water. Determine the mass of sucrose present in that solution.

For this calculation we need the molar mass of sucrose, which we find to be 342.30 g/mol.

Next we use the density equation to get the mass of water, using the density of water, 0.998 g/mL at room temperature. We also convert the mass to kilograms while we are at it because the molality equation requires the solvent mass to be in kilograms.

$$\rho = \frac{m}{V}$$

$$m = \rho \cdot V = 0.998 \ \frac{g}{mL} \cdot 750 \ mL = 748.5 \ g \cdot \frac{1 \ kg}{1000 \ g} = 0.7485 \ kg$$

Now we use the molality equation to determine how many moles of sucrose we have. We begin with the molality equation:

$$\text{molality } (m) = \frac{\text{moles solute}}{\text{solvent mass (kg)}}$$

We solve this equation for the moles of solute and calculate the number of moles, showing the units for molality explicitly to make it clear how the units cancel:

moles solute = molality · solvent mass (kg)

$$= 3.0 \ \frac{\text{mol } C_{12}H_{22}O_{11}}{\text{kg } H_2O} \cdot 0.7485 \ \text{kg } H_2O$$

$$= 2.25 \ \text{mol } C_{12}H_{22}O_{11}$$

Finally, we use the molar mass to convert this mole quantity into grams.

$$2.25 \ \text{mol } C_{12}H_{22}O_{11} \cdot \frac{342.30 \ g \ C_{12}H_{22}O_{11}}{1 \ \text{mol } C_{12}H_{22}O_{11}} = 770 \ g \ C_{12}H_{22}O_{11}$$

This final result is rounded to two significant digits, as required by the given information.

10.4 Compounds in Aqueous Solution

10.4.1 Ionic Equations and Precipitates

In Chapter 7, we saw that double replacement reactions, also called *exchange* or *metathesis* reactions, can lead to precipitation. (Metathesis, which is the Greek word meaning "to transpose," is pronounced meh-TATH-esis.) Earlier in this chapter, in Table 10.2, we reviewed the guidelines for determining which compounds are soluble in water and which are insoluble. Now we put these together and look more closely at how we can describe what happens when compounds in solution are mixed. As we go, keep this basic principle in mind: *in a mixture of soluble ionic compounds—electrolytes—if nothing precipitates, no reaction occurs.*

Consider a possible reaction between lead(II) nitrate and potassium iodide. From Table 10.2 we see that all nitrate compounds and all compounds with alkali metal cations are soluble. This means that lead(II) nitrate and potassium iodide are both strong electrolytes. When separate solutions of these electrolytes are prepared, their dissociation equations are as follows:

$$Pb(NO_3)_2(s) \rightarrow Pb^{2+}(aq) + 2NO_3^-(aq)$$

$$KI(s) \rightarrow K^+(aq) + I^-(aq)$$

There are three important things to notice in these dissociation equations. First, coefficients are placed on ions in the equations where required. The law of conservation of mass in chemical reactions still applies, even when we are just dissolving things in water. The formula for lead(II) nitrate indicates two nitrate ions for each lead ion, so the NO_3^- ion has a coefficient of 2. Second, not only is mass conserved, but charge is conserved as well. The reactant compounds are neutral, so the total numbers of positive and negative charges in the ions must add up to zero. Third, the equations still show the relative mole ratios of substances involved, just as other balanced chemical equations do. In other words, one mole of lead nitrate, $Pb(NO_3)_2$, dissociates to form one mole of Pb^{2+} ions, and two moles of NO_3^- ions. Likewise, one mole of potassium iodide, KI, dissociates to form one mole of K^+ ions and one mole of I^- ions.

Now assume that we combine solutions of these two soluble compounds. The left side of the chemical equation is

$$Pb(NO_3)_2(aq) + KI(aq) \rightarrow$$

To determine if a reaction occurs when these are combined, we consider the possible product compounds that can form. In addition to the reactants we started with, the possibilities are PbI_2 and KNO_3. From Table 10.2 we see that the KNO_3 is soluble. Compounds formed with the I^- anion are generally soluble too, but PbI_2 is an exception, shown in Figure 10.16. *When solutions of soluble compounds are mixed, if an insoluble compound can form and precipitate, it does.* The soluble products remain as ions in solution. So now we see that the formula equation is as follows:

Figure 10.16. Yellow lead iodide precipitates out of solution when lead nitrate and potassium iodide react. The yellow crystals are hexagonal in shape, and the precipitation is sometimes called "golden rain."

$$Pb(NO_3)_2(aq) + KI(aq) \rightarrow PbI_2(s) + KNO_3(aq)$$

In this equation, PbI_2 is identified as a solid (s), which means it precipitates. The balanced equation is

$$Pb(NO_3)_2(aq) + 2KI(aq) \rightarrow PbI_2(s) + 2KNO_3(aq)$$

This familiar form of the reaction equation is called the *molecular equation*. But when studying compounds in aqueous solution, it is helpful to write an *ionic equation* showing the ions present before and after the precipitation occurs. The ionic equation is written by writing all the reactant and product ions, and showing insoluble precipitates as solids. The reactant ions for this equation are shown in the dissociation equations above. The product ions are the K^+ and NO_3^- that remain in solution. Thus the ionic equation is

$$Pb^{2+}(aq) + 2NO_3^-(aq) + 2K^+(aq) + 2I^-(aq) \rightarrow PbI_2(s) + 2K^+(aq) + 2NO_3^-(aq)$$

Notice that all the coefficients from the molecular equation have been preserved on the ions in the ionic equation. This is necessary to show that the law of conservation of mass in chemical reactions is still in force, that charge neutrality is preserved, and to indicate the relative molar quantities involved.

10.4.2 Net Ionic Equations and Spectator Ions

Look again at the ionic equation just above. Some of the ions appear on both sides of the equation, namely, NO_3^- and K^+. Since these ions are in solution before the reaction and remain in solution, they are not part of the precipitation reaction that occurs. Ions present in the same form in both the reactants and products of an ionic equation are called *spectator ions*.

Spectator ions play no direct role in the reaction. To see only the ions and substances that participate in the reaction, we eliminate the spectator ions from both sides of the ionic equation to form the so-called *net ionic equation*. For the reaction above, the net ionic equation is

$$Pb^{2+}(aq) + 2I^-(aq) \rightarrow PbI_2(s)$$

Notice that in the net ionic equation charge and mass are both conserved, just as in the dissociation equations.

There are many different soluble ionic compounds—electrolytes—and thus there are many different dissociation reactions. Net ionic equations are important because they make it easier to compare them. After the illustration we just worked through, we know that any time solutions of ionic compounds containing lead and iodine are mixed, we always get a lead(II) iodide precipitate. For example, magnesium iodide, MgI_2, is soluble. This means that combining aqueous solutions of lead nitrate and magnesium iodide leads to results similar to those of the case we just examined.

In summary, to develop the net ionic equation for a precipitation reaction, follow these steps:

1. Identify which potential product compounds are soluble and insoluble.
2. Write the balanced molecular equation.
3. Write the ionic equation.
4. Eliminate the spectator ions and write the net ionic equation.

▼ Example 10.6

Consider the potential outcomes of combining aqueous solutions of ammonium carbonate and copper(II) chloride. Write the ionic equation, determine if a reaction occurs, identify any spectator ions, and write the net ionic equation if a reaction does occur.

According to Table 10.2, carbonates are generally insoluble, but an exception occurs when carbonate is coupled with the ammonium cation. Also, chlorides are soluble with a few exceptions that aren't relevant here, so both our reactant compounds form aqueous solutions. This means the left side of the formula equation is

$$(NH_4)_2CO_3(aq) + CuCl_2(aq) \rightarrow$$

We see that when these electrolytes dissociate, the ions formed will be NH_4^+, CO_3^{2-}, Cu^{2+}, and Cl^-. Thus potential product compounds are NH_4Cl and $CuCO_3$. From Table 10.2, it is clear that ammonium chloride is soluble and copper carbonate is not. So a reaction does occur with the formation of a copper carbonate precipitate. The full formula equation is

$$(NH_4)_2CO_3(aq) + CuCl_2(aq) \rightarrow NH_4Cl(aq) + CuCO_3(s)$$

Balancing this equation gives us the complete molecular equation, which is

$$(NH_4)_2CO_3(aq) + CuCl_2(aq) \rightarrow 2NH_4Cl(aq) + CuCO_3(s)$$

From here we can write the ionic equation, taking care to preserve quantities by introducing coefficients.

$$2NH_4^+(aq) + CO_3^{2-}(aq) + Cu^{2+}(aq) + 2Cl^-(aq) \rightarrow 2NH_4^+(aq) + 2Cl^-(aq) + CuCO_3(s)$$

The spectator ions are those that appear in the same form on both sides of the ionic equation, namely, NH_4^+ and Cl^-. Deleting these, we have the net ionic equation.

$$CO_3^{2-}(aq) + Cu^{2+}(aq) \rightarrow CuCO_3(s)$$

If you've never seen the copper carbonate precipitate, it has a lovely blue color, as you can see from Figure 10.17.

10.5 Colligative Properties of Solutions

Several important properties of solutions depend on the concentration of solute in a solution but not on the particular chemical species present in the solution. These are called *colligative properties*. The three colligative properties we address here all relate to the vapor pressure, so this is where we begin.

Figure 10.17. Formation of the pretty blue copper carbonate precipitate.

10.5.1 Vapor Pressure Lowering

As you recall from Chapter 8, the vapor pressure of a liquid at a given temperature is the pressure at which that temperature is the boiling point. The curve on a P-T phase diagram separating the vapor and liquid states is the curve representing the vapor pressure of the liquid. In Figure 10.18, the generic phase diagram we examined in Chapter 8 is shown again to illustrate the effect of a nonvolatile solute such as sugar or salt in solution in the liquid. As before, the black curves separate the states of the pure solvent. The red curves in the figure represent the new positions of the curves separating the states when the solute is present in solution.

The upper horizontal dashed line represents atmospheric pressure. This line crosses the phase boundaries at the freezing point (f.p.) and boiling point (b.p.) of the pure solvent. The lower horizontal dashed line indicates the vapor pressure of the solution at the boiling point of the pure solvent. The change in pressure, ΔP, is the amount the vapor pressure (v.p.) has been

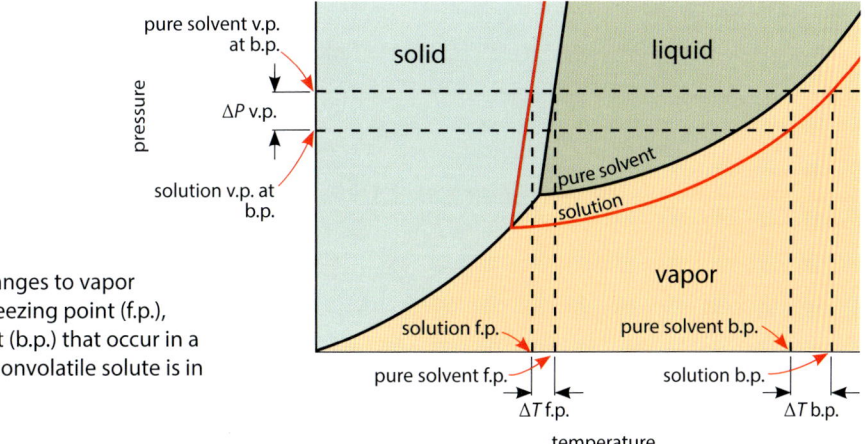

Figure 10.18. Changes to vapor pressure (v.p.), freezing point (f.p.), and boiling point (b.p.) that occur in a solvent when a nonvolatile solute is in solution.

lowered at the boiling point of the pure solvent. Not surprisingly, this effect is called *vapor pressure lowering*.

The P-T phase diagram clearly shows that the lowering of the vapor pressure curve has an effect on the freezing point and boiling point of the solution. Two of the vertical dashed lines in Figure 10.18 indicate the freezing point and boiling point of the pure solvent. The other two are drawn from the points where the atmospheric pressure line crosses the new phase boundaries (shown in red). The changes in temperature, ΔT, between the two sets of lines are the amounts the freezing and boiling points changed with the solute in solution. As you see, the freezing point decreases, an effect known as *freezing point depression*. The boiling point increases, an effect known as *boiling point elevation*.

There is a straightforward physical explanation for the lowering of vapor pressure when a nonvolatile solute is in solution in a liquid, illustrated in the aqueous solution of Figure 10.19. The solute particles (ions in this instance) with their little cages of water molecules are highlighted with yellow circles to make them more easily visible. These clusters of particles are distributed throughout the solvent, including at the liquid surface, where they reduce the surface area available to the water molecules for evaporation. Simply put, they are *in the way* of some of the higher-energy water molecules that might otherwise make it to the surface and evaporate.

The illustration in Figure 10.19 is based on ions in water, but the same effect occurs with non-electrolyte solutes such as sugar, just so long as the solute is nonvolatile so that it is not itself trying to evaporate. As mentioned at the very beginning of Section 10.5, the factor governing colligative properties like vapor pressure lowering is the concentration of the solute, not the identity of the solute. The species of solute doesn't matter; what matters is how many particles there are—the molar concentration. There more particles there are, the more the surface area available to the solvent molecules for evaporation is effectively reduced.

Notice one more important detail in Figure 10.19. When we refer to the concentration of solute particles as the key factor producing the colligative properties of solutions, we are referring to the total number of moles of *particles* in solution, regardless of what those particles are. Sugar molecules do not ionize in water, so one mole of sugar in solution results in one mole of particles. However, NaCl ionizes to form Na^+ ions and Cl^- ions—a pair of them for every unit of NaCl. This means that one mole of NaCl produces two moles of solvent particles in solution. Lithium sulfate, Li_2SO_4, ionizes to form two Li^+ ions and one SO_4^{2-} ion. So every mole of Li_2SO_4 produces three moles of particles in solution. We might expect that a solute that produces two or three moles of particles in solution would have two or three times the effect on the vapor pressure compared to a solute that produces one mole of particles.

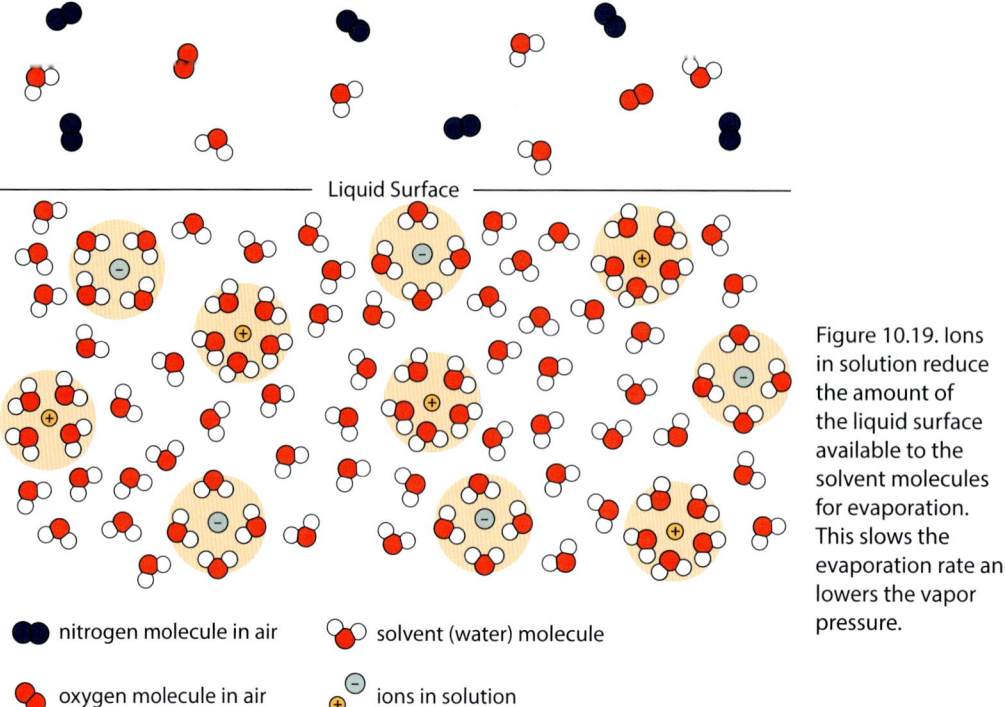

Figure 10.19. Ions in solution reduce the amount of the liquid surface available to the solvent molecules for evaporation. This slows the evaporation rate and lowers the vapor pressure.

This is, in fact, *approximately* what happens. However, there are factors that prevent this analysis from working out so neatly in the real world. For example, the more clustered the solute particles are, the more their motion in the solvent is hindered. Such clustering could be caused by hydration, as in the illustration of Figure 10.19, or by attractions to other ions of the opposite polarity. These clusters encumber the ions, making them less mobile, and if the mobility of particles is impaired due to such clustering, then the solvent particles get around them more easily and evaporate. Thus, we should expect to see something less than the doubling or tripling of the effect that our simple analysis leads us to predict.

What happens in practice depends on whether hydrogen bonding is present, the precise nature of the other intermolecular forces involved, and in the case of ions, their charge. As far as intermolecular forces go, the net result in a solution could be close to, higher than, or lower than what we would predict. As far as ions are concerned, the charge on the ions has a noticeable effect on what happens in the solution. Since intermolecular forces cause the clustering of particles that reduces the effect, we expect that ions with higher charges, and thus greater attractions for oppositely charged species, exhibit less of an effect than ions with lower amounts of charge. As shown below, this is indeed the case.

10.5.2 Freezing Point Depression and Boiling Point Elevation

In Figure 10.18, we see that two side effects of vapor pressure lowering are freezing point depression and boiling point elevation. The graph in Figure 10.18 indicates generally why these effects follow from the vapor pressure lowering, and the discussion of Figure 10.19 provides a physical explanation for why vapor pressure lowering occurs. In this section, we examine a simple quantitative model chemists developed to help predict the size of the effect on the freezing and boiling points for different solvents and different molal concentrations of solute.

Before we proceed to the technical material, I will describe two very common examples of freezing point depression—examples that many students are already familiar with. The first is the fact that in icy winter conditions, many cities use trucks to spread salt on the roads to prevent ice formation or to melt ice that has already formed. Likewise, consumers can purchase salt to sprinkle on their porches and sidewalks. Figure 10.20 shows the effect on some steps in cold weather in the northern U.S.

To be effective, the salt must dissolve in liquid water; salt laying atop solid ice does nothing to melt the ice. But on the surface of the ice there is usually a very thin layer of water. Also, recall from Chapter 8 that if the temperature of the ice is not far below the freezing point, an increase of pressure causes the ice to melt. If the temperature is not too low, the pressure of cars and trucks on roadways causes a thin film of water on the surface of the ice. Once the salt dissolves in this liquid film, the freezing point of the water drops. At the boundary between the water layer and the ice beneath, the water and ice are close to phase transition equilibrium, except that ice molecules are entering the liquid state at a greater rate than the other way around because once liquid, the dissolved ions interfere with movement back toward the ice layer. So the salt water on top of the ice gradually leads to the melting of all the ice, allowing the water to evaporate and leading to a dry road surface.

Figure 10.20. Salt for consumer use on icy steps and walkways. The bare places on the steps were salted just a few minutes before.

Salting roads is only effective down to the new freezing point of the ice water solution. Sodium chloride works down to about 15°F (–9°C). Magnesium chloride produces more ions than sodium chloride, and can be used down to about 5°F (–15°C), and calcium chloride works down to about –20°F (–29°C). The salt product used for the photo in Figure 10.20 is shown at the top of the figure—calcium chloride, the one that provides the most freezing point depression.

Another example of freezing point depression with salt water is fun to be a part of: making ice cream outside at home. A typical home ice cream maker is shown in Figure 10.21. The mixture of cream, sugar, and flavorings is in a metal canister inside the wooden bucket. A mixture of ice and rock salt (sodium chloride) is placed in the space between the canister and the walls of the wooden bucket. The salt mixed in with the ice lowers the freezing point, forcing the ice to melt. For solid H_2O to become liquid, energy must be supplied (the molar heat of fusion). This energy comes from the surrounding ice and water, lowering the temperature of the salt-ice-water mixture. Thus, the ice cream mixture inside the metal canister is sitting in a bath that is colder than 0°C, so the ice cream becomes very cold and freezes. The fun comes in taking turns turning the crank to mix the ice cream as it freezes, and eating the ice cream afterwards! If your family has never made ice cream—try it!

Now we do the math. Using ΔT_f to represent the change in freezing point, and ΔT_b to represent the change in boiling point, the two mathematical models are as follows:

Figure 10.21. Simple ice cream maker for home use.

Chapter 10

$$\Delta T_f = K_f m \qquad (10.3a)$$

$$\Delta T_b = K_b m \qquad (10.3b)$$

In these two equations, m is the molality of the solution, the molal concentration of solute in moles of solute per kilogram of solvent. K_f and K_b are called the *molal freezing-point constant* and *molal boiling-point constant*, respectively. The units for these constants are °C/m, so that when multiplied by the molality (with units m) the result is the temperature change in °C. The presence of a solute always lowers the freezing point and raises the boiling point, and the results of calculations based on Equations (10.3) are interpreted accordingly.

Table 10.3 lists freezing and boiling points and molal constants for a few common solvents. As we go through the following examples, keep in mind that ionizing solutes produce more than one mole of particles in solution for each mole of solute. So to use Equations (10.3) with ionizing solutes, one must multiply the molality, which gives moles of solute per kilogram of solvent, by a factor relating moles of solute to moles of dissolved particles.

▼ Example 10.7

Determine the freezing point of a solution composed of 50.0 g sugar, $C_{12}H_{22}O_{11}$, in 2.00 kg of water.

To calculate the molality of the solution, we first determine the molar mass of sugar, which is 342.3 g/mol. Converting the given mass of sugar to moles, we have

$$50.0 \text{ g } C_{12}H_{22}O_{11} \cdot \frac{1 \text{ mol } C_{12}H_{22}O_{11}}{342.3 \text{ g } C_{12}H_{22}O_{11}} = 0.1461 \text{ mol } C_{12}H_{22}O_{11}$$

With this we calculate the molality:

$$m = \frac{\text{mol solute}}{\text{kg solvent}} = \frac{0.1461 \text{ mol } C_{12}H_{22}O_{11}}{2.00 \text{ kg } H_2O} = 0.07305 \text{ } m$$

Solvent	Freezing Point (°C)	Molal f.p. Constant, K_f (°C/m)	Boiling Point (°C)	Molal b.p. Constant, K_b (°C/m)
acetic acid (p)	16.6	3.63	117.9	3.22
acetone (p)	−94.7	2.67	56.1	1.80
benzene (np)	5.5	5.07	80.1	2.64
ethanol (p)	−114.4	1.959	78.3	1.23
methanol (p)	−97.5	2.56	64.6	0.86
phenol (p)	40.9	6.84	181.9	3.54
toluene (np)	−95.0	3.55	110.6	3.40
water (p)	0.00	1.86	99.974	0.513

Table 10.3. Molal freezing point and boiling point constants for a few solvents (at 1 atm). (p) = polar solvent; (np) = nonpolar solvent

Now we are ready to calculate ΔT_f, using Equation (10.3a) and the K_f value for water from Table 10.3 to calculate the change in the freezing point.

$$\Delta T_f = K_f m = 1.86 \; \frac{°C}{m} \cdot 0.07305 \; m = 0.136 °C$$

Finally, this ΔT_f value tells us how much the freezing point of water was lowered. To get the freezing point of the solution, we need to subtract this value from the freezing point for the pure solvent.

$$T_{f, \text{solution}} = T_{f, \text{normal}} - \Delta T_f = 0.00°C - 0.136°C = -0.14°C$$

In the next example, we treat a case of an ionizing solvent that produces more than one mole of particles in solution for each mole of solute.

▼ Example 10.8

A solution of 180.0 g LiCl in 1.25 L water is prepared. Determine the boiling point of this solution.

We first determine the number of moles of LiCl, based on the molar mass of 42.39 g/mol.

$$180.0 \text{ g LiCl} \cdot \frac{1 \text{ mol LiCl}}{42.39 \text{ g LiCl}} = 4.246 \text{ mol LiCl}$$

Next we determine the mass of the water in kilograms, for which we use the density equation and the density of water, 0.998 g/mL.

$$\rho = \frac{m}{V}$$

$$m = \rho V = 0.998 \; \frac{g}{mL} \cdot 1.25 \text{ L} \cdot \frac{1 \text{ kg}}{1000 \text{ g}} \cdot \frac{1000 \text{ mL}}{L} = 1.248 \text{ kg}$$

Now we compute the molality of the solution:

$$m = \frac{\text{mol solute}}{\text{kg solvent}} = \frac{4.246 \text{ mol}}{1.248 \text{ kg}} = 3.402 \; m$$

This molality is for the solute LiCl. But when LiCl ionizes, two ions are produced (Li⁺ and Cl⁻) for each unit of LiCl. We form a mole ratio from this information to compute the molality of the ions in solution. (We are simply doubling the molal concentration.)

$$3.402 \; m \text{ LiCl} \cdot \frac{2 \; m \text{ ions}}{1 \; m \text{ LiCl}} = 6.804 \; m \text{ ions}$$

This is the molality we need to use for calculating ΔT_b. Next, we compute the ΔT_b value for this solution, using the value of K_b from Table 10.3.

$$\Delta T_b = K_b m = 0.513 \, \frac{°C}{m} \cdot 6.804 \, m = 3.49°C$$

Finally, we calculate the new boiling point by adding the ΔT_b value to the boiling point of water.

$$T_{b,\text{solution}} = T_{b,\text{normal}} + \Delta T_b = 99.97°C + 3.49°C = 103.46°C$$

Notice here that the correct number of significant digits for the ΔT_b value is three. But once this value is added to the boiling point of water using the addition rule, we end up with five significant digits in our final result.

▼ Example 10.9

A solution of benzene and an unknown nonpolar, non-ionizing solute is found to have a freezing point of 1.1°C. Determine the molal concentration of the solution.

To solve this problem, we again use Equation (10.3a). Instead of solving for ΔT_f, we use the given freezing point and the information from Table 10.3 to determine the ΔT_f value. From there we calculate m.

ΔT_f is the difference between the given freezing point and the one in Table 10.3:

$$\Delta T_f = T_{f,\text{normal}} - T_{f,\text{solution}} = 5.5°C - 1.1°C = 4.4°C$$

We use this value with the molal freezing point constant from Table 10.3 to solve for m. Starting with Equation (10.3a), we solve for m, insert the values, and compute the result.

$$\Delta T_f = K_f m$$

$$m = \frac{\Delta T_f}{K_f} = \frac{4.4°C}{5.07 \, \frac{°C}{m}} = 0.87 \, m$$

We have seen that for electrolytes, the degree of freezing point depression or boiling point elevation depends on the molal concentration of ions, and not on the molality of the solute itself. If the solute produces two moles of ions, we expect the ΔT to be twice as large; if three moles, we expect the ΔT to be three times as large. Table 10.4 lists data that demonstrate how closely these predictions agree with empirical results. Three solutes are shown, each producing either two or three moles of ions in solution, and four different concentrations are shown for each. The right column of the table shows the ratio of the experimental ΔT_f to the predicted ΔT_f value for a non-ionizing solute such as sugar (where one mole of solute produces one mole of particles in solution).

HCl produces two moles of ions for each mole of solute, so we expect to see a doubling of the ΔT_f value compared to that predicted for a non-ionizing solute. As you see, the experimental data agree quite well with this prediction, and the data are consistent at various concentrations. Similarly, we expect to see a tripling of the ΔT_f value for $CaCl_2$, and this also agrees fairly well with the empirical data.

Solute	Concentration (m)	Experimental ΔT_f (°C)	ΔT_f Prediction for Nonionizing Solute (°C)	Ratio of Experimental value to non-ionizing prediction
HCl	0.05	0.18	0.093	1.9
	0.25	0.90	0.47	1.9
	0.50	1.86	0.93	2.0
	0.75	2.90	1.40	2.1
MgSO$_4$	0.05	0.13	0.093	1.4
	0.25	0.55	0.47	1.2
	0.50	1.01	0.93	1.1
	0.75	1.50	1.40	1.1
CaCl$_2$	0.05	0.25	0.093	2.7
	0.25	1.27	0.47	2.7
	0.50	2.66	0.93	2.9
	0.75	4.28	1.40	3.1

Table 10.4. Empirical f.p. depression results compared to predictions using molal f.p. constant.

However, the MgSO$_4$ values are significantly less than the doubling we expect for MgSO$_4$. One explanatory factor is in the charges on the ions: the charge on the cation (Mg^{2+}) is higher than the H$^+$ cation from the HCl, and the charge on the anion (SO$_4^{2-}$) is higher than the Cl$^-$ anion from both the other solutes listed. In our discussion a few pages back, we noted that higher ionic charges tend to reduce the freezing-point depression and boiling-point elevation effects because of greater intermolecular forces and resulting particle clustering. Another causal factor may be the size of the SO$_4^{2-}$ ion. This ion is a molecule with five atoms in it, compared to the single-atom Cl$^-$ ions. We expect larger ions to form larger clusters during hydration, becoming less mobile and offering less interference to higher-energy H$_2$O molecules working their way towards the liquid surface.

Chapter 10 Exercises

SECTION 10.1

1. Describe in detail the process of dissolution of an ionic crystal in a polar solvent, such as water.
2. What determines whether the dissolution of a given ionic solute is exothermic or endothermic? Explain the different *sources* or *sinks* of energy in such processes. (A sink is the opposite of a source. Energy is supplied by a source and taken in by a sink.)
3. Why should one never add water to a concentrated acid? Explain the details behind why this action is hazardous.
4. Define heat of solution.
5. Explain the concept of entropy.

Chapter 10

6. Consider the four processes described in the accompanying table and address these questions:

   a. Which process always happens spontaneously?

   b. Which process never happens spontaneously?

   c. Two of the processes may or may not happen in a given instance. Which two are these, and what determines whether the processes occur?

1. Energy increases; entropy increases	2. Energy increases; entropy decreases
3. Energy decreases; entropy increases	4. Energy decreases; entropy decreases

7. Explain what an electrolyte is and distinguish between strong and weak electrolytes.

8. How does the concept of equilibrium apply to solutions of weak electrolytes such as acetic acid?

SECTION 10.2

9. Classify the following compounds as soluble or insoluble in water:

   a. cobalt(II) hydroxide
   b. calcium nitrate
   c. ammonium phosphate
   d. potassium sulfate
   e. iron(III) oxide
   f. magnesium hydroxide
   g. lead(II) acetate
   h. strontium oxide
   i. copper(II) chloride

10. Imagine that you have some sort of sticky residue on the side of your arm and want to clean it off. You try rinsing it off in water, but the water doesn't remove the substance. Explain why it would be pointless to try cleaning off the sticky substance with rubbing alcohol, and suggest a common product that will remove it (other than soap and water).

11. Ammonia is commonly mixed with water as a cleaning agent for mopping floors. Explain why ammonia dissolves in water.

12. Explain why oil and vinegar salad dressing separates into layers of oil and vinegar.

13. Consider the equilibrium that exists in a solution of pressurized gas in a liquid.

    a. Describe the equilibrium.
    b. Describe the effect of pressure on the amount of dissolved gas.

14. Describe the equilibrium that exists in a saturated solution of an ionic solid with excess solid at the bottom of the container. Speculate on what happens to the equilibrium if additional water is added to the solution.

15. Referring to Figures 10.13 and 10.14, respond to these questions:

    a. At 50°C, what is the maximum percentage of an aqueous solution's mass that can be attributed to dissolved cesium chloride?

    b. For a solution of chlorine dissolved in water at 20°C, what are the mole fractions in the solution of chlorine and of water?

    c. Compare the solubilities of KCl and NH_4Cl as functions of temperature.

    d. By what percentage does the solubility of CO_2 decrease if the temperature is changed from 20°C to 30°C?

16. The accompanying table contains solubility data for $Al_2(SO_4)_3$ in aqueous solution. As an example of how to read the solubility values, at a temperature of 40°C, the mass percent of $Al_2(SO_4)_3$ in a saturated solution is 29.2%. This means that if the entire solution has a mass

of 100.0 g, there are 29.2 g Al$_2$(SO$_4$)$_3$ and 70.8 g H$_2$O in the solution. Plot the data from the table on graph paper with the solubility on the vertical axis (resolution of vertical scale = 0.1), and connect the data points with a smooth curve. Then address the following questions:

Temperature (°C)	Al$_2$(SO$_4$)$_3$ Solubility (mass percent of solution mass)
30	28.2
40	29.2
50	30.7
60	32.6
70	34.9
80	37.6

a. Describe the relationship between Al$_2$(SO$_4$)$_3$ solubility and temperature.

b. Estimate the solubility of Al$_2$(SO$_4$)$_3$ at 55°C.

c. In a saturated solution with a mass of 250.0 g at 70°C, how many grams of Al$_2$(SO$_4$)$_3$ are present?

d. In a saturated solution with a mass of 45 g at 30°C, how many moles of both Al$_2$(SO$_4$)$_3$ and water are present? Compare the mass percentages to the mole percentages for Al$_2$(SO$_4$)$_3$ and water at this temperature.

SECTION 10.3

17. Determine the mass of solute required to make each of the following solutions:

   a. an 8.00 M solution of NaOH, with water added to bring the solution volume to 2.25 L.

   b. a 15.0 M solution of acetic acid, CH$_3$COOH, with water added to bring the solution volume to 1.00 L.

18. A solution is prepared by dissolving 65.11 g (NH$_4$)$_2$SO$_4$ in enough water to bring the solution volume to 125 mL. Determine the molarity of this solution.

19. Determine the mass in grams required to make a 1.25 M solution of Ca(NO$_3$)$_2$, with a solution volume of 3.10 L.

20. To extract 22.5 g silver nitrate from a 1.2 M aqueous solution of silver nitrate, what volume of solution is required?

21. If 54.0 mL of a CuSO$_4$ solution reacts with enough iron to produce 7.7 g Cu in a single replacement reaction, what is the molarity of the CuSO$_4$ solution?

22. Solutions of 0.60 M Na$_2$SO$_4$ and 1.00 M Ba(NO$_3$)$_2$ are mixed.

   a. Write the balanced equation for the reaction that occurs.

   b. If 475 mL Na$_2$SO$_4$ solution are combined with excess Ba(NO$_3$)$_2$ solution, what is the mass of the precipitate that forms?

23. Determine the volume in mL of 12.0 M H$_2$SO$_4$ required for a complete reaction with 655 mL of 6.00 M NaOH. The products of this reaction are Na$_2$SO$_4$ and water.

24. Excess iron is reacted with 2.00×10^3 L of 2.0 M HCl to produce hydrogen gas and iron(II) chloride. Determine the volume of H$_2$ produced at STP.

25. Determine the mass of solute required to make each of the following solutions:

   a. a 10.0 m solution of HCl in 1.00 kg H$_2$O.

   b. a 5.25 m solution of NaOH in 1.75 L H$_2$O.

26. If you prepare a solution of 225 g sucrose, C$_{12}$H$_{22}$O$_{11}$, in 5.00×10^2 mL water, what is the molality of the solution?

27. Determine the quantity in milliliters of water that must be added to 235 g of Ca(NO$_3$)$_2$ to produce a 2.0 m solution.

Chapter 10

SECTION 10.4

28. Solutions of Na_2CO_3 and $Ca(NO_3)_2$ are combined.
    a. Use Table 10.2 to identify the precipitate that forms.
    b. Write the balanced molecular equation.
    c. Write the ionic equation and the net ionic equation for the process.
    d. Identify the spectator ions.

29. In a reaction between aqueous solutions of $AgNO_3$ and LiCl, identify the spectator ions and the precipitate compound in the reaction.

30. Aqueous solutions of magnesium chloride and lead(II) nitrate are combined. The magnesium chloride solution is 0.50 molar and 40.0 mL of it are used.
    a. Write the balanced molecular equation, the ionic equation, and the net ionic equation.
    b. Identify the spectator ions and the precipitate.
    c. If all the magnesium chloride is used in the reaction, determine the mass of the precipitate that forms.

31. Aqueous solutions of sodium nitrate and copper(II) chloride are combined. Describe in detail what occurs.

SECTION 10.5

32. Why is the vapor pressure of an aqueous solution of calcium chloride lower than the vapor pressure of water?

33. Determine the boiling point of a solution formed by dissolving 0.4000 kg of table sugar, $C_{12}H_{22}O_{11}$, in 1.100 L of water.

34. Glycerol, $C_3H_8O_3$, is a soluble nonelectrolyte used widely in the food industry. (Also, recall that glycerol is the molecule that remains after the three strands of a triglyceride have been separated during saponification.) Determine the molal concentration of a glycerol and water solution for boiling points of 101.75°C, 100.99°C, and 103.17°C.

35. Ethylene glycol, $C_2H_4(OH)_2$, is the compound in automotive antifreeze. It is a nonelectrolyte, is nonvolatile, and is miscible in water. Determine the molality of ethylene glycol in water required to prevent the solution from freezing at −18°C.

36. The freezing point of a solution of urea, $CO(NH_2)_2$, a nonelectrolyte, in water is found to be −2.93°C. If the mass of the water in the solution is 775 g, determine the mass of the urea in the solution.

37. The *specific gravity* of a substance is defined as the ratio of the substance's density to the density of water. For example, a substance with a specific gravity of 0.750 is 75.0% as dense as water, and thus has a density of 0.750 · 0.998 g/mL = 0.749 g/mL. A solution is prepared by adding carbon tetrachloride, CCl_4, to 2.50 L of benzene. The specific gravity of benzene is 0.878. If the boiling point of the solution is found to be 85.34°C, determine the mass in grams of the CCl_4 in the solution.

38. Predict the freezing point of a solution of 155 g LiCl in 575 g water.

39. Predict the freezing point of a solution of 456 g ethanol and 75.0 g KCl.

40. Assume you are given 0.040 *m* solutions in acetic acid of KI, $MgCl_2$, and $CaSO_4$. Determine the expected change to the boiling point of each.

Solution Chemistry

41. Aqueous solutions with a molality of 0.020 m are prepared from Na_3PO_4, KCl, $C_6H_{12}O_6$ (glucose), and $CaCl_2$. Determine the expected freezing point of each solution.

42. The *CRC Handbook of Chemistry and Physics* indicates that the freezing point of a 2.00 m solution of nitric acid is lowered by 8.34°C compared to pure water. Compute the expected value of the freezing point lowering for this solution. Comment on whether the value listed in the *CRC* is consistent with the models we have studied in this chapter.

GENERAL REVIEW EXERCISES

43. Determine the molar mass of a gas, given that 1.30 g of the gas occupies 2.24 L at 6.85°C and 1.14 atm.

44. If 17.5 L of methane, CH_4, combust completely with excess oxygen at STP, determine the volumes of each of the products formed.

45. What is surface tension and how does it relate to capillary action?

46. As illustrated to the right, a steel support column in a building is supporting a load of 12,500 lb and has a steel plate at the bottom that is 10.0 inches square. What is the pressure in atm under this plate? Note that necessary conversion factors are in Appendix A.

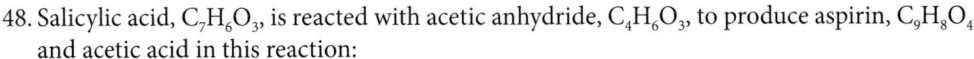

47. The process of coal gasification is used to convert coal, which is pure carbon, into methane gas according to the following reaction:

$$2C(s) + 2H_2O(l) \rightarrow CH_4(g) + CO_2(g)$$

If the reaction has 82% yield, determine the mass of methane produced from 100.0 kg coal.

48. Salicylic acid, $C_7H_6O_3$, is reacted with acetic anhydride, $C_4H_6O_3$, to produce aspirin, $C_9H_8O_4$, and acetic acid in this reaction:

$$C_7H_6O_3(s) + C_4H_6O_3(l) \rightarrow C_9H_8O_4(s) + CH_3COOH(l)$$

If 125.0 mol $C_7H_6O_3$ are used in this reaction, determine the mass in kg of aspirin produced.

49. Determine the percentage composition of salicylic acid.

50. Determine the oxidation state of chlorine in the chlorate ion.

Chapter 11
Acids and Bases

The beautiful flowers of the hydrangea are a natural indicator of soil pH. The flowers are blue in acidic soil and pink in alkaline (basic) soil. Unfortunately for chemistry students, this color palette is just the opposite of *litmus*, one of the oldest laboratory pH indicators. Litmus is a mixture of various dyes extracted from lichens, and turns red in acids and blue in bases. Although other pH indicators are used more commonly now, the use of litmus was so common in chemistry labs for so long that we now associate its colors red and blue with acids and bases, respectively.

Objectives for Chapter 11

After studying this chapter and completing the exercises, you should be able to do each of the following tasks, using supporting terms and principles as necessary.

SECTION 11.1

1. List six general properties of acids and five general properties of bases.

2. Name at least four acids that are important industrial chemicals, stating at least two industrial uses for each.

3. Define *binary acid*, *oxyacid*, and *carboxylic acid*, and give at least three examples of each.

4. Given chemical formulas, state correct names for binary acids and oxyacids.

SECTION 11.2

5. State definitions for *acid*, *base*, and *acid-base neutralization reaction* according to the Arrhenius and Brønsted-Lowry acid-base theories.

6. Explain why water is amphiprotic under the Brønsted-Lowry acid-base theory.

7. Explain why NH_3 acts as a base according to both acid-base theories.

8. Given a chemical equation, identify compounds or ions that act as acids or bases under each of the acid-base theories.

9. Define *conjugate acid* and *conjugate base*, give examples of each, and identify each in a given chemical equation.

10. State the names of at least three of each of the following: strong acid, strong base, weak acid, weak base.

11. Explain the factor that determines whether an acid or base is weak or strong.

12. Write ionic and net ionic equations for neutralization reactions.

13. Identify which direction the equilibrium lies in a Brønsted-Lowry acid-base reaction.

14. Relate acid strength to the number of oxygen atoms bound to the central atom in an acid molecule, and explain why acid strength goes up with the number of oxygen atoms.

SECTION 11.3

15. Explain the concept of the self-ionization of water.

16. Use the ionization constant of water to compute $[OH^-]$ from $[H_3O^+]$, and vice versa.

17. Given $[H_3O^+]$ or $[OH^-]$, compute both pH and pOH.

18. Define *pH* and *pOH*, and compute one given the other.

19. Given pH or pOH, compute both $[H_3O^+]$ and $[OH^-]$.

20. Describe the features of the pH scale, and give five examples of common substances and their pH values.

21. Explain what pH indicators are and what their two main purposes are.

22. Explain how pH indicators function in aqueous solution to produce different colors.

23. State the names and approximate transition intervals for three different pH indicators, one each with its transition interval below 7, including 7, and above 7.

24. Explain the purpose of titration, and give a general explanation of titration procedure.

25. Determine $[H_3O^+]$ and $[OH^-]$ from titration data.

Chapter 11

11.1 Properties and Nomenclature of Acids and Bases

11.1.1 Introduction

Acids and bases are of supreme importance in chemistry—a great deal of chemistry concerns reactions between acids and bases or their reactions with other substances. We are surrounded by compounds that act as acids or bases in daily life, in the foods we eat and the products we use. There are thousands of compounds that exhibit acidic or basic properties, and in this chapter we only scratch the surface of this vast subject.

Figure 11.1. Acids in common foods.

Acids often give flavors to foods, including the sour or tart flavors we are familiar with in fruits. Several examples of these are displayed in Figure 11.1. Many bases are used in products commonly found in the home. Several of these products are shown in Figure 11.2. Tums, the antacid shown in the figure, contains only $CaCO_3$. But $Al(OH)_3$ and $Mg(OH)_2$ are also bases commonly found in antacids.

Acids are also of crucial importance in industry, where their uses are limitless. Some of the major industrial acids and their uses are summarized in Table 11.1.

11.1.2 Properties of Acids and Bases

It is important to be familiar with the well-known properties of acids and bases. For acids, these include the following:

1. In concentrated form, acids are often corrosive, which means they react vigorously with other substances. Because of this property, acids in the laboratory must always be treated with great care. Always wear eye protection, gloves, and—if you value your clothing—a laboratory apron.

2. Acids have a sour taste. This is why several of them appear so often in tart foods. However,

Figure 11.2. Common household bases.

Acids and Bases

Acid	Industrial Uses and Other Notes
hydrochloric acid, HCl historical name: *muriatic acid*	• manufacture of PVC pipe and polyurethane • swimming pool pH maintenance • leather processing • metal cleaning and descaling (pickling) • the acid in the human stomach, known as gastric acid
nitric acid, HNO_3 historical name: *aqua fortis*	• manufacture of ammonium nitrate fertilizer by reaction with ammonia • manufacture of explosives • precursor to manufacture of other compounds, including nylon
sulfuric acid, H_2SO_4 historical name: *oil of vitriol*	• most produced industrial chemical in the world • manufacture of fertilizer • metal cleaning and descaling (pickling) • precursor to manufacture of other compounds • the acid used in car batteries
phosphoric acid, H_3PO_4	• rust removal • food additive for delivering tangy taste • used in dentistry and toothpastes to clean teeth
acetic acid, CH_3COOH	• main ingredient in vinegar (other than water) • source for acetate in photographic film • manufacture of synthetic fabrics • food additive

Table 11.1. Important industrial acids.

note that you should never taste acids in a laboratory. They can be poisonous and extremely damaging to human tissue.

3. Acids react with bases to produce a *salt* and water. This type of chemical reaction is called *neutralization* because the corrosive properties of acids and bases are neutralized when they exchange ions to form the salt. A chemical salt is defined as a compound formed in an acid-base reaction, composed of the cation from the base and the anion from the acid.

4. Acids that form ions in aqueous solution are electrolytes. As we have seen, this means they conduct electricity.

5. Acids cause a color change in *pH indicators*. A pH indicator is a substance used to detect whether a solution is acidic or basic by its color. We study indicators in detail later in this chapter, including the use of *pH* as a measure of the strength of acids and bases. The juice from red cabbage is a natural pH indicator. Its amazing range of color is shown in Figure 11.3.

6. Some acids react vigorously with metals, releasing hydro-

Figure 11.3. The colors exhibited by red cabbage in solutions with the following pH values (L to R): 1–3–5–7–8–9–10–11–13.

gen gas in the process. For this reason, acids are not stored in metal containers or metal cabinets. The photos in Figure 7.17 of magnesium reacting with hydrochloric acid are a typical example of this property.

Bases also have some common properties, many of which correlate to the properties of acids. These properties include:

1. In concentrated form, bases are often corrosive, which means they react vigorously with other substances. Bases in the laboratory must always be treated with great care. Always wear eye protection and gloves.
2. Bases have a bitter taste and slippery feel. Soap exhibits both these characteristics, as you know. However, note that you should never taste bases in a laboratory. They can be poisonous and extremely damaging to human tissue.
3. Bases react with acids to produce a salt and water.
4. Bases that form ions in aqueous solution are electrolytes. As with acids, this means they conduct electricity.
5. Bases cause a color change in pH indicators.

11.1.3 Acid Names and Formulas

We briefly touched on naming acids in the discussion of polyatomic ions in Chapter 5. Now it is time to become more familiar with these terms.

The first major class of acids are the *binary acids*, formed between hydrogen and one of the more electronegative nonmetals. Several of these are *hydrogen halides*—hydrogen with a halogen—HF, HCl, HBr, and HI. Another binary acid is H_2S.

In their pure forms, all these compounds are colorless gases with names that begin with *hydrogen* and end with *–ide*. In this state, they are sometimes referred to as *acid gases*. But in aqueous solution they are proper acids, and their names begin with *hydro–* and end with *acid*. This information is summarized in Table 11.2.

Formula	Name of Pure Compound	Name in Aqueous Solution
HF	hydrogen fluoride	hydrofluoric acid
HCl	hydrogen chloride	hydrochloric acid
HBr	hydrogen bromide	hydrobromic acid
HI	hydrogen iodide	hydroiodic acid
H_2S	hydrogen sulfide	hydrosulfuric acid.

Table 11.2. Binary acids.

Formula	Oxyacid Name	Oxyanion
HIO_3	iodic acid	IO_3^-, iodate
HClO	hypochlorous acid	ClO^-, hypochlorite
$HClO_2$	chlorous acid	ClO_2^-, chlorite
$HClO_3$	chloric acid	ClO_3^-, chlorate
$HClO_4$	perchloric acid	ClO_4^-, perchlorate
$HMnO_4$	permanganic acid	MnO_4^-, permanganate
HNO_2	nitrous acid	NO_2^-, nitrite
HNO_3	nitric acid	NO_3^-, nitrate
H_2CO_3	carbonic acid	CO_3^{2-}, carbonate
H_2CrO_4	chromic acid	CrO_4^{2-}, chromate
H_2SO_3	sulfurous acid	SO_3^{2-}, sulfite
H_2SO_4	sulfuric acid	SO_4^{2-}, sulfate
H_3PO_3	phosphorous acid	PO_3^{3-}, phosphite
H_3PO_4	phosphoric acid	PO_4^{3-}, phosphate

Table 11.3. Common oxyacids and oxyanions.

The second major class of acids is the *oxyacids*. An oxyacid contains one or more oxygen atoms, one or more hydrogen atoms, and at least one other element, usually a nonmetal. We have already encountered many of these in this text. Table 11.3 lists only a *few* of the most common ones; there are many more.

As a brief review of the naming convention for the oxyacids, Table 11.4 lists the four Cl—O oxyacids. The convention for naming the acids is to retain anion prefixes and change *-ate* and *-ite* suffixes to *-ic* and *-ous* in the acid names. This convention is reflected in the names of the oxyacids listed in Table 11.3.

Acid Formula	Acid Name	Anion
$HClO_4$	perchloric acid	ClO_4^-, perchlorate
$HClO_3$	chloric acid	ClO_3^-, chlorate
$HClO_2$	chlorous acid	ClO_2^-, chlorite
$HClO$	hypochlorous acid	ClO^-, hypochlorite

Table 11.4. Illustration of oxyacid naming convention.

A third major category of acids is *carboxylic acids*. These contain the *carboxyl group*, —COOH. Acetic acid, CH_3COOH, which we have encountered before, is a simple carboxylic acid. The number of carboxylic acids is enormous, and they are of great importance in organic chemistry and biochemistry. We do not consider the carboxylic acids further here, except to mention two of the more well-known ones. Formic acid, the simplest of all carboxylic acids, is famous for being in the venom of many ant species. These species bite the victim and spray the formic acid on the wound.[1] Benzoic acid is widely used as a food preservative. The Lewis structures for these three carboxylic acids are shown in Figure 11.4.

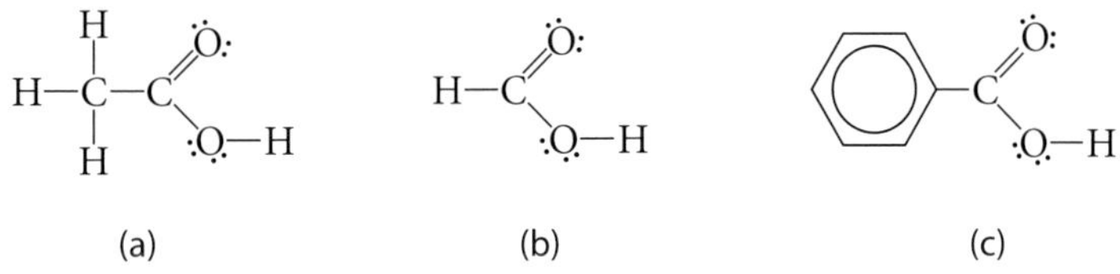

Figure 11.4. Lewis structures for acetic acid (a), formic acid (b), and benzoic acid (c).

Finally, note that some acids do not fit any of the broad acid classes described here. An example of this that we have already seen is hydrocyanic acid, HCN, formed from the cyanide ion, CN^-.

11.2 Acid-Base Theories

Recall once again our discussion in the Introduction that chemistry is all about modeling—developing theories that describe the natural world as accurately as possible. This frame of mind is particularly valuable in the study of acids and bases. The reason for this is that when we try to define what acids and bases are, and then understand their behavior in light of those definitions, nature does not much want to cooperate. The result is three major acid-base theories, each one defining acids and bases differently from the others. The theories do overlap, and substances like the acid H_2SO_4 and the base NaOH are identified as acid or base in all three theoretical systems. But there are differences as well, and in order to engage in the conversation about what happens when acids and bases react, students must be familiar with the different theoretical systems. In this text, we focus on two of them, giving only a brief mention to the third.

One of the first acid theories was developed by French chemist Antoine Lavoisier. Recall from Chapter 7 that in 1774 Lavoisier first identified oxygen and gave the element its name. The

---

[1] However, fire ants, one of the downsides of living in Texas, are different. They only bite to get a grip, and then sting the victim with poison from the abdomen, like a wasp.

name *oxygen* means "acid former," and Lavoisier chose the name because at the time scientists thought that all acids contained oxygen. As it became clear that many acids did not contain oxygen, the task of defining exactly what acids and bases are became pressing.

11.2.1 Arrhenius Acids and Bases

As you recall from the previous chapter, Swedish chemist Svante Arrhenius was a groundbreaking contributor to our understanding of the behavior of substances in solution. The acid-base theory put forward by Arrhenius in 1884 is the first theory to learn. We begin with definitions of acids and bases according to Arrhenius' model.

An Arrhenius acid is a substance that increases the concentration of hydrogen ions, H^+, in aqueous solution. Acids do this because they are electrolytes containing hydrogen, and when they ionize, they contribute H^+ cations to the solution. The solution becomes electrically conductive as a result.

An Arrhenius base is a substance that increases the concentration of hydroxide ions, OH^-, in aqueous solution. Thus, like acids, bases are electrolytes, compounds that ionize in aqueous solution. Bases contain the OH^- anion (or produce them in solution, as NH_3 does). When they ionize, the presence of the OH^- anions in solution makes the solution electrically conductive.

Arrhenius' definitions are very powerful. Thinking about his definition for acids, computer models of three common acids are shown in Figure 11.5. Hydrogen atoms are shown in white in these models, and the hydrogens that ionize in solution are the ones connected to oxygen atoms (red). As you see, nitric acid, HNO_3, contributes one H^+ cation when it ionizes. Such acids are referred to as *monoprotic* (one proton). Sulfuric acid, H_2SO_4, first ionizes as H^+ and HSO_4^-. Then, some of the HSO_4^- anions dissociate further, producing more H^+ cations and SO_4^{2-} anions in solution. The resulting solution contains all these species simultaneously. Acids that can contribute more than one H^+ cation in solution are referred to as *polyprotic*. Those like H_2SO_4 that can contribute two hydrogen ions are termed *diprotic*; an acid such as phosphoric acid, H_3PO_4, that can contribute three hydrogen ions is *triprotic*.

The third acid in Figure 11.5, acetic acid, contains four hydrogen atoms, but those attached to the carbon atom do not dissociate in solution; they stay with the molecule. Only the hydrogen atom connected to the oxygen atom dissociates.

As mentioned in Chapter 10, some acids ionize completely in water and some ionize to a very limited degree. Hydrochloric acid ionizes completely:

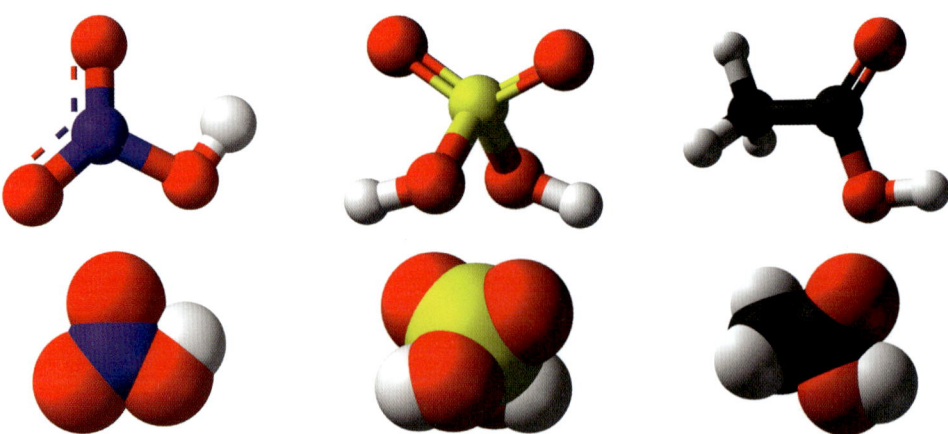

Figure 11.5. Ball-and-stick models (top) and space-filling models (bottom) of three acids: nitric acid (left), sulfuric acid (middle), and acetic acid (right).

$$HCl(aq) \rightarrow H^+(aq) + Cl^-(aq)$$

Acetic acid, on the other hand, ionizes very little in water; most of the molecules remain intact in the solution. Accordingly, the equation shows the ionization of acetic acid as a bidirectional process:

$$CH_3COOH(aq) \rightleftharpoons H^+(aq) + CH_3COO^-(aq)$$

In solution, CH_3COOH reaches an equilibrium in which molecules of CH_3COOH are dissociating at the same rate that H^+ and CH_3COO^- ions are recombining into molecules.

The more completely an acid ionizes, the stronger an electrolyte it is, and the stronger an acid it is. The same holds for bases: when a base ionizes completely, it is a strong electrolyte and a strong base. Those that ionize only to a limited extent are weak electrolytes and weak bases. Table 11.5 identifies a number of common acids and bases, classifying them as strong or weak.

While Arrhenius' definition of an acid is very helpful for understanding acids, we now know that the H^+ ions—protons—contributed to a solution by an acid do not remain in solution as individual protons. Instead, each H^+ ion attaches to one of the nonbonding pairs on a water molecule, forming the H_3O^+ *hydronium ion*. For this reason, it is now common to say that *an Arrhenius acid is a substance that increases the concentration of hydronium ions in solution*. The concepts of increasing H^+ ions (Arrhenius' original concept) and increasing H_3O^+ ions (which is where the H^+ ions end up) are used interchangeably.

Strong Acids	Weak Acids	Strong Bases	Weak Bases
HCl	CH_3COOH	NaOH	NH_3
HNO_3	HCN	KOH	$NaCH_3COO$
H_2SO_4	HF	RbOH	(sodium acetate)
$HClO_4$	H_2CO_3	CsOH	CH_3NH_2
HBr	H_3PO_4	$Ca(OH)_2$	(methylamine)
HI	H_2S	$Sr(OH)_2$	
	HSO_4^-	$Ba(OH)_2$	
	HCO_3^-		

Table 11.5. Classification of some common acids and bases.

As an illustration, the formation of hydronium ions when nitric acid dissolves in water is represented by the following equation:

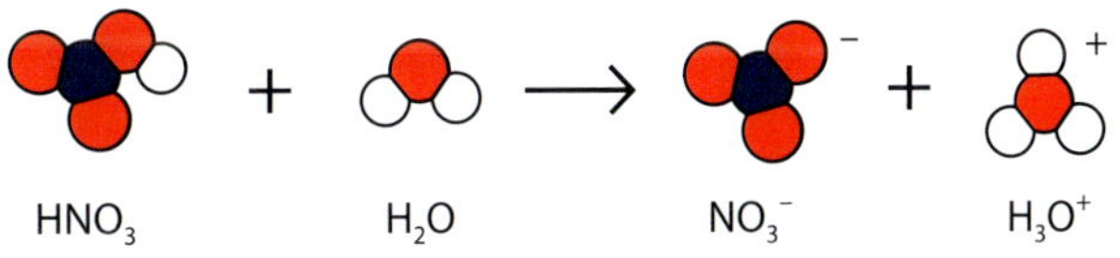

Figure 11.6. Nitric acid dissociating in water to produce nitrate and hydronium ions.

$$HNO_3(l) + H_2O(l) \rightarrow H_3O^+(aq) + NO_3^-(aq)$$

This process is further illustrated by Figure 11.6.

We show the ionization of bases in equations in the same way. For example, the ionization of the strong base sodium hydroxide in water is represented this way:

$$NaOH(s) \xrightarrow{H_2O} Na^+(aq) + OH^-(aq)$$

Chapter 11

The weak base ammonia, NH_3, also dissolves in water, producing OH^- ions in the process. However, ammonia doesn't dissociate. To show what happens in this reaction, we have to show the H_2O molecules in the equation. Ammonia does not contain the OH^- ion, but OH^- ions are produced when hydrogen atoms leave water molecules to bond to ammonia molecules. Recall that the electron domain geometry of the ammonia molecule is tetrahedral, with one nonbonding electron pair on the nitrogen atom. This is where a hydrogen atom can attach, producing an ammonium ion, NH_4^+, as shown in this equation:

$$NH_3(aq) + H_2O(l) \rightarrow NH_4^+(aq) + OH^-(aq)$$

Not very many hydrogen atoms are going to make the jump from water molecules to ammonia molecules. For this reason, not many OH^- ions are produced, and thus ammonia is a weak base.

Arrhenius gave us a good starting place for defining acids and bases. But it gradually became apparent that there are many cases his definitions don't address, primarily because they were articulated in terms of ions in aqueous solution. Many acid-base reactions can occur apart from aqueous solution. For example, HCl and NH_3 can react as gases to produce solid particles of NH_4Cl, as this equation shows:

$$HCl(g) + NH_3(g) \rightarrow NH_4Cl(s)$$

Another example is the acid-base reaction between acetic acid and ammonia:

$$CH_3COOH + NH_3 \rightarrow CH_3COO^- + NH_4^+$$

In this reaction, the acetic acid molecule donates a proton, but the reaction cannot be described in terms of Arrhenius model because the proton does not form hydronium. Instead, it is transferred to the ammonia molecule to form ammonium.

As scientists learned about occasions when substances act as acids or bases apart from aqueous solutions, the door was open for a new, more all-encompassing definition.

11.2.2 Brønsted-Lowry Acids and Bases

In 1923, Danish chemist Johannes Brønsted (Figure 11.7) and English chemist Thomas Martin Lowry (Figure 11.8) independently recognized that an acid-base reaction can be modeled as a *proton transfer*, and they developed definitions for acids and bases based on this fundamental concept. The Brønsted-Lowry theory is now the one most commonly used for general analysis of acid-base interactions.

The definitions for acids and bases according to the Brønsted-Lowry model are as follows: *A Brønsted-Lowry acid is a proton donor, and a Brønsted-Lowry base is a proton acceptor*. These definitions sound simple, but they have far-reaching implications for what we call an acid and what we call a base, as we will now see. As we proceed, keep in mind that a proton is identical to a hydrogen ion.

Figure 11.7. Danish chemist Johannes Brønsted (1879–1947).

Figure 11.8. English chemist Thomas Martin Lowry (1874–1936).

Hmm... Interesting.   What is an alkali?

The beautiful photo below is an image from the salt flats at Badwater Basin in Death Valley, California—the lowest place in North America (282 feet below sea level). Salt flats are sometimes called *alkali flats* because of the carbonates of alkali metals or alkaline-earth metals (chiefly Na_2CO_3) deposited there. Outside the field of chemistry, the term *alkali salt* is sometimes applied to hydroxide compounds, such as NaOH (called *lye* or *caustic soda*), KOH (called *caustic potash*), $Mg(OH)_2$, and $Ca(OH)_2$. These compounds can form when the carbonate compounds react with water. However, chemists refer to the hydroxide compounds as *bases*, reserving the term *salt* for the compounds that form when a base reacts with an acid.

There are also many *alkali lakes* around the world, where living things are rare because of the high concentrations of salts, often seen encrusted along their shores. But the salts in alkali lakes are typically carbonates such as Na_2CO_3, not hydroxides.

The meanings of the terms *alkali* and *alkaline* are tough to chase down because of varying usages in different contexts. When describing natural features like flats and lakes, soils and waters are typically referred to as *alkaline* any time the soil or water pH is high, regardless of the particular base that is causing it.

In the context of chemistry, we also find the terms used in several different ways. The most general use is in referring to any base or basic solution. But some sources consider this usage obsolete, and other sources specifically distinguish between an *alkali* and a *base*, and between *alkalinity* and *basicity*. Other usages apply the term *alkali* to a subset of the bases, but there are several different ways to identify the subset the term applies to, including:

- to describe only bases that include the OH⁻ anion
- to describe any basic salt of the alkali metals or alkaline earth metals
- to describe any water-soluble base that forms OH⁻ ions (which would include NH_3 but not $Mg(OH)_2$, because magnesium hydroxide is not very water soluble)

The term *alkali* derives from the Arabic *al-qaly*, and originally referred to wood ashes containing K_2CO_3 and other potassium compounds, known generally as *potash*. *Caustic potash*, KOH, is a much stronger base, and is the substance English scientist Sir Humphry Davy used when he first isolated potassium in 1807. The chemical symbol K for potassium is from the Medieval Latin word *kalium*—plant ashes—which derives from the same Arabic term that gave us *alkali*. All this is interesting, but doesn't really help us much in establishing a specific contemporary meaning for the term *alkali*.

With so many different definitions in current use, my hunch is that there is no hope of the term regaining any specific meaning, if it ever had one. So use it as you see fit.

Chapter 11

To begin, let's first look again at the reaction between acetic acid and ammonia:

$$CH_3COOH + NH_3 \rightarrow CH_3COO^- + NH_4^+$$

In this reaction, acetic acid is a proton donor, so it is a Brønsted-Lowry acid; ammonia is a proton acceptor, so it is a Brønsted-Lowry base.

However, let's now look again at the dissolving of nitric acid in water:

$$HNO_3(l) + H_2O(l) \rightarrow H_3O^+(aq) + NO_3^-(aq)$$

Here the nitric acid donates a proton, so it is a Brønsted-Lowry acid. But the proton acceptor in this case is water, so according to the Brønsted-Lowry model, water is a Brønsted-Lowry base. After the reaction, notice that the HNO$_3$ has become NO$_3^-$, and as such it can now *accept* a proton. This makes it a base according to the Brønsted-Lowry model. Likewise, H$_2$O has become H$_3$O$^+$, so it is now a potential proton donor. This situation is always the case with the Brønsted-Lowry model: once a proton has been donated, the species that remains is able to accept a proton, and once a proton has been accepted, the new species is able to donate it. Special terms are used to describe these chemical species, illustrated in Figure 11.9.

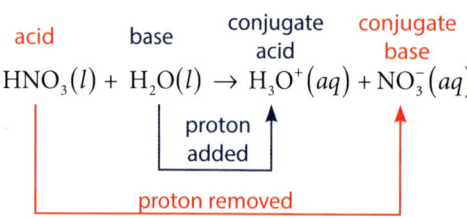

Figure 11.9. Conjugate acids and bases in the Brønsted-Lowry model.

When a Brønsted-Lowry acid donates a proton, the new species created is called the *conjugate base* of that acid. Likewise, when a Brønsted-Lowry base accepts a proton, the new species created is called the *conjugate acid* of that base.

As another example, let's look again at the reaction between ammonia and water:

$$NH_3(aq) + H_2O(l) \rightarrow NH_4^+(aq) + OH^-(aq)$$

Figure 11.10 indicates the proton transfer and identifies the acid, base, conjugate base, and conjugate acid for this reaction. As before, the Brønsted-Lowry base, ammonia, accepts a proton, and the new species created, NH$_4^+$, is the conjugate acid of that base. However, now we see that in this reaction, the H$_2$O molecule is the proton donor, which means that in this reaction water is a Brønsted-Lowry acid. The hydroxide ion formed is the conjugate base of that acid.

In the two examples we have just reviewed, water is a Brønsted-Lowry base in one of them and a Brønsted-Lowry acid in the other. We use the term *amphoteric* to describe any species that can act as either an acid or a base in a general sense. A subset of amphoteric species are molecular compounds that can act as either a proton donor or a proton acceptor. The term *amphiprotic* is used to describe these. Water is an amphiprotic (and amphoteric) compound, and can act as either a Brønsted-Lowry acid or a Brønsted-Lowry base.

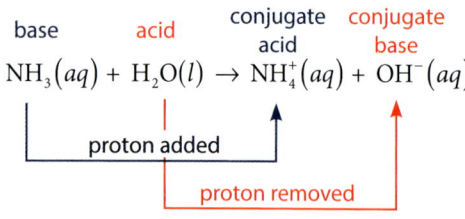

Figure 11.10. Another example of conjugate acids and bases in the Brønsted-Lowry model.

The Brønsted-Lowry model is certainly helpful in understanding what goes on when an acid or base dissolves in water. Let's now see what happens when we attempt to analyze a standard Arrhenius acid-base reaction in terms of the Brøn-

300

sted-Lowry model. Consider the acid-base neutralization reaction between hydrochloric acid and sodium hydroxide:

$$HCl(aq) + NaOH(aq) \rightarrow NaCl(aq) + H_2O(l)$$

The reaction is often written this way, and demonstrates a widely understood fact about acid-base reactions: they produce a salt and water.

But HCl is a strong acid, and dissociates completely in water. Likewise, NaOH is a strong base and dissociates completely in water. So to analyze this reaction further in terms of both the Arrhenius and Brønsted-Lowry models, it is more helpful to write this equation in terms of the ions that appear in solution:

$$HCl(aq) + NaOH(aq) \xrightarrow{H_2O} Na^+(aq) + Cl^-(aq) + H_3O^+(aq) + OH^-(aq)$$

Now it is clear that in terms of the Arrhenius model, HCl is an Arrhenius acid because the hydrogen ion from HCl goes to create a new hydronium ion. Likewise, NaOH is an Arrhenius base because it contributes an OH⁻ ion to the solution.

HCl is also identified as an acid in terms of the Brønsted-Lowry model because it donates the proton that formed the hydronium ion. However, NaOH does not fit the definition of a base according to the Brønsted-Lowry model because it does not accept the proton. The water molecule accepts the proton, and so H_2O is the Brønsted-Lowry base in this reaction. The OH⁻ ion created in the process is now able to receive a proton, and thus is ready to act as a Brønsted-Lowry base in the future. Even though the Brønsted-Lowry model does not identify NaOH as a base, hydroxides such as NaOH are commonly referred to as bases, so one has to get used to mentally switching back and forth from the Arrhenius model to the Brønsted-Lowry model, depending on the reaction in question. Note also that NaOH *produces* a Brønsted-Lowry base (OH⁻) just as HCl *produces* a Brønsted-Lowry acid (H_3O^+). Thus, we can call NaOH a base and HCl an acid in this sense as well.

As mentioned previously, there are also reactions that do not take place in water where the Brønsted-Lowry model is very helpful. The reaction between ammonia and hydrochloric acid is one of these:

$$HCl(g) + NH_3(g) \rightarrow NH_4Cl(s)$$

To see what is happening in this case, the Lewis structures are helpful:

Figure 11.11. Vapors from HCl (left) and NH_3 (right) react in the air to form solid particles of NH_4Cl.

Here, HCl is a Brønsted-Lowry acid, donating a proton and creating a Cl⁻ ion, which is the conjugate base. NH_3 is a Brønsted-Lowry base, accepting the proton to form the NH_4^+ ion, which is the conjugate acid. As mentioned above, this reaction can take place in the vapor state, in the air, as you see from the photo in Figure 11.11. When it does, the Cl⁻ and NH_4^+ ions combine to form solid, airborne particles of the salt NH_4Cl.

Chapter 11

11.2.3 Lewis Acids and Bases

In 1923, the great American chemist G.N. Lewis proposed a third acid-base model. As you know, Lewis originally conceptualized the covalent bond, describing it in terms of the formation of a shared electron pair. Lewis' acid-base theory is expressed in the same terms—electron pairs. Since Lewis expressed his model in terms of electron pairs rather than protons, a substance without any hydrogen atoms can act as a Lewis acid.

A Lewis acid is any species that accepts an electron pair to form a covalent bond. A Lewis base is any species that donates an electron pair to form a covalent bond. A Lewis acid-base reaction occurs when one or more covalent bonds are formed between an electron pair donor and an electron pair acceptor.

The vapor-state reaction between boron trifluoride and ammonia is commonly used to illustrate the Lewis acid-base reaction:

$$BF_3(g) + NH_3(g) \rightarrow F_3BNH_3(g)$$

Represented with Lewis structures, this reaction is as follows:

```
    F           H              F  H
    |           |              |  |
F—B      +  :N—H      →    F—B—N—H
    |           |              |  |
    F           H              F  H
```

Recall that boron acts in exception to the octet rule and forms compounds such as BF_3 that furnish boron with six electrons in its valence shell rather than eight. As the Lewis structures show, the lone pair on the ammonia molecule attaches to the boron atom, giving boron a full octet. In this reaction, BF_3 is a Lewis acid because it accepts the electron pair from NH_3 to form a covalent bond. NH_3 is a Lewis base because it donates the electron pair to form the covalent bond.

Each of the three acid-base theories is useful. The Lewis theory is especially useful in organic chemistry, where it is one of the dominant tools for modeling chemical reactions. As we have seen, sometimes the three theories classify substances in different ways. In other cases, all three models classify a compound the same way. We have seen that NH_3 acts as a base under all three models. It is an Arrhenius base because it causes OH^- ions to form in water, it is a Brønsted-Lowry base because it accepts a proton to form NH_4^+, and it is a Lewis base because it donates an electron pair to form a covalent bond.

▼ Example 11.1

For each of the reactions shown, identify the acids and bases according to the Arrhenius and Brønsted-Lowry models.

$$HBr(aq) + H_2O(l) \rightarrow Br^-(aq) + H_3O^+(aq)$$

$$CH_3COOH + NaHCO_3 \rightarrow CH_3COONa + H_2O + CO_2$$

In the first reaction, HBr ionizes and donates a proton, so it is an Arrhenius acid and a Brønsted-Lowry acid. The H_2O accepts the proton, making it a Brønsted-Lowry base. There is no Arrhenius base.

The second reaction does not have to occur in solution. Pure acetic acid (called glacial acetic acid) is a liquid, and sodium bicarbonate is a solid. The acetic acid donates a proton, so it is a Brønsted-Lowry acid. An OH^- ion from the dissociated $NaHCO_3$ acts as the Brønsted-Lowry

Acids and Bases

base, receiving the proton and forming a molecule of water. This reaction can also occur in solution. In that case, the CH_3COOH contributes H^+ ions (and thus H_3O^+ ions) to the solution, and thus acts as an Arrhenius acid. The $NaHCO_3$ contributes OH^- ions to the solution and thus acts as an Arrhenius base.

▲

11.2.4 Strength of Acids and Bases

Here we use the Brønsted-Lowry model to elaborate further on the concept of acid and base strength. We find that the Brønsted-Lowry model allows us to make general predictions about the equilibrium that occurs when an acid or base ionizes in water.

Table 11.6 lists several Brønsted-Lowry acids and bases in conjugate pairs. In each row, an acid is listed side-by-side with its conjugate base. Water appears twice, and the two placements of water divide the list into three major sections. The strong acids are compounds that ionize 100% in water. The conjugate base for each of these exhibits negligible properties as a base. For example, HCl ionizes completely, and is a strong acid. But Cl^- ions in solution show little tendency to accept a proton, so Cl^- is classified as having negligible base strength.

In the center of the chart are the weak acids and bases. These are weak because they dissociate only partially in water. In acetic acid, for example, only about 1.4% of the molecules in 0.10 M CH_3COOH ionize. But the more an acid dissociates, the stronger it is and the higher it is in the scale of acid strength.

Notice the pattern among the diprotic and triprotic acids. H_2SO_4 is a strong acid, and ionizes in steps. The first step occurs immediately as almost 100% of the H_2SO_4 ionizes to become HSO_4^-. But HSO_4^- is a weak acid, and only partially ionizes. This second step of ionization only occurs to about 29% of the molecules in

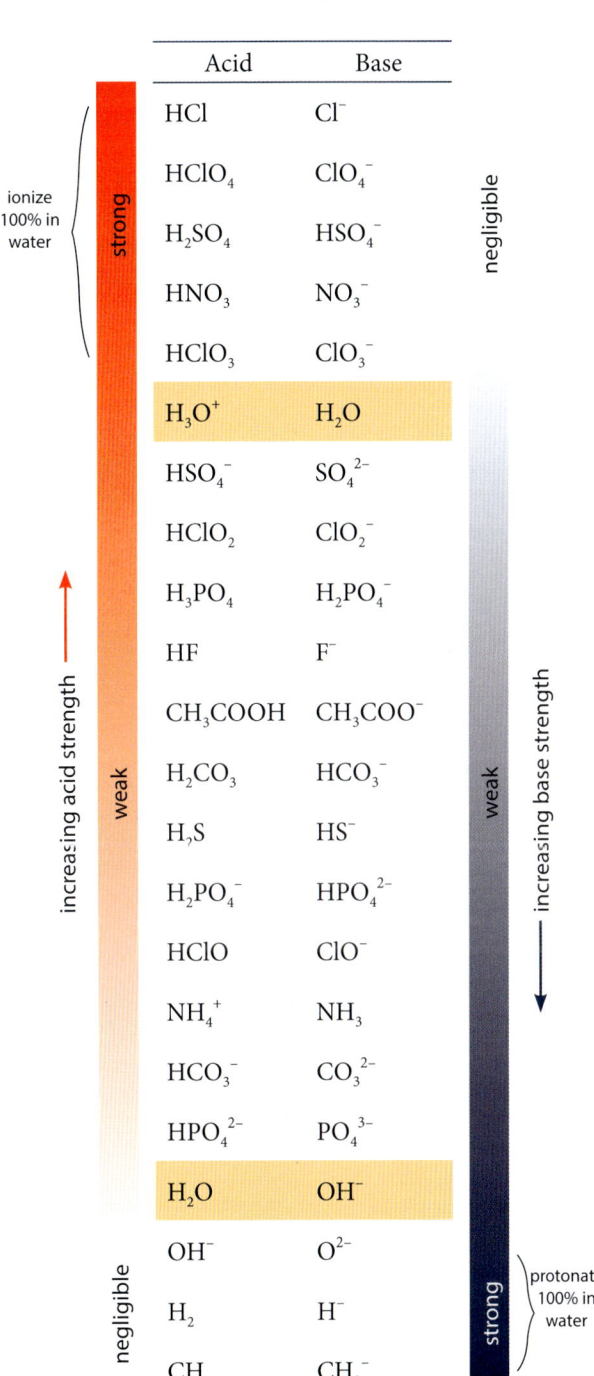

Table 11.6. Relative strengths for some conjugate acid and base pairs.

303

0.10 M H_2SO_4. All polyprotic acids ionize in steps like this, and after each step of ionization, the acid that remains is weaker and ionizes less readily.

In the lower section of the chart are substances that contain hydrogen but display negligible tendency to donate a hydrogen ion, so these compounds have negligible acid strength. Their conjugate bases are strong bases, which means 100% of these molecules accept a proton in aqueous solution, a process called *protonation*.

The first thing we notice when the conjugate acid-base pairs are arranged this way is that *the stronger an acid is, the weaker its conjugate base is, and the stronger a base is, the weaker its conjugate acid is*. In the Brønsted-Lowry model, acid and base strength are determined by a compound's readiness to donate or accept a proton. The strongest acids are those most ready to donate a proton, and the strongest bases are those most ready to accept a proton.

In any Brønsted-Lowry acid-base reaction, there are two conjugate pairs: an acid and its conjugate base, a base and its conjugate acid. In the reaction, a weak base does not compete with a strong base for protonation. Likewise, a weak acid does not compete with a strong acid for donating protons. These considerations lead to this important generalization: *In a Brønsted-Lowry acid-base reaction, the reaction favors the transfer of protons from the strong acid to the strong base, producing larger quantities of the weak acid and weak base*. The way we describe this verbally is to reference the written reaction equation and say that the reaction "favors the right (or left) side," or "the equilibrium point lies to the right (or left)."

To illustrate, consider once again the dissolution of acetic acid in water:

$$CH_3COOH(aq) + H_2O(l) \rightleftharpoons CH_3COO^-(aq) + H_3O^+(aq)$$

In this reaction, acetic acid is a Brønsted-Lowry acid, and water is a Brønsted-Lowry base. The conjugate base is the acetate ion, CH_3COO^-, and the conjugate acid is the hydronium ion, H_3O^+. Referring to Table 11.6, we find that H_3O^+ is a stronger acid than CH_3COOH. We also find that CH_3COO^- is a stronger base than H_2O. So in the equation, the stronger acid and the stronger base are on the right and the weaker acid and weaker base are on the left. Since the reaction favors the production of the weaker acid and the weaker base, we say that the equilibrium lies to the left in this reaction. This is simply a way of saying that the equilibrium point in the solution is such that the production of CH_3COOH and H_2O molecules is favored more than the production of H_3O^+ and CH_3COO^- molecules. In the solution, concentrations of CH_3COOH and H_2O molecules are much higher than the concentrations of H_3O^+ and CH_3COO^- molecules.

Now consider again the dissociation of nitric acid:

$$HNO_3(aq) + H_2O(l) \rightleftharpoons NO_3^-(aq) + H_3O^+(aq)$$

In this reaction, nitric acid is a Brønsted-Lowry acid, and water is a Brønsted-Lowry base. The conjugate base is the nitrate ion, NO_3^-, and the conjugate acid is the hydronium ion, H_3O^+. Referring to Table 11.6, we find that HNO_3 is a stronger acid than H_3O^+. We also find that H_2O is a stronger base than NO_3^-. So in the equation, the stronger acid and the stronger base are on the left and the weaker acid and weaker base are on the right. Since the reaction favors the production of the weaker acid and the weaker base, we say that the equilibrium lies to the right in this reaction. That is, the equilibrium favors the production of NO_3^- and H_3O^+ in the solution.

▼ Example 11.2

For the proton-transfer reaction shown, identify the Brønsted-Lowry conjugate pairs, and predict whether the equilibrium point lies to the right (favoring the production of SO_4^{2-} and HCO_3^-) or to the left (favoring the production of HSO_4^- and CO_3^{2-}).

$$HSO_4^-(aq) + CO_3^{2-}(aq) \rightleftharpoons SO_4^{2-}(aq) + HCO_3^-(aq)$$

We begin by identifying the conjugate acid-base pairs, as shown in the next equation.

$$HSO_4^-(aq) + CO_3^{2-}(aq) \rightleftharpoons SO_4^{2-}(aq) + HCO_3^-(aq)$$

   acid   base   conjugate   conjugate
                                     base   acid

The two acids are HSO_4^- and HCO_3^-. Referring to the acid column in Table 11.6, HSO_4^- is a stronger acid than HCO_3^-. We also see in the table that CO_3^{2-} is a stronger base than SO_4^{2-}. Thus, the stronger acid and the stronger base are on the left side of the equation. The reaction favors the transfer of the proton from the stronger acid to the stronger base, resulting in the production of the species that are the weaker acid and the weaker base. Thus, the equilibrium in this reaction lies to the right.

We have seen that according to both Arrhenius' model and the Brønsted-Lowry model, the strength of an acid depends directly on how readily the acid molecule donates a proton, H^+. When we look at the actual structure of oxyacid molecules, there is a straightforward physical basis for understanding what makes one molecule more ready to donate a proton than another. Figure 11.12 shows space-filling molecular models for the four chlorine oxyacids, arranged in order of increasing acidity. As you see, the only difference between these molecules is the number of oxygen atoms attached to the central chlorine atom.

As you know, oxygen is strongly electronegative. As such, it pulls the electron density of the bonding electron pair of the hydrogen atom toward itself, away from the hydrogen atom. This makes the O—H bond polar and enables the hydrogen atom to separate more easily from the rest of the molecule. This polarity becomes magnified with each additional oxygen atom in the molecule. The more there are, the more strongly the electron density of the electron pair bonding the hydrogen atom is pulled toward the oxygen atoms and away from the hydrogen atom. This makes the molecule more polar, so that less and less energy is required to separate the hydrogen atom from the molecule. As indicated in Table 11.6, the chlorine oxyacids with three and four oxygen atoms are strong acids because they ionize 100% in water.

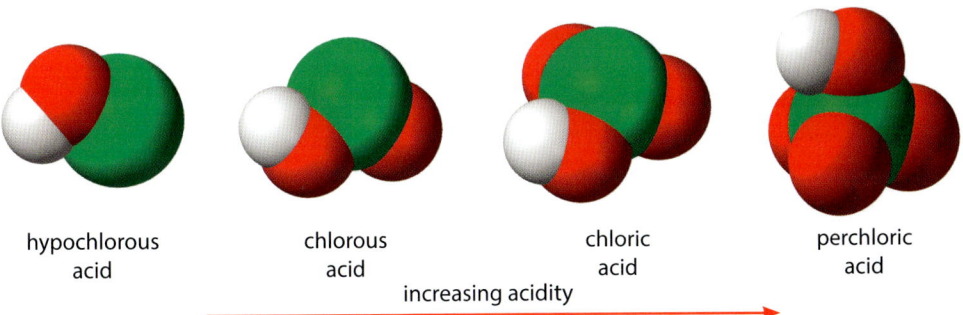

hypochlorous acid   chlorous acid   chloric acid   perchloric acid

increasing acidity

Figure 11.12. The more oxygen atoms there are on the central atom in an acid, the stronger the acid is.

11.3 Aqueous Solutions and pH

11.3.1 The Self-ionization of Water

We have seen that water is amphiprotic; it can donate or accept a proton in a Brønsted-Lowry acid-base reaction. So it may come as no surprise that pure water can and does ionize all by itself, transferring protons from one water molecule to another to create hydronium and hydroxide ions in the water. This process is called the *self-ionization of water*. Any time one water molecule donates a proton to become an OH$^-$ ion, another water molecule accepts the proton to become an H$_3$O$^+$ ion. This is represented in equation form as follows:

$$H_2O(l) + H_2O(l) \rightleftharpoons H_3O^+(aq) + OH^-(aq)$$

The molar concentrations of OH$^-$ and H$_3$O$^+$ ions in pure water, represented as [OH$^-$] and [H$_3$O$^+$], are the same. At 25°C, these values are

$$[H_3O^+] = [OH^-] = 1.0 \times 10^{-7} \, M \tag{11.1}$$

The product of these two concentrations is called the *ionization constant of water*, K_w. The ionization constant K_w is equal to

$$K_w = [H_3O^+] \cdot [OH^-] = 1.0 \times 10^{-7} \, M \cdot 1.0 \times 10^{-7} \, M = 1.0 \times 10^{-14} \, M^2 \tag{11.2}$$

At 25°C, the product of [OH$^-$] and [H$_3$O$^+$] is always equal to this value, even when other ions are present in solution. The ionization constant of water does vary with temperature: the higher the temperature, the more water ionizes, causing K_w to increase. But at any given temperature, K_w is a constant. For our purposes, you may assume that at any ordinary laboratory temperature, K_w is equal to $1.0 \times 10^{-14} \, M^2$.

11.3.2 Calculating [H$_3$O$^+$] and [OH$^-$]

Strong acids are those that readily donate protons to form H$_3$O$^+$ ions. Thus, the measure of the acidity of an aqueous solution is the molar concentration of these ions in the solution, [H$_3$O$^+$]. Pure water, in which [H$_3$O$^+$] = $1.0 \times 10^{-7} \, M$, is considered to be neither acidic nor basic but *neutral*. In fact, in any solution if Equation (11.1) holds, the solution is said to be neutral.

Any aqueous solution in which [H$_3$O$^+$] > $1.0 \times 10^{-7} \, M$ is said to be acidic. And since K_w always equals $1.0 \times 10^{-14} \, M^2$, if the concentration of hydronium ions is higher than its value in pure water, [H$_3$O$^+$] > $1.0 \times 10^{-7} \, M$, the concentration of hydroxide ions is lower than it is in pure water, or [OH$^-$] < $1.0 \times 10^{-7} \, M$. If we know one of these two values, we can use Equation (11.2) to calculate the other.

Suppose we have a solution of 0.010 M NaOH. NaOH is a strong base and thus ionizes 100% in water. As a result, when NaOH dissolves in water, each mole of NaOH in the solution supplies one mole of sodium ions and one mole of hydroxide ions in the solution, as shown by the coefficients in the equation:

$$NaOH(s) \xrightarrow{H_2O} Na^+(aq) + OH^-(aq)$$

This implies that the molar hydroxide ion concentration, [OH$^-$], is the same as the molar concentration of NaOH. Thus, the concentration of hydroxide ions in a 0.010 M NaOH solution is

$[OH^-] = 1.0 \times 10^{-2}$ M. This value for $[OH^-]$ is five orders of magnitude greater than it is for pure water, so this solution is very basic.

From the $[OH^-]$ value we can compute $[H_3O^+]$ using Equation (11.2). The product of $[H_3O^+]$ and $[OH^-]$ is 1.0×10^{-14} M^2. So

$$[H_3O^+] \cdot [OH^-] = 1.0 \times 10^{-14} \ M^2$$

$$[H_3O^+] = \frac{1.0 \times 10^{-14} \ M^2}{[OH^-]} = \frac{1.0 \times 10^{-14} \ M^2}{1.0 \times 10^{-2} \ M} = 1.0 \times 10^{-12} \ M$$

The following example illustrates a slightly more complex situation.

▼ Example 11.3

Calcium hydroxide, $Ca(OH)_2$, is a strong base that dissociates completely in water. Determine the values of $[H_3O^+]$ and $[OH^-]$ for a 0.0012 M solution of $Ca(OH)_2$.

When $Ca(OH)_2$ dissociates, two moles of OH^- ions are produced for each mole of $Ca(OH)_2$. Thus, $[OH^-]$ is

$$[OH^-] = 2 \cdot 0.0012 \ M = 0.0024 \ M$$

From this, we calculate $[H_3O^+]$ using the ionization constant of water.

$$[H_3O^+] \cdot [OH^-] = 1.0 \times 10^{-14} \ M^2$$

$$[H_3O^+] = \frac{1.0 \times 10^{-14} \ M^2}{[OH^-]} = \frac{1.0 \times 10^{-14} \ M^2}{0.0024 \ M} = 4.2 \times 10^{-12} \ M$$

The calculation technique used in the preceding example depends on the knowledge that $Ca(OH)_2$ dissociates completely in water. The same procedure may be applied to solutions of $Ba(OH)_2$ and $Sr(OH)_2$ because these hydroxides are also strong bases that dissociate completely. (The solubility of these compounds in water is quite limited, but as much as dissolves also dissociates.) You may be tempted to assume that the same technique could be applied to H_2SO_4 to determine $[H_3O^+]$. However, as Table 11.6 indicates, only the first step in the ionization of H_2SO_4 is complete. The second step involves HSO_4^-, which is a weak acid and does not ionize completely. For this reason, we cannot simply double the molar concentration for H_2SO_4 and obtain $[H_3O^+]$. There are methods of determining $[H_3O^+]$ for polyprotic acids, but they lie beyond the scope of this text.

11.3.3 pH as a Measure of Ion Concentration and Acidity

As we have seen in these examples, molar concentrations of H_3O^+ and OH^- are very small values. To save ourselves the trouble of always having to write concentrations using scientific notation, the convenient *pH scale* was developed. The pH scale is used widely for specifying the acidity or basicity of a solution. The pH of a solution is defined as

$$pH = -\log[H_3O^+] \tag{11.3}$$

Now, I have a hunch that at least a few of the students reading this text may be a bit fuzzy about how logarithms work and what they mean. So let's pause here and work on this equation for a moment. *A logarithm is an exponent.* To demonstrate what this means, consider the expression

$$\log X = Y$$

In this expression, $\log X$ is an exponent, and Y is an exponent. (They are equal—if one is an exponent, the other is as well.) So exactly what exponent is $\log X$? $\log X$ *is the exponent you have to put on a 10 to get X*. (See footnote.[2]) In other words, the expression $\log X = Y$ means the same thing as

$$10^Y = X$$

So, the expression "$\log[H_3O^+]$" is an exponent. Let's say the numerical value of $[H_3O^+]$ is 1×10^{-4}, which is simply 10^{-4}. The expression $\log[H_3O^+]$ means $\log(10^{-4})$. This expression is a logarithm, and a logarithm is an exponent. So what exponent is this logarithm? It is the exponent you have to put on a 10 to get 10^{-4}. Obviously, that value is -4. That is how logarithms work. (Note: There is an issue here with significant digits, which we will get to in a moment.)

Now let's look again at the definition of pH. By definition, pH is equal to the negative of the log of the hydronium ion concentration. So consider again the illustration in the preceding paragraph. In that illustration, the numerical value of $[H_3O^+]$ is 1×10^{-4}. The value of $\log(10^{-4})$ is -4. The pH is the negative of this value, or pH = 4.

When dealing with the mathematics of logarithms, we have to be careful with the significant digits. When we calculate expressions like $\log(1 \times 10^{-4})$, the number to the left of the decimal in the result does not count in the significant digits; it is an exponent and only *locates the decimal*. The significant digits in a logarithm are the digits to the right of the decimal. This means we must have the same number of significant digits *to the right of the decimal* in the result as we have in the original value that we found the logarithm of. The value 1×10^{-4} has one significant digit, which means we must have one digit to the right of the decimal in the logarithm of it, so $-\log(1 \times 10^{-4}) = 4.0$.

▼ Example 11.4

Determine the pH of pure water.

From Equation (11.1), for pure water, $[H_3O^+] = 1.0 \times 10^{-7}$ M. Thus,

$$pH = -\log[H_3O^+] = -\log(1.0 \times 10^{-7}) = 7.00$$

Note that since the given hydronium ion concentration has two significant digits, we must have two digits to the right of the decimal in the logarithm, that is, in our result.

Figure 11.13 is a representation of the pH scale, with a number of common substances located on it. Solutions with pH values below 7 are acidic, and the lower the pH, the more acidic the solution. Notice that since pH values are powers of 10, a solution with a pH of 4.0 has 10

---

2 This is the way *common* logarithms are defined. The common logarithm, symbolized $\log X$, is a base-10 logarithm. The *natural* logarithm, symbolized $\ln X$, is a base-e logarithm. So the expression $\ln X$ means "the exponent you have to put on e to get X." The number e is, of course, a transcendental number, like π. The numerical value of e begins 2.71828....

times the concentration of hydronium ions compared to a solution with a pH of 5.0. Thus, the solution with pH = 4.0 is 10 times as strong an acid as the solution with pH = 5.0. A pH of 7.0 represents the same hydronium ion concentration (and hydroxide ion concentration) as pure water, and is a neutral solution, neither acidic nor basic. Solutions with pH values above 7.0 are basic. Each higher value means $[H_3O^+]$ is 10 times less, and that means $[OH^-]$ is 10 times more because the product of the two is always 1.0×10^{-14} for aqueous solutions at room temperature.

Sources differ on the original meaning of the term "pH." The term was coined in 1909 by a chemist in Copenhagen, Denmark. This was in Arrhenius' day, when scientists assumed that an ionizing acid added H^+ ions to a solution. That was before we knew that hydrogen ions didn't exist in water, and that they formed hydronium ions instead. So because the term arose in the time when scientists thought that acidity was all about having hydrogen ions in the water, some say the term derives from one of the French phrases *pouvoire hydrogène* or *piussance hydrogène*, both of which mean "power of the hydrogen." Others speculate that the letters derive from the Latin *pondus hydrogenii* (weight of hydrogen) or *potentia hydrogenii* (hydrogen potential). There are other possibilities as well. The fact is, we are not sure what the designation pH meant when it was first used. We do know that in its first use in Copenhagen it was written p_H.

In the same way that pH is defined, we can define the quantity pOH as

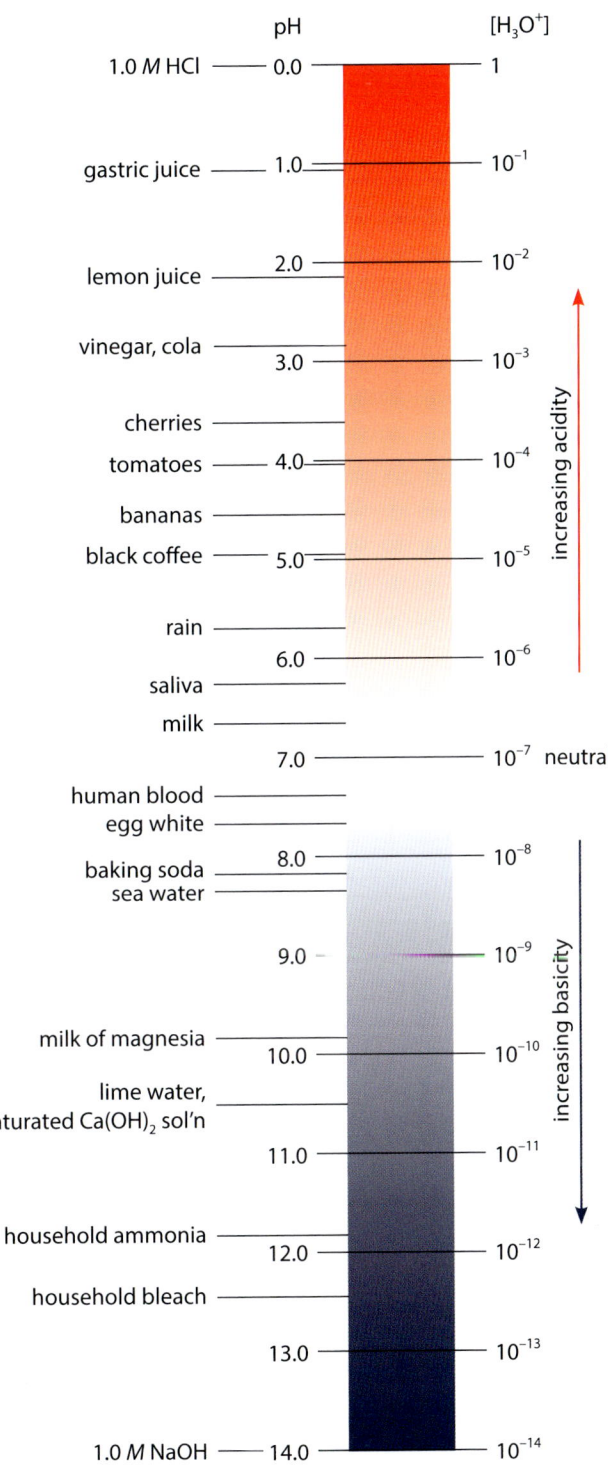

Figure 11.13. The pH scale with approximate values for common substances.

Chapter 11

$$\text{pOH} = -\log[\text{OH}^-] \tag{11.4}$$

Since $[\text{H}_3\text{O}^+]\cdot[\text{OH}^-] = 1.0 \times 10^{14}$, the rules of logarithms give us

$$\text{pH} + \text{pOH} = 14.0 \tag{11.5}$$

If you know either the pH or the pOH of a solution, you can calculate the other from Equation (11.5).

▼ Example 11.5

Given a solution of 3.0×10^{-3} M NaOH, determine $[\text{H}_3\text{O}^+]$, $[\text{OH}^-]$, the pH, and the pOH.

When NaOH ionizes in water, one mole of hydroxide ions are formed for each mole of NaOH in the solution. This means

$$[\text{OH}^-] = 3.0 \times 10^{-3} \, M$$

From this value, we use the ionization constant of water to determine $[\text{H}_3\text{O}^+]$:

$$[\text{H}_3\text{O}^+]\cdot[\text{OH}^-] = 1.0 \times 10^{-14} \, M^2$$

$$[\text{H}_3\text{O}^+] = \frac{1.0 \times 10^{-14} \, M^2}{[\text{OH}^-]} = \frac{1.0 \times 10^{-14} \, M^2}{3.0 \times 10^{-3} \, M} = 3.3 \times 10^{-12} \, M$$

Now we calculate pH and pOH using the definitions for each given in Equations (11.3) and (11.4).

$$\text{pH} = -\log[\text{H}_3\text{O}^+] = -\log(3.3 \times 10^{-12}) = 11.48$$

$$\text{pOH} = -\log[\text{OH}^-] = -\log(3.0 \times 10^{-3}) = 2.52$$

Note that pH + pOH = 14.00.

When working in a research laboratory, pH is typically measured with a *pH meter*, illustrated in Figure 11.14. Once the pH is known, the chemist can then calculate the values of $[\text{H}_3\text{O}^+]$ and $[\text{OH}^-]$. Such a calculation requires using the definition of a logarithm discussed previously to rearrange the logarithmic Equations (11.3) and (11.4) into base-10 power functions.

To illustrate, if you know $[\text{H}_3\text{O}^+]$ and wish to compute the pH, you use Equation (11.3):

$$\text{pH} = -\log[\text{H}_3\text{O}^+]$$

This is as simple as entering the molar hydronium ion concentration into your calculator and hitting the log$x$ button. But now we are talking about the inverse calculation: given the pH, we desire to calculate $[\text{H}_3\text{O}^+]$.

To do this, we must re-express the logarithmic equation as a base-10 power function. Remember, the expression

$$\log X = Y$$

means the same thing as

$$10^Y = X$$

This is a base-10 power function.[3] In the same way, the equation

$$pH = -\log[H_3O^+]$$

can be re-expressed as a base-10 power function, as follows:

$$pH = -\log[H_3O^+]$$

$$\log[H_3O^+] = -pH$$

$$10^{-pH} = [H_3O^+]$$

This gives us

$$[H_3O^+] = 10^{-pH} \qquad (11.6)$$

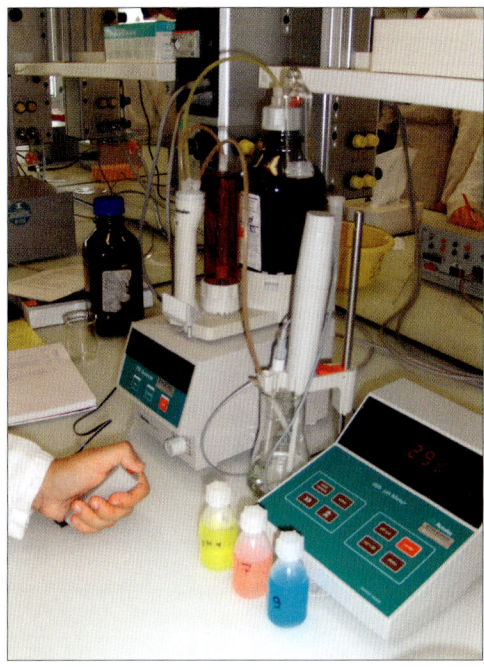

Figure 11.14. Using an electronic laboratory pH meter to measure the pH of various solutions.

A similar expression applies for calculating [OH⁻] from the pOH value:

$$[OH^-] = 10^{-pOH} \qquad (11.7)$$

We now illustrate the use of this technique in the following example.

▼ Example 11.6

A technician measures the pH of a solution with a pH meter and finds that pH = 4.72. Determine the molar hydronium ion concentration and molar hydroxide ion concentration for this solution.

We use the pH to find $[H_3O^+]$ using Equation (11.6):

$$[H_3O^+] = 10^{-pH} = 10^{-4.72} = 1.9 \times 10^{-5} \, M$$

Note here that the two digits to the right of the decimal in the pH value allow us two significant digits in the molar hydronium concentration. Now that we have $[H_3O^+]$, we use the ionization constant of water in Equation (11.2) to calculate $[OH^-]$.

$$[H_3O^+] \cdot [OH^-] = 1.0 \times 10^{-14} \, M^2$$

$$[OH^-] = \frac{1.0 \times 10^{-14} \, M^2}{[H_3O^+]} = \frac{1.0 \times 10^{-14} \, M^2}{1.9 \times 10^{-5} \, M} = 5.3 \times 10^{-10} \, M$$

---

3 The base-10 power function is sometimes referred to as an *antilog*.

Chapter 11

We note that this solution is acidic, since the pH is less than 7.0.

There is a significant reason why this type of calculation is important. If you prepare a solution of a strong acid such as HCl or HNO_3, the molar concentration of hydronium ions can be determined directly from the molarity of the solution. The fact that the acid completely dissociates in water means that the molar hydronium ion concentration is equal to the molarity of the acid (for monoprotic acids). As Example 11.3 shows, alkaline-earth hydroxides dissociate completely, and the molar hydroxide concentration of an alkaline-earth hydroxide is twice the molar concentration of the base.

However, referring again to Table 11.6, most of the acids and bases listed are weak, which means they do not ionize completely in water. Without complete ionization, we cannot determine $[H_3O^+]$ or $[OH^-]$ from the solution molarity. Instead, we must determine the values of $[H_3O^+]$ and $[OH^-]$ experimentally. One way to do this is to measure the pH and perform the calculation above. Another way is to use the technique known as *titration* to determine $[H_3O^+]$ and $[OH^-]$. We consider titration procedure in detail soon. First though, we need to become familiar with the world of pH measurement and pH indicators, our next topic.

11.3.4 pH Measurement, pH Indicators, and Titration

A *pH indicator* is a substance that changes color in solutions of different pH. Some pH indicators, such as *litmus*, exhibit two basic colors. Litmus is an extract from lichens, and is one of the oldest known pH indicators. Illustrated in Figure 11.15, litmus paper turns red when it comes in contact with an acidic solution, and blue when in contact with a basic solution. The reason we associate red with acidity and blue with basicity (see again Table 11.6 and Figure 11.13) is because of litmus paper. Litmus paper has been around a long time.

We have already encountered two other natural pH indicators. The opening page of the chapter illustrates the two colors the hydrangea takes on in acidic and basic soils (and as the caption says, the colors are the opposite of litmus paper). Another natural pH indicator is the juice from red cabbage, which exhibits an entire spectrum of colors across the range of the pH scale, shown in Figure 11.3. A fourth type of indicator, popular now in school science laboratories, is indicator paper that uses various colors to indicate the pH more precisely. One brand of this type of indicator paper is shown in Figure 11.16. Indicator strips like this are convenient to use and inexpensive.

Electronic pH meters are now rapidly rendering paper-strip type indicators obsolete. Figure 11.14 shows an expensive bench pH meter, for use in a research laboratory. Figure 11.17 shows one of the inexpensive portable models available today. pH meters work by inserting two electrodes into the solution and reading the voltage between them. Acids and bases are electrolytes, and the meter uses the voltage measurement to determine the conductivity of the solution. The conductivity depends on ion concentration, and so does the pH.

There are two basic reasons for using a pH indicator. The first is to estimate the pH of a solution. The pH indicator strips shown in Figure 11.16 are suitable for this purpose. But again, inexpensive pH meters are rapidly taking over this task from indicator papers.

A second use of indicators arises in a titration, a procedure used to ascertain the molar concentration of an unknown acid

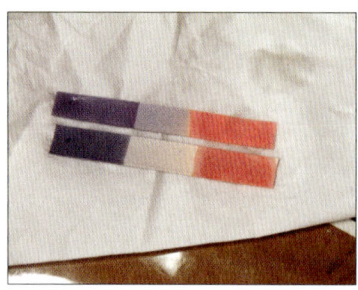

Figure 11.15. Strips of litmus paper. The paper turns red when in contact with an acidic solution, and blue when in contact with a basic solution. In a neutral solution, the paper remains a dull color.

or base solution based on the molar quantity of solution of a known molarity required to achieve neutralization. I am reserving the details of titration procedure for Section 11.3.5, but in general, here's how a titration works: during titration, an acid of precisely known concentration, called a *standard solution*, is gradually added to a base of unknown concentration (or vise versa) in a measured fashion until the point is reached when neutralization occurs, indicating that chemically equivalent amounts of the acid and base exist in the solution. From the molar amount of the standard solution and the volumetric amount of the solution of unknown concentration, the molar concentration of the unknown may be calculated.

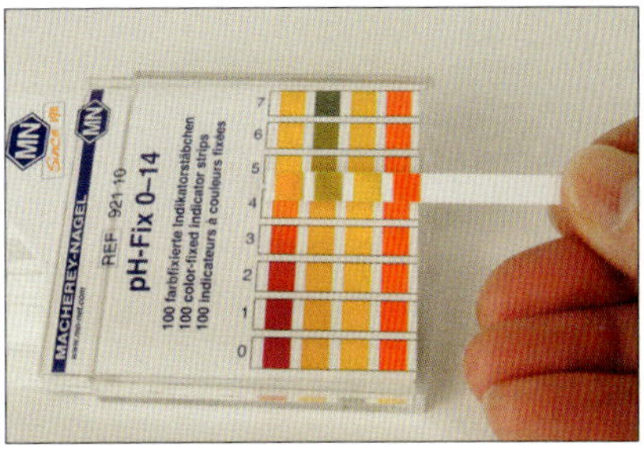

Figure 11.16. With four colored bars on each strip, this indicator paper can give a fairly accurate indication of pH. The strip shown has been dipped in a solution with a pH of 5, or possibly a bit lower as evidence by the color on the left end. The second color spot is much closer to a pH of 5 than a pH of 4.

A liquid indicator in the solution changes color when chemically equivalent amounts of acid and base are present. There are a number of different liquid indicators that may be used for this purpose, and a few of these are shown in Figure 11.18 on the next page. At the top is *bromothymol blue* (also called *bromthymol blue*). This indicator changes color in the pH range of 6.0–7.6. This range is called the indicator's *transition interval*.

The next indicator shown is *bromophenol blue* (also called *bromphenol blue*), with a transition interval in the pH range of 3.0–4.6. The third indicator shown is a solution called *phenolphthalein*, which has its transition interval in the pH range of 8.2–10.0. There are many other indicators as well, with transition intervals in different regions of the pH scale. A few of the most common indicators are listed in Table 11.7.

It may appear to you that the transition intervals for the pH indicators in Figure 11.18 are rather wide, and that an accurate pH measurement from the color would be next to impossible, but that's not really the point. First, in a titration, we are not actually trying to measure the pH; we only want to know when chemically equivalent amounts of acid and base are present in the solution. This point in the titration is called the *equivalence point*. The indicator changes color at a point called its *endpoint*, which we use as an approximation for the equivalence point. Second, when the equivalence point is reached in a titration, the pH and the indicator color both change very fast with only a minute amount of additional acid or base in the solution. So even though the color change takes place over a relatively wide range of pH, this does not really affect the accuracy with which the equivalence point is known. We return to these matters shortly when we discuss titration procedure.

There are three different basic scenarios in a titration, summarized in Table 11.8. These three scenarios depend on whether the acid and base in the titration are strong or weak. As before, the terms strong and weak refer to how fully the acid or base ionizes in water, per Table 11.6.

When a strong acid is neutralized with a strong base, the resulting salt is neutral. This means the entire solution is neutral at the equivalence point, which occurs close to pH = 7.0. The indicator used in the titration needs to have its transition in-

Figure 11.17. An inexpensive, portable electronic pH meter.

4.5   5.9   8.2
bromothymol blue (transition interval 6.0–7.6)

2.5   3.0   4.0
bromophenol blue (transition interval 3.0–4.6)

7.9   8.4   9.0
phenolphthalein (transition interval 8.2–10.0)

Figure 11.18. Colors for three pH indicators at different pH values. In each case, colors on the left and right are the colors just below and above the transition. The color in the center is right at the transition. The difference between each photo in each series is a single drop of the standard solution added to the solution being titrated.

terval in the pH region where the equivalence point is expected to occur. Of the three indicators pictured in Figure 11.18, the transition interval for bromothymol blue is in the interval where the equivalence point is expected for a strong acid/strong base titration, so this indicator is an appropriate choice for this titration.

When a strong acid is neutralized by a weak base, the resulting salt solution is slightly acidic. This means that the equivalence point occurs at a pH below 7.0. For this titration, an indicator such as bromophenol blue, methyl red, or malachite green should be used, since their transition intervals are below 7.0. Malachite green has two color transitions, one on the very acidic end of the pH scale and one on the very basic end of the scale.

When a weak acid is neutralized with a strong base, the resulting salt is slightly basic, so the equivalence point occurs above pH 7.0. For a titration like this, an indicator such as phenolphthalein should be used, since its transition interval is above 7.0. As Table 11.7 shows, malachite green might also be used. Finally, the weak acid/weak base combination is not used for titration because the equivalence point can occur at just about any pH, depending on the specific compounds in solution. The relationships between acid and base strength and their impact on titration are summarized in Table 11.8.

pH indicators are weak acids or weak bases. In a solution, an indicator partially ionizes and is present in equilibrium. For example, an acidic indicator in solution is represented by this equation:

$$HX \rightleftharpoons H^+ + X^-$$

In this equation, X represents the anion of the indicator (usually a large, complicated molecule). The anion of an indicator like this acts as a Brønsted-Lowry base in solution. If an acid is present, the X^- ions accept protons from the acid to form HX molecules, and thus the indicator is mostly in its non-ionized form in the solution. In this form, the indicator takes on its acid-indicating color, illustrated in Figure 11.19. In a basic solution, there are more OH^- ions present than H^+ ions, prompting the indicator to ionize and give up its proton. In this case, the indicator is present in its ionized form, and the solution takes on a different color.

As mentioned in the caption to Figure 11.18, the color transition at the titration end point occurs with the addition of a single drop or so of the standard solution. The addition of another drop drives the color change to completion. Again, the endpoint in a titration is when the color change occurs, and this is used as an estimate for the equivalence point—the moment when chemically equivalent amounts of acid and base are present in the solution.

As the equivalence point is approached, the pH of the combined solution during a titration becomes very sensitive to additional amounts of the standard solution. Figure 11.20 illustrates how the pH changes when a strong acid of unknown concentration is titrated with a standard solution of a strong base. At first the pH increases very gradually. Then addition of smaller and smaller quantities of the base causes the pH to increase by larger and larger amounts. Then when the equivalence point is reached, just a drop of the base increases the pH dramatically and initiates the color change of the indicator. One more drop of the standard solution again increases the pH dramatically, and the color transition of the indicator goes to completion. (See the pH values listed under the colored solutions in Figure 11.18.)

Indicator	Transition Interval
malachite green (first transition)	0.0–2.0
bromophenol blue	3.0–4.6
methyl orange	3.1–4.4
methyl red	4.4–6.2
bromothymol blue	6.0–7.6
phenol red	6.8–8.2
phenolphthalein	8.2–10.0
malachite green (second transition)	11.6–14.0

Table 11.7. pH indicators and their transition intervals.

Titration	Resulting Salt	Example Indicator
strong acid—strong base	neutral	bromothymol blue
strong acid—weak base	acidic	bromophenol blue
weak acid—strong base	basic	phenolphthalein

Table 11.8. The three types of titrations lead to different types of salts, and require different pH indicators.

Figure 11.19. The color of a pH indicator in solution depends on whether the indicator is in its non-ionized form or its ionized form.

The fact that the pH of the combined solution is so sensitive at the equivalence point allows the equivalence point to be determined with high precision. The pH indicator in the solution is a critical factor and provides the experimenter with a visual notification of when the equivalence point is being approached and when it has been achieved.

11.3.5 Titration Procedure

Titration makes use of an item of laboratory apparatus called a *buret*, pictured in Figure 11.21. Two burets are required, along with a stand, a dual buret clamp, and an *Erlenmeyer flask*.

Before the titration, a standard solution of precisely known molarity must be prepared. This is done by first preparing a volume of the standard solution with the desired concentration. The standard solution is then titrated with a *primary standard*, a solid, highly purified compound that is used to determine the precise concentration of the standard solution. Then the standard solution is ready to be used for titrating the solution of unknown concentration.

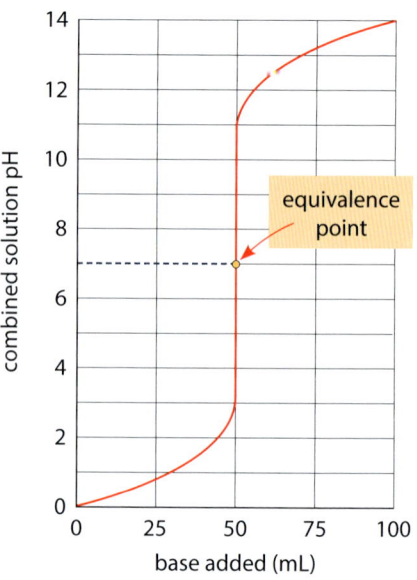

Figure 11.20. Change in the pH of the combined solution as a strong acid of unknown concentration is titrated with a standard solution of strong base.

The procedure for titration is described below. This procedure describes the titration of an acid solution of unknown concentration with a basic standard solution.

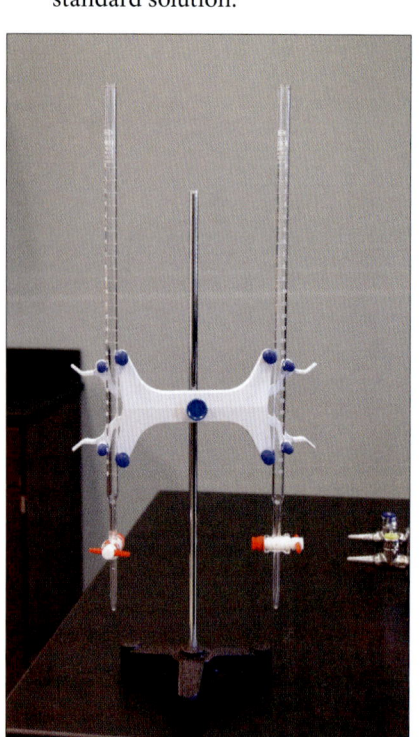

Figure 11.21. Dual buret setup for titration.

1. Set up the stand with two clean burets and assign one of the burets to hold the acid and the other to hold the base. Rinse the acid buret three times with the acid solution to be titrated. Rinse the base buret three times with the base standard solution. Note that these rinses

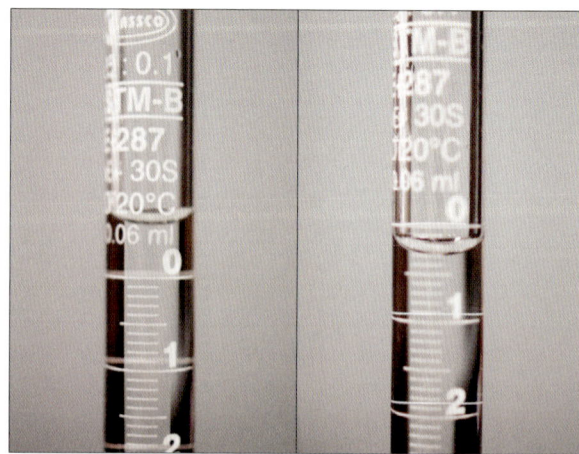

Figure 11.22. Filling the acid solution above the calibrated scale (left), then releasing acid until the liquid is within the calibrated scale (right).

cannot be performed with water because even a few drops of water remaining in the buret compromises the accuracy of the titration.

2. Fill the acid buret to above the calibrated scale with the solution to be titrated. To eliminate any air bubbles in the stopcock, release some acid into a waste beaker until the top of the liquid falls down into the region of the calibrated scale on the buret. Record the reading in the buret to the nearest 0.01 mL. (Remember to read the buret at the bottom of the meniscus.) These steps are illustrated in Figure 11.22.

3. Release acid into a clean, dry Erlenmeyer flask. Your instructor or lab procedure will indicate the approximate volume of acid to use. Record the new buret reading to the nearest 0.01 mL. The difference between the two buret readings is the precise amount of acid solution in the flask.

4. Add three drops of the pH indicator to the flask. The rest of this description titration assumes phenolphthalein is being used.

5. Fill the base buret with the standard solution to above the calibrated scale. To remove air bubbles from the stopcock, release some base into a waste beaker until the buret level falls to within the calibrated scale. Record the buret reading to the nearest 0.01 mL.

6. Place the Erlenmeyer flask containing the acid solution under the base buret. Adjust the height of the buret so the tip of the buret extends slightly into the flask, as shown in Figure 11.23.

7. Slowly release standard base solution into the flask, swirling the flask as you do, as illustrated in Figure 11.24. As the concentration of base in the flask increases, flares of pink color appear where the base solution enters the titration solution. The colored liquid dissipates and becomes colorless as the flask is swirled.

Figure 11.23. The tip of the buret should extend slightly into the flask.

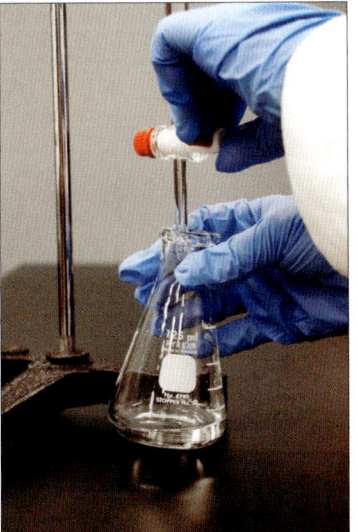

Figure 11.24. Titration.

Figure 11.25. Flares of pink color begin lasting for a second or two when the indicator endpoint is approaching.

Figure 11.26. A single drop of the standard solution eventually turns the indicator pink, indicating that the equivalence point has been reached.

8. When the pink color begins to stay in the solution for a second or two before disappearing, the endpoint is being approached, shown in Figure 11.25. After this, begin adding the standard base slowly, drop by drop.

9. When a single drop of base causes the solution to become light pink and stay that way even after 30 seconds of swirling, the endpoint has been reached, as shown in Figure 11.26. Record the reading on the standard base buret to the nearest 0.01 mL. The difference between the initial and final base buret readings is the amount of standard solution required to neutralize the volume of acid solution in the flask.

11.3.6 Determining [H₃O⁺] or [OH⁻] from Titration Data

Determining the concentration of the solution of unknown concentration with titration data is essentially a stoichiometric calculation. The starting point is the chemical equation showing the mole ratios in the reaction. For example, if titrating HCl with NaOH, the equation is

$$HCl(aq) + NaOH(aq) \rightarrow NaCl(aq) + H_2O(l)$$

The coefficients on the HCl and NaOH in this equation tell us that 1 mol HCl neutralizes 1 mol NaOH. Thus, chemically equivalent amounts of acid and base are equal mole quantities.

Next, determine the number of moles of standard solution present at the titration endpoint. For example, if 24.50 mL are used of a standard solution of 1.100×10^{-3} M NaOH, then the mole quantity of NaOH is

$$24.50 \text{ mL} \cdot \frac{1 \text{ L}}{1000 \text{ mL}} \cdot \frac{1.100 \times 10^{-3} \text{ mol}}{\text{L}} = 0.02695 \text{ mol}$$

Since the mole ratio of HCl to NaOH is 1 : 1, there are 0.02695 mol HCl in the solution as well. Now supposing that 21.00 mL HCl are initially released into the flask, the concentration of the acid solution is computed as

$$\frac{0.02695 \text{ mol}}{21.00 \text{ mL}} \cdot \frac{1000 \text{ mL}}{\text{L}} = 1.283 \text{ M}$$

The following example illustrates a case when the mole ratios of [OH⁻] and [H₃O⁺] in the titration solution are not the same.

▼ **Example 11.7**

Suppose 22.73 mL of a standard solution of 0.0995 M Sr(OH)₂ are required to neutralize 23.21 mL of HNO₃ solution. Determine the concentration of the HNO₃ solution.

Recall from Example 11.3 that the three soluble alkaline-earth hydroxides each dissociate completely in water. For Sr(OH)₂, there are two OH⁻ ions in each formula unit of Sr(OH)₂, and thus the molarity of the OH⁻ is twice the molarity of the Sr(OH)₂. From this information we know that

$$[OH^-] = 2 \cdot 0.0995 \text{ M} = 0.199 \text{ M}$$

Using this with the volume given for the standard solution, we find the number of moles of OH⁻ to be

$$22.73 \text{ mL Sr(OH)}_2 \cdot \frac{1 \text{ L}}{1000 \text{ mL}} \cdot \frac{0.199 \text{ mol OH}^-}{\text{L Sr(OH)}_2} = 4.523 \times 10^{-3} \text{ mol OH}^-$$

The ionic equation for this neutralization reaction is

$$2HNO_3(aq) + Sr(OH)_2(aq) + 2H_2O(l)$$
$$\rightarrow 2NO_3^-(aq) + Sr^{2+}(aq) + 2OH^-(aq) + 2H_3O^+(aq)$$

This equation indicates that the mole ratio of hydroxide ions to hydronium ions is 2 : 2, which is the same as 1 : 1. This is also the mole ratio of OH⁻ to HNO₃. Thus, the number of moles of HNO₃ is 4.523×10^{-3} mol.

Now we calculate the HNO₃ concentration with the number of moles and the volume:

$$\frac{4.523 \times 10^{-3} \text{ mol HNO}_3}{23.21 \text{ mL HNO}_3} \cdot \frac{1000 \text{ mL}}{\text{L}} = 0.195 \text{ } M$$

Chapter 11 Exercises

SECTION 11.1

1. List the general properties of acids and bases.
2. What properties do acids and bases have in common?
3. Distinguish between oxyacids and carboxylic acids.
4. Name the following acids, and identify each as monoprotic, diprotic, or triprotic.

   a. HF
   b. HClO
   c. HNO₂
   d. H₂SO₃
   e. HBr
   f. H₃PO₄
   g. H₂S
   h. H₂CrO₄
   i. CH₃COOH

5. Is formic acid a binary acid? Explain your response.
6. Write the formulas for the following acids:

   a. chlorous acid
   b. perbromic acid
   c. hydrofluoric acid
   d. hydroiodic acid

7. Describe three industrial uses for three different major industrial acids.
8. Draw the Lewis structures for formic acid and benzoic acid.
9. Explain why acetic acid is not a "tetraprotic" acid.

SECTION 11.2

10. Describe the Arrhenius and Brønsted-Lowry acid-base theories in detail.
11. Give examples of two acids and two bases according to the Arrhenius and Brønsted-Lowry acid-base theories. For each one, write a short explanation of why the species acts as you have described under the given theory.
12. Explain why water is amphiprotic under the Brønsted-Lowry acid-base model.

Chapter 11

13. For the reactions shown, identify the Brønsted-Lowry acid-base conjugate pairs.

   a. $HNO_2(aq) + CN^-(aq) \rightleftharpoons NO_2^-(aq) + HCN(aq)$

   b. $HCO_3^-(aq) + H_2O(l) \rightleftharpoons H_2CO_3(aq) + OH^-(aq)$

   c. $HSO_4^-(aq) + C_2O_4^{2-}(aq) \rightleftharpoons SO_4^{2-}(aq) + HC_2O_4^-(aq)$

   d. $NH_4^+(aq) + HPO_4^{2-}(aq) \rightleftharpoons NH_3(aq) + H_2PO_4^-(aq)$

14. Explain why ammonia is classified as a base under both the Arrhenius and Brønsted-Lowry acid-base theories.

15. Explain the factor that determines whether an acid or base is classified as strong or weak.

16. Explain why the chlorine oxyacids get more acidic as the number of oxygen atoms in the molecule increases.

17. Based on your knowledge of molecular bonding, formulate a reason for the stepwise ionization and subsequent weakening of the acid for polyprotic acids.

18. Sodium carbonate is an important industrial base, even though it contains no hydroxide anion. Write an equation that demonstrates the character of sodium carbonate as a base.

19. For each of the following acid-base neutralization reactions, complete the formula equation and balance it. Then write the ionic equation and the net ionic equation.

   a. $HCl(aq) + KOH(aq) \rightarrow$

   b. $HNO_3(aq) + NaOH(aq) \rightarrow$

   c. $H_2SO_4(aq) + Ca(OH)_2(aq) \rightarrow$

   d. $H_2SO_3(aq) + KOH(aq) \rightarrow$

20. In each of the following reactions, an acid reacts with a metal to form a salt, releasing hydrogen gas. Complete the formula equation and balance it. Then write the ionic equation and the net ionic equation.

   a. $Al(s) + H_2SO_4(aq) \rightarrow$

   b. $Zn(s) + HCl(aq) \rightarrow$

21. In each of the following reactions, an acid reacts with a carbonate base. Complete the formula equation and balance it.

   a. $HNO_3(aq) + CaCO_3(aq) \rightarrow$

   b. $HCl(aq) + BaCO_3(aq) \rightarrow$

   c. $H_2SO_4(aq) + Na_2CO_3(aq) \rightarrow$

   d. $HNO_3(aq) + MgCO_3(aq) \rightarrow$

Acids and Bases

22. For each acid-base reaction given, complete the proton-transfer equation for the transfer of a single proton. Then, using Table 11.6, predict the direction in which the equilibrium is favored.

    a. $H_3PO_4(aq) + F^-(aq) \rightarrow$
    b. $HSO_4^-(aq) + CO_3^{2-}(aq) \rightarrow$
    c. $ClO_2^-(aq) + H_3O^+(aq) \rightarrow$
    d. $HClO_3(aq) + HCO_3^-(aq) \rightarrow$
    e. $H_3O^+(aq) + HS^-(aq) \rightarrow$
    f. $HPO_4^{2-}(aq) + NH_3(aq) \rightarrow$

23. A 196-g sample of sodium carbonate is dissolved in nitric acid. Assuming excess acid, determine the following:

    a. the mass and volume at STP of dry CO_2 gas produced by the reaction.
    b. the volume of 6.00 M nitric acid required for this reaction.

24. According to the activity series of metals (Table 7.2), copper does not react with sulfuric acid. However, if the acid is hot enough and concentrated enough, copper reacts with H_2SO_4 in a single-replacement reaction. If excess copper reacts with 125 mL of 8.0 M H_2SO_4, determine the following:

    a. the mass in grams of copper sulfate produced.
    b. the volume in liters of hydrogen gas released at STP.

25. Some forms of marble are made of calcium carbonate. In a reaction with hydrochloric acid, $CaCl_2$, CO_2, and H_2O are produced. Determine the following:

    a. the mass of marble required to produce 2.5×10^4 mL CO_2 at STP.
    b. the corresponding volume of 2.00 M HCl required.

SECTION 11.3

26. Explain the significance of the small value of the ionization constant of water, K_w, and relate this to the electrical conductivity of water.

27. Explain what the pH of a solution indicates. Select three different pH values to use as examples in your explanation.

28. For each of the following, indicate whether the solution is acidic, basic, or neutral.

    a. $[OH^-] = 1.0 \times 10^{-4}$ M
    b. $[H_3O^+] = 1.0 \times 10^{-6}$ M
    c. $[H_3O^+] = 1.0 \times 10^{-9}$ M
    d. $[H_3O^+] = [OH^-]$
    e. pH = 4.91
    f. $[H_3O^+] < [OH^-]$
    g. pOH = 4.91
    h. $[OH^-] = 1.0 \times 10^{-7}$ M

29. Distinguish between equivalence point and endpoint.

30. There are four possible combinations of strong/weak acid/base that could be used in a titration. For each combination, state in general what the expected pH is at the equivalence point, and suggest an appropriate pH indicator.

31. Calculate $[H_3O^+]$ and $[OH^-]$ for each of the following:

    a. 1.56×10^{-3} M $HClO_4$
    b. 0.0110 M $Sr(OH)_2$
    c. 2.50 M NaOH
    d. 8.911×10^{-5} M $HClO_3$

Chapter 11

32. Calculate the pH and pOH for each of the following:

   a. $0.0250\ M$ NaOH
   b. $6.0\ M$ HNO$_3$
   c. $12\ M$ HCl
   d. $2.03 \times 10^{-2}\ M$ Ba(OH)$_2$

33. For each of the following values, determine [H$_3$O$^+$] and [OH$^-$].

   a. pH = 1.00
   b. pH = 3.5
   c. pOH = 10.0
   d. pH = 11.88
   e. pOH = 2.1
   f. pOH = 7.0
   g. pOH = 4.00
   h. pH = 14.0

34. In each of the following items, one of these values is given: pH, pOH, [H$_3$O$^+$], or [OH$^-$]. In each case, calculate the other three.

   a. pH = 5.6
   b. [OH$^-$] = $5.5 \times 10^{-10}\ M$
   c. [H$_3$O$^+$] = $1.0 \times 10^{-9}\ M$
   d. pOH = 11.5
   e. [H$_3$O$^+$] = $4.5 \times 10^{-3}\ M$
   f. pOH = 1.00
   g. pOH = 7.22
   h. pH = 13.50
   i. pH = 2.0
   j. [OH$^-$] = $3.2 \times 10^{-2}\ M$

35. Listed below are several acid-base combinations that could be used in a titration. Determine the molarity of the first substance listed that is chemically equivalent with the amount given for the second substance.

   a. 18.55 mL HNO$_3$ with 24.05 mL of $1.65 \times 10^{-3}\ M$ NaOH
   b. 23.78 mL KOH with 21.00 mL of $1.500\ M$ HClO$_3$
   c. 22.94 mL Ca(OH)$_2$ with 20.87 mL of $4.077 \times 10^{-2}\ M$ HClO$_4$
   d. 22.00 mL HCl with 20.05 mL of $3.050\ M$ LiOH

36. A chemist uses 23.09 mL of a $5.06 \times 10^{-3}\ M$ standard solution of HCl to neutralize 20.07 mL of KOH. Determine the molar concentration of the KOH solution.

37. In a titration, a student adds 22.01 mL of $0.0231\ M$ Ba(OH)$_2$ to 27.0 mL of an HCl solution of unknown concentration. Determine the molarity of the acid solution.

38. A $2.5 \times 10^{-3}\ M$ standard solution of NaOH is used to titrate a $0.04098\ M$ standard solution of a monoprotic weak acid. If 23.90 mL of the base are required to neutralize 20.00 mL of the acid, what is the percent ionization of the acid?

GENERAL REVIEW EXERCISES

39. The solubility of strontium hydroxide is 1.77 g Sr(OH)$_2$/100 mL water. With this information, verify that it is possible to prepare the Sr(OH)$_2$ solution referred to in Example 11.7.

40. A solution is prepared by dissolving 38.95 g sucrose, C$_{12}$H$_{22}$O$_{11}$, in enough water to bring the solution volume to 250.0 mL. Determine the molar concentration of the solution.

41. An ice cube with a mass of 37.05 g is in equilibrium at −18.0°C. Determine the amount of heat required to raise the ice cube's temperature to 0°C, melt it, and warm the resulting water to 20.0°C. (The properties of water are in Appendix A.)

Acids and Bases

42. An aqueous solution is prepared with 35.55 g NaOH. To this is added 157.0 mL of 4.777 M HNO$_3$.

    a. Determine the limiting reactant.
    b. Determine the amount of sodium nitrate produced.
    c. Determine the amount of water produced.
    d. Determine how many moles of the excess reactant remain unconsumed by the reaction.

43. Draw the Lewis structures for the following molecules and polyatomic ions:

    a. CCl$_4$
    b. CO$_3^{2-}$
    c. NO$_2^-$
    d. C$_2$H$_2$
    e. O$_3$
    f. H$_2$S
    g. CO$_2$
    h. H$_3$O$^+$

44. Aqueous solutions of sodium carbonate and barium nitrate are combined. The sodium carbonate solution is 0.050 molar and 25.50 mL of it are used.

    a. Write the balanced molecular equation, the ionic equation, and the net ionic equation.
    b. Identify the spectator ions and the precipitate.
    c. If all the sodium carbonate is used in the reaction, determine the mass of the precipitate that forms.

45. When copper reacts with sulfuric acid, the reaction products are copper sulfate and hydrogen gas. Assume that 109 mL of H$_2$ is collected over water at a barometric pressure of 765 Torr and a temperature of 25°C. Determine how many grams of copper are consumed in the reaction.

46. A 2.500-L tank contains oxygen gas at 22.0°C and 13.11 atm.

    a. Determine the number of moles of gas in the tank.
    b. Determine the mass of the gas.

47. A 25.00 g sample of iron metal is placed in an aqueous solution containing 35.8 g CuSO$_4$.

    a. Determine the limiting reactant.
    b. Determine the mass of iron(II) sulfate produced.
    c. Determine the mass of the excess reactant that remains.

48. At 30.0°C, a 2.00-L container contains a gas sample with a mass of 8.755 g at a pressure of 1.05 bar. Determine the molar mass of the gas.

49. Determine the energy in electron volts of an X-ray photon with a wavelength of 0.22 nm.

Chapter 12
Redox Chemistry

This is a 1940s-vintage photo of a phosphate smelting furnace in Alabama. The smelter produces elemental phosphorus. The specific redox reaction to produce phosphorus depends on the particular mineral from which the phosphorus is being extracted. Calcium phosphate mined in Florida is one possibility, for which the redox reaction is $2Ca_3(PO_4)_2 + 6SiO_2 + 10C \rightarrow 6CaSiO_3 + 10CO + P_4$. The calcium phosphate is mixed with sand (SiO_2) and coke (carbon) and heated to around 1,500°C. The carbon is oxidized to form carbon monoxide; the phosphorus is reduced to form elemental phosphorus vapor (P_4).

In the process shown above, the heat is being supplied by electricity, and the large cables are electrical conductors carrying the current into the furnace. You can be confident that where the worker is standing, the heat is so intense it is difficult to remain there.

Objectives for Chapter 12

After studying this chapter and completing the exercises, you should be able to do each of the following tasks, using supporting terms and principles as necessary.

SECTION 12.1

1. Define *oxidation* and *reduction* and write example reaction equations to represent each.
2. Identify the oxidation states of elements in elemental molecules and in compounds.
3. Compare relative strengths of oxidizing and reducing agents, and use a table of such agents to predict if one substance is able to oxidize another.

SECTION 12.2

4. Given a substance and its oxidized or reduced form, complete and balance the half-reaction and indicate whether the half-reaction is an oxidation or reduction.
5. Given substances in a redox reaction, write the oxidation and reduction half-reactions.
6. Balance redox reaction equations using the half-reaction method.
7. After balancing a redox reaction equation, combine and/or add ions to show the original compounds used to produce the reaction.

SECTION 12.3

8. Explain the essential principles of electrochemistry.
9. Describe the basic components of an electrochemical cell.
10. Describe and differentiate between voltaic and electrolytic cells.
11. On a diagram of a voltaic or electrolytic cell, identify the anode, cathode, direction of electron flow, salt bridge or porous barrier, and direction of ion movement.
12. Draw a diagram of a voltaic or electrolytic cell, indicating the anode, cathode, direction of electron flow, salt bridge or porous barrier, and direction of ion movement.
13. Explain the purpose of a salt bridge and describe how a salt bridge is physically constructed.
14. Describe the processes of oxidation and reduction in electrochemical cells, including where they occur and what kinds of species (atoms or ions) are involved in each.
15. Define *electrode potential* and explain how standard reduction potential values are established.
16. Calculate cell potentials from a table of standard electrode potentials.
17. Identify several technologies that use electrochemical redox reactions.

Chapter 12

12.1 Oxidation and Reduction

12.1.1 Introduction to Redox Reactions

Chemical reactions in which substances are oxidized and reduced are *oxidation-reduction reactions,* or *redox reactions.* These reactions form a very large class, including many of the reactions we have studied. All combustion reactions are redox reactions, illustrated by the beautiful launch of Apollo 15 in Figure 12.1. All reactions in which coatings such as rust or verdigris form on metals are also redox reactions, illustrated by the verdigris on a copper bay window in Figure 12.1.

Reactions between metals and acids or bases that result in corrosion or destruction of the metal are also redox reactions. Alkali batteries, such as those shown in Figure 12.2, contain potassium hydroxide, a strong base. When the battery seals fail and the KOH leaks out, a redox reaction corrodes the metal. The batteries themselves also supply electrical power by means of redox reactions, a topic we address later in this chapter.

The single replacement reactions we studied in Section 7.2.4 are also redox reactions. The sequence of photos in Figure 12.3 illustrates several redox reactions. The top three images show an aluminum wire placed into a solution of copper(II) chloride. The aluminum replaces the copper in the solution, and the copper comes out of solution, forming a fluffy coating on the aluminum. In the lower image, copper hydroxide (blue) and copper carbonate (blue green) have formed on the copper—both of these being redox reactions between the copper and the air. The copper compounds then leached back into the water containing the aluminum ions.

Oxidation reactions and reduction reactions always occur together in pairs—when one substance is oxidized, another substance is reduced. Recall from Section 7.1.5 that redox reactions may be generally thought of as electron transfer reactions, particularly in the case of the oxidation or reduction of elements in ionic compounds. (And remember the mnemonic: *LEO the lion says GER*—Lose Electrons Oxidize; Gain Electrons Reduce.)

Figure 12.1. The combustion of rocket fuel at the Apollo 15 launch (above) and the formation of verdigris on a copper-clad bay window in Brooklyn (below) are examples of redox reactions.

In the case of molecular compounds, it is not very helpful to think in terms of electron transfer. Instead it is better to think in terms of the oxidation states of the elements involved in the reaction. We review this information below, but it would be a good idea at this point for you to reread Section 7.1.5 if you have not done so recently.

12.1.2 Oxidation States

When a metal atom loses one or more electrons, it becomes an ion with a positive charge. Its oxidation state as a pure metal is 0. As an ion, its oxidation state is the same as its ionic charge: +1 for alkali metals, +2 for alkaline-earth metals, and so on.

Figure 12.2. Metal corrosion from the strong base potassium hydroxide contained inside alkali batteries.

Redox Chemistry

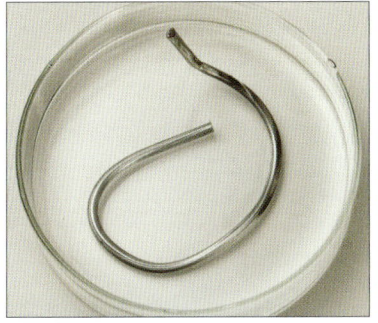

Figure 12.3. More redox: Aluminum wire (top); aluminum wire in CuCl$_2$ solution (second from top); copper metal out of solution on the aluminum (third from top); Cu(OH)$_2$ and CuCO$_3$ form on the copper, then leach into the water (bottom).

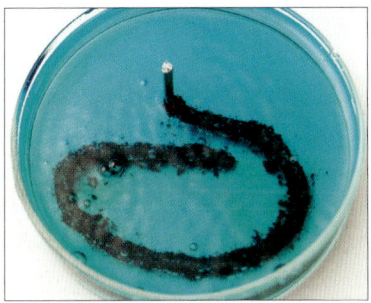

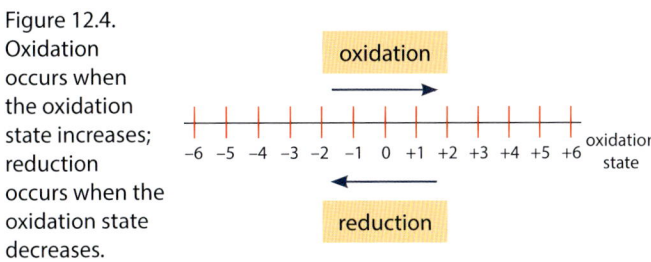

Figure 12.4. Oxidation occurs when the oxidation state increases; reduction occurs when the oxidation state decreases.

When a nonmetal atom gains one or more electrons, it becomes an ion with a negative charge. Its oxidation state as a pure element is 0 (including elements that form diatomic molecules). As an ion, its oxidation state is equal to its ionic charge: –1 for the halogens, –2 for the chalcogens, and so on.

In general, *any time the oxidation state of an element increases, that element is said to have been oxidized.* Conversely, *any time the oxidation state of an element decreases, that element is said to have been reduced.* These definitions are illustrated in Figure 12.4. Thinking in terms of electron transfer, an element that loses electrons is oxidized. The electrons must have somewhere to go, and the element they go to is the element that is reduced. The element that is oxidized causes the reduction of the element that is reduced. For this reason, the oxidized element is called the *reducer* or *reducing agent* or *reductant*. Conversely, the element that is reduced causes the oxidation of the element that is oxidized. For this reason, the reduced element is called the *oxidizer* or *oxidizing agent* or *oxidant*. These relationships are illustrated in Figure 12.5.

For monatomic ions and elements in ionic compounds, the oxidation state is equal to the value of the ion's charge. However, in molecules, the oxidation state of an element in the molecule is a hypothetical or artificial number assigned to each element in the molecule as if bonding electrons were held completely by one atom or another. In molecules, the oxidation state is associated with the strength of the element's attraction to the bonding electrons, and thus is based on the element's electronegativity relative to the other elements in the molecule. The oxidation state in a molecule is not based on any actual charge.

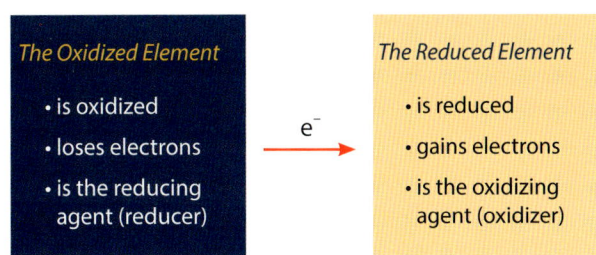

Figure 12.5. Relationships and nomenclature for the oxidizing agent and the reducing agent.

327

Chapter 12

Analyzing redox reactions requires knowledge of how to establish the oxidation state of any element in any compound. As indicated above, this is easy for ions. For molecules, we have to go by the set of rules you learned in Section 7.1.5. For your convenience, here again are the rules we follow to determine the oxidation state of elements in covalent compounds:

1. The operating rule is this: *in a molecular compound, the oxidation states of the elements involved add up to the charge on the molecule.* In a neutral molecule, the oxidation numbers add up to zero. In a polyatomic ion, they add up to the charge on the ion.

2. In pure elements in their natural state, the oxidation state is 0. This also applies to diatomic molecules such as H_2 and the other elements that naturally make molecules with themselves, S_8 and P_4.

3. The most electronegative element in the molecule has the oxidation state it would have if it were an anion. Since fluorine has the highest electronegativity, its oxidation state is always –1. Oxygen has the second-greatest electronegativity, so it is usually assigned an oxidation state of –2. One exception is when oxygen forms peroxides in its form as the molecule O_2^{2-}, in which case its oxidation state is –1. A common peroxide is hydrogen peroxide, H_2O_2. Another exception is when oxygen bonds with fluorine to form OF_2, in which case its oxidation state is +2 (since F is always –1).

4. The oxidation state for halogens is –1. The exception is when they bond with oxygen to form polyatomic ions with negative charge (oxyanions). In that case, they have positive oxidation states (except for fluorine).

5. When hydrogen forms a compound with an element that is more electronegative than itself (a nonmetal), its oxidation state is +1. In compounds with metals, the oxidation state of H is –1.

Let's look at some examples of reactions and identify the oxidation states we find in the reactants and products. The equation for the combustion of hydrogen is

$$2H_2(g) + O_2(g) \rightarrow 2H_2O(g)$$

The elements in pure gases, hydrogen and oxygen, have oxidation states of 0. In water, the oxidation state of oxygen and hydrogen are –2 and +1, respectively. Writing in the oxidation states, we have:

$$\overset{0}{2H_2}(g) + \overset{0}{O_2}(g) \rightarrow \overset{+1\ -2}{2H_2O}(g)$$

Hydrogen is oxidized and oxygen is reduced in this reaction.

When zinc metal is placed in a solution of hydrochloric acid, the hydrogen ions react with the zinc to produce zinc ions and hydrogen gas. This reaction is represented as follows:

$$2H^+(aq) + Zn(s) \rightarrow H_2(g) + Zn^{2+}(aq)$$

Here hydrogen ions are forming hydrogen gas and zinc is being ionized. The oxidation states on the ions are the same as their ionic charges; the oxidation states of the pure elements are zero. Writing these in, we have:

$$\overset{+1}{2H^+}(aq) + \overset{0}{Zn}(s) \rightarrow \overset{0}{H_2}(g) + \overset{+2}{Zn^{2+}}(aq)$$

328

Redox Chemistry

Now we see that hydrogen is reduced and zinc is oxidized. Since hydrogen is reduced, it is the oxidizing agent—hydrogen oxidizes the zinc. The zinc is oxidized, so it is the reducing agent—zinc reduces the hydrogen.

As one more example, from Table 7.2 we see that zinc is a more active metal than copper. This means zinc replaces copper in solution in a single replacement reaction, such as the following:

$$Zn(s) + Cu(NO_3)_2(aq) \rightarrow Cu(s) + Zn(NO_3)_2(aq)$$

When we write in the oxidation states on this more complex equation, we have:

$$\overset{0}{Zn}(s) + \overset{+2\ -2}{Cu(NO_3)_2}(aq) \rightarrow \overset{0}{Cu}(s) + \overset{+2\ -2}{Zn(NO_3)_2}(aq)$$
$${+5}{+5}$$

Note here that the charge on the nitrate ions is –1, and the sum of the oxidation states of the elements in that polyatomic ion add up to the charge on the ion. As usual, the oxidation state of oxygen is –2. Since there are three oxygen atoms, the oxidation state of nitrogen is +5 so that 5 + 3(–2) = –1. In this reaction, the oxidation states of the elements in the nitrate ions do not change. The redox reaction is the oxidation of zinc and the reduction of copper.

▼ Example 12.1

"Nicad" batteries are common now as AA and AAA size rechargeable dry cell batteries. These batteries contain nickel and cadmium, hence their name. The redox reaction that allows the batteries to produce electricity is shown below. Write the oxidation states above each element in the reaction. Identify the substances that are oxidized and reduced, and the oxidizing and reducing agents.

$$Cd(s) + NiO_2(s) + H_2O(l) \rightarrow Cd(OH)_2(s) + Ni(OH)_2(s)$$

Using the guidelines for assigning oxidation states, we write in these numbers as follows:

$$\overset{0}{Cd}(s) + \overset{+4\ -2}{NiO_2}(s) + \overset{+1\ -2}{H_2O}(l) \rightarrow \overset{+2\ -2\ +1}{Cd(OH)_2}(s) + \overset{+2\ -2\ +1}{Ni(OH)_2}(s)$$

Cadmium is oxidized, since its oxidation state increases from 0 to +2. Nickel is reduced, since its oxidation state decreases from +4 to +2. Since cadmium is oxidized, cadmium is the reducing agent. Since nickel is reduced, nickel is the oxidizing agent.

▼ Example 12.2

When sulfur dioxide gas is dissolved in water, small amounts of sulfurous acid form according to the reaction shown below. Write the oxidation states above each element in the reaction. Identify the substances that are oxidized and reduced, and the oxidizing and reducing agents.

$$SO_2(g) + H_2O(l) \rightarrow H_2SO_3(aq)$$

Using the guidelines for assigning oxidation states, we write in these numbers as follows:

$$\underset{+4}{\overset{-2}{SO_2}}(g) + \overset{+1\ -2}{H_2O}(l) \rightarrow \underset{+4}{\overset{+1\ -2}{H_2SO_3}}(aq)$$

None of the oxidation states change. Thus, this is not a redox reaction.

12.1.3 Strengths of Oxidizing and Reducing Agents

In Section 11.2.4, we studied Table 11.6, which charts the relative strengths of acids and bases. By comparing their relative strengths, we are able to predict the direction of the equilibrium in an acid-base reaction. In a similar way, Table 12.1 lists a few representative substances according to their relative strengths as oxidizing and reducing agents. Some of the information in this table anticipates topics we have not yet addressed. So for now, attend only to the first two columns. The significance of the "half-reactions" in the third column will become clear in the next section, and we discuss the numerical "electrode potentials" in the fourth column later in the chapter.

From our studies so far in this text, you already know that lithium is an extremely active metal. It reacts vigorously to lose its single $2s$ electron, which is very loosely held in the second shell. When lithium loses its electron in reactions, it is oxidized, increasing its oxidation state from 0 to +1. The substance lithium is reacting with is the oxidizing agent. In such a reaction, lithium is the reducing agent, and the other substance in the redox reaction is reduced, accepting the electron from lithium.

Each substance listed in Table 12.1 appears in two columns side-by-side. The red column shows the relative strength as an oxidizing agent; the green column shows the relative strength as a reducing agent. As you see, lithium is the strongest reducing agent. And once lithium loses its electron, it shows no tendency to want to take it back, which makes Li^+ the weakest oxidizing agent.

Another strong reducer is sodium metal. If you have never seen the reaction that occurs when a lump of sodium metal is tossed into a basin of water, you need to—it's fantastic. (Ask your chemistry teacher about demonstrating it.)[1]

The strongest oxidizing agent is fluorine gas. Fluorine oxidizes just about anything, which makes it a very dangerous substance. It is interesting to compare the relative strengths of some of the oxidizing agents you are familiar with. The strength of ozone, O_3, is just below that of fluorine. In the discussion of "Why nitrates and nitros blow up" (page 179), we saw that hydrogen peroxide is a very aggressive oxidizer that is used for rocket fuel. Here you can see that it is the third strongest oxidizer listed. And note that oxygen, the most familiar oxidizer of all, is a fairly strong oxidizer but not as aggressive as some others. This has some important practical consequences. Oxygen is readily available in the atmosphere. As a strong oxidizer, it reacts readily with fuels to power the machines in the industrial world we live in. This is obviously of immense value for contemporary society. But O_2 is not such a strong oxidizer that it reacts with everything in sight the way F_2 does. If it did, the world would be a radically different place—everything would be destroyed by being oxidized. So you might say that oxygen's strength as an oxidizer is in the "Goldilocks zone": not too hot, not too cold, but just right. The downside of oxygen's strength as an oxidizer is that it oxidizes many metals, causing rust on iron and steel and corrosion of other unprotected metals.

---

[1] Note to teachers: Instructions on how to perform the sodium demonstration safely are in my book, *Favorite Experiments for Physics and Physical Science*. See novarescienceandmath.com.

Redox Chemistry

Oxidizing Agent	Reducing Agent	Reduction Half-Reaction	$E°$ (V)
Li^+ (weakest)	Li (strongest)	$Li^+ + e^- \rightarrow Li(s)$	–3.0401
K^+	K	$K^+ + e^- \rightarrow K(s)$	–2.931
Ca^{2+}	Ca	$Ca^{2+} + 2e^- \rightarrow Ca(s)$	–2.868
Na^+	Na	$Na^+ + e^- \rightarrow Na(s)$	–2.71
Mg^{2+}	Mg	$Mg^{2+} + 2e^- \rightarrow Mg(s)$	–2.372
Al^{3+}	Al	$Al^{3+} + 3e^- \rightarrow Al(s)$	–1.662
Zn^{2+}	Zn	$Zn^{2+} + 2e^- \rightarrow Zn(s)$	–0.7618
Fe^{2+}	Fe	$Fe^{2+} + 2e^- \rightarrow Fe(s)$	–0.447
Cd^{2+}	Cd	$Cd^{2+} + 2e^- \rightarrow Cd(s)$	–0.4030
Co^{2+}	Co	$Co^{2+} + 2e^- \rightarrow Co(s)$	–0.28
Ni^{2+}	Ni	$Ni^{2+} + 2e^- \rightarrow Ni(s)$	–0.257
Sn^{2+}	Sn	$Sn^{2+} + 2e^- \rightarrow Sn(s)$	–0.1375
Pb^{2+}	Pb	$Pb^{2+} + 2e^- \rightarrow Pb(s)$	–0.1262
H^+ (H_3O^+)	H_2	$2H^+ + 2e^- \rightarrow H_2(g)$	0
Cu^{2+}	Cu	$Cu^{2+} + 2e^- \rightarrow Cu(s)$	0.3419
I_2	I^-	$I_2(s) + 2e^- \rightarrow 2I^-$	0.5355
Fe^{3+}	Fe^{2+}	$Fe^{3+} + e^- \rightarrow Fe^{2+}$	0.771
Ag^+	Ag	$Ag^+ + e^- \rightarrow Ag(s)$	0.7996
NO_3^-	NO	$NO_3^- + 4H^+ + 3e^- \rightarrow NO(g) + 2H_2O$	0.957
Br_2	Br^-	$Br(l) + 2e^- \rightarrow 2Br^-$	1.066
O_2	H_2O	$O_2(g) + 4H^+ + 4e^- \rightarrow 2H_2O$	1.229
$Cr_2O_7^{2-}$	Cr^{3+}	$Cr_2O_7^- + 14H^+ + 6e^- \rightarrow 2Cr_3^+ + 7H_2O$	1.36
Cl_2	Cl^-	$Cl_2 + 2e^- \rightarrow 2Cl^-$	1.3583
MnO_4^-	MnO_2	$MnO_4^- + 4H^+ + 3e^- \rightarrow MnO_2(s) + 2H_2O$	1.679
H_2O_2	H_2O	$H_2O_2(aq) + 2H^+ + 2e^- \rightarrow 2H_2O$	1.776
O_3	O_2	$O_3(g) + 2H^+ + 2e^- \rightarrow O_2(g) + H_2O$	2.076
F_2 (strongest)	F^- (weakest)	$F_2(g) + 2e^- \rightarrow 2F^-$	2.866

Table 12.1. Relative strengths of oxidizing and reducing agents, with reduction half-reactions and standard reduction potentials.

Chapter 12

Which metals will be oxidized by oxygen? This table answers that question. *Any reducing agent will be oxidized by any of the oxidizing agents below it.* And the farther apart the oxidizer and reducer are in the table, the likelier it is that a redox reaction will take place between them.

You may be wondering why aluminum, which is far above oxygen, doesn't exhibit the same tendency to be corroded (oxidized) by the oxygen in the atmosphere as other common metals do. After all, we are surrounded by aluminum products nowadays—just look around. It is probable that the window frames in the building you are in (and maybe the outer case of your laptop or tablet) are made of aluminum. The reason is that aluminum *is* oxidized by oxygen, and very quickly. However, in aluminum's case the oxide formed remains as a thin surface layer, protecting the interior metal from reacting further with oxygen. This process is called *passivation*, because the rest of the metal becomes "passive" (i.e., it doesn't react) because of its protective coating. This handy property makes aluminum an attractive structural material.

Notice that chlorine gas is one of the strongest oxidizing agents. Chlorine gas was the first chemical weapon used in World War I, deployed by Germany against the French in 1915. Chlorine gas is devastating to human tissue. In large water treatment facilities, chlorine gas is dissolved in water to produce hypochlorous acid (HOCl), a relatively weak acid. Both Cl_2 and HOCl kill microorganisms and algae (by oxidizing the carbon in them, and thus destroying them), which is why chlorine is so important for water treatment. Because Cl_2 is such a strong oxidizer, it is too dangerous for consumer use. Thus, treatment for home swimming pools uses HOCl only. The HOCl forms when solid cakes of $Ca(ClO)_2$ or NaClO are added to the water.

Finally, if the list of metals in the column of reducing agents looks familiar, it should. This is where the ordering of activity series of metals comes from (Table 7.2). The explanation for the activity series lies in the strengths of the metals as reducers and oxidizers. Why are copper ions in solution replaced by aluminum ions in a single replacement reaction (Figure 12.3)? Because aluminum metal is a strong reducing agent that is above Cu^{2+} in the table. So a redox reaction occurs between these two, the Al metal being oxidized to become Al^{3+} ions, and the Cu^{2+} ions being reduced to metallic copper. Figure 7.16 shows that iron metal and copper ions do the same thing. Table 12.1 helps us to see why.

We return to Table 12.1 later in the chapter after we get into the details of redox chemistry. But as we go through the examples of redox reactions in the coming pages, you should refer back to the table to see how the strengths of the oxidizing agent and reducing agent in the reactions compare.

12.2 Redox Reaction Equations

12.2.1 Redox Half-Reactions

Redox reactions may be thought of as electron-transfer reactions. Just as acid-base reactions may be thought of as a process of proton transfer, redox reactions may be thought of as a process of electron transfer. The analytical methods used for redox reactions amount to a bookkeeping system that keeps track of the electrons. The first tool you need in order to do this is to be able to establish elements' oxidation states. The second tool is to be able to write separate equations for the oxidation and reduction processes. These separate equations are called *half-reactions*.

Consider the reaction between solid sodium metal and chlorine gas to form sodium chloride. If we write the equation with the oxidation states above the elements in the compounds, we have:

$$\overset{0}{2Na}(s) + \overset{0}{Cl_2}(g) \rightarrow \overset{+1\ -1}{2NaCl}(s) \tag{12.1}$$

In this reaction, sodium is oxidized because sodium loses one electron and its oxidation state increases from 0 to +1. Chlorine is the oxidizing agent, so it is reduced; its oxidation state goes from 0 to –1. The half-reactions for this redox reaction are two separate ionic equations showing the oxidation and reduction processes. The oxidation half-reaction is written as follows:

$$\overset{0}{2\text{Na}(s)} \rightarrow \overset{+1}{2\text{Na}^+(s)} + 2e^- \tag{12.2}$$

This equation shows the two atoms of sodium ionizing to become Na^+ ions. Since their oxidations states increase from 0 to +1, the sodium atoms are oxidized. In the half-reaction, the electrons the sodium atoms give up are shown as a reaction product.

Like all reaction equations, half-reactions like Equation (12.2) must include coefficients to indicate the conservation of mass. But since half-reactions show electron transfer, they have a second requirement: they must also indicate the conservation of charge. The two sodium ions with charges of +1 are accompanied by two electrons with charges of –1. Thus, the total charge on each side of the equation is zero.

The reduction half-reaction is as follows:

$$\overset{0}{\text{Cl}_2(g)} + 2e^- \rightarrow \overset{-1}{2\text{Cl}^-(s)}$$

Sodium enters the reaction as individual atoms of sodium leaving the crystal lattice of the metal. But chlorine enters the reaction as molecules of gas which decompose to become separate Cl^- ions. As before, the charges on both sides of the equation must balance. Since this is a reduction half-reaction, the electrons are shown on the reactant side because they are combining with the chlorine atoms. The total charge of –2 on the left (from the two electrons) balances the total charge of –2 on the right (from the two Cl^- ions).

The two half-reactions may be added together to produce the complete net equation for the reaction:

oxidation	$2\text{Na}(s) \rightarrow 2\text{Na}^+(s) + \cancel{2e^-}$
reduction	$\text{Cl}_2(g) + \cancel{2e^-} \rightarrow 2\text{Cl}^-(s)$
	$2\text{Na}(s) + \text{Cl}_2(g) \rightarrow 2\text{NaCl}(s)$

Notice that in the sum, the electrons cancel out because there are two electrons on each side.

Many redox reactions occur between ions in aqueous solution. Two important features of reactions in solution need mentioning here because of their bearing on how the half-reactions are written. First, to make it easier to keep track of the ions, hydrogen ions from acids are written simply as H^+ rather than as the hydronium ions they actually are in solution. Writing the ions as H_3O^+ is more convenient for acid-base reactions, but in redox reactions, it is more convenient to treat ionized hydrogen atoms as H^+ ions (even though, in reality, the positive ions in the water are hydronium ions, not hydrogen ions).

A second feature of redox reactions in solution is that once compounds ionize in solution, spectator ions appear that are not actually part of the redox reaction process. Since they are not part of the redox reaction, the spectator ions are not included in either half-reaction. In such cases, adding the half-reactions together shows the redox part of the reaction that occurs, but it does not show all the ions that came from the original compounds in solution unless they are added back in. We see in the next section how we handle this in the more general scheme of

Chapter 12

things. For now, just note that when we write the half-reactions for a redox reaction that occurs in solution, the sum of the two half-reactions does not include the spectator ions.

To illustrate these two points, consider the reaction shown in Figure 7.17 between magnesium and a solution of hydrochloric acid. When we are not concerned with the ions and the redox chemistry, we write this reaction this way:

$$Mg(s) + 2HCl(aq) \rightarrow MgCl_2(aq) + H_2(g) \tag{12.3}$$

Adding the oxidation states, we have the following:

$$\overset{0}{Mg}(s) + 2\overset{+1}{H}\overset{-1}{Cl}(aq) \rightarrow \overset{+2}{Mg}\overset{-1}{Cl_2}(aq) + \overset{0}{H_2}(g)$$

In this reaction, magnesium metal is oxidized to become Mg^{2+} ions, and hydrogen ions are the oxidizing agent. Hydrogen ions are reduced to become H_2 gas and magnesium is the reducing agent.

Now, from our studies in prior chapters, you know that the HCl and $MgCl_2$ shown in this equation are actually ions in solution. Let's rewrite the equation to show the ions explicitly. As we do so, we leave the hydrogen molecule intact because it represents an evolving gas in the reaction, not an ionized substance in solution.

$$\overset{0}{Mg}(s) + 2\overset{+1}{H^+}(aq) + 2\overset{-1}{Cl^-}(aq) \rightarrow \overset{+2}{Mg^{2+}}(aq) + 2\overset{-1}{Cl^-}(aq) + \overset{0}{H_2}(g)$$

With the equation written as ions this way, it is easier to pick out the oxidation and reduction half-reactions. When magnesium is oxidized to become the Mg^{2+} ion, two electrons are set loose. Thus, the half reaction for the oxidation of magnesium is as follows:

$$\overset{0}{Mg}(s) \rightarrow \overset{+2}{Mg^{2+}}(aq) + 2e^-$$

The two electrons on the right are necessary to make the total charge on the right zero, as it is on the left. The reduction of hydrogen involves two H^+ ions receiving the two electrons to become hydrogen gas:

$$2\overset{+1}{H^+}(aq) + 2e^- \rightarrow \overset{0}{H_2}(g)$$

The two electrons on the left are necessary to make the total charge on the left zero, as it is on the right.

If we now add these two equations together, we have the complete redox reaction:

oxidation: $Mg(s) \rightarrow Mg^{2+}(aq) + \cancel{2e^-}$

reduction: $2H^+(aq) + \cancel{2e^-} \rightarrow H_2(g)$

$Mg(s) + 2H^+(aq) \rightarrow Mg^{2+}(aq) + H_2(g)$

As you can see, the reaction we obtain by adding the half-reactions together is the actual redox part of the original reaction equation. The chlorine ions go into solution and remain there as spectator ions; they do not participate in the redox reaction itself.

Example 12.3

The single replacement reaction shown in the first three images of Figure 12.3, and again in Figure 12.6, is as follows:

$$2\text{Al}(s) + 3\text{CuCl}_2(aq) \rightarrow 3\text{Cu}(s) + 2\text{AlCl}_3(aq)$$

For this reaction,

Figure 12.6. Redox reaction between aluminum metal and aqueous copper(II) chloride.

a. Write the oxidations states above the elements in the given equation.
b. Write the ionic equation with oxidation states shown.
c. Write the two half-reactions, showing the oxidation states.
d. Add the two half-reactions together to show the redox reaction that occurs.

From the chlorine subscripts, we infer the oxidation states on the copper and aluminum in the two salts. In an ionic compound, chlorine's oxidation state is −1. The aluminum and copper metals (s) have oxidation states of 0, as all pure elements do. This gives us the following for the answer to part (a):

$$\overset{0}{2\text{Al}(s)} + \overset{+2\ -1}{3\text{CuCl}_2(aq)} \rightarrow \overset{0}{3\text{Cu}(s)} + \overset{+3\ -1}{2\text{AlCl}_3(aq)}$$

The copper chloride and aluminum chloride are both in aqueous solution. Dividing these compounds into aqueous ions gives us the answer to part (b). Notice that there are six Cl⁻ ions on each side of the ionic equation.

$$\overset{0}{2\text{Al}(s)} + \overset{+2}{3\text{Cu}^{2+}(aq)} + \overset{-1}{6\text{Cl}^-(aq)} \rightarrow \overset{0}{3\text{Cu}(s)} + \overset{+3}{2\text{Al}^{3+}(aq)} + \overset{-1}{6\text{Cl}^-(aq)}$$

The aluminum metal is oxidized to Al^{3+} ions. Two aluminum atoms are shown in the equation. Thus we write the half reaction as

$$\overset{0}{2\text{Al}(s)} \rightarrow \overset{+3}{2\text{Al}^{3+}(aq)} + 6e^-$$

Six electrons are added to the right side so that the total charge on the right is zero, as it is on the left.

(A comment is in order here about half-reactions such as this one. It is equally legitimate to write the half-reaction without the coefficients, as follows:

$$\overset{0}{\text{Al}(s)} \rightarrow \overset{+3}{\text{Al}^{3+}(aq)} + 3e^-$$

I am writing the half-reactions with the coefficients simply because they are given in the original equation. If we were working with an unbalanced equation, we would not know what coefficients to use. In such a case, the half-reaction is written without coefficients. Don't worry about this just yet. We sort out these coefficients in the next section.)

From the ionic equation above, we see that the copper is being reduced as Cu^{2+} ions are converted to copper metal. The reduction half-reaction is written as

$$3\overset{+2}{Cu^{2+}}(aq) + 6e^- \rightarrow 3\overset{0}{Cu}(s)$$

For the answer to item (d), we add the half-reactions together, cancelling out the electrons as follows:

oxidation $\qquad 2Al(s) \rightarrow 2Al^{3+}(aq) + \cancel{6e^-}$

reduction $\qquad \underline{3Cu^{2+}(aq) + \cancel{6e^-} \rightarrow 3Cu(s)}$

$$\qquad\qquad 2Al(s) + 3Cu^{2+}(aq) \rightarrow 2Al^{3+}(aq) + 3Cu(s)$$

From here we can see the actual redox reaction that takes place: aluminum is oxidized and copper is reduced. As a final note about the coefficients, note that the coefficients in the two half-reactions are different. If we had written the half-reactions without coefficients, we would have to apply coefficients here for the sum. We would simply select the coefficients so that the electrons cancel out when the half-reactions are added together. The coefficients of 2 and 3 give us the least common multiple of 6 for the number of electrons involved.

12.2.2 Balancing Redox Equations

Balancing simpler equations may be done by inspection, as we learned in Chapter 7 and have been doing throughout this text. However, balancing the more complex equations involved in redox reactions is much trickier. For this reason, a methodical step-by-step procedure is necessary. The method we present here for balancing redox reaction equations is called the *half-reaction method*. The general idea builds on what you learned in the previous section: we write an unbalanced ionic equation for the reaction, extract the half-reactions from it, and add them together to produce the balanced equation.

To use this method, there are two important items to add to what you learned in the previous section. First, in the examples we encountered there, we didn't have to worry about balancing the masses in the half-reactions. All the elements we needed were present and the only issue we encountered was the coefficients. However, in the more general cases we consider now, the half-reactions are often missing oxygen or hydrogen atoms necessary to balance the equation. This occurs when reactions take place in acidic or basic solutions. As shown in the examples coming up, we handle this as follows:

- In acidic solutions, we add water molecules as necessary to one side to balance the oxygen, then we add H^+ ions as necessary to one side to balance the hydrogen.
- In basic solutions, we add water molecules as necessary to one side to balance the oxygen, then we add H^+ ions as necessary to one side to balance the hydrogen, then we add OH^- ions as necessary to both sides to "neutralize" the H^+ ions.

The second new item is that we must make sure that the number of electrons set loose in the oxidation half-reaction matches the number of electrons taken in by the reduction half-reaction. Doing this may require multiplying one or both of the half-reactions by coefficients.

Redox Chemistry

We now go through the method. For this example, we use the redox reaction that occurs when hydrogen sulfide gas is combined with nitric acid in solution. The products of the reaction are sulfuric acid, nitrogen dioxide, and water.

Step 1 *Write the formula equation and the ionic equation.*

formula equation:

$$H_2S(g) + HNO_3(aq) \rightarrow H_2SO_4(aq) + NO_2(g) + H_2O(l)$$

ionic equation:

$$H_2S(g) + H^+(aq) + NO_3^-(aq) \rightarrow 2H^+(aq) + SO_4^{2-}(aq) + NO_2(g) + H_2O(l)$$

Step 2 *Assign oxidation states to each element. Identify all species containing only elements that do not change oxidation state in the reaction, and delete them from the equation.*

$$\overset{+1\,-2}{H_2S}(g) + \overset{+1}{H^+}(aq) + \overset{+5\,-2}{NO_3^-}(aq) \rightarrow 2\overset{+1}{H^+}(aq) + \overset{+6\,-2}{SO_4^{2-}}(aq) + \overset{+4\,-2}{NO_2}(g) + \overset{+1\,-2}{H_2O}(l)$$

Sulfur and nitrogen both change oxidation states, so species containing S and N are retained. All others are deleted.

$$\overset{+1\,-2}{H_2S}(g) + \overset{+5\,-2}{NO_3^-}(aq) \rightarrow \overset{+6\,-2}{SO_4^{2-}}(aq) + \overset{+4\,-2}{NO_2}(g)$$

Step 3 *Write the oxidation reaction terms, then balance the atoms, then balance the charges.*

This is the reaction involving sulfur, which is oxidized in the reaction.

$$\overset{-2}{H_2S}(g) \rightarrow \overset{+6}{SO_4^{2-}}(aq)$$

Sulfur is already balanced. This is an acidic solution, so to balance the oxygen and hydrogen we use H_2O and H^+. We first add four H_2O molecules to the left side to balance the oxygen.

$$H_2S(g) + 4H_2O(l) \rightarrow SO_4^{2-}(aq)$$

Now we see that we have 10 hydrogen atoms on the left, so we must add 10 H^+ ions to the right to balance the hydrogen.

$$H_2S(g) + 4H_2O(l) \rightarrow SO_4^{2-}(aq) + 10H^+(aq)$$

Now we have to add electrons to make the charges balance. On the right side, the sulfate ion has a charge of −2. This is added to 10 positive charges on the hydrogen ions, giving a total charge on the right side of −2 + 10(+1) = +8. The total charge on the left side is zero. To make the charges on the two sides match, we must add eight electrons to the right side to cancel out the +8 charge on that side.

Chapter 12

$$H_2S(g) + 4H_2O(l) \rightarrow SO_4^{2-}(aq) + 10H^+(aq) + 8e^-$$

This is the complete oxidation half-reaction.

Step 4 *Write the reduction reaction terms, then balance the atoms, then balance the charges.*

Nitrogen is reduced. These terms are:

$$\overset{+5\ -2}{NO_3^-}(aq) \rightarrow \overset{+4\ -2}{NO_2}(g)$$

Nitrogen is already balanced. Again, this is an acidic solution, so we balance oxygen and hydrogen with H_2O and H^+. We first add one water molecule to the right side to balance the oxygen.

$$NO_3^-(aq) \rightarrow NO_2(g) + H_2O(l)$$

Now we must add two H^+ ions to the left side to balance the hydrogen.

$$NO_3^-(aq) + 2H^+(aq) \rightarrow NO_2(g) + H_2O(l)$$

Next, we have to add electrons to make the charges balance. The charges shown on the left are –1 from the nitrate ion and two +1 charges from the H^+ ions for a total charge of $-1 + (2)(+1) = +1$. The total charge on the right side is zero, so to make the left side balance with the right side, we must add one electron to the left side to cancel out the +1 charge there.

$$NO_3^-(aq) + 2H^+(aq) + e^- \rightarrow NO_2(g) + H_2O(l)$$

This is the complete reduction half-reaction.

Step 5 *Add the half-reactions together. Before adding, multiply one or both equations by a constant as necessary so that the numbers of electrons in the two half-reactions are equal. (The appropriate number of electrons is the least common multiple of the coefficients on the electrons in the two half-reactions.)*

First, we write the two half-reactions.

$$H_2S(g) + 4H_2O(l) \rightarrow SO_4^{2-}(aq) + 10H^+(aq) + 8e^-$$
$$NO_3^-(aq) + 2H^+(aq) + e^- \rightarrow NO_2(g) + H_2O(l)$$

With eight electrons in the first reaction and one electron in the second reaction, we multiply the second reaction by eight to make the numbers of electrons in the two equations match. The goal here is that the electrons cancel out when the two half-reactions are added together.

$$H_2S(g) + 4H_2O(l) \rightarrow SO_4^{2-}(aq) + 10H^+(aq) + 8e^-$$
$$8[NO_3^-(aq) + 2H^+(aq) + e^- \rightarrow NO_2(g) + H_2O(l)]$$

Now we multiply this value into the second equation and add the two equations together. The electrons cancel out. H₂O and H⁺ appear on both sides, so some of these also cancel out. H₂O and H⁺ each appear on only one side of the sum.

oxidation $H_2S(g) + \cancel{4}H_2O(l) \rightarrow SO_4^{2-}(aq) + \cancel{10}H^+(aq) + \cancel{8e^-}$

reduction $8NO_3^-(aq) + \cancel{16}H^+(aq) + \cancel{8e^-} \rightarrow 8NO_2(g) + \cancel{8}H_2O(l)$

with 6 above 16, 4 above 8.

$$H_2S(g) + 8NO_3^-(aq) + 6H^+(aq) \rightarrow SO_4^{2-}(aq) + 8NO_2(g) + 4H_2O(l)$$

Some texts regard the problem as complete at this point. The equation as it stands shows the ions and molecules involved in the redox chemistry in their proper proportions. However, I like to add the next step because it allows us to see again the compounds the ions all came from.

Step 6 *Combine the ions to form the compounds in the original formula equation. If you need extra ions of any sort, add the same number to both sides of the equation.*

Nitric acid, HNO₃, is in the formula equation on the left side, so we need to combine the H⁺ ions with the NO₃⁻ ions to form HNO₃. However, we only have six H⁺ ions. Thus, we add two more H⁺ ions to each side of the equation. The two added to the right side combine with the SO_4^{2-} ion to make the H₂SO₄ that appears on the right side of the formula equation. First, we'll add the two H⁺ ions:

$$H_2S(g) + 8NO_3^-(aq) + 8H^+(aq) \rightarrow SO_4^{2-}(aq) + 8NO_2(g) + 4H_2O(l) + 2H^+(aq)$$

Now we'll combine the ions:

$$H_2S(g) + 8HNO_3(aq) \rightarrow H_2SO_4(aq) + 8NO_2(g) + 4H_2O(l)$$

This is the complete, balanced equation showing the original compounds that provided the species in the redox reaction.

Step 7 *Check the final equation to make sure that all elements are correctly balanced and that there is no common multiple that can be divided out of all the coefficients.*

A final check of our equation indicates all is correct.

▼ Example 12.4

Potassium permanganate, KMnO₄, reacts with hydrocyanic acid, HCN, in aqueous solution to form fulminic acid, HCNO, manganese oxide, MnO₂, and potassium hydroxide, KOH. Note that manganese oxide is insoluble and precipitates out of solution. Because of the production of excess hydroxide ions, this reaction occurs in basic solution.

Figure 12.7. Fulminic acid, HCNO. (H and O are white and red, as usual. C is black and N is blue.)

Determine the complete, balanced equation for this redox reaction, showing the original compounds involved.

As a side note, a space-filling model of the fulminic acid molecule is shown in Figure 12.7.

Chapter 12

Step 1 From the description above, the formula equation is

$$KMnO_4(aq) + HCN(aq) \rightarrow HCNO(aq) + MnO_2(s) + KOH(aq)$$

Each of the compounds is soluble and ionizes in solution, except MnO_2. Even without the tip-off in the problem statement, we know that manganese oxide is insoluble from the solubility rules of Table 10.2. So leaving MnO_2 as a non-ionized solid, the ionic equation is

$$K^+(aq) + MnO_4^-(aq) + H^+(aq) + CN^-(aq) \rightarrow$$
$$H^+(aq) + CNO^-(aq) + MnO_2(s) + K^+(aq) + OH^-(aq)$$

Step 2 Our next task is to write in the oxidation states.

$$\overset{+1}{K^+}(aq) + \overset{+7\ -2}{MnO_4^-}(aq) + \overset{+1}{H^+}(aq) + \overset{+2\ -3}{CN^-}(aq) \rightarrow$$
$$\overset{+1}{H^+}(aq) + \overset{+4\ -2}{\underset{-3}{CNO^-}}(aq) + \overset{+4\ -2}{MnO_2}(s) + \overset{+1}{K^+}(aq) + \overset{-2\ +1}{OH^-}(aq)$$

Carbon is oxidized and manganese is reduced. The manganese and carbon compounds contain these elements. Eliminating everything else, we have

$$\overset{+7\ -2}{MnO_4^-}(aq) + \overset{+2\ -3}{CN^-}(aq) \rightarrow \overset{+4\ -2}{\underset{-3}{CNO^-}}(aq) + \overset{+4\ -2}{MnO_2}(s)$$

Step 3 Next we write the terms involved in the oxidation, then balance the atoms, and then balance the charges.

The carbon is oxidized, so writing the carbon terms gives

$$CN^-(aq) \rightarrow CNO^-(aq)$$

Carbon and nitrogen are already balanced. We add one H_2O to the left to balance the oxygen:

$$CN^-(aq) + H_2O(l) \rightarrow CNO^-(aq)$$

Now we add two H^+ ions to the right to balance the hydrogen:

$$CN^-(aq) + H_2O(l) \rightarrow CNO^-(aq) + 2H^+(aq)$$

We have one extra step to perform as part of the mass balancing. Recall that for redox reactions in basic solution, we must "neutralize" the H^+ ions by adding an equal number of OH^- ions to each side of the equation. Adding two OH^- ions to each side gives us

$$CN^-(aq) + H_2O(l) + 2OH^-(aq) \rightarrow CNO^-(aq) + 2H_2O(l)$$

Before we continue, notice that we have water molecules on both sides. We can simplify the equation here a bit by subtracting one H_2O molecule from each side. This is not strictly necessary; the cancellations occur when we add the half-reactions. So if you forget this step, it's no problem. But I like to keep things simplified, so now we have:

$$CN^-(aq) + 2OH^-(aq) \rightarrow CNO^-(aq) + H_2O(l)$$

Next we balance the charges. The total charge on the left side is –3; on the right side it is –1. To make these balance, we add two electrons to the right side:

$$CN^-(aq) + 2OH^-(aq) \rightarrow CNO^-(aq) + H_2O(l) + 2e^-$$

This is the completed oxidation half-reaction.

Step 4 Now we write the terms involved in the reduction, then balance the atoms, and then balance the charges.

The manganese is reduced. Writing the manganese terms gives

$$MnO_4^-(aq) \rightarrow MnO_2(s)$$

Manganese is already balanced. We add two H_2O molecules to the right side to balance the oxygen:

$$MnO_4^-(aq) \rightarrow MnO_2(s) + 2H_2O(l)$$

Now we add four H^+ ions to the left to balance the hydrogen:

$$MnO_4^-(aq) + 4H^+(aq) \rightarrow MnO_2(s) + 2H_2O(l)$$

Again, this reaction occurs in basic solution. So we have the extra step here of adding enough OH^- ions to both sides to neutralize those H^+ ions. Adding four OH^- ions to each side we have

$$MnO_4^-(aq) + 4H_2O(l) \rightarrow MnO_2(s) + 2H_2O(l) + 4OH^-(aq)$$

As before, there are now water molecules on both sides. We can subtract two from each side to simplify a bit:

$$MnO_4^-(aq) + 2H_2O(l) \rightarrow MnO_2(s) + 4OH^-(aq)$$

The last piece of this step is to balance the charges. The total charge on the left is –1; on the right it is –4. We add three electrons to the left side to make these match.

$$MnO_4^-(aq) + 2H_2O(l) + 3e^- \rightarrow MnO_2(s) + 4OH^-(aq)$$

This is the completed reduction half-reaction.

Step 5 We can now write the two half-reactions in preparation for adding them. But before we add them together, we need to make sure the electrons will cancel.

oxidation $\quad CN^-(aq) + 2OH^-(aq) \rightarrow CNO^-(aq) + H_2O(l) + 2e^-$

reduction $\quad MnO_4^-(aq) + 2H_2O(l) + 3e^- \rightarrow MnO_2(s) + 4OH^-(aq)$

There are two electrons in the oxidation half-reaction and three in the reduction half-reaction. To make the electrons cancel, we need six electrons in each half-reaction (six being the least common multiple of two and three). We need to multiply the oxidation half-reaction by three and the reduction half-reaction by two. Multiplying in these constants gives us the equations below. I have also marked out the items that cancel and added the equations together.

oxidation $\quad 3CN^-(aq) + \cancel{6OH^-}(aq) \rightarrow 3CNO^-(aq) + \cancel{3H_2O(l)}^{1} + \cancel{6e^-}$

reduction $\quad 2MnO_4^-(aq) + \cancel{4H_2O(l)}^{2} + \cancel{6e^-} \rightarrow 2MnO_2(s) + \cancel{8OH^-}(aq)$

$\quad\quad\quad 3CN^-(aq) + 2MnO_4^-(aq) + H_2O(l) \rightarrow 3CNO^-(aq) + 2MnO_2(s) + 2OH^-(aq)$

This is the completed redox equation. However, to show the original compounds involved as the problem requires, we need to add in some ions in the next step.

Step 6 We need three H^+ ions on each side to join with the carbon-containing ions. We also need to add two K^+ ions to make potassium permanganate and potassium hydroxide. Adding these ions, we have:

$3H^+(aq) + 3CN^-(aq) + 2K^+(aq) + 2MnO_4^-(aq) + H_2O(l) \rightarrow$
$\quad\quad\quad 3H^+(aq) + 3CNO^-(aq) + 2MnO_2(s) + 2K^+(aq) + 2OH^-(aq)$

And combining the ions gives us finally

$3HCN(aq) + 2KMnO_4(aq) + H_2O(l) \rightarrow 3HCNO(aq) + 2MnO_2(s) + 2KOH(aq)$

Step 7 Our final step is to double check the mass balance by counting elements on both sides. This checks.

12.3 Electrochemistry

12.3.1 Copper and Zinc Redox

We have seen several single replacement reactions with metals in copper sulfate. You now know that these are all redox reactions. Here is another one, this time involving zinc. The plating known as *galvanizing* that is often used to protect steel from rusting is made of zinc. Figure 12.8 shows a beaker of copper sulfate solution and a galvanized steel plate purchased from a hardware store. (I used a galvanized plate for this demo because I didn't have any solid zinc metal handy. The galvanized part from the hardware store is completely covered in zinc, so it works just fine.) The zinc-coated plate is placed into the copper sulfate solution, and five minutes later the zinc plate has a nice copper coating.

Redox Chemistry

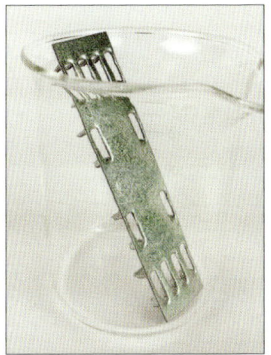

Figure 12.8. The spontaneous redox reaction between zinc and copper. Copper ions in the blue $CuSO_4$ solution are reduced by the zinc, so the copper ions come out of solution. The copper metal adheres to the zinc plate. (The plate was removed from the solution before the reaction was complete, so the solution was still blue.)

This redox reaction occurs spontaneously. In the reaction, the electron transfer from the atoms of zinc metal (which are being oxidized) to the copper ions (which are being reduced) occurs directly between the zinc atoms at the surface of the metal and the copper ions in solution right next to the zinc metal.

The introduction to electrochemistry in this section involves an apparatus using the same redox reaction, but we are going to arrange things so the copper and zinc are not in direct contact. For the redox reaction to occur, the electrons take a different route to get from the zinc to the copper. We provide this electron pathway in the form of an electrical conductor (a copper wire) connected from the zinc to the copper. So before we go any further with the redox reaction, let's discuss the benefits of doing this.

12.3.2 Electricity Instead of Heat

Every chemical reaction involves a transfer of energy. As you now know, heat flows into a chemical system during an endothermic reaction and out of a chemical system in an exothermic reaction. But since redox reactions involve electron transfer, there is another way energy can flow into or out of a chemical system. Electric current is the flow of electrons. Since we have electrons in motion anyway in redox reactions, we use the electric current of the electrons being transferred as a means for getting useful electrical energy out of a reaction, or as a means of forcing reactions to occur by putting electrical energy in. This is what electrochemistry is all about.

Instead of being released as heat, the energy released during an exothermic redox reaction is released in the form of an electric current. All we have to do is rig up a system so that the electrons being transferred in the redox reaction flow through an external electric circuit. When we do this, we have made what everyone calls a *battery*.

The importance of redox chemistry should now be obvious to you. We are surrounded now by battery-operated devices. Not only do we have millions of different products that run on standard batteries such as the common AAA, AA, C, and D cells (and others), but we now also have rechargeable batteries everywhere—every mobile phone, mobile device, and laptop computer uses a rechargeable battery.

A battery—actually, the proper term is *cell*; a *battery* is a set of two or more cells connected together—is an electrochemical system that uses a redox reaction to create an electric current composed of the electrons being transferred in the reaction. When the cell is connected to an external circuit, the electrons flow and the reaction occurs. When the external circuit is interrupted, by opening a switch in the external circuit or by the cell being disconnected from the device it is connected to, the electron flow ceases and the redox reaction ceases as well. The energy produced by the redox reaction flows as electrical energy to the device being powered by the cell.

Electrochemical cells also operate in reverse—instead of energy flowing out of the cell, energy can flow into the cell from another source of power. The obvious application of this we are all familiar with is when rechargeable cells are being recharged. Another very important application is the use of cells for industrial processes such as smelting and electroplating. We will look at these a bit more after we dive into the details of how electrochemical cells work.

12.3.3 Electrochemical Cells

The diagram in Figure 12.9 shows the basic components of an *electrochemical cell*, a container in which the two half-reactions of an electrochemical redox reaction take place. The cell shown makes use of zinc and copper metal as the two *electrodes*, the sites of the oxidation and reduction half-reactions.[2] The zinc electrode is immersed in a zinc sulfate solution. This solution conducts electricity, and is thus an electrolyte.[3] The copper electrode is immersed in a solution of copper sulfate, also an electrolyte.

The zinc and copper solutions are divided into separate *half-cells* where the redox half-reactions take place. The zinc and copper solutions are separated so that the half-reactions occur in separate compartments, but there must be a means for ions to migrate from one cell to another. (We are setting up an electric circuit in which charges—electrons or ions—flow in a closed loop. If charges can't get from one half-cell to the other, the cell won't work.) In the simplified cell pictured, ion migration occurs through a *porous barrier* separating the half cells. This barrier keeps the solutions from randomly mixing, but it is made of a material with microscopic holes large enough to allow ions to migrate through from one side of the cell to the other. A better cell design allows for ion transfer while keeping the zinc and copper ions completely separated from each other. Such a cell allows ion transfer to occur through a *salt bridge*. We look more closely at these details a bit later.

The final component of the electrochemical cell is a conductor connected between the two electrodes. With this conductor between the electrodes, we now have a complete closed electrical circuit. Electrons flow up one electrode,

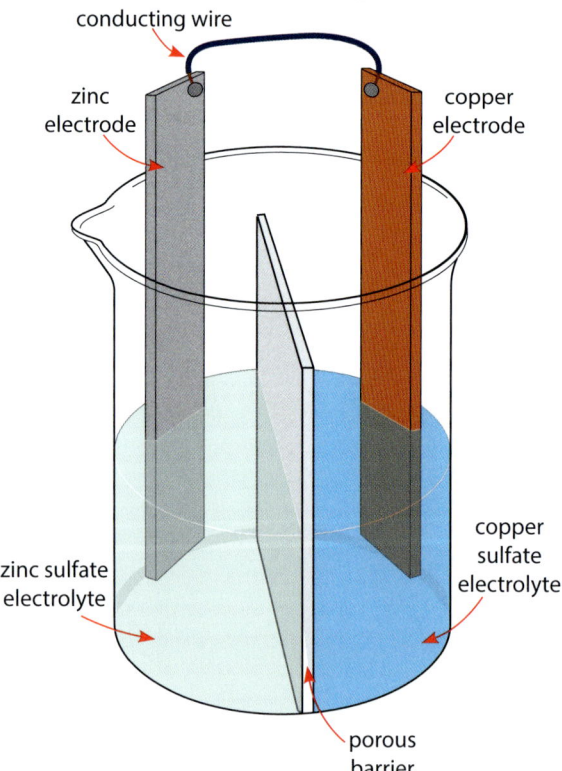

Figure 12.9. The essential parts of an electrolytic cell. The cell shown uses zinc and copper for the electrodes.

---

2 Many other metals may be used in electrochemical cells. The copper/zinc cell is the usual one used for demonstration and discussion of the basic principles of the cell's operation.

3 In our studies in previous chapters, we use the term *electrolyte* to refer to a compound that ionizes in water to produce an electrically conductive solution. In the context of electrochemistry, it is customary to refer to the solution itself as the electrolyte.

Redox Chemistry

through the wire, and down the other electrode. To complete the circuit, ions flow through the porous barrier. The movement of charge represented by the electrons in the wire is balanced by ion transfer through the barrier. The direction of the electron flow depends on whether the cell is producing electrical energy or consuming it, as I explain below. Note that a simple wire connecting the electrodes wouldn't actually allow the cell to be used for anything. Instead, that connecting wire needs to connect to an outside device if the cell is producing electricity, or have a power source inserted in it if the cell is consuming electricity. Details on these connections are also part of the discussion below.

The cell shown in Figure 12.9 contains liquid, and is accordingly called a *wet cell*. The cells we use to power portable electronics contain moist solids (pastes or gels), but no liquids, and are called *dry cells*. Cells powering mobile phones, laptops, and flashlights are dry cells; conventional car batteries are wet cells.

There are two basic configurations of the electrochemical cell, based on whether the cell is producing electricity or consuming it, and equally based on the direction of electron flow in the redox process. The *voltaic cell* or *galvanic cell* (named after Alessandro Volta or Luigi Galvani, depending on what you call it) is used to provide electrical power to an external device. Cells in mobile devices, power tools, and flashlights are voltaic cells. The other configuration is the *electrolytic cell*. Electrolytic cells are those that consume electricity instead of producing it. We will now examine the details of what is going on in each type of cell.

Voltaic Cells The basic arrangement of the voltaic cell is shown on the left side of Figure 12.10. If you have not yet studied DC circuits in one of your science classes, just focus on the diagram on the left. But for those students who are familiar with DC circuits, the electrochemical cell is represented by the "battery" (cell) in the electrical schematic diagram shown on the right side of the figure. If you are familiar with DC circuits, the schematic will help you understand how the cell functions electrically. Recall from your study of DC circuits that the way the flow of positive current (I) is defined, the current, I, and the electrons flow in opposite directions in the circuit.

The figure shows a generic electrical device (the light bulb) being powered by the voltaic cell. The connections between the voltaic cell and the electrical device are shown so you can get an idea of how a voltaic cell is put to use. Naturally, we would not actually use a wet cell made of two beakers to power anything; we would use the dry cells and batteries we are all familiar with. But the voltaic cell shown might be used in a laboratory for experimental purposes and is convenient for demonstrating how electrochemical cells work.

We are going to examine this diagram in detail, but first, here's the big picture in brief: The voltaic cell is divided into two half-cells containing the two electrodes, each immersed in a salt solution of the same metal. For our demonstration example, the two electrolytes are zinc sulfate and copper sulfate. The cell and the device it is connected to form a complete electrical circuit—a closed path in which electrons or other forms of charge flow in a closed loop. Beginning at the bottom of the zinc electrode, the electrons flow up the zinc electrode, out of the cell in the blue wire, through the electrical device (the light bulb or whatever it is), through another blue wire, into the copper electrode, and down the copper electrode to where it is submerged in the electrolyte. At this point the movement of charge around the circuit is taken over by ions in the two half-cells and in the salt bridge connecting them together.

We begin our tour around the loop in the zinc half-cell. At the lower end of the zinc electrode, down in the electrolyte, oxidation of the zinc metal is occurring: zinc atoms are ionizing and the Zn^{2+} ions are separating from the zinc metal and entering the zinc sulfate electrolyte. (Thus, the zinc metal is gradually dissolving.) Note this: *In any electrochemical cell, the electrode*

Chapter 12

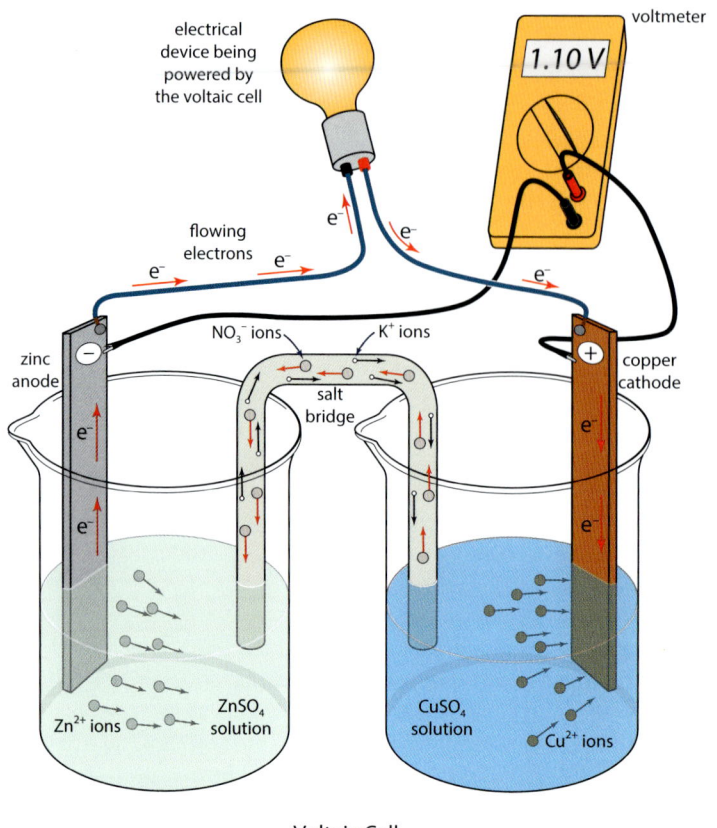
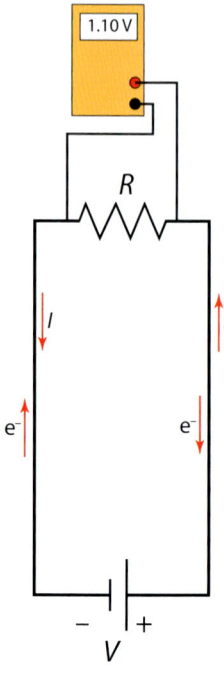

Voltaic Cell

The cell is powering an electrical device, while a voltmeter measures the voltage the cell is producing.

Schematic

An electrical representation of the voltaic cell. The cell is driving current into an external device, represented in the schematic by a resistor.

Figure 12.10. Diagram of the voltaic cell showing flow of charge (left). The cell is providing electrical energy to a generic electrical device shown. A voltmeter is connected to the cell to show the voltage the cell is producing. On the right is the equivalent electrical schematic representation of the system. (Note that the cell can only be used this way if the 1.10 V it produces is a high enough voltage to run the device.)

where the oxidation is taking place is called the anode. Thus, in our demonstration voltaic cell, the zinc electrode is the anode. The half-reaction for the oxidation of the zinc is

$$Zn(s) \rightarrow Zn^{2+}(aq) + 2e^-$$

The electrons turned loose by all the oxidizing zinc atoms (two per atom, according to the half-reaction) flow up the anode and out of the cell in the blue wire, as indicated by the red arrows. This flow of electrons goes through an electrical device, providing energy to operate the device. The redox reaction is releasing energy, and this is where the energy is going. The electrons returning to the cell from the electrical device enter the copper electrode. Down under the surface of the $CuSO_4$ electrolyte, Cu^{2+} ions in the electrolyte are being reduced and are attaching to the crystal lattice of the copper electrode. Thus the copper electrode is where the reduction is taking place. *In any electrochemical cell, the electrode where the reduction is taking place is called the cathode*. Thus, the copper electrode in this voltaic cell is the cathode. The electrons flowing into the upper end of the cathode travel down the copper metal and reduce the copper ions as they attach to the cathode and become copper metal. (As a result, the copper ions are gradually being depleted from the electrolyte, and the cathode is getting larger and heavier as additional copper atoms continue to attach to it.) The reduction half-reaction is

$$Cu^{2+}(aq) + 2e^- \rightarrow Cu(s)$$

(To aid you in remembering which electrode goes with which half-reaction, note that *anode* and *oxidation* both begin with vowels; *cathode* and *reduction* both begin with consonants.)

We return to Figure 12.10 to discuss the salt bridge in a moment. But first, Figure 12.11 shows hypothetical extreme close-up depictions of the two electrodes. (Remember that atoms and ions are much smaller than the wavelength of light, and no amount of magnification would allow us actually to see these things. These images are visual models of entities that cannot be perceived with our eyes.) At the anode, oxidation of zinc atoms sends them into the $ZnSO_4$ electrolyte as Zn^{2+} ions, while the electrons from the oxidized atoms flow up the anode and out of the cell. The electrons return to the cell at the cathode, where they travel down to where the reduction of Cu^{2+} ions is occurring, which adds copper metal to the cathode.

Now we return to our discussion of Figure 12.10. For an electric circuit to work, charge must be able to flow in a path that forms a complete, closed loop. Without a way for charge to flow between the two half-cells, nothing happens and no current flows. In a demonstration cell like the one we are considering here, the connection allowing charge to flow between the cells is

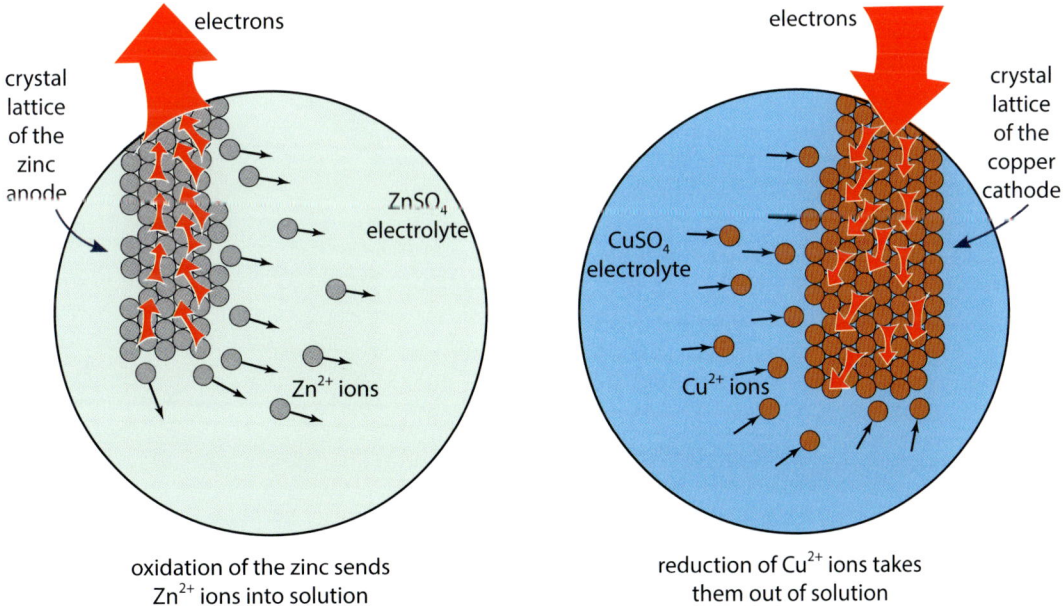

Figure 12.11. A close-up view of the oxidation occurring at the anode (left) and the reduction occurring at the cathode (right) in the voltaic cell. The red arrows represent the electrons streaming away from the oxidized zinc ions on the left, and flowing down to reduce the copper ions on the right. (Note that the ions are shown on a colored background for clarity. If we could really "see" ions, we would not see the rest of the solution as a colored background. We would instead see the particles of which the solution is composed—a dense mixture of water molecules, metal ions, sulfate ions, and of course, a few hydronium and hydroxide ions as well.)

often provided by a pathway called a *salt bridge*. The salt bridge is essentially a tube filled with concentrated salt solution. The salt is chosen so that the particular cations and anions, which are dissolved in solution, do not react with the substances of which the electrodes and electrolytes are composed. Potassium nitrate is a salt commonly used for this purpose. The anions in the salt solution, NO_3^- in our example, are negatively charged just like the electrons are. Thus, they move around the electrical circuit in the same direction the electrons are moving. This means that the salt anions migrate through the salt bridge toward the half-cell containing the anode. The cations in the salt solution (K^+) migrate in the opposite direction. To summarize, in the electrodes and in the external electric circuit, the flowing charge is electrons. In the salt bridge connecting

Hmm... Interesting.   How are salt bridges made?

If you are the curious type, you want to know what a salt bridge *is*—that is, what they are *made of*. We know the function of the salt bridge in the electrochemical cell—to allow charge migration between the half-cells. This can be done with a simple salt solution. But the trick is, we need some way to keep the salt solution inside the glass tube of the salt bridge. This is done in a clever way that makes use of the porous properties of gels. In Chapter 2, we looked at some types of heterogeneous mixtures, one of which was gels. I mention there that in a gel, the solid particles link together in long molecules known as polymers, and these linked solids form a network or matrix throughout the substance. What this means is that a gel is like a sponge, but with pores too small to see. The solid material is the network of polymer strands and between these strands is a connected matrix of micro-sized passageways.

So here's how the salt bridge is prepared. You mix up a 1-*M* solution of KNO_3. Then you mix up a batch of a gel called *agar*. This is just like making Jell-O, which is a gel (gelatin). The gelatin is initially a solid powder, which you mix into hot water. As the gelatin mixture cools, it turns into a gel, like Jell-O. For salt bridges, and for biological work such as culturing bacteria, agar is used because it does not enable the growth of bacteria the way other gelatins do. The photo to the right shows a petri dish of agar (blue) in which is growing a species of moss.

For the salt bridge, the salt solution is combined with the hot gelatin mixture and mixed. Then the empty glass salt bridge tube is inverted and filled with the salt/gelatin mixture. When the mixture cools, it solidifies like Jell-O, with the salt solution inside it occupying the spaces between the gelatin molecules. In this way, the ions are able to migrate through the salt bridge, and the agar gel holds the salt solution in the salt bridge so it doesn't run out.

The salt bridge described here is typically used for demonstration cells like the one described in this chapter. But a simple strip of paper towel soaked in salt water and draped between the beakers also works. And instead of a salt bridge, some demonstration cells have their compartments connected together at the bottom by a tube containing some kind of porous barrier. The porous material is chosen to allow the sulfate and zinc ions to migrate directly from one half-cell to another without the use of an extra salt. In such a case, the negatively charged SO_4^{2-} ions migrate toward the zinc anode and the Zn^{2+} ions migrate toward the cathode.

the two half-cells, the flowing charge is the ions from the salt solution. The interesting details of how a salt bridge is actually made are in the accompanying box.

With the salt bridge connecting the two half-cells of the cell, charge can flow all the way around the circuit. As the cell operates, nitrate ions from the salt bridge become concentrated in the anode half-cell with the zinc sulfate solution. This solution also becomes increasingly concentrated with Zn^{2+} ions from the decomposing anode. The potassium cations in the salt bridge migrate into the cathode half-cell and become concentrated there in the copper sulfate solution. The concentration of copper ions in this solution is decreasing, as copper ions come out of solution and join the other copper metal atoms in the cathode. For each electron that leaves the anode, another electron from the external circuit enters the cathode. Together, these charge movements make the anode one unit more positive and the cathode one unit more negative. For each of these electron exchanges, one nitrate ion enters the anode compartment from the salt bridge, replacing the negative charge lost as the electron leaves. Also, one potassium ion enters the cathode compartment from the salt bridge, its positive charge cancelling out the negative charge of the electron that enters. The result is that charge remains balanced between the two half-cells as the two half-reactions proceed.

Finally, note that the reaction in a voltaic cell occurs spontaneously and delivers energy to the outside world. You may recall from our discussion in Section 10.1.3 that when the change in Gibbs free energy for a process is negative, the process occurs spontaneously. In this case, the redox reaction generating the flow of electric charge happens all by itself once the outside circuit is connected to the cell. This is because the change in the Gibbs free energy for the reaction is negative.

Electrolytic Cells

Electrolytic cells are essentially voltaic cells running in reverse, with all reactions proceeding in the opposite direction. But the change in the Gibbs free energy for a system operating this way is positive, so the process is not spontaneous. An energy source must be supplied to force the electrical current to flow in the opposite direction.

The components and arrangement in the electrolytic cell are the same as for a voltaic cell, with one difference. Figure 12.12 is a diagram showing the electrode arrangement for the electrolytic cell. An electrical power source is now connected into the external circuit and electron flow is in the opposite direction. Instead of the cell producing power to operate an external device, an external source of power drives the redox reaction in the cell. This is essentially what is going on when "batteries" are recharged. The forward redox reaction concentrates ions in one region until they are so thick the cell can no longer operate. The recharging puts all these ions back in the other direction so the forward reaction can begin afresh.

It is important for you to note in Figure 12.12 that *the zinc electrode is now the cathode*. If you refer again to Figure 12.10, you see that in the voltaic cell the cathode is the copper electrode. The electrons are flowing *into* this electrode because this electrode is the site of the re-

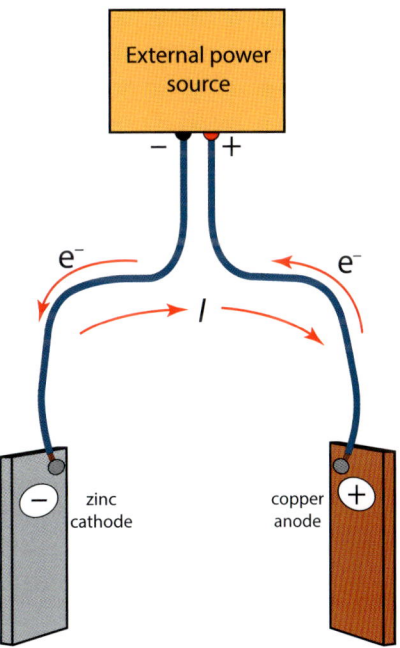

Figure 12.12. Connection of a power source to the electrodes of an electrolytic cell. The power supply drives the redox reaction occurring in the cell. (The rest of the cell is not shown.)

Chapter 12

duction half-reaction. For reduction to occur, a continuous electron supply is needed. When reduction occurs, ions are combining with electrons, reducing their oxidation states, coming out of solution, and becoming metal atoms in solid form. The incoming electrons are joining with the ions, enabling the reduction to happen.

Compare that situation to the sketch here in Figure 12.12. With the power source connected between the electrodes, the electrons are being forced by the power source to flow in the opposite direction. The electrode where the electrons are entering is the one where the reduction is taking place, and thus it is the cathode.

To identify the anode and cathode in any electrochemical cell, simply remember this: when reduction occurs, electrons are being supplied to metal ions to convert them from ions to atoms of pure metal. So to identify the cathode, you can simply look at the direction of the electron flow. The electrode where the electrons are entering the cell is the electrode where the reduction is taking place, and is thus the cathode. And again, as a final reminder to those who have studied DC circuits, in any electrical circuit, the direction of the flow of positive current, I, is the opposite of the direction of electron flow. The same applies to the electrolytic cell.

Figure 12.13 summarizes the nomenclature and arrangement of the two types of electrochemical cells. In this sketch, the two types of cells are shown side-by-side to aid you in studying and remembering the details. The voltaic cell has a resistor in the external circuit. This resistor represents whatever the cell is connected to (even if it is only connected to a voltmeter). The electrolytic cell has an external source of DC current in the wire between the electrodes.

12.3.4 Electrode Potentials

Look once again at Figure 12.10. In the diagram, the voltmeter connected to the electrodes reads 1.10 volts (V). The voltage produced by a voltaic cell depends on the materials the two electrodes are made of. (It also depends on the concentration of the electrolytes, a detail we will not concern ourselves with here). Each different electrode material exhibits a certain tendency to engage in a reduction half-reaction. For a given reducing agent in a particular reduction half-reaction, we quantify this tendency with a constant called the *standard reduction potential*, symbolized as $E°$ and measured in volts. The standard reduction potential for a material used as an electrode is usually just called the *electrode potential*.

The overall potential for a complete cell made of two electrodes is referred to as the *cell potential*. The cell potential is calculated as

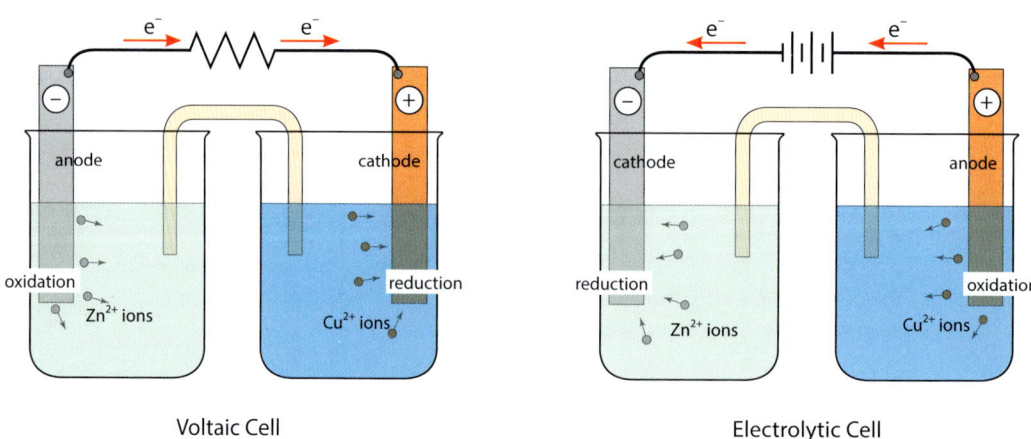

Figure 12.13. Summary of configurations and ion flow for voltaic and electrolytic cells. The key to remembering everything is that the electrode where the electrons are entering is where the reduction is taking place, and is thus the cathode.

$$E^\circ_{cell} = E^\circ_{cathode} - E^\circ_{anode} \qquad (12.4)$$

For the redox reaction in any cell to occur spontaneously, the cell potential must be greater than zero. Ideally, the voltage produced by a voltaic cell is equal to this cell potential. However, nothing is ever ideal, so in practice the actual voltage the cell produces while electrical current is flowing between the electrodes is a bit less and drops slowly over time as the concentrations of the electrolytes change.

It is now time to refer back to Table 12.1, which lists the standard reduction potentials for a number of different reduction half-reactions.[4] The standard reduction potential is a measure of the readiness of the reducing agent to engage in a particular reduction half-reaction. The more negative the standard electrode potential for the half-reaction is, the stronger the reducing agent is. The table begins with the most negative half-reaction potential, which is associated with the strongest reducing agent, and ends with the most positive potential, associated with the strongest oxidizing agent. For any pair of substances selected to be the electrodes in a voltaic cell, *the substance with the most positive electrode potential is the cathode.*

With the data from Table 12.1 and Equation (12.4), we can calculate the cell potential for the copper-zinc cell (denoted as Cu | Zn) we have been considering. The electrode potential for copper is listed as 0.3419 V and the electrode potential for zinc is listed as –0.7618 V. Inserting these values into Equation (12.4), the cell potential for our Cu | Zn cell is

$$E^\circ_{cell} = E^\circ_{cathode} - E^\circ_{anode} = 0.3419 \text{ V} - (-0.7618 \text{ V}) = 1.1037 \text{ V}$$

This is why the voltmeter in Figure 12.10 reads 1.10 V.

If you look once again at Figure 12.10, you see that the copper cathode is labeled positive (⊕) and the zinc anode is labeled negative (⊖). In a voltaic cell, the electrode with the positive (or more positive) electrode potential, which is the cathode, is labeled positive. Now you know what the + and – symbols on a dry cell or battery actually mean! In an electrolytic cell, represented by Figure 12.12, the reaction processes operate in reverse, and the electrode potentials are reversed as well. This makes the anode the more positive electrode in an electrolytic cell.

▼ Example 12.5

A voltaic cell is to be constructed using cadmium and tin electrodes. The standard reduction half-reactions for these metals are as follows:

$Cd^{2+} + 2e^- \rightarrow Cd(s)$

$Sn^{2+} + 2e^- \rightarrow Sn(s)$

Identify the anode and cathode, write the two redox half-reactions for this cell, and determine the cell potential for the resulting voltaic cell.

The standard reduction potentials listed in Table 12.1 for Cd^{2+}/Cd and Sn^{2+}/Sn are as follows:

---

[4] Any oxidation potential for an oxidation half-reaction (the reverse of the reduction half-reaction) is the negative of the reduction potential. Thus, there is no need to list both in tables such as Table 12.1. Chemists deal in standard reduction potentials by common consent. There is no reason to prefer reduction potentials over oxidation potentials, except for the convenience of having a standard way to refer to data.

Cd²⁺/Cd: −0.4030 V
Sn²⁺/Sn: −0.1375 V

Tin is the material with the most positive electrode potential, and thus tin is the cathode in this cell. Since reduction occurs at the cathode, the standard reduction half-reaction given above for tin applies. The cadmium electrode is the anode, where the oxidation occurs. Accordingly, the reduction half-reaction for cadmium must be reversed to convert it to an oxidation half-reaction. This gives us the following two half-reactions for the cell:

oxidation $Cd(s) \rightarrow Cd^{2+} + 2e^-$

reduction $Sn^{2+} + 2e^- \rightarrow Sn(s)$

The cell potential is obtained by inserting the electrode potentials into Equation (12.4):

$$E^\circ_{cell} = E^\circ_{cathode} - E^\circ_{anode} = -0.1375 \text{ V} - (-0.4030 \text{ V}) = 0.2655 \text{ V}$$

The cell potential is positive, so when this cell is assembled and connected to an outside circuit, the redox reaction commences and current flows in the external circuit. The voltage of the cell is ideally 0.2655 V, but in an actual cell is slightly less.

▲

At this point, the meaning of the standard reduction potentials listed in Table 12.1 may still be a bit of a mystery to you. To understand what the standard reduction potential means, it helps to know how it is measured. Standard reduction potentials are relative voltage values, as are all electrical voltages. This is analogous to the fact that the height of any object is relative to our reference for measuring height, which could be the ground, or sea level, or the floor in a room. The reference standard for measuring reduction potentials is the reduction of hydrogen ions to form hydrogen gas. In this half-reaction, hydrogen ions are the oxidizing agent:

$$2H^+ + 2e^- \rightarrow H_2(g) \quad (12.6)$$

A standard reduction potential is the voltage a voltaic cell reads when a given reducing agent is being oxidized in the anode half-cell while a special reference electrode reduces H^+ to H_2 in the cathode half-cell.

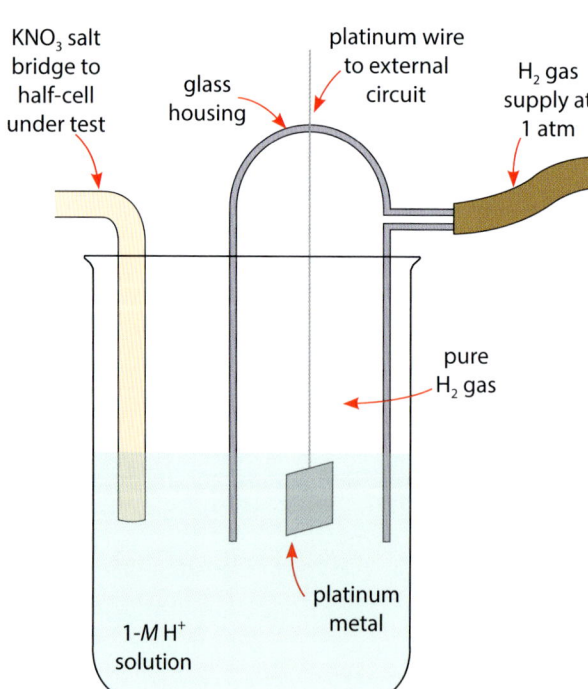

Figure 12.14. A half-cell containing a standard hydrogen electrode for testing standard reduction potentials of other electrode materials.

Redox Chemistry

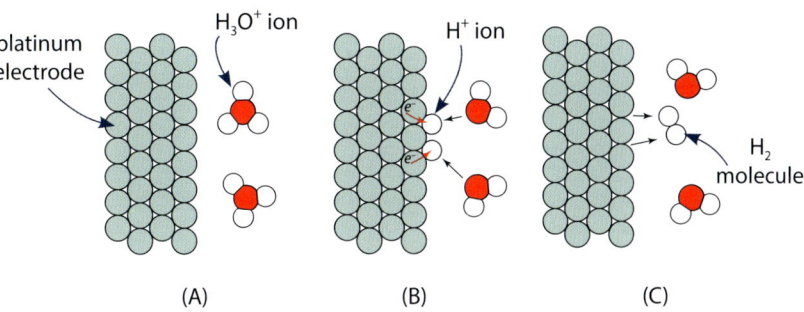

Figure 12.15. The sequence A–B–C shows how the platinum electrode acts as a catalyst for the reduction of H^+ ions to H_2. H^+ ions separate from H_3O^+ ions, adsorb onto the platinum where they are reduced, and then combine to form H_2.

The reference electrode is called a *standard hydrogen electrode* (SHE), or *normal hydrogen electrode* (NHE). The standard reduction potential for the half-reaction shown in Equation (12.6) is defined as 0 V and has been established as the reference standard for measuring all other standard reduction potentials. All other standard reduction potentials are measured relative to an SHE.

Figure 12.14 shows generally how an SHE is constructed. A platinum electrode is lowered into a pure solution with an H^+ concentration of $[H^+] = 1\ M$ (pH = 0). Platinum is used because it is a fairly unreactive metal. The platinum electrode is suspended by a platinum wire that serves as the electrical connection to the external circuit. The entire platinum assembly is contained inside a glass housing filled with H_2 gas at a pressure of 1 atm. A conventional KNO_3 salt bridge connects the hydrogen half-cell to the half-cell containing the electrode under test.

The mechanics of the reduction half-reaction that forms the H_2 gas in the SHE are illustrated in Figure 12.15. As you know, the H^+ ions come from H_3O^+ ions in the water. The platinum electrode acts as a *catalyst* for the reduction half-reaction. A *catalyst* is a substance that expedites or facilitates a chemical reaction without itself being consumed by the reaction. Metals often act as catalysts because of a phenomenon known as *adsorption*. Adsorption is the collection of molecules on a surface, attracted there by electrostatic attraction or Van der Waals forces. When reactant molecules adsorb onto the surface of a catalyst, the surface forces cause reactant molecules to separate into individual atoms, at which time they are free to bond to other atoms to form new molecules. In the SHE, the H^+ ions separate from H_3O^+ ions and adsorb onto the platinum surface. There they are reduced and combine to form molecules of H_2 gas.

To summarize, the standard reduction potentials are relative values, as are all electrical voltages. The SHE has been established as the reference-standard half-cell. The electrode potentials of all other reducing agents are measured by installing them in an anode half-cell connected to an SHE cathode half-cell. The overall potential of the voltaic test cell—and thus the electrode potential of the reducing agent under test—is the difference between the $E°$ of the SHE cathode (0 V) and the $E°$ of the anode, according to Equation (12.4).

12.3.5 Electrochemical Applications

We conclude this chapter with a brief look at a few of the hundreds of real-world applications of electrochemistry in technology.

Dry Cells Beginning with the most familiar application of all, Figure 12.16 is a diagram of a typical alkaline dry cell. The materials shown as "zinc/KOH anode" and "manganese oxide cathode" are typically made of a gel or paste that allows movement of ions throughout the mate-

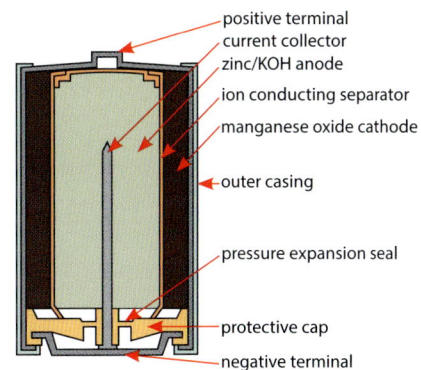

Figure 12.16. Typical alkaline dry cell construction.

353

rial. The anode is actually a zinc powder mixed with KOH, which is where the cell gets the name *alkaline*. The cathode material is in direct contact with the positive terminal of the cell at the top. Electrical connection to the negative terminal at the bottom is made with a metal rod that extends up into the anode material. The half-reactions for alkaline dry cells are quite complex, but can be approximated as follows:

oxidation $\quad\quad\quad\quad Zn(s) + 2OH^-(aq) \rightarrow Zn(OH)_2(s) + 2e^-$

reduction $\quad 2MnO_2(s) + 2H_2O(l) + 2e^- \rightarrow 2MnO(OH)(s) + 2OH^-(aq)$

Wet Cell Batteries

Figure 12.17 shows the inside of a small six-cell automotive battery. The battery shown came out of a four-wheeler, but larger batteries for cars are similar. Typical automotive batteries use plates made of lead and lead oxide immersed in an H_2SO_4 electrolyte. For this reason they are often called *lead-acid* batteries. There are six cells connected in series inside the battery, so this device is truly a *battery*, unlike dry cells. Automotive batteries are designed to deliver a very large current (hundreds of amps) for a very brief time (typically under one second, just long enough to start the car). This requirement is quite different from the way dry cells typically operate, which is to provide a very small current to an electronic device for an extended period of time.

Figure 12.17. A small, six-cell automotive battery cut open to show the cells inside.

Each of the six cells has a cell potential of just over 2.0 V, and six of them connected in series produces an overall voltage for the battery of 12 V. The half-reactions for this common battery are as follows:

oxidation $\quad\quad\quad\quad Pb(s) + HSO_4^-(aq) \rightarrow PbSO_4(s) + H^+(aq) + 2e^-$

reduction $\quad PbO_2(s) + HSO_4^-(aq) + 3H^+(aq) + 2e^- \rightarrow PbSO_4(s) + 2H_2O(l)$

Electroplating

Electroplating is a very widely used instance of electrochemistry. The basic idea behind plating is to form a tightly adhering metallic coating on a metal or plastic object. In electroplating, the part being plated is the cathode. The anode is made of the metal that is plating the part. The reaction takes place with anode and cathode together in the same electrolyte, which contains a high concentration of ions of the same metal as the anode. Once the current is switched on, the ions in the electrolyte begin plating onto the cathode piece.

Figure 12.18. Chrome plating on a Velocette Viper, a British motorcycle manufactured in the 1950s and 60s.

Electroplating technology is used for a wide variety of purposes. One use is to coat an object made of a metal that rusts or corrodes (such as steel) with a metal that is more resistant to corrosion, such as chrome, nickel, or gold. A second purpose is for aesthetic appeal. The chrome plating on the motorcycle parts in Figure 12.18 protects the steel parts from rust, and at the same time makes the machine shiny and attractive. Some coins are made this way as well. Nearly all coins are made of one metal plated onto another. Since 1982, the U.S. penny has been made of a zinc disk with an electroplated copper coating. The penny is 97.5% zinc and 2.5% copper.

Figure 12.19. Printed circuit boards (PCBs) hanging in an electroplating machine.

A third example of electroplating technology is to adhere a copper layer to a fiberglass board in the manufacturing of printed circuit boards (PCBs). Electronic devices such as computers, televisions, and microwave ovens all contain printed circuit boards where the electronic components are mounted and interconnected. Instead of being connected together with individual wires, the electronic components on the PCB are connected together with copper "traces" plated onto the board. Figure 12.19 shows some PCBs hanging in an electroplating machine during the manufacturing process. The plated copper is visible on the PCBs.

Anodizing

Anodizing is an electrochemical process commonly used on aluminum parts, although other metals may be anodized as well. Recall from Section 12.1.3 that aluminum parts form a passivation layer of aluminum oxide that protects the aluminum from further corrosion. The purpose of anodizing is to increase the thickness of the autopassivated layer (the layer the aluminum part forms on itself) by an electrochemical redox reaction. The anodizing also allows the metal part to be dyed. Anodized aluminum objects are quite common these days. Figure 12.20 displays a couple of familiar examples. The purpose of the anodizing on the cooking pan at the top is to protect the aluminum. With the inexpensive "not-for-climbing" carabiners below, the purpose is clearly for the colorful dye that makes the objects attractive.

In anodizing, the part to be anodized is the anode—hence the name of the process. Recall from Figure 12.12 that in electrolytic cells, the anode is the positive electrode. The electrons flow through the electrolyte from the cathode to the anode. The electric current releases hydrogen at the cathode and oxygen at the anode. The oxygen-rich environment at the anode is what causes the aluminum oxide layer to form.

Figure 12.20. Examples of anodized aluminum objects.

Cathodic Protection

Underground steel pipes and ships with steel hulls or steel propellers have this in common: they are subject to the merciless processes of corrosion—oxidation—that destroy the metal. If not controlled, corrosion can damage steel pipes so badly that they rupture.

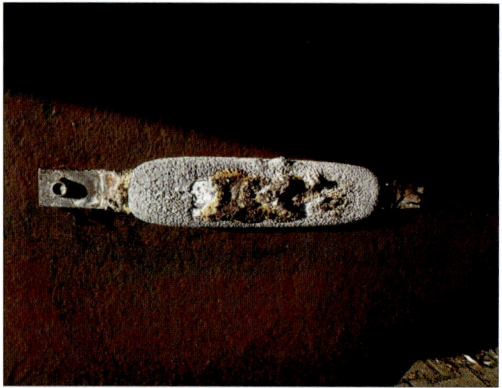

Figure 12.21. At the top, the lozenge-shaped object on the bottom of this boat is a sacrificial anode. The middle image shows what the sacrificial anode looks like after it has corroded during use. The metallic rectangles on the large ship in the bottom image are zinc sacrificial anodes. Metal propellers are also protected from corrosion as long as there is a solid electrical contact between the propeller and the cathode, which is usually the steel hull of the ship.

The steel hulls of ships can simply corrode away to the point where only costly repair work can save the ship from the scrap yard.

This type of passive corrosion of steel parts is a redox reaction. Recall from our discussion of Figure 12.10 that the anode in a voltaic cell dissolves away during the redox reaction. This means that it is the anode that is corroded in these corrosion reactions—when active metals like steel (which is mostly iron) are buried in the ground or pushed around in salt water, the salts in the ground and in the water allow the steel parts to become the anodes—as if they were in voltaic cells—in redox reactions that gradually destroy those parts.

Cathodic protection, illustrated in Figure 12.21, makes use of a *sacrificial anode* to protect other parts, such as pipes, ship hulls, and ship propellers. The idea is to convert the parts to be protected from anodes to cathodes by attaching a chunk of metal that is a more active reducing agent than the steel parts are. This chunk of active metal is called the sacrificial anode, because it is specifically chosen to corrode (i.e., sacrifice itself) in lieu of the corrosion of the protected parts.

For cathodic protection to work, the anode must have a more negative electrode potential than the metal to be protected. Typical metals for this purpose are zinc, aluminum, or magnesium, or alloys of these. The electrode potentials of these metals are all more negative than the electrode potential of steel.

Referring again to Figure 12.10, keep in mind that electrochemical processes that run by themselves are basically voltaic cells. In a voltaic cell, the electrons flow out of the anode into the external circuit, and charge in the form of ions flows from the cathode half-cell toward the anode half-cell. This is essentially what is happening when a pipe corrodes underground or a ship hull corrodes in the water. In the case of a pipe without cathodic protection, oxidation and reduction both occur all along the pipe. Iron atoms in the pipe oxidize to form iron oxide—rust. Atmospheric oxygen is reduced at the same time.

But a sacrificial anode changes the scenario, as illustrated in Figure 12.22. With the addition of the sacrificial anode, a metal that is a stronger reducing agent than the iron in the steel is, the

Redox Chemistry

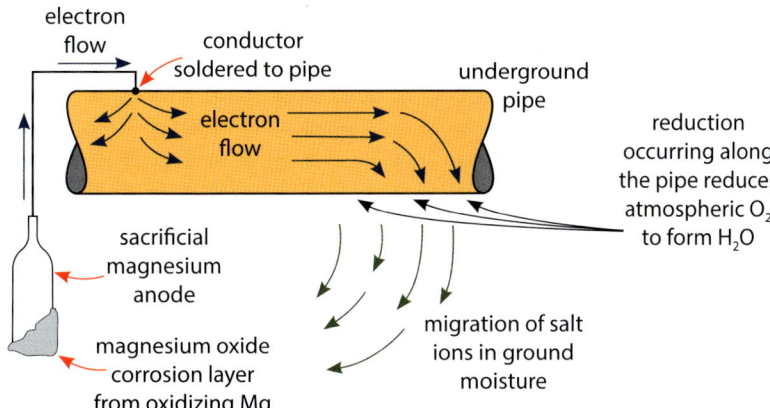

Figure 12.22. Instead of oxidation occurring all along the pipe, the sacrificial anode alone is destroyed by oxidation. Reduction still occurs all along the pipe, as atmospheric oxygen is reduced to form H_2O.

oxidation all occurs at the sacrificial anode, so it gets corroded instead of the pipe. The electrons from the oxidation flow into the pipe, and reduction still occurs all along the pipe walls, wherever there is oxygen to be reduced to form water. But the reduction doesn't hurt the pipe. The corrosion is not the result of the reduction half-reaction; it is from the oxidation half-reaction at the anode. With a sacrificial anode, the anode can simply be replaced from time to time after it corrodes away, and the pipe remains in good condition. The same idea works in the case of ship hulls underwater.

Smelting   I have mentioned a few times in this text that *smelting* is another important example of redox technology at work. Of all the technologies we have reviewed here, smelting is *by far* the oldest. Evidence now persuades historians and anthropologists that lead smelting was in use in Asia Minor by 6500 BC. Apparently, copper smelting was known by 5500 BC in the region of the Balkans.

The point of smelting is to reduce the metal ions in ores to pure metal. You can't simply heat the ore up and melt the metal out. (At least, not unless you heat the ore to the extremely high temperature needed to melt the rock.) The metal ions are bonded in the crystal lattice of the mineral, so to convert them into metal, they must be reduced.

Pure metals have always been valuable as tools, weapons, sculptural materials, and structural materials, so people have smelted for thousands of years. In the old days, the redox reactions required for smelting were generated strictly from heat. Thus, smelters prior to the 20th century were essentially large furnaces. The classic reducing agent for smelting is carbon monoxide, CO, which is easily produced by an oxygen-starved fire. In the thermal process, metal ores are heated to the necessary temperature in the presence of CO. The carbon in the CO is oxidized and forms CO_2 (which increases the oxidation state of carbon from +2 to +4). In the process, the metal from the ore is reduced and forms a pure liquid metal, which is cast into ingots.

Once electrochemical theory was developed and industrial-grade electrical supplies were available, smelting by electrochemical processes became as common as smelting by strictly thermal processes. Today, electrochemical smelting of some metals (zinc, for example) is more common than thermal processes. The facilities for smelting metals like aluminum and zinc can have cells the size of large automobiles connected in series circuits of over 100 cells. The electrical currents involved can be well over 100,000 amps. Alas, the manufacturers all guard their technologies closely, so photos of such facilities are not to be had. However, I can tell you that the magnetic fields surrounding conductors carrying these kinds of currents are huge. If you wear a watch, carry credit cards, or take a mobile phone into such an area, any of them will cease functioning—either permanently erased or magnetized so strongly that they won't work. (Hmm... Interesting.)

Chapter 12

Chapter 12 Exercises

SECTION 12.1

1. Distinguish between oxidation and reduction.

2. For the reaction equations listed below, write the oxidation states for each element in each compound, identify elements that are oxidized or reduced, and identify all oxidizing agents and reducing agents.

   a. $PBr_3(l) + 3H_2O(l) \rightarrow H_3PO_3(aq) + 3HBr(aq)$

   b. $2KClO_3(s) \rightarrow 2KCl(s) + 3O_2(g)$

   c. $2H_2SO_4(aq) + 2NaBr(s) \rightarrow Br_2(l) + SO_2(g) + Na_2SO_4(aq) + 2H_2O(l)$

   d. $NaCl(aq) + AgNO_3(aq) \rightarrow AgCl(aq) + NaNO_3(aq)$

   e. $NaI(aq) + 3HClO(aq) \rightarrow 3HCl(aq) + NaIO_3(aq)$

   f. $3SO_2(g) + 2HNO_3(aq) + 2H_2O(l) \rightarrow 3H_2SO_4(aq) + 2NO(g)$

3. Use Table 12.1 to identify the strongest and weakest reducing agent in each of the following groups of substances:

   a. I^-, Ca, NO, Zn
   b. Fe^{2+}, Li, Br^-, Cu
   c. Al, Mg, Cr^{3+}, Ag
   d. Sn, O_2, F^-, MnO_2

4. Use Table 12.1 to address the following questions:

   a. Can I_2 be used to oxidize nickel metal?
   b. Will cadmium metal reduce nitrate ions?
   c. Can silver metal act as a reducing agent to reduce Ca^{2+} ions?
   d. Will Fe^{2+} ions oxidize copper metal?

5. What is the significance of one reducer being stronger than another?

SECTION 12.2

6. For each of the following half-reactions, complete and balance the half-reaction and indicate whether it is an oxidation or reduction.

   a. $SO_3^{2-}(aq) \rightarrow SO_4^{2-}(aq)$ (basic solution)

   b. $N_2(g) \rightarrow NH_3(g)$ (basic solution)

   c. $N_2(g) \rightarrow NH_4^+(aq)$ (acidic solution)

   d. $Sn^{2+}(aq) \rightarrow Sn^{4+}(aq)$ (acidic or basic solution)

   e. $ClO_3^-(aq) \rightarrow Cl^-(aq)$ (acidic solution)

   f. $OH^-(aq) \rightarrow O_2(g)$ (basic solution)

   g. $H_2SO_3(aq) \rightarrow SO_4^{2-}(aq)$ (acidic solution)

h. $O_2(g) \rightarrow H_2O(l)$ (basic solution)

7. For each of the following redox reactions, complete and balance the equation using the half-reaction method. Identify the oxidizing and reducing agents. Combine and/or add ions to produce a balanced equation showing the reactants and products given.

   a. In aqueous solution, ferric chloride, $FeCl_3$, combines with hydrogen sulfide, H_2S, to form ferrous chloride, $FeCl_2$, hydrochloric acid, and solid sulfur. Everything dissociates except the solid sulfur, and the reaction takes place in acidic solution.

   b. Solid iodine, I_2, combines with aqueous hypochlorous acid, HClO, to produce aqueous iodic acid, HIO_3, and aqueous hydrochloric acid, HCl. In this reaction, everything ionizes except the I_2. The reaction takes place in acidic solution.

   c. Aqueous potassium permanganate, $KMnO_4$, combines with aqueous methanol, CH_3OH, to produce aqueous manganese(II) hydroxide, aqueous formic acid, HCOOH, and aqueous potassium hydroxide, KOH. Everything ionizes except the methanol, and the reaction takes place in acidic solution. (Tip 1: $KMnO_4$ ionizes to K^+ and permanganate, MnO_4^-; formic acid ionizes to H^+ and $COOH^-$. Tip 2: You need to add in potassium and hydroxide ions at the very end.)

   d. Aqueous hydrogen peroxide, H_2O_2, and aqueous chlorine dioxide, ClO_2, combine to produce chlorous acid, $HClO_2$, and oxygen gas. Only the acid dissociates. The reaction takes place in basic solution.

   e. Solid aluminum combines with aqueous lithium chloride and aqueous lithium nitrite to produce aqueous lithium aluminate, $LiAlO_2$, and aqueous ammonium chloride. Dissociations should be obvious. The reaction takes place in basic solution.

   f. In aqueous solution, hydrochloric acid combines with permanganic acid, $HMnO_4$, to produce chlorine gas and manganese(II) chloride.

   g. Solid sulfur combines with aqueous nitric acid to produce aqueous sulfurous acid and nitrous oxide gas, N_2O. (N_2O is sometimes called *laughing gas* because inhaling it induces euphoria. Formerly, it was used as an anesthesia. Today, it is added to racing fuel because it is a powerful oxidizer.)

   h. In aqueous solution, hydrobromic acid, HBr, combines with permanganic acid, $HMnO_4$, to produce bromic acid, $HBrO_3$, and solid manganese(IV) oxide.

8. Why do free electrons always appear on the left side of a reduction half-reaction?

9. A *thermite reaction* occurs when a metal fuel powder is combined with a powdered metal oxide. When heated, the powdery mixture briefly emits intense light and heat. The reaction is not explosive, but generates enough heat that it is used for a process called *thermite welding*. Thermite reactions are redox reactions. The following thermite reaction is used for welding steel rails together in train tracks:

$$Fe_2O_3(s) + 2Al(s) \rightarrow 2Fe(s) + Al_2O_3$$

In the image at the right, a small quantity (about 110 g) of the iron(III) oxide/alumi-

Chapter 12

num powder is heated. The sparks are molten particles of pure iron. The reaction is taking place in a cast iron skillet (destroyed during the reaction). Address the following questions about this thermite reaction.

a. Explain why this reaction occurs. (Use data from Table 12.1 to support your answer.)
b. Identify the oxidizing agent and reducing agent.
c. What mass of powdered aluminum should be mixed with 65 g of powdered Fe_2O_3 to make the thermite powder described by the reaction above?
d. If 65 g Fe_2O_3 is consumed in the reaction, what mass of iron is produced?

SECTION 12.3

10. Explain the basic principles of electrochemistry. In formulating your response, consider addressing the question of why electrons spontaneously flow in an external electric circuit connected to a voltaic cell.

11. Explain the difference between the anode and the cathode in an electrochemical cell. Does your answer hold for both voltaic and electrolytic cells?

12. Draw a diagram of a voltaic cell made with cobalt and iron metals, $CoCl_2$ and $FeCl_2$ electrolytes, and sodium chloride salt bridge. On your diagram, identify the following: anode, cathode, positive electrode, negative electrode, direction of electron flow, direction of chloride ion migration in the salt bridge, direction of sodium ion migration in the salt bridge. Finally, determine the cell potential.

13. Explain why electrolytic cells require an outside source of electrical power to drive the reaction.

14. Draw a diagram of a voltaic cell made with Ag in $AgNO_3$ and Ni in $NiSO_4$, with a potassium nitrate salt bridge. On your diagram, identify the following: anode, cathode, positive electrode, negative electrode, direction of electron flow, direction of nitrate ion migration in the salt bridge, direction of potassium ion migration in the salt bridge. Finally, determine the cell potential.

15. Describe what you expect to happen if an aluminum wire is dipped into a solution of tin(II) sulfate. Base your response on the values of standard reduction potentials.

16. When a voltaic cell operates with a potassium nitrate salt bridge, potassium ions are drawn from the salt bridge into the cathode half-cell. Explain why this happens.

17. Explain why the copper electrode in a Cu | Zn electrolytic cell is the positive electrode.

18. Explain why $E°_{cell}$ must be greater than zero for a voltaic cell to work.

19. For the following pairs of half-cells, determine which is the anode and which is the cathode if the redox reaction is to occur spontaneously. Then determine the overall cell potential, $E°_{cell}$.

a. Sn^{2+}/Sn, Ag^+/Ag
b. Co^{2+}/Co, Fe^{3+}/Fe^{2+}
c. Na^+/Na, Br_2/Br^-
d. I_2/I^-, Br_2/Br^-

20. Explain why some standard reduction potentials are positive and some are negative.

21. Explain the general idea behind cathodic protection. In your response, address the question of why cathodic protection works—that is, why the protected object is protected from corrosion.

22. Explain what is meant by the phrase, "reduction occurs at the cathode." Include the details about what is actually happening at the atomic level at the cathode.

23. Explain how a rechargeable dry cell incorporates the electrochemistry of both the voltaic cell and the electrolytic cell.

GENERAL REVIEW EXERCISES

24. Nitrogen dioxide reacts with ozone to produce dinitrogen pentoxide and oxygen according to the following equation:

$$2NO_2(g) + O_3(g) \rightarrow N_2O_5(s) + O_2(g)$$

   a. Determine the percent yield for a reaction in which 1.27 g NO_2 reacts and 1.20 g N_2O_5 is recovered.

   b. Determine the mass of oxygen produced if 4.0 mol ozone react and the percent yield is 76.0%.

   c. Determine the limiting reagent if 4.304 g NO_2 are available to react with 2.268 g O_3.

25. A steel tank of O_2 gas is filled to a gauge pressure of 62.4 atm at 22.00°C. The temperature then increases to 35.00°C. Determine the new pressure in the tank and state your result as a gauge pressure in atm. Assume the volume of the tank does not change.

26. A chemist collects a sample of ethane by water displacement at a temperature of 16.0°C and a total pressure of 102,360 Pa. The collection bottle used to draw the sample has a volume of 275 mL. Determine the number of moles of ethane collected in the sample.

27. The molecular formula for aspirin, acetylsalicylic acid, is $C_9H_8O_4$. How many carbon atoms are contained in a bottle of aspirin, if the bottle contains exactly 200 aspirin tablets with a mass of 325 mg per tablet?

28. Ethanol burns to produce water and CO_2. Determine the volume of ethanol consumed in producing 1.00 L of water. The density of ethanol is 0.789 g/cm^3.

Glossary

A

absolute zero	The theoretical lowest temperature that can exist. At absolute zero all atomic and molecular motion ceases and the entropy of crystals is zero.
acid-base neutralization reaction	A chemical reaction in which an acid and base react to produce a salt, water, and sometimes carbon dioxide. In the process, the low-pH solution of the acid and the high-pH solution of the base are neutralized to form a solution with a neutral pH of approximately 7.
acid dissociation constant	The equilibrium constant for an acid in aqueous solution.
acid rain	Rain with a low pH due to the presence of sulfuric acid. The acid forms in the atmosphere from sulfur dioxide and moisture.
activity series of metals	A list of metals in order of their chemical activity. More active metals replace less active metals in single replacement reactions.
actual yield	The amount of a given product in a chemical reaction that is produced in an actual experiment or chemical process.
adsorption	The collection of molecules onto a solid surface, attracted there by electrostatic attraction or Van der Waals forces.
alkali metal	Any element in Group 1 of the Periodic Table of the Elements.
alkaline-earth metal	Any element in Group 2 of the Periodic Table of the Elements.
alloy	A homogeneous mixture of metals formed by mixing the molten metals together and allowing the solution to cool and solidify.
amphiprotic species	Any species that can act either as a proton acceptor or a proton donor, depending on the conditions.
amphoteric species	Any species that can act either as an acid or as a base, depending on the conditions.
angular momentum quantum number	See azimuthal quantum number.
anion	A negatively charged ion.
anode	The electrode in an electrochemical cell where the oxidation half-reaction takes place.
atomic mass	The average mass of the atoms of a given element in unified atomic mass units, u.
atomic mass unit	An obsoleted unit used for relative atomic masses; replaced by the unified atomic mass unit, u.
atomic number	A number, denoted by Z, that designates the number of protons contained in the nucleus of atoms of a given element. The elements are arranged in the periodic table in order by their atomic numbers from 1 to 118.
atomic spectrum	The set of wavelengths of electromagnetic radiation emitted by the excited atoms of a particular element.
Aufbau principle	The principle that in the ground state electrons occupy orbitals in atoms starting with the lowest energy orbital and going sequentially up from there to fill orbitals associated with higher energy as required.
azimuthal quantum number	A number denoted by $l = 0, 1, 2, 3$ or 4 and used to designate specific subshells within the shells of an atom. The values of l are usually denoted by the letters $s, p, d, f,$ and g.

B

barometer	A device that measures atmospheric pressure.

Glossary

barometric pressure	Atmospheric pressure as measured by a barometer.
base dissociation constant	The equilibrium constant for a base in aqueous solution.
binary acid	An acid formed from hydrogen and one of the more electronegative nonmetals.
binary formula	A chemical formula for a compound consisting of two parts, such as the symbols for cation and anion.
birefringence	The property, exhibited by some crystals, of refracting light in two directions as the light passes through the crystal.
body-centered cubic lattice	A close-packed metallic lattice with layers in the sequence A–B–A–B, all A layers being aligned together, all B layers aligned together, and the atomic centers of A layers aligned with the interstitial centers of the B layers, and vice versa.
boiling	The phase transition that occurs when a liquid is at the boiling point temperature and additional heat is added, causing the liquid to transition to the vapor state; vaporization.
boiling point elevation	The increase in boiling point that occurs in a solution relative to the boiling point of the pure solvent; a colligative property.
bond energy	The energy holding two atoms together in a covalent bond; the energy that must be supplied to separate two atoms held together by a covalent bond.
bonding domain	An electron domain between two atoms covalently bonded together.
bond length	The distance between nuclei in the chemical bond between two atoms.
bond number	A measure of covalent bond strength between two atoms. A single, double, or triple bond between exactly two atoms has a bond number of 1, 2, or 3, respectively. If a bond is also included as part of a resonance structure, then the bond number is the sum of the single dedicated bond (1) and the bond's fractional share of the resonance structure bond.
Brownian motion	The jittery, random motion exhibited by particles such as pollen grains when the particles are suspended in water and observed through a microscope. The phenomenon was first observed by Robert Brown. In 1905, Albert Einstein demonstrated that the phenomenon is caused by collisions of water molecules with the suspended particles.
buret	A tall, thin article of glassware, fitted with a stopcock at the bottom, and marked along the side with a volume scale. Burets are used for dispensing precise amounts of liquid.

C

calorimeter	A device used for making temperature measurements on a sample, while keeping the contents of the calorimeter thermally isolated. Calorimeters are used for determining molar heat capacity, specific heat capacity, heat of combustion, and other thermal parameters.
capillary action	The tendency of a liquid to be drawn upward against gravity into a thin tube or solid porous substance. Capillary action is a result of the forces involved in surface tension, and the polar character of the particles in the solid surface.
carboxylic acid	A compound incorporating the carboxyl functional group, the group having a hydrogen atom or alkyl group attached.
catalysis	The process of using a catalyst to expedite the rate of a reaction.
catalyst	A substance that expedites or facilitates a chemical reaction without itself being consumed in the reaction.
cathode	The electrode in an electrochemical cell where the reduction half-reaction takes place.
cathodic protection	The technique of protecting ship hulls, underground pipes and the like from corrosion due to redox reactions by attaching a sacrificial anode to metal parts.
cation	A positively charged ion.

Glossary

chalcogen	Any element in Group 16 of the Periodic Table of the Elements. (The synthetic element 116 is not generally included with the chalcogens at this time, but is expected to exhibit chalcogen properties as well.)
charge	A fundamental property of subatomic particles. Each proton carries one unit of positive charge, each electron carries one unit of negative charge, and each neutron carries no charge. An electrical current is a moving stream of charged particles.
chemical activity	A reference to how readily or aggressively a substance engages in various chemical processes.
chemical change	A change involving a chemical reaction, in which new chemical species are formed, with chemical identities and chemical properties different from those of the original substances.
chemical equation	An equation written with the chemical formulas of the compounds involved in a chemical reaction that shows the reactants and the products, their ratios, and (optionally) the forms they are in.
chemical equilibrium	A state in which chemical species are transforming into each other at equal rates in an equilibrium system.
chemical formula	An expression of the elements in a compound and their ratios in which the elements are denoted by their chemical symbols.
chemical property	A property of pure substances related to a substance's tendency to engage in specific chemical reactions or processes; a property related to a substance's tendency to form or break chemical bonds under specific conditions. Examples of quantifiable chemical properties are number of valence electrons, oxidation state(s), electronegativity, ionization energy, and electron affinity. Examples of qualitative chemical properties include flammability, susceptibility to corrosion, and the types of bonds the element forms (ionic or covalent).
chemistry	The study of the elements, how they combine to form mixtures and compounds, the properties of these substances, and the processes involved.
colligative properties	Properties of solutions that depend only on the concentration of solute and not on the identity of the solute.
colloid	A colloidal dispersion.
colloidal dispersion	A heterogeneous mixture formed when microscopic particles, ranging in size from 1 to 1,000 nm, are dispersed throughout a dispersing medium.
combustion reaction	A chemical reaction in which a substance combines with oxygen, releasing energy in the form of heat and light; burning.
compound	A substance, either molecular or crystalline, in which the atoms of two or more different elements are chemically bonded together.
condensation	The phase transition that occurs when a vapor is at the boiling point temperature and heat is removed, causing the vapor to transition to the liquid state.
condensed electron configuration	A shorter form of the electron configuration for an atom of a given element in which the noble gas of the preceding period in the periodic table is used to represent the atom's core electrons.
conduction electron	The valence electrons from the metal atoms in a metallic lattice that are free to move about in the overlapping orbitals in the lattice under the influence of an electric potential. Collectively, the conduction electrons are referred to as the electron sea.
conjugate acid	According to Brønsted-Lowry acid-base theory, the species formed when a base accepts a proton.
conjugate base	According to Brønsted-Lowry acid-base theory, the species that remains when an acid donates a proton.
core electron	As distinguished from valence electrons, any of the electrons in the inner shells of an atom that do not participate in bonding.

Glossary

covalent bond	A chemical bond between two or more nonmetal atoms in which the atoms bond together in individual molecules through the sharing of electron pairs.
critical point	The point located on a pressure-temperature phase diagram by the critical temperature and critical pressure.
critical pressure	The lowest pressure at which a substance can exist as a liquid at the critical temperature.
critical temperature	The highest temperature at which a substance can exist as a liquid.
crystal lattice	A regular, three-dimensional, geometric array of positive and negative ions bound together by electrical attraction; one of the two fundamental atomic structures of matter (the other being molecules).
crystallogen	Any element in Group 14 of the Periodic Table of the Elements. (The synthetic element 114 is not generally included with the crystallogens at this time, but is expected to exhibit crystallogen properties as well.)

D

Debye force	The force of electrical attraction between molecules, one of which is a permanent dipole, and the other of which is a temporary, induced dipole due to the presence of the permanent dipole; one of the three Van der Waals forces.
decomposition reaction	A chemical reaction in which a compound is broken down into other compounds or elements.
delocalization	The condition of electrons that cannot be associated with the orbitals involved in a single pair of atoms, but are instead occupying an orbital that spreads across several atoms in a molecule or lattice. An example of delocalized electrons is the conduction electrons in a metallic lattice.
deposition	The phase transition in which a gas changes into a solid.
diamond	An allotrope of carbon with a strong crystal lattice.
diatomic molecule	A molecule consisting of two atoms, often referring to one of the elements that form diatomic molecules with themselves, such as hydrogen, fluorine, chlorine, bromine, and iodine.
diffusion	The tendency of molecules in fluids spontaneously to mix and mingle due to their random motions and collisions. Foreign atoms in a solid crystal lattice are also known diffuse or migrate through the lattice.
dipole	A molecule with one end or region more electrically positive than the other due to the bonding electrons being located more densely in one part of the molecule than another. Dipoles can be permanent dipoles, induced dipoles, or randomly occurring temporary dipoles. More generally, any object with one end more electrically positive than the other.
diprotic acid	An acid with two hydrogen atoms in the molecule, both capable of dissociating.
dissociation	The process of separating into independent ions in solution. This process is exhibited by soluble ionic compounds, and by acids and bases.
dissolution	The process of dissolving.
double bond	A covalent bond in which two atoms share two pairs of electrons.
double replacement reaction	A chemical reaction in which the cations and anions of two binary compounds switch places.
dry cell	An electrochemical cell that contains no liquid, such as cells used to power portable electronics.
ductility	The ability of a substance to deform under tension without snapping, thus permitting the substance to be drawn into a wire. This property is typical of many metals.

Glossary

E

effective nuclear charge	A value representing the ability of an atomic nucleus to attract electrons, roughly calculated as the difference between the total number of electrons in an atom and the number of core electrons in the atom.
effusion	The process of a gas passing from one contained area to another through a hole so small that only one particle of gas can pass through at a time. Effusion occurs when the hole diameter is less than the mean free path of the gas particles. If the hole diameter is larger than the mean free path, multiple particles pass through simultaneously and diffusion occurs.
electrical conductivity	The ability of a substance to conduct electricity; the flow of electric charge (usually electrons).
electric potential	Voltage.
electrochemical cell	A container in which an electrochemical redox reaction can take place. The cell includes two half-cells, each of which contains an electrode in an electrolyte, and a means of ion transfer between the half-cells.
electrode	One of the two sites—anode or cathode—in an electrochemical cell where the redox half-reactions take place. The oxidation half-reaction occurs at the anode; the reduction half-reaction occurs at the cathode.
electrolyte	A substance that dissociates in water to produce an electrically conductive solution. In electrochemistry, the electrically conductive solution itself.
electrolytic cell	An electrochemical cell that consumes electricity. Electrochemical cells are used for industrial processes such as electroplating and smelting.
electromagnetic spectrum	The range of wavelengths of electromagnetic radiation found in nature, including radio waves, microwaves, infrared light, visible light, ultraviolet light, X-rays, and gamma rays.
electron	One of the three basic subatomic particles in atoms. Electrons carry the atom's negative charge and reside in the orbitals surrounding the atomic nucleus. The activity of electrons determines an atom's chemical behavior.
electron affinity	The amount of energy released when an electron is added to a ground-state, isolated, gaseous atom.
electron configuration	A shorthand notation that lists the orbitals containing electrons in the ground-state atoms of a given element.
electron domain	The region occupied by a shared pair of electrons in a covalent bond or around a central atom in a molecule.
electronegativity	A metric invented by Linus Pauling that indicates the relative attraction of atoms for bonding electron pairs in molecules. The electronegativity scale ranges from a low of 0.7 for the element francium to a high of 3.98 for the element fluorine.
electron sea	The vast number of conduction electrons in a metal.
element	A pure substance consisting entirely of atoms possessing the same number of protons, and listed in the Periodic Table of the Elements.
empirical formula	A formula that represents the ratios—rather than the actual numbers—of elements in a compound. The actual numbers of atoms of different elements in one molecule of a molecular compound are indicated by the molecular formula.
endpoint	The pH at which a pH indicator changes color, used in titration as an estimate for the equivalence point.
energy level	Any one of the energy states available to electrons according to the Bohr model of the atom.
entropy	A thermodynamic variable associated with the degree of disorder present in a system. Entropy is also associated with the number of microstates available to particles in a system.

Glossary

equilibrium	A state in which chemical or physical species are transforming into each other at equal rates in an equilibrium system.
equivalence point	The point during a titration at which chemically equivalent amounts of acid and base are present in the titration solution.
evaporation	The spontaneous phase transition in which particles in a liquid transition to the vapor state, while the temperature of the liquid is below the boiling point.
evaporative cooling	The cooling that occurs during evaporation, when the most energetic liquid particles enter the vapor state, leaving behind particles in the liquid state with lower average energy and thus lower temperature.
exchange reaction	A double replacement reaction.
excitation	The increased energy state an atom enters upon absorbing a quantum of energy from a photon or through collision with another atomic or subatomic particle.
excited state	Any energy state of an atom that is higher than the ground state.
experiment	A test designed to confirm or disconfirm a particular hypothesis.
extensive property	Any property of a substance (such as mass and volume) that depends on the amount of the substance that is present.

F

face-centered cubic lattice	A close-packed metallic lattice with layers in the sequence A–B–C–A–B–C, all A layers being aligned together, all B layers aligned together, and all C layers aligned together. The atomic centers of B layers are displaced slightly from those of layer A, and the C layers are displaced slightly again. The fourth layer is displaced yet further and is again at the A alignment.
formula mass	The mass, in unified atomic mass units (u), obtained by adding together the atomic masses of the elements in a compound, taking into account the number of atoms of each element represented in the chemical formula of the compound.
formula equation	An unbalanced chemical equation that lists reactants and products.
formula unit	Typically used in reference to crystalline substances, a formula unit is one set of the atoms represented by the chemical formula of the compound. In molecular substances, the formula unit is the molecule.
free radical	Any atom, molecule, or ion that has a lone, unpaired electron.
freezing	The phase transition that occurs when a liquid is at the freezing (melting) point temperature and heat is removed, causing the liquid to transition to the solid state.
freezing point depression	The reduction in freezing point that occurs in a solution relative to the freezing point of the pure solvent; a colligative property.

G

galvanic cell	A voltaic cell.
gas	One of the four basic states of matter; the non-ionized state characterized by the highest internal energy, in which the particles of a substance are completely free from one another.
gauge pressure	A pressure measurement relative to atmospheric pressure; a pressure measurement that uses atmospheric pressure as the zero reference.
Gibbs free energy	A thermodynamic variable based on changes of energy, temperature, and changes of entropy in chemical processes. The algebraic sign of the change in Gibbs free energy for a given chemical process indicates whether the process occurs spontaneously. Also called *free energy*.
ground state	The lowest possible energy state of an atom, when all electrons in the atom are in orbitals representing the lowest possible total energy.
group	A column in the Periodic Table of the Elements.

Glossary

H

half-cell	One of the two compartments in an electrochemical cell. Each half-cell contains one of the electrodes immersed in an electrolyte.
half-reaction	An equation showing separately the oxidation or reduction process in a redox reaction.
half-reaction method	A method of balancing chemical equations in oxidation-reduction reactions that makes use of the oxidation and reduction half-reactions.
halogen	Any element in Group 17 of the Periodic Table of the Elements. (The synthetic element 117 is not generally included with the halogens at this time, but is expected to exhibit halogen properties as well.)
heat of solution	The net amount of heat energy absorbed or released during the process of dissolution.
heterogeneous catalyst	A catalyst in a different phase or state than the reactants in a chemical system.
heterogeneous mixture	A mixture of two or more pure substances in which the different substances appear in lumps containing many atoms or molecules. Typically the individual substances in a heterogeneous mixture may be discerned with the naked eye or with a microscope, because the individual particles in the mixture are large enough to reflect light.
heterogeneous reaction	A reaction in which more than one state of matter is present among the reactants.
homogeneous catalyst	A catalyst in the same phase or state as the reactants in a chemical system.
homogeneous mixture	A solution.
homogeneous reaction	A reaction in which all reactants are in the same phase or state.
Hund's rule	The principle that in an unfilled subshell, electrons occupy separate orbitals singly until each orbital contains one electron, after which the electrons begin doubling up in the orbitals.
hydrate	An ionic compound that contains water molecules trapped in the crystal lattice.
hydration	The process of polar water molecules being attracted to and surrounding ions in aqueous solution.
hydride	A hydrogen atom that has ionized by gaining an electron to acquire a charge of −1.
hydrocarbon	Any compound containing only carbon and hydrogen.
hydrogen bond	The attraction between nonbonding electron pairs on the central atom of a molecule and hydrogen atoms that are attached to oxygen, nitrogen, or fluorine atoms.
hygroscopic	Having a strong affinity for water.
hypothesis	An informed prediction about what will happen in certain circumstances. Every hypothesis is based on a particular theory. It is hypotheses that are tested in scientific experiments.

I

icosagen	Any element in Group 13 of the Periodic Table of the Elements. (The synthetic element 113 is not generally included with the icosagens at this time, but is expected to exhibit icosagen properties as well.)
ideal gas	A gas modeled according to the assumptions of the kinetic-molecular theory of gases.
induced dipole	A molecule with temporary dipole character due to the proximity of a permanent dipole.
inner transition metal	Any one of the f-block elements, usually shown in two separate rows of elements under the main body of the Periodic Table of the Elements. Also known as rare-earth elements.

intensive property	Any bulk property of substances (such as electrical conductivity or density) that does not depend on the amount of the substance present.
intermolecular force	The forces of electrical attraction between molecules, relatively weak compared to the forces involved in ionic and covalent bonds. Intermolecular forces consist of hydrogen bonds and Van der Waals forces. The Van der Waals forces consist of Keesom forces between permanent dipoles, Debye forces between a dipole and an induced dipole, and London dispersion forces between instantaneously formed temporary dipoles. The presence and strength of intermolecular forces governs properties in molecular substances such as melting point, boiling point, and solubility.
internal energy	The total energy represented by the sum of the kinetic energies of each of the particles in a substance.
ion	(1) An atom that has gained or lost one or more electrons, thus acquiring a net electrical charge. (2) A molecule with a net electrical charge.
ionic bond	A chemical bond between atoms in which positive and negative ions are bound together by electrical attraction in a crystal lattice. Ions form when electrons are transferred from metal atoms to nonmetal atoms, so ionically bonded substances typically involve metal and nonmetal atoms. Ionic bonds also form between metals and polyatomic ions.
ionic equation	A chemical equation showing ionizable reactant and product compounds as individual ions rather than as compound formula units or molecular units.
ionization constant of water	K_w, the product of the hydronium ion and hydroxide ion concentrations in aqueous solution, a constant at any given temperature. At 25°C this product is equal to $K_w = 1.0 \times 10^{-14}\ M^2$.
ionization energy	The amount of energy required to remove a ground-state electron from an isolated, gaseous atom. The energy required to remove one electron is called the first ionization energy. The additional energy required to remove a second electron is the second ionization energy, and so on.
ionization equation	A chemical equation describing ionization and specifying the energy involved.
ionizing radiation	Particles of matter or electromagnetic radiation (photons) with sufficient energy to remove an electron from an atom, thus ionizing it. Types of ionizing radiation include high-speed subatomic particles and electromagnetic radiation at wavelengths equal to or shorter than those of the high-energy end of the ultraviolet region of the electromagnetic spectrum
isotope	Any of the atoms of a given element, distinguished from one another by the different numbers of neutrons the atoms can have in the nucleus.

K

Keesom force	The force of electrical attraction between molecules that are permanent dipoles; one of the three Van der Waals forces.
kinetic energy	The energy associated with any object in motion. Kinetic energy is directly proportional to the object's mass and proportional to the square of the object's speed.

L

lattice energy	The total energy binding the ions together in the crystal lattice of one mole of an ionically bonded substance.
Lewis structure	A molecular representation in which atoms are represented by their chemical symbols and bonds between atoms are represented by line segments.
Lewis symbol	A representation of an atom and its valence electrons in which the atom is represented by its chemical symbol and the electrons are represented by dots placed around the element symbol.
limiting reactant	The reactant that is consumed first in a chemical reaction, thereby determining the quantities of products produced by the reaction.

Glossary

limiting reagent	A limiting reactant.
liquid	One of the four basic states of matter; the state in which the particles of a substance have sufficient kinetic energy to remain separated from each other enough to move around, but not enough to break free entirely.
London dispersion forces	The forces of electrical attraction between molecules exhibiting instantaneous, temporary dipole character.
lone pair	A nonbonding electron domain; an unshared pair of electrons.
luster	Shininess.

M

Madelung rule	The mathematical rule that specifies the sequence in which an atom's orbitals fill with electrons in the ground state as they fill up from low-energy orbitals first and then to higher energy orbitals according to the Aufbau principle.
magnetic quantum number	A number denoted by m_l that denotes orbitals of specific shape and orientation within a given subshell in an atom.
main group element	Any element in Groups 1–2 or 13–18 of the Periodic Table of the Elements.
malleability	The ability of a substance to be shaped (rather than shattered) by pounding. This property is typical of many metals.
manometer	A thin, U-shaped tube filled with liquid and used to make measurements of small differences in pressure by measuring the height difference between the liquid columns in the two sides of the tube.
mass number	The total number of protons and neutrons in the nucleus of an atom.
melting	The phase transition that occurs when a solid is at the melting point temperature and heat is added, causing the solid to transition to the liquid state.
metal	An element exhibiting the physical and chemical properties of metals. The metals are located in Groups 1–12 (except hydrogen), in the lower parts of Groups 13–15, and in the region of the inner transition metals in the Periodic Table of the Elements.
metallic bond	The bonding arrangement that binds atoms together in the crystal lattice of a pure metal. In the lattice, the atomic shells and orbitals overlap and the valence electrons of the atoms occupy orbitals in more than one atom simultaneously. The energy that holds the electron in multiple orbitals at once has the effect of holding the atoms together in the crystal lattice.
metalloid	Any of the elements located in the region between the metals and nonmetals in the Periodic Table of the Elements, usually assumed to include elements 5, 14, 32, 33, 51, 51, and 84. The metalloids possess properties that are not clearly those of the metals nor those of the nonmetals.
metathesis	A double replacement reaction.
microstate	For an atom or molecule, one of the possible combinations of position and velocity.
miscible	Able to mix together in any proportions to form a solution.
mixture	A combination of two or more pure substances in which the individual substances retain their respective chemical identities and may be separated by physical means. The two main classes of mixtures are homogeneous mixtures, which are solutions, and heterogeneous mixtures.
molality	A measure of solution concentration, defined as the number of moles of solute per kilogram of solvent.
molar heat capacity	For a given substance, the amount of heat that must be added or removed to change the temperature of one mole of the substance by 1°C.
molar heat of fusion	For a given substance, the amount of heat per mole of substance that must be added to effect a phase transition from solid to liquid (melting) or removed to effect a phase transition from liquid to solid (freezing).

Glossary

molar heat of vaporization	For a given substance, the amount of heat per mole of substance that must be added to effect a phase transition from liquid to vapor (vaporization or boiling) or removed to effect a phase transition from vapor to liquid (condensation).
molarity	A measure of solution concentration, defined as the number of moles of solute per liter of solution.
molar mass	The mass of one mole of a substance in units of grams per mole.
mole	The amount of a pure substance that contains Avogadro's number of particles of the substance.
molecular equation	A chemical equation showing the reactant and product compounds as formula units or molecular units.
molecular formula	A formula indicating the actual numbers of atoms of different elements in one molecule of a molecular compound.
molecular mass	For compounds that exist as molecules, the mass, in unified atomic mass units (u) obtained by adding together the atomic masses of the elements in the compound, taking into account the number of atoms of each element represented in the chemical formula of the compound. May also be expressed in grams (g) by adding together the gram masses of the atoms in the molecule, or by converting a molar mass to grams with the use of the Avogadro constant.
molecular orbital	A single orbital formed by the melding together of individual orbitals in individual atoms as the atoms bond together in a molecule.
molecule	A cluster of atoms chemically bound together by covalent bonds; one of the two fundamental atomic structures of matter (the other being the crystal lattice).
monoprotic acid	An acid with one hydrogen atom in the molecule.
monovalence	The tendency exhibited by some elements, especially those in Groups 1–4, 16, 17, and parts of 15, always to acquire the same charge and oxidation state during ionization.
multivalence	The ability of many elements, particularly the transition metals, to ionize to any one of two or more different oxidation states, acquiring various ionic charges, depending on the circumstances.

N

net ionic equation	An ionic equation with the spectator ions removed.
neutral	Neither acidic nor basic, and thus having a pH of 7.0.
neutralization	An acid-base reaction, in which the solution after the reaction is neither acidic nor basic but neutral.
nitration	A substitution reaction in which a nitro group (NO_2^+) is added to a hydrocarbon molecule in place of a hydrogen atom that is removed from the molecule.
noble gas	Any element in Group 18 of the Periodic Table of the Elements. (The synthetic element 118 is not generally included with the noble gases at this time, but is expected to exhibit noble gas properties as well.)
nonbonding domain	An electron domain containing an unshared pair of electrons around a central atom in a molecule that does not bond the central atom to any other atom.
nonmetal	Elements on the extreme right end of the periodic table, to the right of the metalloids, exhibiting the physical and chemical properties of nonmetals.
nucleon	A proton or neutron, so called because these two particles are contained in the atomic nucleus.
neutralization	A chemical reaction in which an acid and base react to produce a salt, water, and sometimes carbon dioxide. In the process the low-pH solution of the acid and the high-pH solution of the base are neutralized to form a solution with a neutral pH of approximately 7.

371

Glossary

neutron	One of the three basic subatomic particles in atoms. Neutrons carry no charge and have mass slightly higher than the mass of protons. Neutrons reside in the atomic nucleus.
nucleus	The tiny central region in an atom that contains the atom's protons and neutrons.
nuclide	Any isotope of any element.

O

octet	The eight electrons that occupy the s and p subshells in main group elements.
octet rule	The principle that when forming chemical bonds, main group elements seek to obtain a full octet by gaining, losing, or sharing electrons.
orbital	One of the various cloud-like regions surrounding the atomic nucleus where an atom's electrons are held. The orbitals are arranged in subshells and shells associated with different levels of energy. Each orbital can hold up to two electrons.
orbital diagram	A notation in which atomic orbitals are represented as boxes and electrons as arrows. The notation is used to provide a graphical representation of where the electrons are in the orbitals of an atom of a given element.
osmosis	The process in which particles of solvent pass through a semipermeable membrane in greater quantities in the direction toward the side of the membrane with the highest solute concentration.
osmotic pressure	The pressure required on the higher-concentration side of a semipermeable membrane to stop the process of osmosis.
oxidant	An oxidizing agent.
oxidation	A chemical change in which the oxidation state of an element increases.
oxidation number	The number (positive or negative) representing an element's oxidation state.
oxidation-reduction reaction	A chemical reaction in which elements are oxidized and reduced. Oxidation-reduction reactions are also called redox reactions.
oxidation state	For ions or elements in ionic compounds, the oxidation state is equal to the charge on a monatomic ion. In molecules, the oxidation state of an element in the molecule is a hypothetical or artificial number assigned to each element in the molecule as if bonding electrons were held completely by one atom or another. In molecules, the oxidation state is associated with the strength of the element's attraction to the bonding electrons, and thus is based on the element's electronegativity relative to the other elements in the molecule. The oxidation state in a molecule is not based on any actual charge.
oxidizer	An oxidizing agent.
oxidizing agent	An element that receives electrons to cause the oxidation of another element.
oxyacid	An acid containing oxygen and at least one other element, usually a nonmetal.
oxyanion	A polyatomic ion containing oxygen.

P

passivation	The phenomenon, exhibited by aluminum and some other metals, of forming a thin oxide coating on the surface of the metal which then protects the interior of the metal from further oxidation. Also called autopassivation.
Pauli exclusion principle	The principle that no two electrons in an atom can occupy the same quantum state.
percent composition	The percentages, by mass, of each element in a compound.
percent yield	In a chemical process, the ratio of actual yield to theoretical yield, multiplied by 100%.
periodicity	The cyclic variation of the physical and chemical properties of elements in the Periodic Table of the Elements.

Glossary

periodic law	The fact that the elements, when arranged in order according to atomic number, exhibit periodicities in many physical and chemical properties. Notably, elements in the same group exhibit similar properties.
permanent dipole	A molecule with permanent dipole character due to differences in the electronegativities of the atoms in the molecule.
petroleum	A natural mixture of hydrocarbon molecules of many different sizes.
pH	A measure of acidity in a solution, calculated as the negative of the common logarithm of the hydronium ion concentration.
phase diagram	A diagram showing the phases or states of a particular substance with respect to pressure, temperature, and/or energy.
phase (of matter)	State (of matter).
phase transition	The change of state of a substance from one state or phase to another.
pH indicator	A weak acid or weak base that changes color as the pH of a solution increases or decreases through its transition interval. pH indicators are used in titrations to indicate when the equivalence point has been reached. The indicator in the solution changes color at its endpoint, and the endpoint is used as an estimate for the equivalence point. Some pH indicators are also used to measure the pH of a solution.
photon	A single packet of electromagnetic energy; a quantum of electromagnetic radiation; a particle of light.
physical change	A change of any kind in which the substances involved retain their chemical identities and chemical properties.
physical equilibrium	A state in an equilibrium system in which particles are transforming between two or more physical states at equal rates. Examples include different phases of a substance at the triple point, and gas molecules evaporating out of solution and dissolving into solution in a sealed container.
physical property	Properties of pure substances other than those that describe the substance's chemical behavior. Physical properties of substances include color, density, boiling point, melting point, electrical conductivity, malleability, vapor pressure, thermal conductivity, and many others.
plasma	A state of matter consisting of a gas with ionized particles.
pnictogen	Any element in Group 15 of the Periodic Table of the Elements. (The synthetic element 115 is not generally included with the pnictogens at this time, but is expected to exhibit pnictogen properties as well.)
pOH	A measure of basicity similar to pH for acidity. In an aqueous solution, pH + pOH = 14.0.
polarity	An imbalance in the distribution of electron electrical charge in a molecule due to the different electronegativities of the atoms in the molecule. Polarization causes one region of the molecule to be electrically positive relative to another region that is electrically negative.
polyatomic ion	A molecule formed by covalent bonds between atoms of nonmetals that possesses a positive or negative net electrical charge.
polyprotic acid	An acid with more than one hydrogen atom in the molecule, each capable of dissociating.
precipitate	(1) (noun) A solid that forms when reactants in solution combine to form an insoluble product. (2) (verb) The process of an insoluble solid coming out of solution.
precipitation	The process of forming a precipitate.
precursor	A compound used as the starting point for the manufacture of another compound.
pressure	The average amount of force exerted per unit area on a surface by collisions from the moving particles (atoms or molecules) in a substance.

Glossary

principle quantum number	A number denoted by $n = 1, 2, 3$, etc., which designates one of the shells in an atom that contains subshells and orbitals in which electrons are contained.
probability distribution	A mathematical description resulting from a solution to the Schrödinger equation of where electrons are likely to be found in the orbitals of atoms.
proton	One of the three basic subatomic particles in atoms. Protons carry the atom's positive charge. The number of protons in an atom determines which element the atom represents. Protons have mass slightly lower than the mass of neutrons. Protons reside in the atomic nucleus.
protonation	The process of a molecule accepting a proton in aqueous solution (and thus acting like a Brønsted-Lowry base).
pure substance	Any element or compound.

Q

quantum	The smallest possible single unit of energy possessed by any subatomic particle or photon.
quantum model	A model of the atom based on the theory of energy quantization and the solutions to the Schrödinger equation.
quantum number	One of the four values that define the state of an electron in an atom. The four quantum numbers are the principle quantum number, the azimuthal (or angular momentum) quantum number, the magnetic quantum number, and the spin projection quantum number.
quantum state	The state of an electron in an atom defined by the four quantum numbers known as the principle quantum number, the azimuthal (or angular momentum) quantum number, the magnetic quantum number, and the spin projection quantum number.

R

rare-earth element	Any one of the f-block elements, usually shown in two separate rows of elements under the main body of the Periodic Table of the Elements. Also known as the inner transition metals.
real gas	A gas modeled according to the theory in the Van der Waals equation, a theoretical gas model based on the kinetic-molecular theory of gases, but with particle volume and intermolecular forces taken into account.
redox reaction	A chemical reaction in which elements are oxidized and reduced.
reducer	A reducing agent.
reducing agent	An element that loses electrons to cause the reduction of another element.
reductant	A reducing agent.
reduction	A chemical change in which the oxidation state of an element decreases.
representative element	A main group element.
resonance structure	A molecular structure in which one or more covalent bonds formed by electron pairs are shared or spread out over multiple bonding locations in a molecule. In such structures the electrons are said to be *delocalized*.

S

salt	One of the products of an acid-base reaction, formed from anion of the acid and the cation of the base.
salt bridge	A tube filled with salt ions embedded in a gel matrix that allows migration of charge (the ions) between the two half-cells of an electrochemical cell.
saturated solution	A solution containing the maximum amount of solute that will dissolve in the solvent at a particular temperature.

Glossary

Schrödinger equation	A complex equation that provides through its solutions information about the probable locations of electrons in the orbitals of atoms.
scientific fact	A proposition based on a large amount of scientific data that is correct so far as we know.
shield effect	The forming of an electrical screen by core electrons around an atomic nucleus, which has the effect of reducing the nuclear electrical attraction for valence electrons (also called atomic shielding).
shell	A term used to describe the major collections of subshells and orbitals in atoms where electrons are contained. The shells are denoted by the principle quantum number n, which takes on the integer values, 1, 2, 3, etc.
single replacement reaction	A chemical reaction in which an element in a compound is removed and replaced by a different element.
solid	One of the four basic states of matter; the state characterized by the lowest internal energy, in which the particles of a substance are bound together by electrical or bonding forces and do not have sufficient energy to break free. Particles in solids vibrate in place, but do not flow in bulk.
soluble	Able to dissolve in a given solvent to an appreciable degree.
solute	The solid substance dissolved into a liquid solvent to form a solution.
solution	A mixture of two or more pure substances with uniform composition all the way down to the atomic or molecular level; a homogeneous mixture. Solutions may be solid (alloys), liquid, or gaseous. Individual particles in solutions are the size of molecules or atoms (typically < 1 nm in diameter), and thus are orders of magnitude too small to be seen with the eye, even under magnification.
solvation	The process of polar solvent molecules being attracted to and surrounding ions in solution. In aqueous solution, solvation is referred to as hydration.
solvent	The liquid into which a solid dissolves to form a solution.
sparingly soluble	Essentially insoluble except to a minute degree, typically far less than approximately 0.05 mol/L or 0.1 g/100 mL.
specific gravity	The ratio of the density of a substance to the density of water, symbolized as s.g. Substances with a specific gravity of s.g. < 1 will float on water.
spectator ion	An ion that appears on both the reactant side and the product side in an ionic equation, and thus takes no part in the chemical reaction represented by the equation.
spectroscopy	The science of identifying the presence of elements by means of the electromagnetic spectra they emit.
spin projection quantum number	The quantum number used to differentiate between the two possible spin states (spin up and spin down) electrons can possess in a single orbital in an atom. Each orbital can hold up to two electrons which must have opposite spin states to fulfill the Pauli exclusion principle.
standard hydrogen electrode	A specially designed electrode configuration made of platinum wire and hydrogen gas, established as the reference standard for measuring all standard reduction potentials.
standard reduction potential	The voltage a voltaic cell reads when a given reducing agent is oxidized in one half-cell while a standard hydrogen electrode reduces H^+ to H_2 in the other half-cell.
standard solution	A solution of precisely known molar concentration. Standard solutions are used in titration to determine the unknown concentration of an acid or base.
standard temperature and pressure	A set of conditions defined as 0°C and 100 kPa.
state (of matter)	Any of the solid, liquid, gas, or plasma states defined by the internal energy, thermal properties, and (in the case of plasmas) ionization of a substance.

Glossary

Stock system	A system for naming ionic compounds in which the oxidation state of the cation is indicated by upper-case Roman numerals in parentheses after the name of the cation and before the name of the anion.
stoichiometry	The calculations associated with chemical equations, in which given quantities of reactants or products are used to determine the required amounts of other reactants or the amounts produced of other products.
stopcock	A valve in an article of laboratory glassware.
strong electrolyte	A substance that dissociates completely, or nearly so, in aqueous solution.
structural formula	Diagrams similar to Lewis structures used to represent organic compounds.
subatomic particle	A proton, neutron, or electron, or any of the smaller particles of which protons and neutrons are composed.
sublimation	The phase transition in which a solid changes into a gas.
subshell	The term used to denote the various ordered clusters of orbitals associated with a given principle quantum number (shell) and available in an atom for electrons to be contained in.
substitution reaction	A chemical reaction in which one or more atoms in a molecule are replaced by others. Halogenation is a typical example.
supercritical fluid	A substance at conditions above the critical temperature and critical pressure. At these conditions distinct liquid and vapor states do not exist. Instead, the fluid will be more vapor-like or more liquid-like depending on its pressure and temperature.
surface tension	A characteristic of liquids caused by intermolecular forces between particles at the liquid surface. Because of these forces, some liquids—notably water—form a meniscus, allowing objects denser than the liquid to float (such as an insect walking on the surface), and exhibit capillary action.
suspension	A heterogeneous mixture formed when particles of size approximately 1 μm or larger are dispersed in a fluid medium. Particles in suspensions eventually settle out.
synthesis reaction	A chemical reaction in which two or more elements or compounds combine to form a single compound.

T

temporary dipole	A molecule with a temporary dipole character, either due to electrical induction from a nearby permanent dipole, or due to random fluctuations of electrons in the molecular orbitals.
temperature	A variable defined by thermodynamic theory that correlates directly with the internal energy of a substance.
theoretical yield	The calculated amount of a given product produced in a chemical reaction, determined by stoichiometric calculation based on the available quantity of the limiting reactant and 100% reaction efficiency.
theory	A mental model that accounts for the data (facts) in a certain field of research, and attempts to relate them together, interpret them and explain them.
thermal conductivity	The ability of a substance to allow heat to flow by conduction.
thermochemical equation	A chemical equation that shows the physical states of all reactants and products, as well as the enthalpy of reaction.
titration	A procedure used to determine the unknown concentration of an acid or base, in which an acid or base of precisely known concentration, called a standard solution, is gradually added to a base or acid until chemically equivalent amounts of the acid and base are present in the solution, a condition called the equivalence point. The equivalence point is indicated by a pH indicator in the solution that changes color at its endpoint, and the endpoint is used as an estimate for the equivalence point.

Glossary

TNT	2,4,6-trinitrotoluene.
transition interval	The range of pH values in which a pH indicator changes color, depending on whether the pH of the solution is higher or lower than the pH at the indicator's endpoint.
triple bond	A covalent bond in which two atoms share three pairs of electrons.
triple point	The particular combination of unique temperature and pressure that enables a substance to exist in equilibrium simultaneously as solid, liquid, and vapor.
triprotic acid	An acid with three hydrogen atoms in the molecule, each capable of dissociating.

U

unified atomic mass unit	A unit of relative atomic mass, defined as exactly 1/12 the mass of an atom of carbon-12.

V

valence electron	Any electron in the outer, unfilled shell of an atom. The valence electrons are the electrons that participate in ionization and chemical bonding.
valence shell	The outer, unfilled shell of an atom where the valence electrons are contained.
Valence Shell Electron Pair Repulsion Theory	The theory that bonding domains and nonbonding domains around central atoms in molecules repel each other, providing an explanation for three-dimensional molecular structure.
Van der Waals forces	Intermolecular forces, not including hydrogen bonding. Van der Waals forces include the Keesom force, the Debye force, and London dispersion forces.
vapor	Gas.
vaporization	The phase transition in which a liquid changes into a vapor.
vapor pressure	The pressure the vapor of a liquid assumes at a given temperature due to evaporation in a closed container; the pressure at which a liquid boils at a given temperature.
vapor pressure lowering	The reduction in vapor pressure that occurs in a solution relative to the vapor pressure of the pure solvent; a colligative property.
visible spectrum	The range of wavelengths in the electromagnetic spectrum that are visible to the human eye, approximately 700 nanometers to 400 nanometers. Wavelengths in the visible spectrum appear to humans as the colors in the rainbow, including red, orange, yellow, green, blue and violet.
volumetric flask	A flask marked with a specific volume and used to prepare solutions with a given molarity.
voltaic cell	An electrochemical cell that produces electricity, such as the dry cells used to power electronic devices.

W

weak electrolyte	A substance that dissociates only to a limited degree in aqueous solution.
wet cell	An electrochemical cell containing liquid, such as a car battery.
work	A form of energy, specifically, energy transferred from one object, machine, or person to another by means of applying a force to an object and moving the object through a distance. When the force applied and distance moved point in the same direction, the amount of work done, and thus energy transferred, is equal to the product of force and distance.

Answers to Selected Exercises

There are always errors in any technical writing project of this scope, and regrettable as it is, there are almost certainly errors in this list of answers to problems. If you find an answer you believe to be erroneous, first check the errata file on our website (novarescienceandmath.com/catalog/free-resources) to see if it is an error we know about and have posted the correction for. If you don't see it there, we would greatly appreciate it if you would send an email to us about it at info@novarescienceandmath.com.

Chapter 1

5. a. 8 Pa b. 5 cm c. 3 MA
   d. 2 km e. 4 ms f. 6 kN
   g. 8 kg
   h. 7 μL i. 1 mJ
8. a. 4.183 yd b. 2352.6 ft
   c. 12.296 ft d. 7.363 ft^3
   e. 1,610,000 in^3
   f. 19,907,000 gal/hr
   g. 109,000 hr h. 0.80623 mi
   i. 16.17863 qt/s
   j. 41,980 in/hr
9. a. 0.0354 m b. 76,991 μL
   c. 0.03444 L d. 63.3 kg/m^2
   e. 0.00935 mm/ms^2
   f. 0.5422 J/s g. 0.0566 ms
   h. 44.19 cm^3 i. 0.532 μm
   j. 96,963 m^3/s
   k. 2,956,000 μL
   l. 7,873 mL m. 0.00875 m^2
   n. 0.871 m/s^2
   o. 0.01575 g/cm^3 p. 875 m
   q. 16,056,000 kPa
   r. 7.845 mA

10.

	°F	°C	K
a.	431.1	221.7	494.9
b.	−69.0	−56.1	217.1
c.	−431.0	−257.2	16.0
d.	0.00	−17.7$\overline{7}$	255.37$\overline{2}$
e.	−107	−77.0	196.2
f.	6,744	3,729	4,002
g.	−32.0	−35.6	237.6
h.	149.5	65.25	338.40
i.	3,556	1,958	2,231
j.	1,337	724.8	998.0

18. a. 2%, 0.59%, 4.96%
    b. 3% c. 0.008%
    d. 7.7%, 1.0 × 10^1%

19. Standard Notation

a.	–
b.	0.00220 kg
c.	591,000 μL
d.	400,000 mi
e.	–
f.	0.00075 m^3
g.	–
h.	605,000 s
i.	5.57 g/cm^3
j.	–
k.	–
l.	0.0052 m^3
m.	0.876 in^2
n.	9.802 m/s^2
o.	0.000417 in
p.	1,105.6 kg/m^3
q.	–
r.	–
s.	29 m/s
t.	0.0000249 in
u.	33,640,000 mi/hr
v.	6,010 J/mol
w.	466 psi
x.	–
y.	–
z.	0.000385 J/(mg·K)
aa.	0.013 ft^3
ab.	–
ac.	44,400 ft^2/s

Answers to Selected Exercises

19. Scientific Notation
   a. 5.700×10^6 ft
   b. 2.20×10^{-3} kg
   c. 5.91×10^5 μL
   d. 4×10^5 mi
   e. 5.302×10^{-35} ft
   f. 7.5×10^{-4} m^3
   g. 6.7061×10^8 mi/hr
   h. 6.05×10^5 s
   i. –
   j. 1.0×10^1 m^3/min
   k. 6×10^{10} L/hr
   l. 5.2×10^{-3} m^3
   m. 8.76×10^{-1} in^2
   n. –
   o. 4.17×10^{-4} in
   p. 1.1056×10^3 kg/m^3
   q. 1.36×10^{10} mg/m^3
   r. 1.5×10^{13} cm
   s. 2.9×10^1 m/s
   t. 2.49×10^{-5} in
   u. 3.364×10^7 mi/hr
   v. 6.01×10^3 J/mol
   w. 4.66×10^2 psi
   x. 1.16×10^{-8} cm
   y. 3.99×10^{-25} in^3
   z. 3.85×10^{-4} J/(mg·K)
   aa. 1.3×10^{-2} ft^3
   ab. 6.95622×10^{17} mm^2
   ac. 4.44×10^4 ft^2/s

Chapter 2

12. 28.086 u
14. 92p, 92e, 146n; 1p, 1e, 0n
15. 1.9342 u, or 3.2118×10^{-24} g
16. 1.96×10^{-3} g/mL
17. 580 mL
18. 0.00295 m^3, or 2,950 cm^3
19. 7.23 g/cm^3
20. 210 kg
21. 1.6 cm
22. 13,600 kg/m^3
23. 1.1×10^7 kg
24. 5,800 kg
26. a. 6.94×10^{23}
    b. 8.13×10^{23}
    c. 4.9133×10^{24}
27. a. 39.10 g
    b. 3×10^{-20} g
    c. 0.131 g d. 14 g
    e. 336.56 g f. 24.04 g
28. a. 0.34 mol
    b. 0.71216 mol
    c. 28.32 mol
    d. 0.5255 mol
    e. 0.313 mol
    f. 3.40×10^{-6} mol
29. a. 17.0304 g/mol
    b. 44.010 g/mol
    c. 70.9054 g/mol
    d. 159.610 g/mol
    e. 132.089 g/mol
    f. 342.30 g/mol
    g. 46.069 g/mol
    h. 44.096 g/mol
    i. 60.0843 g/mol
30. a. 95.2104 u
    b. 164.088 u
    c. 96.064 u
    d. 159.610 u
    e. 67.806 u
    f. 153.822 u
31. 382 g
32. a. 14.6 mol
    b. 8.81×10^{24} atoms
    c. 9.50×10^{22} atoms
33. a. 2.10×10^2 mol
    b. 2.53×10^{26} atoms
    c. 2.91×10^{22} atoms

Chapter 3

4. $E = 3.66 \times 10^{-19}$ J
5. $\lambda = 874.39$ nm, IR
6. $\lambda = 91.1$ nm, UV
7. 410 nm: 4.8×10^{-19} J
   434 nm: 4.58×10^{-19} J
   486 nm: 4.09×10^{-19} J
   656 nm: 3.03×10^{-19} J
19. a. Cl: $1s^2 2s^2 2p^6 3s^2 3p^5$
    b. O: $1s^2 2s^2 2p^4$
    c. Ru: $1s^2 2s^2 2p^6 3s^2 3p^6 4s^2 3d^{10} 4p^6 5s^1 4d^7$
    d. K: $1s^2 2s^2 2p^6 3s^2 3p^6 4s^1$
    e. V: $1s^2 2s^2 2p^6 3s^2 3p^6 4s^2 3d^3$
    f. Br: $1s^2 2s^2 2p^6 3s^2 3p^6 4s^2 3d^{10} 4p^5$
20. a. Cl: $[Ne]3s^2 3p^5$
    b. N: $[He]2s^2 2p^3$
    c. Al: $[Ne]3s^2 3p^1$
    d. Y: $[Kr]5s^2 4d^1$
    e. Sr: $[Kr]5s^2$
    f. W: $[Xe]6s^2 4f^{14} 5d^4$
    g. Cs: $[Xe]6s^1$
    h. I: $[Kr]5s^2 4d^{10} 5p^5$
    i. Nd: $[Xe]6s^2 4f^4$
23. Yb: $[Xe]6s^2 4f^{14}$
    Es: $[Rn]7s^2 5f^{11}$
    No: $[Rn]7s^2 5f^{14}$
24. CH_3O, $C_2H_6O_2$
25. a. 92.261% C, 7.739% H
    b. CH c. C_6H_6
26. a. 27.367% Na, 1.1998% H, 14.298% C, 57.136% O
    b. 74.1858% Na, 25.8142% O
    c. 69.943% Fe, 30.0567% O
    d. 63.49928% Ag, 8.24539% N, 28.2553% O
    e. 25.339% Ca, 30.375% C, 3.8234% H, 40.4620% O
    f. 60.002% C, 4.4756% H, 35.5226% O
27. 43.854%
28. Na_2SO_4
29. $C_4H_8O_4$
30. a. $C_4H_5N_2O$, $C_8H_{10}N_4O_2$
    b. $C_{13}H_{18}O_2$ c. C_3H_8
    d. $C_{14}H_{18}N_2O_5$ e. CH, C_2H_2
31. 91.2% C, 8.75% H, C_7H_8
32. 98.079 g/mol, 98.079 u
33. 8.913×10^{23}
34. 2.49×10^6 kg
35. 3.441×10^{-10} μg
36. 4.977×10^{24} molecules
37. 1.24×10^{26} molecules
38. 30.62 cm^3
46. a. 0.003688 gal b. 220.6 K
    c. 139.75 in^3 d. 12.509 μm

Answers to Selected Exercises

e. 1300.05 bar
f. 10,000,000,000 cm
g. 1.50×10^2 °F h. 10.6 L
i. 2.317×10^{-5} s
j. 37.0°C
k. 1.600076×10^{-4} cm
l. 0.01002 ng

Chapter 4

8. Xe < Ag < Y < Rb
9. Mg < Na < Ba < Cs
10. Ca > Mg > Mg^{2+} > Be^{2+}
15. Cu^{2+}: $[Ar]3d^9$
As^{5+}: $[Ar]3d^{10}$
Ag^+: $[Kr]4d^{10}$
Au^{3+}: $[Xe]4f^{14}5d^8$
17. Ba < Sr < Mg < P < Ar
19. Rb < S < Br
26. Cs < Ta < Ni < Se < Cl < F
30. E = 13 eV
31. λ = 390 nm (UV)
32. a. P: *p*, 3, 15
b. Tl: *p*, 6, 13 c. Mo: *d*, 5, 6
d. V: *d*, 4, 5
33. 16, 32
34. 1.53×10^{24}
35. 2.0553% H, 32.694% S, 65.251% O; H: 4.30×10^{23} atoms; S: 2.15×10^{23} atoms; O: 8.60×10^{23} atoms
37. 1.20×10^2 g
38. WCl_6

Chapter 5

5. a. $SrBr_2$, b. MgO, c. $CuCl_2$,
d. Li_2Se, e. Ni_3P_2, f. K_3N,
g. $BeCl_2$ h. NaI i. Cr_2O_3
j. Ag_2Se k. MnF_4 l. $SbBr_3$
m. Rb_2S n. Sc_2O_3 o. AlF_3
p. MgF_2 q. $ZnCl_2$ r. Li_2S
s. Co_2O_3 t. VI_5
6. a. strontium bromide
b. magnesium oxide
c. copper(II) chloride; cupric chloride
d. lithium selenide
e. nickel phosphide
f. potassium nitride
g. beryllium chloride
h. sodium iodide
i. chromium(III) oxide; chromous oxide
j. silver selenide
k. manganese(IV) fluoride; manganic fluoride
l. antimony(III) bromide; antimonous bromide
m. rubidium sulfide
n. scandium oxide
o. aluminum fluoride
p. magnesium fluoride
q. zinc chloride
r. lithium sulfide
s. cobalt(III) oxide; cobaltic oxide
t. vanadium(V) iodide; vanadic iodide

13.
a. H—C≡C—Cl
b. $[H-NH_2-H]^+$ (ammonium)
c. CCl_4 structure
d. NO_3^- structure
e. PCl_5 structure
f. NO_2^- structures
g. C≡O
h. $SiCl_4$ structure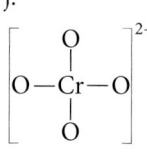
i. O_3 structures
j. $[CrO_4]^{2-}$ structure
k. AlF_3 structure
l. NOCl structure
m. $[PO_4]^{3-}$ structure
n. H_2S structure
o. N_2O_4 structures

Answers to Selected Exercises

p.

```
    H H
    | |
H—C—C—H
    | |
    H H
```

q.

$$\left[\begin{array}{c} H-\ddot{O}-H \\ | \\ H \end{array}\right]^+$$

r.

```
      Cl
      |
Cl—B
      |
      Cl
```

s.

```
    O
    ‖
O=Cl—O—H
    ‖
    O
```

t.

$$\left[\begin{array}{c} :\ddot{Cl}-O \\ | \\ O \end{array}\right]^-$$

u.

```
O=C—O—H
    |
    O—H
```

v.

```
Cl
  \
   C=S
  /
Cl
```

w.

```
     H H
     | |
Cl—C—C—F
     | |
     H H
```

x.

```
    H O—H
    | |
H—C—C—O
    | |
    H H H
```

14. c. carbon tetrachloride
e. phosphorus pentachloride
g. carbon monoxide
h. silicon tetrachloride
k. aluminum trifluoride

n. dihydrogen sulfide (common name hydrogen sulfide)
o. dinitrogen tetroxide
r. boron trichloride

15. a.
```
    H
    |
H—Al
    |
    H
```
(exception)

b. S=C=S

c.
```
      Cl
       |    Cl
Cl—Sb
       |   \Cl
      Cl
```
(exception)

d. $\ddot{N}=N=\ddot{N}$ (free radical)

e.
```
    Cl
    |
Cl—C—H
    |
    H
```

f.
```
O—S—O—H
    ‖
    O
    |
    H
```

g. ·Ö—H (free radical)

h. $\ddot{O}=\ddot{N}-\ddot{O}\cdot$
  $\cdot\ddot{O}-\ddot{N}=\ddot{O}$ (free radical)

17. a. sulfuric acid
b. hydrofluoric acid
c. carbonic acid
d. perchloric acid
e. nitrous acid
f. bromous acid
g. hydrobromic acid
h. phosphoric acid

18. a. $MgCO_3$ magnesium carbonate
b. Na_2CrO_4 sodium chromate
c. $Ca_3(PO_4)_2$ calcium phosphate
d. $(NH_4)_3PO_4$ ammonium phosphate
e. $CuSO_4$ copper(II) sulfate
f. $NaHCO_3$ sodium bicarbonate
g. $Fe(ClO_3)_3$ iron(III) chlorate
h. $Al(OH)_3$ aluminum hydroxide

i. $Cr(SO_3)_3$ chromium(VI) sulfite
j. $Ca(CH_3COO)_2$ calcium acetate

20. a. O—F: nonpolar covalent, least polar; C—F: polar covalent; Be—F: ionic, most polar

b. F—F: least polar, nonpolar covalent; S—O: polar covalent; B—F: ionic, most polar

c. O—Cl: nonpolar covalent, least polar; C—P: nonpolar covalent; S—Cl: polar covalent, most polar

21. red: a, yellow: b, orange: c
22. 1.8051×10^{24} atoms of O
23. S: 21.955%, F: 78.0457%
24. Cl: $1s^2 2s^2 2p^6 3s^2 3p^5$

Sc: $1s^2 2s^2 2p^6 3s^2 3p^6 4s^2 3d^1$

Cd: $1s^2 2s^2 2p^6 3s^2 3p^6 4s^2 3d^{10}$ $4p^6 5s^2 4d^{10}$

Gd: $1s^2 2s^2 2p^6 3s^2 3p^6 4s^2 3d^{10}$ $4p^6 5s^2 4d^{10} 5p^6 6s^2 4f^7 5d^1$

25. 2.94×10^{22} Cl atoms
27. $Mg^{2+} < Ca^{2+} < K^+ < Rb^+ < Rb$
28. $CuSO_4 \cdot 5H_2O$; copper sulfate pentahydrate
29. 533 nm

Chapter 6

3.

a.
```
H—P̈—H
    |
    H
```

tetrahedral/pyramidal

b.
```
      Cl
      |
Cl—C—Cl
      |
      Cl
```

tetrahedral/tetrahedral

c.
```
    S̈
   ⫽ ⟍
  O     O
```

trigonal/bent

381

Answers to Selected Exercises

d.

Cl—S—Cl (with lone pairs on S)

tetrahedral/bent

e.

H—N(..)—H
    |
    Cl

tetrahedral/pyramidal

f.

$\left[\begin{array}{c} H \\ | \\ H-N-H \\ | \\ H \end{array} \right]^+$

tetrahedral/tetrahedral

g.

$[O-Cl(..)-O]^-$

tetrahedral/bent

h.

O=O(..)—O

trigonal/bent

i. H—C≡C—H
linear/linear

j. S=C=S
linear/linear

k.

O=S(=O)—O

trigonal/trigonal

l.

Cl—O(..)—Cl

tetrahedral/bent

m. H—Cl:
linear/linear

n. $\left[\begin{array}{c} H-O(..)-H \\ | \\ H \end{array} \right]^+$

tetrahedral/pyramidal

o. H—C≡N
linear/linear

7. a. 109.5° b. 104.5°
c. 109.5° d. 109.5° e. 120°
f. 104.5° g. 107°
9. a. tetrahedral/bent
b. tetrahedral
c. tetrahedral/pyramidal
d. trigonal bipyramidal
e. octahedral/square planar
13. a. tetrahedral
b. tetrahedral
c. tetrahedral/pyramidal

d. trigonal planar
e. tetrahedral/bent
27. 89.094 g/mol, 40.444% C, 15.721% N, 35.916% O, 7.9189% H
28. 2.06 eV
29. a. polar covalent b. ionic
c. nonpolar covalent
d. polar covalent e. nonpolar covalent f. polar covalent
30. b, c, d, f
31. $F < F^- < Cl^- < I^- < Te^{2-}$
28. Hg: $[Xe]6s^2 4f^{14} 5d^{10}$
Br: $[Ar]4s^2 3d^{10} 4p^5$
Ba: $[Xe]6s^2$
Mn: $[Ar]4s^2 3d^5$
S: $[Ne]3s^2 3p^4$
35. Sr
37. a. 3.64 mol
b. 0.00718 mol c. 0.002 mol
d. 138 mol e. 10.5 mol
f. 47.635 mol
38. a. 1.4×10^{23} Hg atoms
b. 2.26×10^{25} Fe atoms
c. 6.0221×10^{23} Ca atoms
d. 1.5×10^{23} Sr atoms
e. 2.409×10^{24} Na atoms
f. 1.92×10^{25} Ca atoms
39. 55.845 u
43. 40.0% C, 6.72% H, 53.28% O; CH_2O; $C_6H_{12}O_6$

Chapter 7

1.
$2C_4H_{10} + 13O_2 \rightarrow$
$8CO_2 + 10H_2O$

2.
$FeS + 2HCl \rightarrow$
$H_2S + FeCl_2$

3.
$2C_2H_2 + 5O_2 \rightarrow$
$4CO_2 + 2H_2O$

4.
$CO + 2H_2 \rightarrow CH_3OH$

5.
$K_2CO_3 + H_2SO_4 \rightarrow$
$K_2SO_4 + CO_2 + H_2O$

6.
$WO_3 + 3H_2 \rightarrow W + 3H_2O$

7.
$2NaN_3 \rightarrow 2Na + 3N_2$

8. a. P: +3, Br: −1
b. O: −2, As: +5
c. O: −2, Cl: +7
d. F: 0 e. F: −1, U: +6
f. O: −2, C: +4
g. O: −2, H: +1, P: +5
h. O: −2, N: +5, Zn: +2
i. H: +1, C: −2.5
j: O: −2, Cr: +6
k. H: −1, K: +1 l. Fe: 0
10. butane: none
FeS: double replacement
C_2H_2: none
CO: synthesis
K_2CO_3: none
WO_3: single replacement
NaN_3: decomposition
11. a. Fe reduced, Al oxidized, single replacement
b. O reduced, S oxidized
c. N reduced, H oxidized, synthesis
d. Cl reduced, I oxidized, single replacement
13.
a.
$Mg(OH)_2 \rightarrow MgO + H_2O$
decomposition
b.
$2NaCl + H_2SO_4 \rightarrow$
$Na_2SO_4 + 2HCl$
double replacement
c.
$CaO + H_2O \rightarrow Ca(OH)_2$
synthesis

Answers to Selected Exercises

d.
$NH_4HCO_3 + NaCl \rightarrow NaHCO_3 + NH_4Cl$
double replacement

14.
a. $2Ca(s) + O_2(g) \rightarrow 2CaO(s)$
b. no reaction
c. no reaction
d. $2Al(s) + 3H_2SO_4(aq) \rightarrow Al_2(SO_4)_3(aq) + 3H_2(g)$
e. $Ni(s) + CuCl_2(aq) \rightarrow NiCl_2(aq) + Cu(s)$
f. $Ba(s) + 2H_2O(l) \rightarrow Ba(OH)_2(aq) + H_2(g)$
g. no reaction
h. $2Al(s) + 3Pb(NO_3)_2(aq) \rightarrow 3Pb(s) + 2Al(NO_3)_3(aq)$
i. $Li(s) + KI(aq) \rightarrow LiI(aq) + K(s)$
j. no reaction

15.
a. $2LiClO_3(s) \rightarrow 2LiCl(s) + 3O_2(g)$
b. $Cu(OH)_2(s) \rightarrow CuO(s) + H_2O(l)$
c. $H_2CO_3(aq) \rightarrow CO_2(g) + H_2O(l)$

16. 660 g
17. 787.5 mol O_2, 1575 mol H_2O
18. 2,379 g $Al(OH)_3$
19. a. 0.029 mol HCl
    b. 0.52 g H_2O
20. 1.75×10^4 kg Fe
21. a. NaOH
    b. 52.47 g Na_2SO_4
    c. 13.31 g H_2O
22. a. 4375 mol O_2
    b. 1.4 kg H_2O
    c. 105 kg O_2
23. a. O_2
    b. 1,310 g NO
    c. 96.9%
24. a. H_2SO_4
    b. 8.5 g $Al(OH)_3$
    c. 37.0 g $Al_2(SO_4)_3$, 11.7 g H_2O
25. a. N_2O_4
    b. 2349 kg H_2O
    c. 1.178×10^{29} N atoms
26. a. 54.1 mol HF
    b. 102 g H_2SiF_6
    c. 1300 mol HF
    d. SiO_2
    e. 2.89 kg H_2SiF_6
    f. 91.9%
27. a. 90.5 g C_6H_5Br b. 69.9%
28. a. 5.70×10^{-6} mol NaI
    b. 18.6 g NaI
    c. NaI, 1232 g I_2
    d. 5.846×10^{24} I atoms
29. a. 67.2% C, 6.94% H, 12.1% N, 13.8% O
    b. $C_{13}H_{16}N_2O_2$
    c. 1.0×10^{20} C atoms
30. 545.9 nm
31. Y: $[Kr]5s^24d^1$
    Sn: $[Kr]5s^24d^{10}5p^2$
    Ti: $[Ar]4s^23d^2$
    I: $[Kr]5s^24d^{10}5p^5$
33. 28.085 u
34. Fr < Ba < Fe < Se < Cl < F

35.
a. $Cl_2C=S$ (Cl, Cl on C double-bonded to S)

b. $[O-P(O)(O)-O]^{3-}$ (phosphate)

c. $[O=N=O]^+$

d. $H-N=N-H$ with lone pairs on each N

e. $O=O$ with lone pairs (ozone-like, bent with third O)

f. SF_6 octahedral (F–S–F with six F around S)

g. $:C{\equiv}O:$

h. $[H-\ddot{O}-H]^+$ with additional H (hydronium H_3O^+)

36. a. trigonal planar
b. tetrahedral
c. linear
d. trigonal planar/bent at each N atom
e. tetrahedral/bent
f. trigonal bipyramidal
g. linear
h. tetrahedral/pyramidal

Chapter 8

3. 0–900 m/s, 350 m/s
5. 10.4 m
6. 12,800,000 Pa, 12,800 kPa, 95,700 Torr, 126 atm, 128 bar
7. 53 atm
16. ethanol
19. a. 7.8 kJ b. 162 kg
    c. 4.00×10^3 kJ d. 176 mL

Answers to Selected Exercises

e. 1920 kJ
20. a. −75°C b. −55°C, 218 K, 4 atm, 3000 Torr
c. −55°C, −35°C d. 40 atm
e. 15 atm, 1000 atm
f. 35 atm
22. 1500 Torr
25. 93°C
26.
Br: $1s^2 2s^2 2p^6 3s^2 3p^6 4s^2 3d^{10} 4p^5$
Y: $1s^2 2s^2 2p^6 3s^2 3p^6 4s^2 3d^{10} 4p^6 5s^2 4d^1$
27. a. 62.5 mol
b. 9170 g
28.
$2NH_4NO_3(s) \rightarrow$
$2N_2(g) + O_2(g) + 4H_2O(g)$
29. a. $AgNO_3$ b. 1.67 g
c. 1.92 g
30. $C_{13}H_{18}O_2$ (empirical and molecular)
31. S: +6, C: +4
32. C—N: nonpolar covalent
I—I: nonpolar covalent
Al—Cl: polar covalent
34.
a. $[O-\overset{O}{\underset{O}{S}}-O]^{2-}$ tetrahedral
b. $H-\overset{..}{\underset{..}{S}}-H$ bent
c. $[O-\overset{O}{\underset{O}{Cl}}-O]^{-}$ tetrahedral

Chapter 9

1. a. 2.52 L b. 463 K
2. 0.77 bar
3. 2010 cm³
9. 2271 L
10. 4950°F
11. 15.5 L
12. 0.698 mol
13. a. 68,300 mol
b. 1.10×10^3 kg
c. 0.707 kg/m³ d. 4.11×10^{28} carbon atoms
14. a. 834 kg b. 4.20×10^2 kg
15. a. 1.876 mol b. chlorine
16. a. 11.2 mol b. 8.20×10^2 L
17. 28 psig
18. a. 58.69 mol b. 1333 L
19. a. 20.8 g/mol
b. 6.87 g/mol
23. $P_{nitrogen} = 1.32$ atm
$P_{argon} = 0.193$ atm
$P_{methane} = 0.097$ atm
24. a. 602 Torr
b. 5.40 mol
25. $X_{propane} = 0.479$,
$X_{methane} = 0.521$,
$P_{propane} = 2.55$ bar,
$P_{methane} = 2.77$ bar,
$P_T = 5.33$ bar
26. a. $n_{oxygen} = 0.128$ mol
$n_{hydrogen} = 0.0432$ mol
b. $P_{oxygen} = 1.03$ atm
$P_{hydrogen} = 0.350$ atm
c. $P_T = 1.38$ atm
27. $n_{nitrogen} = 0.011$ mol
$P_{nitrogen} = 0.97$ atm
$X_{nitrogen} = 0.97$
$n_{water} = 0.000307$ mol
$P_{water} = 0.026$ atm
$X_{water} = 0.026$
28. 0.467 g
29. 6.01 g
30. 2.5 L O_2, 5.0 L CO_2
31. 150.0 m³ O_2, 120.0 m³ H_2O, 90.00 m³ CO_2
32. 4.5 g $Fe(OH)_3$
3.3 g Fe_2O_3
33. 71.4 kg
34. 71.4 kg
35. 0.016 g
36. 0.351 L
37. a. 0.682 mol KI, 0.341 mol Cl_2, 0.682 mol KCl, 0.341 mol I_2
b. 113 g KI, 24.2 g Cl_2, 50.8 g KCl, 86.6 g I_2
38. a. Al
b. 34.80 L
39. a. CO b. 55.0 m³
c. 605.0 m³
40. a. 0.17 L b. 5.0 g
41. 62.1 nm (UV)
42. 15.867% C, 2.2192% H, 18.5040% N, 63.4094% O
43. a. 84.0%
b. 4.85×10^{26} molecules
c. 121 kg
44. a. plutonium b. bromine
c. plutonium d. strontium
e. bromine f. bromine
45. 9180 kJ
46. gas, solid

Chapter 10

9. insoluble: a, e, f, soluble: b, c, d, g, h, i
15. a. 68%
b. 0.0015, 0.9985
d. 23% decrease (from 0.7 to 0.54)
16. b. 31.7
d.

	$Al_2(SO_4)_3$	H_2O
% of total mass	0.282	0.718
% of total moles	0.020	0.980

17. a. 7.20×10^2 g b. 901 g
18. 3.94 M
19. 636 g
20. 110 mL
21. 2.2 M
22. a.
$Na_2SO_4(aq) + Ba(NO_3)_2(aq)$
$\rightarrow BaSO_4(s) + 2NaNO_3(aq)$
b. 67 g
23. 164 mL
24. 45,000 L
25. a. 365 g
b. 367 g
26. 1.32 m
27. 720 mL

28. a. $CaCO_3$
b.
$Na_2CO_3(aq) + Ca(NO_3)_2(aq)$
$\rightarrow 2NaNO_3(aq) + CaCO_3(s)$
c.
$2Na^+(aq) + CO_3^{2-}(aq)$
$+ Ca^{2+}(aq) + 2NO_3^-(aq)$
$\rightarrow 2Na^+(aq) + 2NO_3^-(aq)$
$+ CaCO_3(s)$

$CO_3^{2-}(aq) + Ca^{2+}(aq) \rightarrow$
$CaCO_3(s)$

d. Na^+, NO_3^-
29. precipitate: AgCl
spectator ions: Li^+, NO_3^-
30.
a.
$MgCl_2(aq) + Pb(NO_3)_2(aq)$
$\rightarrow Mg(NO_3)_2(aq) + PbCl_2(s)$

$Mg^{2+}(aq) + 2Cl^-(aq)$
$+ Pb^{2+}(aq) + 2NO_3^-(aq)$
$\rightarrow Mg^{2+}(aq) + 2NO_3^-(aq)$
$+ PbCl_2(s)$

$Pb^{2+}(aq) + 2Cl^-(aq)$
$\rightarrow PbCl_2(s)$

b. spectator ions: Mg^{2+}, NO_3^-, precipitate: $PbCl_2$
c. 5.6 g
33. 100.520°C
34. 3.46 m, 1.98 m, 6.23 m
35. 9.7 m
36. 73.3 g
37. 670 g
38. −23.7°C
39. −123.0°C

40. ΔT_b = +0.26°C, +0.39°C, +0.26°C (this last value is ideal, but is likely to be more like −0.17°C)
41. T_f = −0.15°C, −0.07°C, −0.04°C, −0.11°C
42. −7.44°C
43. 11.7 g/mol
44. 17.5 L $CO_2(g)$, 35.0 L $H_2O(g)$
46. 8.50 atm
47. 55 kg
48. 22.52 kg
49. 60.871% C, 4.3783% H, 34.7503% O
50. +5

Chapter 11

4. a. hydrofluoric acid (monoprotic) b. hypochlorous acid (monoprotic) c. nitrous acid (monoprotic) d. sulfurous acid (diprotic) e. hydrobromic acid (monoprotic) f. phosphoric acid (triprotic) g. hydrosulfuric acid (diprotic) h. chromic acid (diprotic) i. acetic acid (monoprotic)
6. a. $HClO_2$ b. $HBrO_4$ c. HF d. HI
8. a.

H—C(=O)—O—H

b.

(benzene ring)—C(=O)—O—H

13.

	acid	base	conj. acid	conj. base
a	HNO_2	CN^-	HCN	NO_2^-
b	H_2O	HCO_3^-	H_2CO_3	OH^-
c	HSO_4^-	$C_2O_4^{2-}$	$HC_2O_4^-$	SO_4^{2-}
d	NH_4^+	HPO_4^{2-}	$H_2PO_4^-$	NH_3

19.
a.
$HCl + KOH \rightarrow KCl + H_2O$

$H^+ + Cl^- + K^+ + OH^-$
$\rightarrow K^+ + Cl^- + H_2O$

$H^+ + OH^- \rightarrow H_2O$

b.
$HNO_3 + NaOH \rightarrow NaNO_3 + H_2O$

$H^+ + NO_3^- + Na^+ + OH^-$
$\rightarrow Na^+ + NO_3^- + H_2O$

$H^+ + OH^- \rightarrow H_2O$

c.
$H_2SO_4 + Ca(OH)_2$
$\rightarrow CaSO_4 + 2H_2O$

$2H^+ + SO_4^{2-} + Ca^{2+} + 2OH^-$
$\rightarrow Ca^{2+} + SO_4^{2-} + 2H_2O$

$2H^+ + 2OH^- \rightarrow 2H_2O$

d.
$H_2SO_3 + 2KOH$
$\rightarrow K_2SO_3 + 2H_2O$

$2H^+ + SO_3^{2-} + 2K^+ + 2OH^-$
$\rightarrow 2K^+ + SO_3^{2-} + 2H_2O$

$2H^+ + 2OH^- \rightarrow 2H_2O$

20.
a.
$2Al + 3H_2SO_4$
$\rightarrow Al_2(SO_4)_3 + 3H_2$

$2Al + 6H^+ + 3SO_4^{2-}$
$\rightarrow 2Al^{3+} + 3SO_4^{2-} + 3H_2$

$2Al + 6H^+ \rightarrow 2Al^{3+} + 3H_2$

b.
$Zn + 2HCl \rightarrow ZnCl_2 + H_2$

$Zn + 2H^+ + 2Cl^-$
$\rightarrow Zn^{2+} + 2Cl^- + H_2$

Answers to Selected Exercises

$Zn + 2H^+ \rightarrow Zn^{2+} + H_2$

21.
a.
$2HNO_3 + CaCO_3$
$\rightarrow Ca(NO_3)_2 + H_2CO_3$

b.
$2HCl + BaCO_3$
$\rightarrow BaCl_2 + H_2CO_3$

c.
$H_2SO_4 + Na_2CO_3$
$\rightarrow Na_2SO_4 + H_2CO_3$

d.
$2HNO_3 + MgCO_3$
$\rightarrow Mg(NO_3)_2 + H_2CO_3$

Note: These answers all show the formation of carbonic acid, H_2CO_3. This acid is unstable and immediately breaks down to CO_2 and water. Thus, each equation could show the products as: $... + CO_2 + H_2O$.

22.
a.
$H_3PO_4 + F^- \rightarrow H_2PO_4^- + HF$
right
b.
$HSO_4^- + CO_3^{2-} \rightarrow SO_4^{2-} + HCO_3^-$
right
c.
$ClO_2^- + H_3O^+ \rightarrow HClO_2 + H_2O$
right
d.
$HClO_3 + HCO_3^-$
$\rightarrow ClO_3^- + H_2CO_3$
right
e.
$H_3O^+ + HS^- \rightarrow H_2O + H_2S$
right
f.
$HPO_4^{2-} + NH_3 \rightarrow PO_4^{3-} + NH_4^+$
left

23. a. 81.4 g, 42.0 L b. 616 mL
24. a. 160 g b. 22.7 L
25. a. 110 g b. 1.1 L
28. basic: a, c, f, g
acidic: b, e
neutral: d, h

31.
a.
$[H_3O^+] = 1.56 \times 10^{-3}\ M$,
$[OH^-] = 6.4 \times 10^{-12}\ M$
b.
$[H_3O^+] = 4.5 \times 10^{-13}\ M$,
$[OH^-] = 0.0220\ M$
c.
$[H_3O^+] = 4.0 \times 10^{-15}\ M$,
$[OH^-] = 2.50\ M$
d.
$[H_3O^+] = 8.911 \times 10^{-5}\ M$,
$[OH^-] = 1.1 \times 10^{-10}\ M$

32. a. pOH = 1.602, pH = 12.4
b. pH = −0.78, pOH = 14.8
c. pH = −1.08, pOH = 15.1
d. pH = 12.6, pOH = 1.391

33.
a.
$[H_3O^+] = 0.10\ M$,
$[OH^-] = 1.0 \times 10^{-13}\ M$
b.
$[H_3O^+] = 3 \times 10^{-4}\ M$,
$[OH^-] = 3 \times 10^{-11}\ M$
c.
$[H_3O^+] = 1 \times 10^{-4}\ M$,
$[OH^-] = 1 \times 10^{-10}\ M$
d.
$[H_3O^+] = 1.3 \times 10^{-12}\ M$,
$[OH^-] = 7.7 \times 10^{-3}\ M$
e.
$[H_3O^+] = 1 \times 10^{-12}\ M$,
$[OH^-] = 8 \times 10^{-3}\ M$
f.
$[H_3O^+] = 1 \times 10^{-7}\ M$,
$[OH^-] = 1 \times 10^{-7}\ M$
g.
$[H_3O^+] = 1.0 \times 10^{-10}\ M$,
$[OH^-] = 1.0 \times 10^{-4}\ M$
h.
$[H_3O^+] = 1 \times 10^{-14}\ M$,
$[OH^-] = 1\ M$

34.

	pH	pOH	$[H_3O^+]$	$[OH^-]$
a	5.6	8.4	$3 \times 10^{-6}\ M$	$4 \times 10^{-9}\ M$
b	4.7	9.26	$1.8 \times 10^{-5}\ M$	$5.5 \times 10^{-10}\ M$
c	9.00	5.00	$1.0 \times 10^{-9}\ M$	$1.0 \times 10^{-5}\ M$
d	2.5	11.5	$3 \times 10^{-3}\ M$	$3 \times 10^{-12}\ M$
e	2.35	11.66	$4.5 \times 10^{-3}\ M$	$2.2 \times 10^{-12}\ M$
f	13.0	1.00	$1.0 \times 10^{-13}\ M$	$0.10\ M$
g	6.8	7.22	$1.7 \times 10^{-7}\ M$	$6.0 \times 10^{-8}\ M$
h	13.50	0.5	$3.2 \times 10^{-14}\ M$	$0.31\ M$
i	2.0	12.0	$1 \times 10^{-2}\ M$	$1 \times 10^{-12}\ M$
j	12.51	1.49	$3.1 \times 10^{-13}\ M$	$3.2 \times 10^{-2}\ M$

35. a. $2.14 \times 10^{-3}\ M$
b. $1.325\ M$
c. $0.01855\ M$
d. $2.780\ M$
36. $5.82 \times 10^{-3}\ M$
37. $3.77 \times 10^{-2}\ M$
38. 7.3%
39. $0.146\ M$ max
40. $0.4552\ M$
41. 16.8 kJ
42. a. HNO_3 b. 0.7500 mol
c. 0.7500 mol d. 0.1388 mol

43.
a.

$$\begin{array}{c} Cl \\ | \\ Cl-C-Cl \\ | \\ Cl \end{array}$$

b.

$$\left[\begin{array}{c} O \\ \| \\ O-C-O \end{array}\right]^{2-}$$

c.

$$\left[O-\ddot{N}-O \right]^-$$

d.

$H-C\equiv C-H$

e.

$$O=\ddot{O}=O$$

f.

$$H-\ddot{\underset{..}{S}}-H$$

386

Answers to Selected Exercises

g.
O=C=O

h.
$\left[\text{H}-\overset{\cdot\cdot}{\underset{\underset{\text{H}}{|}}{\text{O}}}-\text{H}\right]^+$

44.
a.
$Na_2CO_3 + Ba(NO_3)_2$
$\rightarrow 2NaNO_3 + BaCO_3$

$2Na^+ + CO_3^{2-} + Ba^{2+} + 2NO_3^-$
$\rightarrow 2Na^+ + 2NO_3^- + BaCO_3$

$CO_3^{2-} + Ba^{2+} \rightarrow BaCO_3$

b. spectator ions: Na^+, NO_3^-
precipitate: $BaCO_3$
c. 0.25 g
45. 0.276 g
46. a. 1.353 mol b. 43.29 g
47. a. $CuSO_4$ b. 34.1 g
c. 12.5 g
48. 105 g/mol
49. 5600 eV

Chapter 12

2.
a.
+3 −1   +1 −2   +1+3 −2   +1 −1
$PBr_3 + 3H_2O \rightarrow H_3PO_3 + 3HBr$

Not a redox reaction.
b.
+1+5 −2   +1 −1   0
$2KClO_3 \rightarrow 2KCl + 3O_2$

Cl is reduced; it is the oxidizing agent. O is oxidized; it is the reducing agent.
c.
+1+6 −2   +1 −1
$2H_2SO_4 + 2NaBr \rightarrow$
0   +4 −2   +1+6 −2   +1 −2
$Br_2 + SO_2 + Na_2SO_4 + 2H_2O$

S is reduced; it is the oxidizing agent. Br is oxidized; it is the reducing agent.

d.
+1 −1   +1+5 −2
$NaCl + AgNO_3 \rightarrow$
+1 −1   +1+5 −2
$AgCl + NaNO_3$

Not a redox reaction.
e.
+1 −1   +1 +1 −2   +1 −1   +1+5 −2
$NaI + 3HClO \rightarrow 3HCl + NaIO_3$

Cl is reduced; it is the oxidizing agent. I is oxidized; it is the reducing agent.
f.
+4 −2   +1+5 −2   +1 −2
$3SO_2 + 2HNO_3 + 2H_2O$
+1+6 −2   +2 −2
$\rightarrow 3H_2SO_4 + 2NO$

N is reduced; it is the oxidizing agent. S is oxidized; it is the reducing agent.

3.

	strongest	weakest
a.	Ca	NO
b.	Li	Br⁻
c.	Mg	Cr³⁺
d.	Sn	F⁻

4. a. yes b. yes c. no d. no
6.
a.
$SO_3^{2-} + 2OH^- \rightarrow$
$SO_4^{2-} + H_2O + 2e^-$

b.
$N_2 + 6H_2O + 6e^- \rightarrow$
$2NH_3 + 6OH^-$

c.
$N_2 + 8H^+ + 6e^- \rightarrow 2NH_4^+$

d.
$Sn^{2+} \rightarrow Sn^{4+} + 2e^-$

e.
$ClO_3^- + 6H^+ + 6e^- \rightarrow$
$Cl^- + 3H_2O$

f.
$4OH^- \rightarrow O_2 + 2H_2O + 4e^-$

g.
$H_2SO_3 + H_2O \rightarrow$
$SO_4^{2-} + 4H^+ + 2e^-$

h.
$O_2 + 2H_2O + 4e^- \rightarrow 4OH^-$

7.
a.
$2FeCl_3 + H_2S \rightarrow$
$2FeCl_2 + 2HCl + S$

oxidizing agent: Fe
reducing agent: S
b.
$I_2 + 5HClO + H_2O \rightarrow$
$2HIO_3 + 5HCl$

oxidizing agent: Cl
reducing agent: I
c.
$4KMnO_4 + 5CH_3OH$
$+ H_2O \rightarrow 4Mn(OH)_2$
$+ 5HCOOH + 4KOH$

oxidizing agent: Mn
reducing agent: C
d.
$H_2O_2 + 2ClO_2 \rightarrow$
$2HClO_2 + O_2$

oxidizing agent: Cl
reducing agent: O
e.
$2Al + LiCl + LiNO_2$
$+ 2H_2O \rightarrow 2LiAlO_2 + NH_4Cl$

oxidizing agent: N
reducing agent: Al
f.
$14HCl + 2HMnO_4 \rightarrow$
$5Cl_2 + 2MnCl_2 + 8H_2O$

oxidizing agent: Mn
reducing agent: Cl

Answers to Selected Exercises

g.
$$2S + 2HNO_3 + H_2O \rightarrow$$
$$2H_2SO_3 + N_2O$$
oxidizing agent: N
reducing agent: S

h.
$$2HMnO_4 + HBr \rightarrow$$
$$HBrO_3 + 2MnO_2 + H_2O$$
oxidizing agent: Mn
reducing agent: Br

9. b. oxidizing agent: Fe, reducing agent: Al
c. 22 g
d. 45 g

19.

	cathode	anode	$E°_{cell}$
a.	Ag	Sn	0.9371 V
b.	Fe	Co	1.05 V
c.	Br	Na	3.78 V
d.	Br	I	0.530 V

24. a. 80.5% b. 97 g c. NO_2
25. 65.2 atm
26. 0.0115 mol
27. 1.96×10^{24} carbon atoms
28. 1.08 L

Appendix A

Reference Data

On this page and the next, items marked in yellow are approximate. All others are exact.

Table A.1. SI prefixes.

	Prefix	deca–	hecto–	kilo–	mega–	giga–	tera–	peta–	exa–	zetta–	yotta–
Multiples	Symbol	da	h	k	M	G	T	P	E	Z	Y
	Factor	10	10^2	10^3	10^6	10^9	10^{12}	10^{15}	10^{18}	10^{21}	10^{24}
	Prefix	deci–	centi–	milli–	micro–	nano–	pico–	femto–	atto–	zetto–	yocto–
Fractions	Symbol	d	c	m	μ	n	p	f	a	z	y
	Factor	1/10	$1/10^2$	$1/10^3$	$1/10^6$	$1/10^9$	$1/10^{12}$	$1/10^{15}$	$1/10^{18}$	$1/10^{21}$	$1/10^{24}$

Table A.2. Physical constants.

Quantity	Symbol	Value
speed of light in a vacuum	c	299,792,458 m·s^{-1}
acceleration due to gravity, sea level	g	9.80 m·s^{-2}
atmospheric pressure	P_{atm}	101,325 Pa
		1 atm
		14.7 psi
proton mass	m_p	1.672622 × 10^{-27} kg
		1.007276 u
neutron mass	m_n	1.674927 × 10^{-27} kg
		1.008665 u
electron mass	m_e	9.109382 × 10^{-31} kg
		0.0005486 u
Avogadro constant	N_A	6.02214076 × 10^{23} mol^{-1}
Planck constant	h	6.62607015 × 10^{-34} J·s
Boltzmann constant	k_B	1.380649 × 10^{-23} J·K^{-1}
Rydberg constant	R	1.097373 × 10^7 m^{-1}
unified atomic mass unit	u	1.660539 × 10^{-27} kg
electron charge	e	1.602176634 × 10^{-19} C
ideal gas constant	R	8.31446261815324 J·mol^{-1}·K^{-1}
		0.082057 L·atm·mol^{-1}·K^{-1}
		8.314462 L·kPa·mol^{-1}·K^{-1}
		62.36337 L·Torr·mol^{-1}·K^{-1}

Table A.3. Unit conversion factors.

Pressure

Unit Name	Symbol	Definition
pascal	Pa	1 Pa = 1 N/m^2
kilopascal	kPa	1 kPa = 1000 Pa
atmosphere	atm	1 atm = 101,325 Pa
bar	bar	1 bar = 100,000 Pa
pounds per square inch	psi	1 psi = 6894.757 Pa
torr	Torr	1 Torr = 1/760 atm

Length

1 mi = 5280 ft	1 ft = 12 in	1 in = 2.54 cm	1 cm = 10 mm
1 yd = 3 ft	1 mi = 1609 m	1 ft = 0.3048 m	1 Å = 0.1 nm

Temperature

$$T_C = \frac{5}{9}(T_F - 32°) \qquad T_F = \frac{9}{5}T_C + 32° \qquad T_K = T_C + 273.15 \qquad T_C = T_K - 273.15$$

Volume

1 gal = 3.78541 L	1 L = 1000 cm^3	1 mL = 1 cm^3	1 m^3 = 1000 L
1 tsp = 4.92892 mL	1 Tbsp = 14.7868 mL	1 cp = 236.588 mL	1 pt = 473.176 mL
1 qt = 0.946353 L	1 gal = 4 qt	1 ft^3 = 7.480523 gal	

Force

1 lb = 4.448 N

Energy and Mass

1 eV = 1.60218 × 10^{-19} J   1 u = 1.660539 × 10^{-27} kg

Time

1 min = 60 s	1 hr = 60 min	1 hr = 3600 s	1 dy = 24 hr
1 yr = 365 dy			

Appendix A

Table A.4. Vapor pressure of water.

T (°C)	P (kPa)	P (Torr)	T (°C)	P (kPa)	P (Torr)
0.01	0.6117	4.588	52	13.631	100.2
2	0.7060	5.295	54	15.022	112.7
4	0.8136	6.103	56	16.533	124.0
6	0.9354	7.016	58	18.171	136.3
8	1.0730	8.048	60	19.946	149.6
10	1.2282	9.212	62	21.867	164.0
12	1.4028	10.52	64	23.943	179.6
14	1.5990	11.99	66	26.183	196.4
16	1.8188	13.64	68	28.599	214.5
18	2.0647	15.49	70	31.201	234.0
20	2.3393	17.55	72	34.000	255.0
22	2.6453	19.84	74	37.009	277.6
24	2.9858	22.40	76	40.239	301.8
25	3.1699	23.78	78	43.703	327.8
26	3.3639	25.23	80	47.414	355.6
28	3.7831	28.38	82	51.387	385.4
30	4.2470	31.86	84	55.635	417.3
32	4.7596	35.70	86	60.173	451.3
34	5.3251	39.94	88	65.017	487.7
36	5.9479	44.61	90	70.182	526.4
38	6.6328	49.75	92	75.684	567.7
40	7.3849	55.39	94	81.541	611.6
42	8.2096	61.58	96	87.771	658.3
44	9.1124	68.35	98	94.390	708.0
46	10.099	75.75	100	101.42	760.7
48	11.177	83.83	102	108.87	816.6
50	12.352	92.45	104	116.78	875.9

Table A.5. Constants for water.

Constant	Symbol	Value
molar heat capacity (ice, −10.0°C)	C	$0.0364 \ \frac{kJ}{mol \cdot K}$
molar heat capacity (water)	C	$0.0752 \ \frac{kJ}{mol \cdot K}$
molar heat capacity (steam, 110°C)	C	$0.0369 \ \frac{kJ}{mol \cdot K}$
molar heat of fusion	H_f	$6.01 \ \frac{kJ}{mol}$
molar heat of vaporization	H_v	$40.7 \ \frac{kJ}{mol}$
density (4.0°C)	ρ	$0.9999749 \ \frac{g}{cm^3}$
density (22.0°C)	ρ	$0.9978 \ \frac{g}{cm^3}$
density (25.0°C)	ρ	$0.9970 \ \frac{g}{cm^3}$

Table A.6. Molal freezing point and boiling point constants for representative solvents at P = 1 atm. (p) = polar solvent; (np) = nonpolar solvent

Solvent	Freezing Point (°C)	Molal f.p. Constant, K_f (°C/m)	Boiling Point (°C)	Molal b.p. Constant, K_b (°C/m)
acetic acid (p)	16.6	3.63	117.9	3.22
acetone (p)	−94.7	2.67	56.1	1.80
benzene (np)	5.5	5.07	80.1	2.64
ethanol (p)	−114.4	1.959	78.3	1.23
methanol (p)	−97.5	2.56	64.6	0.86
phenol (p)	40.9	6.84	181.9	3.54
toluene (np)	−95.0	3.55	110.6	3.40
water (p)	0.00	1.86	99.974	0.513

Appendix B

Scientists to Know About

Wolfgang von Goethe once wrote that "the history of science is science." I agree, and this is why knowing about the contributions of significant scientists is an important component in any science class. A number of important scientists make an appearance in this text. Instead of including learning objectives about the scientists on the chapter Objectives lists, the scientists are listed here. This way, teachers can decide what knowledge about the scientists they will require, and students have an outline of names all in one place to help with organizing their studies.

Section	Scientist	Life Dates, Birth Country	Contribution
2.1.2	John Dalton	1766–1844, England	Developed the first scientific atomic model. (See also 9.3.1.)
2.1.2	J.J. Thomson	1856–1940, England	Discovered the electron and developed the plum pudding model of the atom.
2.1.2	Robert Millikan	1868–1953, U.S.A.	Determined the charge on the electron.
2.1.2	Ernest Rutherford	1871–1937, New Zealand	Discovered the atomic nucleus and the proton.
2.1.2	James Chadwick	1891–1974, England	Discovered the neutron.
2.2.2	Robert Brown	1773–1858, Scotland	First observed Brownian motion.
2.4.2	Jean Perrin	1870–1942, France	Determined the value of the Avogadro constant.
2.4.2	Amedeo Avogadro	1776–1856, Italy	Proposed that at a given pressure and temperature, the volume of a gas is proportional to the number of particles of the gas. (See also 5.3.1 and 9.1.3.)
3.1.1	Max Planck	1858–1947, Germany	First proposed the quantization of energy as a mathematical trick.
3.1.1	Albert Einstein	1879–1955, Germany	First theorized that energy and light were actually, physically quantized.
3.1.3	Johann Balmer	1825–1898, Switzerland	Discovered the formula predicting the lines in the hydrogen atom visible spectrum.
3.1.3	Johannes Rydberg	1854–1919, Sweden	Worked out the general formula for wavelengths in the hydrogen spectrum.
3.1.3	Theodore Lyman	1875–1954, U.S.A.	First observed the ultraviolet series in the hydrogen spectrum.
3.1.3	Friedrich Paschen	1865–1947, Germany	First observed the infrared series in the hydrogen spectrum.
3.2	Niels Bohr	1885–1962, Denmark	Proposed the Bohr model of the atom, which includes electrons in fixed energy levels.
3.3.1	Erwin Schrödinger	1887–1961, Austria	Developed the Schrödinger equation describing particles in quantum systems.
3.3.1	Wolfgang Pauli	1900–1958, Austria	Proposed the Pauli exclusion principle.
4.1	Dmitri Mendeleev	1834–1907, Russia	Developed the first fully-formed periodic table.
4.4.4	Linus Pauling	1901–1994, U.S.A.	Developed the electronegativity scale.

Scientists to Know About

5.3.1	Stanislao Cannizzaro	1826–1910, Italy	Demonstrated that Amedeo Avogadro's molecular theory is correct.
5.3.1	Amedeo Avogadro	1776–1856, Italy	First introduced the theory that atoms in gases bond together to form molecules. (See also 2.4.2 and 9.1.3.)
5.3.6	Gilbert N. Lewis	1875–1946, U.S.A.	Developed the theory of covalent bonding. (See also 9.2.2 and 11.2.3.)
5.3.8	Michael Faraday	1791–1867, England	First isolated and identified benzene.
5.3.8	Friedrich August Kekulé	1829–1896, Germany	Discovered the structure of the benzene ring.
7.1.2	Antoine Lavoisier	1743–1794, France	Discovered the law of conservation of mass in chemical reactions.
7.1.2	Joseph Priestley	1733–1804, England	First isolated oxygen.
8.1.2	James Clerk Maxwell	1831–1879, Scotland	Developed the velocity distribution for particles in a gas.
8.1.2	Ludwig Boltzmann	1844–1906, Austria	Developed the explanatory theory behind the Maxwell-Boltzmann velocity distribution.
8.1.4	Blaise Pascal	1623–1662, France	Developed many of the general principles pertaining to pressure in liquids and gases.
8.1.4	Evangelista Torricelli	1608–1647, Italy	Invented the mercury barometer.
9.1.1	Robert Boyle	1627–1691, Ireland	Discovered Boyle's law.
9.1.2	Jacques Charles	1746–1823, France	Discovered Charles' law.
9.1.2	Joseph Gay-Lussac	1778–1850, France	Published Charles' law and credited it to Jacques Charles.
9.1.3	Amedeo Avogadro	1776–1856, Italy	Proposed that at a given pressure and temperature, equal volumes of gases contain equal numbers of particles. (See also 2.4.2 and 5.3.1.)
9.2.2	Johannes Van der Waals	1837–1923 The Netherlands	Developed the first equations to model the behavior of real gases.
9.2.2	Gilbert N. Lewis	1875–1946, U.S.A.	Statements on the distinction between science and truth claims. (See also 5.3.6 and 11.2.3.)
9.3.1	John Dalton	1766–1844, England	Developed the law of partial pressures. (See also 2.1.2.)
10.1.1	Svante Arrhenius	1859–1927, Sweden	Published the first theory of ionic dissociation. (See also 11.2.1.)
11.2.1	Svante Arrhenius	1859–1927, Sweden	Developed Arrhenius acid-base theory. (See also 10.1.1.)
11.2.2	Johannes Brønsted	1879–1947, Denmark	Conceptualized acid-base reactions in terms of proton transfer.
11.2.2	Martin Lowry	1879–1947, England	Conceptualized acid-base reactions in terms of proton transfer.
11.2.3	Gilbert N. Lewis	1875–1946, U.S.A.	Proposed the Lewis acid-base theory. (See also 5.3.6 and 9.2.2.)

Appendix C

Memory Requirements

We list here the various items students must know from memory for courses using this text. Students must also be mindful of the chapter Objectives Lists, where all the tasks students are responsible for are listed. However, the prefixes, constants, and conversion factors/equations below are listed here for convenience, since most of them are not listed in the Objectives Lists. Values shaded in yellow are not exact and thus must be taken into account when dealing with precision in computations (significant digits).

1. The following metric prefixes (including their symbols and meanings):

Fractions			Multiples		
Prefix	Symbol	Factor	Prefix	Symbol	Factor
centi–	c	$1/10^2$	kilo–	k	10^3
milli–	m	$1/10^3$	mega–	M	10^6
micro–	μ	$1/10^6$	giga–	G	10^9
nano–	n	$1/10^9$	tera–	T	10^{12}
pico–	p	$1/10^{12}$			

2. The following physical constants:

Avogadro constant	$6.02214076 \times 10^{23}$ mol^{-1}	number of particles in one mole of substance
ρ_{water}	998 kg/m^3 = 0.998 g/cm^3	density of water at room temperature
g	9.80 m/s^2	acceleration at sea level due to gravity

3. The following conversion factors and equations:

2.54 cm = 1 in
1000 cm^3 = 1 L
1000 L = 1 m^3
1 mL = 1 cm^3

$$T_C = \frac{5}{9}(T_F - 32°) \qquad T_F = \frac{9}{5}T_C + 32° \qquad T_K = T_C + 273.15 \qquad T_C = T_K - 273.15$$

References and Citations

The following works were consulted in the writing of this text:

Brown, Theodore L., LeMay, H. Eugene, Jr., Bursten, Bruce E., Murphy, Catherine J. *Chemistry: The Central Science*, 11th ed., Pearson, 2009.

Cobb, Cathy and Fetterrolf, Monty L. *The Joy of Chemistry*, Prometheus, 2010.

Coffee, Patrick. *Cathedrals of Science: The Personalities and Rivalries that Made Modern Chemistry*, Oxford, 2008.

CRC Handbook of Physics and Chemistry, 91st ed., CRC Press, 2010.

Kean, Sam. *The Disappearing Spoon*, Back Bay Books, 2010.

King, G. Brooks and Caldwell, William E., *The Fundamentals of College Chemistry*, American Book, 1954.

Le Couteur, Penny and Burreson, Jay. *Napoleon's Buttons: 17 Molecules that Changed History*, Tarcher/Penguin, 2004.

McKeague, Charles, and Turner, Mark, *Trigonometry*, 6th ed., Cengage Learning, 2008.

Modern Chemistry, Holt, Rinehart and Winston, 2005.

Pauling, Linus. *College Chemistry*, W. H. Freeman, 1964.

Pauling, Linus. *The Nature of the Chemical Bond*, Cornell UP, 1960.

Peters, Edward I. *Introduction to Chemical Principles*, Saunders, 1986.

Scerri, Eric R. *The Periodic Table: A Very Short Introduction*, Oxford, 2011.

Stwertka, Albert. *A Guide to the Elements*, Oxford, 2012.

Unless otherwise noted, all tabulated data are from the *CRC Handbook of Physics and Chemistry*, 91st ed.

In addition to the sources above, the excellent science pages at en.wikipedia.com were used as a source of general information and for fact checking.

Specific credit is due for the following:

Section 1.3.2	The definition for significant digits listed in Case 1 is quoted from Charles McKeague and Mark Turner, *Trigonometry*, 6th ed.
Section 3.4.1	The metaphor attributed to Wolfgang Pauli is from Kean, *The Disappearing Spoon*, p. 360.
Section 4.1	Atomic mass values for the Periodic Table of the Elements are from Stwertka, *A Guide to the Elements*.
Section 4.1	The arrangement for elements 57, 89, 71, and 103 is based on Scerri, *The Periodic Table: A Very Short Introduction*, p. 134–136.

References and Citations

Section 5.4.2	The color sequence for % ionic character is from Pauling, *The Nature of the Chemical Bond*.
Example 7.10	Some details about the Haber-Bosch process are from Coffee, *Cathedrals of Science*.
Chapter 7 Exercises	Problem 25 is adapted from Holt, *Modern Chemistry*.
Section 9.2.2	The quote by G.N. Lewis is from Coffee, *Cathedrals of Science*, p. 185.
Example 9.8	Adapted from Brown, et al., *Chemistry: The Central Science*.
Chapter 9 Exercises	Problems 24, 25, and 34 are adapted from Brown, et al., *Chemistry: The Central Science*.
Section 10.1.1	The quote, "the wild army of the Ionists" is from Coffee, *Cathedrals of Science*, p. 23.
Section 10.2.2	The solubility guidelines in Table 10.2 are adapted from Brown, et al., *Chemistry: The Central Science*.
Section 10.2	The boxed explanation for how soap works is adapted from Le Couteur and Burreson, *Napoleon's Buttons: 17 Molecules that Changed History*, p. 285–289.
Section 10.5.2	Working temperature data for salting roads is from chemistry.about.com.
Section 11.2.4	Ionizing percentage for HSO_4^- is from Peters, *Introduction to Chemical Principles*.
Section 12.3.5	The half-reactions for alkaline dry cells are from Brown, et al., *Chemistry: The Central Science*.

Image Credits

Cover. Nada Orlic, http://nadaorlic.info/ 2. Dhatfield, public domain. 4. Ben Mills, public domain. 5. water: Ben Mills, public domain. ice: Solid State, licensed under CC-BY-SA-3.0. 6. drop on leaf: tanakawho, licensed under CC-BY-SA-2.0. drops on wood: neekoh.fi, licensed under CC-BY-SA-2.0. ball on slope: John D. Mays. 7. John D. Mays. 8. John D. Mays, based on At09kg, licensed under CC-BY-SA-3.0. 9. proton-electron: John D. Mays. crystal model: Benjah-bmm27, public domain. fluorite: CarlesMilan, licensed under CC-BY-SA-3.0. 11. John D. Mays. 14. John D. Mays, based on Unit_relations_in_the_new_SI.svg.png via https://commons.wikimedia.org/wiki/File:Unit_relations_in_the_new_SI.svg. Author: Emilio Pisanty, licensed under CC-BY-SA 4.0. 16. badger: Public domain. boy: duhoki, licensed under CC-BY-SA-3.0. potassium permanganate: Benjah-bmm27, public domain. laser light: John D. Mays 17. John D. Mays. 19. meter stick and receptacle: John D. Mays. meter bar: Unknown, public domain. 21–39. John D. Mays. 42. kaneiderdaniel, public domain. 44. Vlada Marinković, licensed under CC-BY-SA-3.0. 45. oxygen molecule: Ulflund, public domain. chlorine molecule: Benjah-bmm27, public domain. Dalton: Joseph Allen/William Henry Worthington, public domain. 46. Thomson: Public domain. CRT: Mrjohncummings, licensed under CC-BY-SA-2.0. plum pudding model: John D. Mays. 47. Millikan: Public domain. oil-drop photo: Public domain. oil-drop sketch: Robert Millikan, public domain. 48. Rutherford: Public domain. atoms: John D. Mays, based on Fastfission, public domain. 49. Chadwick: Bortzells Esselte, Nobel Foundation, public domain. substances: John D. Mays. 50–51. John D. Mays. 52. sodium: John D. Mays; chlorine: Greenhorn1, public domain. carbon dioxide: Jynto, public domain. propane and ozone: Benjah-bmm27, public domain. 53. Benjah-bmm27, public domain. 54. John D. Mays. 55. colbalt glass: Public domain. cranberry glass: Schtone, licensed under CC-BY-SA-3.0. 56. Brown: Henry William Pickersgill, public domain. beakers: John D. Mays. 58. top: FK1954, public domain. middle: FK1954, public domain. bottom: Chemicalinterest, public domain. 65. Public domain. 68. John D. Mays. 74. John D. Mays. 76. Gringer, public domain. 77. Planck: Public domain. Einstein: Nobel Prize photograph, public domain. 79. atoms: John D. Mays. EAT: Ctankcycles, licensed under CC-BY-SA-3.0. phonons: John D. Mays, based on image by FlorianMarquardt at the English Language Wikipedia, licensed under CC-BY-SA-3.0. 80. John D. Mays 81. spectrum: Gringer, public domain. 82. John D. Mays. 83. Schrodinger: Public domain. Pauli: Bettina Katzenstein/ETH Zürich, licensed under CC-BY-SA-3.0. 86. Dhatfield, public domain. 87–92. John D. Mays. 39. John D. Mays. 100. top: Dmitri Mendeleev, public domain. middle left: Marco Piazzalunga, licensed under CC-BY-SA-3.0. middle right: Mardeg at en.wikipedia, licensed under CC-BY-SA-3.0. lower left: Bastianow/Lidia/Fourdraine/J. Scholten, licensed under CC-BY-SA-3.0. lower right: Richard Powell (Singinglemon), public domain. 102. Public domain. 103–114. John D. Mays. 115. Pauling: Public domain. water: Ben Mills, public domain. 116. John D. Mays. 117. top: John D. Mays. nebula: NASA, Hui Yang University of Illinois ODNursery of New Stars, public domain. 120. Dover: Immanuel Giel, licensed under CC-BY-SA-3.0. model: Benjah-bmm27, public domain. 122–124. John D. Mays. 125. copper (I) oxide: Licensed under CC-BY-SA-3.0. copper (II) oxide: Walkerma, public domain. models: Benjah-bmm27, public domain. 129. photo: Public domain. models: Benjah-bmm27, public domain. 130. John D. Mays. 131. Arpingstone, public domain. 142. Kekulé: Public domain. models: Benjah-bmm27, public domain. 144. John D. Mays. 145. chlorine: Greenhorn1, public domain. bromine: Jurii, licensed under CC-BY-SA-3.0. 146. water model: Benjah-bmm27, public domain. others: John D. Mays. 147: Benjah-bmm27, public domain. 148. glass model: John D. Mays based on Jdrewitt, public domain. beta quartz model: Benjah-bmm27, public domain. Crystal photo: Rob Lavinsky, iRocks.com, licensed under CC-BY-SA-3.0. 150: Benjah-bmm27, public domain. 153–158 John D. Mays. 3-D models generated using CrystalMaker®, CrystalMaker Software Ltd, Oxford, England (www.crystalmaker.com). 159. model: John D. Mays. photo: Jurii, licensed under CC-BY-SA-3.0. 113. Alchemist-hp, licensed under CC-BY-SA-3.0. 161. water model: John D. Mays. 3-D models: Benjah-bmm27, public domain. 162. shoes: Yarnalgo, licensed under CC-BY-SA-2.0. models: John D. Mays. 163. model: John D. Mays. gecko: w:User:Lpm, licensed under C-BY-SA-3.0. 165. Benjah-bmm27, public domain. 168. Tubifex, licensed under CC-BY-SA-3.0. 170. John D. Mays 171. Lavoisier: Public domain. Priestly: Public domain. mercury oxide: Materialscientist, licensed under CC-BY-SA-3.0. 172. Materialscientist, CC-BY-SA-3.0. 173. sulfur sample: Benjah-bmm27, public domain. model: Benjah-bmm27, public domain. burning sulfur: Johannes 'volty' Hemmerlein, licensed under CC-BY-SA-3.0. 174. Laitr Keiows, licensed under CC-BY-SA-3.0. 175. Martin Walker, public domain. 176.

Image Credits

Tubifex, licensed under CC-BY-SA-3.0. 179. Tomas er, licensed under CC-BY-SA-3.0. 180. calcium hydroxide: Walkerma, public domain. trees: Nipik, public domain. 181. Toby Hudson, licensed under CC-BY-SA-3.0. 182. John D. Mays. 183. magnesium: John D. Mays. aqua regia: Thejohnlei, public domain. 184. Sculptor and engraver Erik Lindberg, public domain. 185. Tubifex, licensed under CC-BY-SA-3.0. 189. Mondalor, licensed under CC-BY-SA-3.0. 193. Drahkrub, licensed under CC-BY-SA-3.0. 197. Třinecké železárny, free use license. 198. 3-D model: Benjah-bmm27, public domain. 200. NASA/GRC/Martin Brown and Quentin Schwinn, public domain. 202. G. J. Stodart, public domain. 203. Boltzmann: unbekannt, public domain. graph: John D. Mays, based on Pdbailey, public domain. 204. Anonymous, public domain. 205. John D. Mays. 206. Torricelli: S. L. Pelaco, public domain. barometer: John D. Mays. 207. John D. Mays. 208–209. John D. Mays. 210. bricks: Hankwang, licensed under CC-BY-SA-3.0. other: John D. Mays. 211–212. John D. Mays. 213. sushi: Justinc, licensed under CC-BY-SA 2.0. nitrogen: Jurii, licensed under CC-BY-SA 3.0. lightning: Unknown, public domain. aurora: NASA, public domain. 214. John D. Mays. 215. Schnobby, licensed under CC-BY-SA-3.0. 216–220. John D. Mays. 221. Henry Gray, Gray's Anatomy, public domain. 222. John D. Mays. 224. John D. Mays. 226. Alchemist-hp, licensed under CC-BY-SA-3.0. 228. Boyle: Johann Kerseboom, public domain. other: John D. Mays. 229. balloon: Antoine Sergent dit Sergent-Marceau (?), public domain. Gay-Lussac: Francois Seraphin Delpech, public domain. 230–231. John D. Mays. 232. C. Sentier, public domain. 234. Unknown, public domain. 237. John D. Mays. 243. Thomas Phillips, public domain. 246, 250, 253, 256. John D. Mays. 258. Arrhenius: Public domain. model: John D. Mays. 259–263. John D. Mays. 264. acid model: Benjah-bmm27, public domain. other: John D. Mays. 266. ethanol, methanol, acetone: Benjah-bmm27, public domain. isopropyl: Jynto, public domain. 268. top: Jynto, public domain. middle: Benjah-bmm27, public domain. bottom: Matt18224, public domain. 269. micelle: John D. Mays, based on Smokefoot, public domain. door knocker: Alfred Löhr, public domain. 270–273. John D. Mays. 276. Der Kreole, licensed under CC-BY-SA-3.0. 278–280. John D. Mays. 281. top: Jacqueline Arneson, used by permission. bottom: Bengt Oberger, licensed under CC-BY-SA-3.0. 289. John D. Mays. 290. left: Riley Huntley (Huntley Photography), licensed under CC-BY-SA-3.0. right: Joanne Bergenwall Aw, licensed under CC-BY-SA-3.0. 292. John D. Mays. 293. Supermartl, altered by Haltopub, licensed under CC-BY-SA-3.0. 296. Benjah-bmm27, public domain. 297. John D. Mays. 298. Brønsted: Peter Elfelt, public domain. Lowry: Public domain. 299. Photographersnature, licensed under CC-BY-SA-3.0. 301, 303. John D. Mays. 305. Benjah-bmm27, public domain. 309. John D. Mays. 311. Martina Steiner, public domain. 312. Chemicalinterest, public domain. 313–317. John D. Mays. 324. Alfred T. Palmer, public domain. 326. Apollo: NASA, public domain. verdigris: Beyond My Ken, licensed under CC-BY-SA-3.0. batteries: Mathieu BOIS, licensed under CC-BY-SA-3.0. 327–335. John D. Mays. 339. Jynto, public domain. 343–347. John D. Mays. 348. Sabisteb, licensed under CC-BY-SA-1.0. 349–352. John D. Mays. 353. top: John D. Mays. bottom: John D. Mays, based on Tympanus, public domain. 354. Top: Sunny. solanki, licensed under CC-BY-SA-3.0. bottom: Ronald Saunders, licensed under CC-BY-SA-2.0. 355. top: Swoolverton, licensed under CC-BY-SA-3.0. middle: FiveRings, public domain. bottom: Polyparadigm, public domain. 356. top: Rémi Kaupp, licensed under CC-BY-SA-3.0. middle: Zwergelstern, licensed under CC-BY-SA-3.0. bottom: Knotnic, public domain. 357. John D. Mays. 359. Schuyler S., licensed under CC-BY-SA-2.5.

Index

absolute pressure, 229, 233, 235
absolute scale, 231
absolute temperature, 19, 231, 233, 235
absolute units, 231, 239
absolute zero, 230
acceleration, 17, 226
accuracy, 28, 29
acid, 58, 182, 184, 290–329, 326, 333; dissociation of, 116, 136, 260, 263, 296, 297, 304, 312; in equilibrium, 263; naming of, 135, 136, 294, 295; properties of, 292, 293; strength of, 303–306
acid-base indicator. *See* pH indicator
acid-base neutralization reaction, 183, 185, 186, 333
acid-base theory, 295–303
acid gas, 294
acid indigestion, 185
acidic solution, 336–338
acidity, 305, 307–309
acid rain, 180
actinide series (aka, actinides), 105
activity series of metals, 181
actual yield, 195
adsorption, 353
agar, 348
air, 185, 188, 193, 202, 204, 206, 220, 326
alchemist, 45
alcohol, 266
alkali, 290, 299
alkali battery, 326
alkali flat, 299
alkali lake, 299
alkali metal, 105, 106, 115, 117, 122, 265, 269, 276, 299, 326
alkaline dry cell, 353, 354
alkaline-earth metal, 105, 123, 124, 177, 299, 312, 326
alkali salt, 299
alloy, 157, 267, 269, 270
alpha particle, 48, 49
amino acid, 150
amorphous structure, 148, 208, 209
amphiprotic, 301, 306
amphoteric, 300
analog instrument, 31, 32
angstrom, 106
angular momentum quantum number, 84
anion, 107, 134; charge of, 178, 285, 328; definition of, 104; in acids, 135, 136, 186, 293, 295; in ionic bonds, 123–126, 134; resulting from ionization, 107, 112, 113, 117, 296; size of, 107, 108
anode, 46, 346–357
anodizing, 355
ant, 295
antacid, 185, 186
antilog, 311
aqua fortis, 293
aqua regia, 183, 184
aqueous solution, 116, 129, 136, 146, 147, 212, 256, 259, 261, 263, 271, 293, 294, 296–329, 335; preparation of, 273; reaction in, 174, 181, 183, 185, 275–278, 333
Aristotle, 207
Arrhenius acid-base theory, 296–298, 301–303, 305
Arrhenius, Svante, 258, 296, 298, 309
atmosphere, 180, 206
atmospheric pressure, 181, 205–207, 214, 217, 229, 233, 235, 247, 270, 279
atom, 3, 5, 10, 42, 45, 46, 49, 51, 52, 59, 64, 65, 74, 76, 110, 113, 350; attraction for electrons, 115, 143; electrons of, 79–81, 83–93; energy in, 78–83, 210; existence of, 9, 65, 203, 232; in covalent bonds, 131, 132, 152; in gases, 202, 203; in chemical equations, 172–176, 187, 337; in ionic substances, 124; in Lewis structures, 138; in metallic lattice, 157, 158; in molecules, 5, 131, 133, 152, 157, 162, 327; ionization of, 107, 109–114, 177; mass of, 44, 64, 69, 70; model of, 10, 45–49, 74; orbitals of, 2, 4, 9, 10, 114; quantum model of, 83–90; size of, 53, 105–108, 347; whole-number ratios of, 45
atomic bomb, 250
atomic mass, 51, 59–62, 64, 66, 67, 69, 70, 96, 232
atomic mass unit, 60
atomic number, 49, 51, 52, 61, 88, 102, 143
atomic radius. *See* atom, size of
atomic shielding. *See* shield effect
atomic spectrum, 74, 88
atomic theory, 9, 10, 44, 78, 132, 258
atomist, 9
Aufbau principle, 89
aurora borealis, 213
austenite, 157
autopassivation, 355
Avogadro, Amedeo, 65, 132, 232

Avogadro constant, 64, 69, 70, 132, 237
Avogadro's law, 232, 234
Avogadro's number, 64, 65, 70
azimuthal quantum number, 84, 85, 111
Badwater Basin, 299
balanced equation, 172–176, 248, 249, 276
balancing equations, 172–176, 187, 336–342
balloon, 226, 229
Balmer, Johann, 81
Balmer series, 81, 82
bar, 205
barometer, 204–207
barometric pressure, 206, 207, 246
base, 290–329, 326; dissociation of, 260, 263, 307; properties of, 294; strength of, 303–306
base unit, 14, 18–22
basicity, 299, 308, 309
basic solution, 339, 341
battery, 326, 343, 351, 354
bent geometry, 153–155
Bible, xiv
Big Bang, xiv
binary acid, 294
binary compound, 143
binary formula, 125, 134
birefringence, 130, 131
body-centered cubic lattice, 157, 158
Bohr Institute, 184
Bohr model, 74, 82–84
Bohr, Niels, 49, 74, 82, 184
boiling point, 145, 214, 216, 279, 281–285; elevation, 279–285
Boltzmann, Ludwig, 202, 203
bond angle, 153, 155–157
bond character, 145–147
bond energy, 143, 144, 202
bonding domain, 152–157
bonding pair, 136, 137, 152, 305
bond length, 143, 144
Bosch, Carl, 193
Boyle, Robert, 228
Boyle's law, 10, 228–230, 233, 239
brass, 269
brittleness, 130, 267
bromophenol blue, 313–315
bromphenol blue, 313
bromothymol blue, 313–315
bromthymol blue, 313
Brønsted, Johannes, 298
Brønsted-Lowry acid-base reaction, 306

401

Index

Brønsted-Lowry acid-base theory, 298–303, 305
bronze, 270
Brownian motion, 55, 56
Brown, Robert, 56
buret, 32, 316–318
caffeine, 215
calculator, 24, 26, 30, 35–37
Cannizzaro, Stanislao, 132, 232
capillary action, 210
carat, 269
carbon steel, 267
carboxyl group, 295
carboxylic acid, 295
cast iron, 267
catalyst, 353
cathode, 46, 346–357
cathode ray tube, 46
cathodic protection, 355–357
cation, 107, 147, 265, 276, 277; charge of, 285; definition of, 104; in bases, 186, 293; in chemical formulas, 123–127; 184; in ionic bonds, 123–126, 133, 134; resulting from ionization, 107, 112, 113, 116, 296; size of, 107, 108
caustic potash, 299
caustic soda, 299
cell (electrochemical), 343–354
cell potential, 350, 351, 354
Celsius scale, 19, 27
cement, 181
Chadwick, James, 49
chalcogen, 105, 124, 327
chalk, 120, 186
charge, 3, 4, 8, 44, 47, 106, 107, 109, 156, 163, 178, 213, 328; balancing of, 276, 277, 337–342, 349; conservation of, 171, 276, 277, 333; flow of, 158, 263, 344–346, 348, 349; in polyatomic ions, 133, 178, 328; ionic, 124, 177, 264, 280, 285, 326, 327
charge-to-mass ratio, 47
Charles, Jacques, 226, 229
Charles' law, 229–231, 233, 234, 239
chemical activity, 181, 182
chemical change, 58, 59
chemical formula, 65, 124
chemical equation, 51, 171–176, 248, 249, 276–278, 318
chemical reaction, 3, 51, 53, 57, 58, 64, 108, 113, 168–195
chemical property, 51, 52, 54, 57–59, 102, 140; of metals, 104
chemical symbol, 51, 90, 92, 93, 107, 138
chemical vapor deposition, 200, 215
chemistry, 3, 8, 10, 12, 44, 64, 83, 108, 115, 127, 146, 213, 233, 292
Christ, xiv
Christian, xiv, 221, 232

close packing, 157, 208
coal, 180
cobalt glass, 55
coefficient, 171–176, 187–191, 248, 276–278, 335, 336
cold pack, 260
collecting gas over water, 246–248
colligative property, 278–285
colloid, 55–57
colloidal dispersion, 54–56, 267
color, in visible spectrum, 76, 81, 88, 145, 212; of acid-base indicators, 290, 293, 294, 312–318; of emitted light, 89; of metals, 159; of solutions, 146, 147
combustion, 170, 192
combustion reaction, 185, 186, 326
compound, 3, 4, 9, 10, 44, 49, 52–54, 65, 93, 186; definition of, 51; in equations, 180–186; ionic, 124, 125; molar mass of, 66; quantities in reactions, 188–195; synthesis of, 195
compressibility, 211
concentration, 272–275, 278, 279, 282, 284, 306–312, 316–319
concrete, 180
condensed electron configuration, 92, 93, 108, 109
conduction electron, 158, 159
conductor, 343, 344, 348
conjugate acid/base, 300, 303
conjugate pair, 303
conversion factor. *See* unit conversion factor
cooking time, 214
Copernicus, Nicolaus, 82
core electron, 93, 108, 109, 111, 112, 115
corrosion, 355–357
covalent bond, 116, 122, 131–147, 158, 160, 209, 302
covalent bonding theory, 152
covalent characteristic, 146
covalent compound, naming of, 143; oxidation states in, 177, 178, 328; physical properties of, 144, 145
cranberry glass, 55
creation, xvi, 3, 8, 82, 161
Creator, 8, 82
critical point, 215
critical pressure, 215
critical temperature, 215
crystal, 5, 52, 53, 65, 131, 202; dissociation of, 183, 258–265
crystal lattice, 52, 65, 122, 144, 148, 183, 259, 264; in ionic substances, 123, 125, 129, 130, 160, 177, 357; in a metal, 64, 157, 269, 333
crystallogen, 105
Cycle of Scientific Enterprise, xvi, 11
cylinder, 228, 231

dalton, 60
Dalton, John, 9, 10, 45, 132, 243, 258
Dalton's law of partial pressures, 243–247
Davy, Sir Humphry, 299
Debye force, 162–164
decaffeinated coffee, 215
decomposition reaction, 181
delocalization, 141, 158
de Hevesy, George, 184
Democritus, 45
density, 62–64, 69, 102, 206, 207, 210; in gas law calculations, 240–242; in solution calculations, 274; of a gas, 211, 236, 237, 240–242; of ionic compounds, 130, 208, 209; of metals, 208, 209; of water, 161, 216
deposition, 214, 215
derived unit, 18–20
descaling, 293
design, 161, 221
diatomic gas, 44, 67, 174
diatomic molecule, 7, 45, 132, 173, 179, 232, 327, 328
diffusion, 210–212, 249, 250
digital instrument, 31
dipole, 145, 161–164, 266
diprotic acid, 296, 304
disorder, 261, 262
dispersing medium, 55
dissociation, 183, 258, 259, 263, 276, 277; of acids and bases, 258, 260, 263, 296, 297, 304, 307, 312
dissociation equation, 276, 277
dissociation theory, 258
dissolution, 258–271
double bond, 132, 137–140, 144
double replacement reaction, 184, 275
Dover cliffs, 120
dry cell, 345, 351, 353, 354
ductility, 57, 159, 267
dynamite, 179
effervescence, 270
effusion, 250
Einstein, Albert, 9, 56, 77, 137
elastic collision, 204
electrical attraction, 4, 5, 7, 115, 262, 353; between atoms, 122, 209; between molecules, 4, 204, 208; due to polarity, 258–261; in ionic bonds, 123, 130, 143, 177
electrical conductivity, 158, 263, 296, 313, 344
electrical force, 5, 123, 130, 163, 164, 204
electrical potential, 350, 351
electric circuit, 343–346, 350, 357
electric current, 158, 177, 263, 343, 345, 346, 349–351, 355
electric field, 263

electricity, 324; conduction of, 258, 263, 293, 294
electrochemical cell, 343–354
electrochemistry, 342–357
electrode, 344–353
electrode potential, 350–352
electrolyte, 261, 263, 264, 266, 276, 277, 284, 293, 294, 296, 297, 313; in electrochemical cell, 344–346, 348, 354, 355
electrolytic cell, 345, 349–351
electromagnetic radiation, 17, 76, 77, 79, 89, 221
electromagnetic spectrum, 76, 80, 81
electron, 3, 7, 8, 44–48, 79, 85, 108, 112, 337; charge of, 4; energy of, 79, 88; flow, 344–346, 348–350, 355; in Bohr model, 74, 82, 83; in bonding, 122; in covalent bonds, 116, 131, 132, 152; in ionic bonds, 123, 124; in Lewis structures, 136–141; in orbitals, 4, 87–93, 105, 113, 131, 143, 157, 158, 160, 163; in plasma, 213; in subshells, 2, 84–93, 106, 143, 302; mass of, 8, 44, 61; octet of, 122; transfer in redox reactions, 326, 327, 332–343; with respect to ionization, 107, 116, 177, 330; with respect to nuclear attraction, 115
electron affinity, 112–114, 122, 127
electron configuration, 90–93, 108, 111
electron domain, 152–157
electronegativity, 4, 115–117, 122, 145–147, 161, 162; 177, 178, 305; table of, 117; with respect to oxidation state, 178, 327, 328
electron pair, 302
electron sea, 158
electron volt, 110, 111
electroplating, 344, 354, 355
electrostatic force, 130, 160
element, 3, 4, 44–46, 52, 65, 93, 187, 327, 328, 337; atomic mass of, 51, 60, 61; atomic number of, 59; atoms of, 10; definition of, 49, 51; in chemical equations, 180–186; in covalent compounds, 178; in the periodic table, 92, 93, 100, 102–117; ionization energies of, 110–112; isotopes of, 60; molar mass of, 187; oxidation state of, 177–180, 326, 335; symbol of, 51; spectrum of, 74
emission spectrum, 76, 82, 212
empirical formula, 96; determination of, 93–95
endothermic process, 8, 259, 260, 262
endothermic reaction, 343
endpoint, 313, 318

energy, 4, 6, 8, 17, 81, 113, 114; in atoms, 202; in covalent bonds, 143, 144; in dissociation, 259–263; in ionic bonds, 127–129; in molecules, 7, 202, 203; in redox reaction, 343, 346, 349; ionization, 109–114, 213; minimization of, 6, 8, 89, 90, 93, 122, 123, 127, 129, 140, 143, 157, 209, 261–263; of electron, 44, 79–83, 88; of photon, 77, 78; potential, 6, 9, 80, 143, 144, 209, 259; supplying of, 213, 281; quantization of, 9, 77, 81; transfer, 343, 346
energy level, 79, 80, 82, 84, 88
energy state, 6, 7, 8, 109, 114, 123
entropy, 8, 261–263; increase in, 261, 267; maximization of, 262
enzyme, 185
epistemology, xiii
equilibrium, 215, 246, 256, 264, 270, 281, 297, 303; acid-base, 303–305
equivalence point, 313–316
Erlenmeyer flask, 316–318
error, 29
evaporation, 220–222, 246, 266, 270, 279, 280
evaporative cooling, 220, 221
evolution (of gas), 181, 183, 184
exchange reaction, 275
excitation, 78, 79, 202
excited state, 79, 202
exothermic process, 127, 259, 260
exothermic reaction, 8, 343
experiment, 11, 12
experimental error, 37
extensive property, 129, 130
face-centered cubic lattice, 157, 158
Fahrenheit scale, 19, 27, 231
Faraday, Michael, 141
ferrite, 157
fertilizer, 179, 193
fight or flight response, 221
fire ant, 295
Fire Diamond, 42
flame test, 89
flammability, 42, 58, 266
fluid, 78, 202
fluorescence, 79, 117
foam, 55
fog, 56
force, 204, 205, 207, 226, 259; on molecules, 210
formula equation, 172–176, 189, 276, 278, 337, 339
formula mass, 66, 67, 95; definition of, 65, 66
formula unit, 65, 129
Franck, James, 184
free energy, 262, 263
free radical, 140

freezing point, 214, 279, 281–285; depression, 279–285
frequency, 77
Galvani, Luigi, 345
galvanic cell, 345
galvanization, 342
gas, 55–57, 132, 202, 267, 333; diffusion of, 212, 249, 250; effusion of, 250; in explosions, 179; in liquid solution, 270–272; kinetic-molecular theory of, 203, 204; particles of, 202; pressure in, 10, 204, 205; state of matter, 211, 212
gaseous diffusion, 250
gas law, 203, 226–250
gastric acid, 185
gauge pressure, 233
Gay-Lussac, Joseph, 229
gecko, 163
gel, 55, 348
gelatin, 348
general relativity, 11
Genesis, xiv
Gibbs free energy, 262, 263, 349
glass, 148, 208, 210
God, xiv, 3, 8, 11, 161, 221, 222, 232
Goethe, Wolfgang, xvi
golden rain, 276
gold foil experiment, 13, 48
gold ruby, 55
graduated cylinder, 32
gravitational potential energy, 6
ground state, 79, 80, 88, 109, 113
group, 104
gunpowder, 179
Haber-Bosch process, 193, 194
Haber, Fritz, 193
half-cell, 344–347, 349, 350, 352, 353
half-reaction, 332–342, 344, 350–354
half-reaction method, 336–342
halogen, 105, 115, 122, 135, 144; behavior of, 116, 184; in acids, 135, 136, 185; oxidation number of, 112, 177, 326, 328
halogenation, 199
heartburn, 185
heat, 216, 217, 219, 260–263, 324, 343, 357; released in a reaction, 7, 8, 127, 128, 170, 183; to create a plasma, 213; to effect a phase transition, 216, 217, 219
heat of solution, 261, 262
heterogeneous mixture, 49, 53–56, 348
Hindenburg disaster, 229
history, xiii, xvi, xix
Hitler, Adolf, 184
homogeneous mixture, 49, 53, 54, 210, 211, 249, 267
hot air balloon, 226
Hubble Space Telescope, 117
human body, 220–222

403

Index

Hund's rule, 89, 90, 108
hydrate, 129
hydration, 259, 280, 285
hydride, 117
hydrogen, uniqueness of, 116, 117
hydrogen atom, energy levels in, 80–82, 88
hydrogen bonding, 5, 160–162, 164, 208, 210, 215, 216, 220, 259, 266, 267, 269, 280
hydrogen halide, 294
hydrogen spectrum, 82
hydronium, 133, 297–312
hydroxyl group, 162, 264, 266, 268
hypothalamus, 221
hypothesis, 11, 12, 207, 232
ice, 5, 208, 281
ice cream, 281
ice skating, 216
icosagen, 105
ideal gas, 204, 233
ideal gas constant, 235, 238, 242
ideal gas law, 230, 233–242, 244–249
incompressibility, 208–210
induced dipole, 163, 164
inertia, 17
infrared light, 77, 81, 89, 221
inner transition metal, 105
instrument, 29, 31, 33
integration, xiii–xv
intensive property, 129, 130, 145, 158
intermolecular force, 5, 160–164, 204, 208–210, 234, 243, 280, 285
internal energy, 202, 204, 210, 216, 220, 232, 259
International System of Units, 18
interstice, 267
ion, 3, 8, 53, 79, 110, 111, 135, 140, 143, 164, 177, 184, 326, 327, 339, 342, 350, 357; charges on, 177, 264, 285, 327; flow, 263, 344; in a crystal lattice, 123, 130, 159; in plasma, 213; in salt bridge, 344–346, 349; in solution, 129, 182, 183, 258–265, 279–281, 313, 333, 348; oxidation state of, 177–180; size of, 107, 108, 347
ionic bond, 122–131, 143, 146, 259
ionic characteristic, 145, 146
ionic compound, 122–130, 160, 326; in solution, 258–265; naming of, 125–127; oxidation state in, 125, 126, 327; physical properties of, 130, 131, 144, 158, 159, 208, 209
ionic equation, 277, 278, 319, 333, 335, 336, 337
ionic radius, 107, 108
ionists, 258
ionization, 112, 116, 117; during redox, 333–342; in solution, 261, 266, 340; of a gas, 212, 213; of acid or base in solution, 296–312

ionization constant, 306
ionization energy, 109–114, 122, 127, 128
ionization equation, 109, 113
ionizing radiation, 213
iron, 130, 267
isotope, 46, 59–62, 64, 70
Karlsruhe Institute of Technology, 193
Keesom force, 162, 164, 204, 266
Kekulé, Friedrich August, 142
Kelvin scale, 19, 27
kinetic energy, 6, 7, 78, 79, 158, 202–204, 208, 209, 220, 259
kinetic-molecular theory, 202, 220
kinetic-molecular theory of gases, 203, 204, 230, 232, 243
lab journal, 33
lab report, xvii, 33
lanthanide series (aka, lanthanides), 105
lattice energy, 127, 128, 130, 144, 160, 202, 259
Lavoisier, Antoine, 170, 171, 177, 296
law of conservation of mass in chemical reactions, 171, 172, 192, 276, 277
law of partial pressures, 243–248
laws of physics, 3
lead-acid battery, 354
Le Châtelier, Henri Louis, 228
Lewis acid-base theory, 302, 303
Lewis, Gilbert N., 137, 152, 234
Lewis structure, 136–142, 152, 153–157, 179, 234, 295, 301, 302
Lewis symbol, 131, 132, 136
lift, 226
light, 17, 53, 131; absorption of, 147; emitted by plasmas, 167; released in a reaction, 7, 127, 128, 170; speed of, 77, 78; wavelength of, 44, 74, 76, 78, 347
limestone, 120, 186
limiting reactant, 192–194
limiting reagent, 192, 193
linear geometry, 153, 154
liquid, 202, 209–211, 214, 216, 217, 222, 249, 278–280
liquid aerosol, 55
liquid emulsion, 55
litmus, 290, 312
litmus paper, 312
logarithm, 308, 310, 311
London dispersion force, 163, 164, 204, 267
lone pair, 152
Lowry, Thomas Martin, 298
luster, 159, 269
lye, 268, 299
Lyman series, 82
Lyman, Theodore, 81

Madelung rule, 89, 90, 92, 93, 108, 159
magnetic quantum number, 84, 85
magnetosphere, 213
main group elements, 104, 122, 125
malachite green, 314, 315
malleability, 57, 159, 267
Manhattan Project, 250
manometer, 237
mass, conservation of, 333; definition of, 14, 16–18; of subatomic particles, 61; units for, 17, 18, 60
mass number, 60
mastery, xiii–xv
mathematical model, 282
mathematical structure in nature, 3, 82
matter, 16–18, 42, 49, 62, 117
Maxwell-Boltzmann distribution, 202, 203
Maxwell, James Clerk, 202, 203
mean free path, 250
measurement, 14–37
melting, 214
melting point, 214, 216; of covalent compounds, 145, 209; of ionic compounds, 130, 209; of metals, 209
Mendeleev, Dmitri, 100, 102, 258
meniscus, 32, 210, 317
mental model, 10, 12
mercury barometer, 206, 207
metabolism, 189
metal, 130, 177, 293, 326, 328, 333, 342, 347, 350, 353; chemical properties of, 104, 181, 182; cleaning, 293; corrosion of, 326, 355–357; elemental, 177; in ionic bonds, 123–127; in the periodic table, 104, 105; ionization of, 107, 116, 123, 177; oxidation state of, 335; physical properties of, 104, 158–160; smelting of, 177, 357
metallic bonding, 122, 157, 158, 160, 208, 269
metallic lattice, 157, 158, 160, 208
metalloid, in the periodic table, 104
metathesis reaction, 275
methyl orange, 315
methyl red, 314, 315
metric prefix, 20–22, 24
metric system, 18–22
Meyer, Lothar, 102
micelle, 269
Millikan, Robert, 47, 48
miscibility, 56, 258, 267
mixture, 3, 49, 53–56
MKS units, 20, 62, 235
model, 10, 11, 141, 233, 281, 282, 347
modeling, xiii, xvi, 10, 83, 122, 146, 203, 234, 295, 302

404

Index

molal boiling point constant, 282
molal freezing point constant, 282
molality, 274, 275, 282–284
molar gas constant. *See* ideal gas constant
molar heat capacity, 218, 219
molar heat of fusion, 219, 281
molar heat of vaporization, 219
molarity, 272–274, 312
molar mass, 69, 95, 96, 273; as a conversion factor, ; calculation of, 65–68; definition of, 65, 66; in concentration calculations, 272–275; in gas law calculations, 240–242; in stoichiometry, 187, 189–194
mole, 128, 187, 188, 218, 234, 243, 283; definition of, 64, 65; in chemical equations, 172; in relation to solution concentration, 272–275, 279, 280, 284, 285
molecular equation, 277, 278
molecular formula, 94, 140; determination of, 95, 96
molecular geometry, 138, 152–157
molecular mass, 95, 96; calculation of, 66, 67, 70, 71, 232; definition of, 65, 66
molecular orbital, 131
molecular solid, 148
molecular structure, 3, 141
molecular theory, 132, 152, 232
molecular theory of gases. *See* kinetic-molecular theory of gases
molecule, 9, 10, 42, 52, 53, 64, 94, 122, 326; adsorption of, 353; atomic attraction in, 115, 162; collision of, 202, 204–206, 209; covalent bonds in, 116, 144, 160, 209; electrons in, 115, 145, 146, 162, 163, 327; existence of, 232; forces between, 160–163, 266; in equilibrium, 263; in gases, 202–205, 333; in ionic bonds, 133, 134; in liquids, 222; in solution, 135, 258–270, 304; ionization of, 297, 306; mass of, 65–67, 69, 70; oxidation states in, 178, 327, 328; polar, 5, 161–163, 258, 266, 269; structure of, 152–157; with respect to Lewis structures, 136–143; with respect to polyatomic ions, 133–135
mole fraction, 243, 245
mole ratio, 188–194, 318
monomer, 162
monoprotic acid, 296, 312
monovalence, 126
Montgolfier balloon, 226
Montgolfier brothers, 226
Mount Sapo, 268
multivalence, 126

muriatic acid, 293
neon sign, 79, 212, 213
net ionic equation, 277, 278
neutral, 306
neutralization, 293, 313, 319, 336, 340, 341
neutron, 3, 44–46, 49, 60, 88, 213; mass of, 44, 61; numbers of in nucleus, 59, 60, 62
newton, 1, 205
NGC604 nebula, 117
nicad battery, 329
nitration, 198
nitro compound, 179
Nobel, Alfred, 179
Nobel Foundation, 179
Nobel Peace Prize, 115
Nobel Prize, 137, 179, 184
Nobel Prize in Chemistry, 48, 115
Nobel Prize in Physics, 46, 48, 49, 65, 77, 83, 234
noble gas, 64, 92, 93, 102, 105, 106, 112–115; in plasmas, 212, 213; particle motion of, 202, 203; octet of, 122
nonbonding domain, 152–157
nonbonding pair, 136, 137, 139, 152, 161, 162, 164, 298
nonmetal, chemical properties of, 131, 143; in covalent bonds, 116, 131, 143, 328; in ionic bonds, 123; in the periodic table, 104, 105; ionization of, 107, 123, 326
nonpolar molecule, 145, 162, 204, 265, 267
nonpolar solvent, 267
normal hydrogen electrode, 353
nuclear decay, 48
nuclear fallout, 115
nucleon, 44, 60, 61
nucleus, 3, 4, 44–46, 48, 49, 59–62, 80, 85, 88, 104, 111; attraction for electrons, 83, 106, 108, 109, 112, 143, 144, 160; charge in, 106, 110, 111; of atoms in metallic lattice, 158; shielding of, 106, 108, 110, 112, 113, 115; size of, 105
nuclide, 59, 60, 70
Oak Ridge, 250
octahedral geometry, 153–155
octahedron, 153
octet, 122, 125, 131, 132, 133, 152, 155, 302; in Lewis structures, 137–141
octet rule, 122, 138, 140, 141, 302; exceptions to, 140, 141
oil-drop experiment, 47, 48
oil of vitriol, 293
Oklahoma City bombing, 179
orbital, 2, 4, 9, 44, 79, 84–93, 105, 113, 114, 122, 143, 147, 157, 158, 159

orbital diagram, 90, 91
order, 261, 262
Ordinary Portland Cement, 181
organic chemistry, 295, 302
oxidant, 177, 327
oxidation, 58, 177–180, 183, 186, 189, 190, 324–357
oxidation number, 177, 178
oxidation potential, 351
oxidation-reduction reaction, 186, 187; 324–357
oxidation state, 112, 122, 124–126, 174, 177–180, 186, 326–330, 332, 333–335, 337, 350; table of, 112
oxidizer, 177, 179, 186, 327
oxidizing agent, 177, 179, 186, 327, 329, 330–334
oxyacid, 136, 181, 295, 305, 306
oxyanion, 134, 135, 178, 328
partial pressure, 243–247
pascal, 204
Pascal, Blaise, 204, 207
Paschen, Friedrich, 81
Paschen series, 82
passivation, 332, 355
Pauli exclusion principle, 83, 85, 90
Pauling electronegativity scale, 115–117
Pauling, Linus, 115
Pauli, Wolfgang, 83, 90
percent composition, 93, 94
percent difference, 37
percent ionic character, 146, 147
percent yield, 195
period, 104
periodicity, 102, 115
periodic law, 102
Periodic Table of the Elements, 2, 4, 49–51, 60, 61, 64, 65, 82, 91, 92, 100–117, 143, 159; an element's position in, 177
permanent dipole, 162, 164, 266
peroxide, 178, 179, 328
Perrin, Jean, 65, 132
pH, 290, 293, 294, 308–319
phase diagram, 214–217, 278, 279
phase transition, 213–223, 281
phases of matter, 208–223, 278–280
phenolphthalein, 313–315, 317, 318
phenol red, 315
pH indicator, 290, 293, 294, 312–318
pH meter, 310–313
phonon, 79
photon, 74, 79, 80, 81, 137; energy of, 77, 78, 88, 89
photosynthesis, 8, 190
pH scale, 308, 309, 313
physical change, 58, 59
physical property, 44, 51, 52, 54, 57, 58, 62, 129; of ionic substances, 130, 131; of metals, 104, 158–160; periodicity of, 105–108

405

Index

physics, 3, 44, 233
pickling, 293
piston, 228, 231
Planck, Max, 74, 77
Planck constant, 77, 78
Planck relation, 77, 78, 80, 88
plasma, 117, 200, 212, 213
plastic, 208
plum pudding model, 46–48
pnictogen, 105
pOH, 310, 311
polarity, 4, 5, 6, 115, 129, 145, 146, 266, 280, 305, 306
polarization of light, 131
polyatomic ion, 58, 133–135, 164; in acids, 135, 136, 185; in chemical equations, 174–176; Lewis structures of, 138–140; naming of, 134, 135; oxidation states in, 178
polymer, 55, 161, 162, 348
polyprotic acid, 296, 307
porous barrier, 344
potash, 268, 299
potential energy. *See* energy, potential
precipitate, 170, 176, 184, 275–278
precipitation, 168, 181, 182, 275–278
precision, 28–35
precursor, 141
prefix, 134–136, 143, 295; metric, 20–22
pressure, 222, 281; in gases, 10, 203–205, 211; in gas laws, 228–249; in liquids, 207; in phase diagrams, 214–217, 278, 279; relation to gas solubility, 270
pressure gauge, 233
Priestly, Joseph, 171, 172, 181
primary standard, 316
principle quantum number, 84, 85, 88, 92, 106, 111
printed circuit board, 355
probability distribution, 87, 105
product (of reaction), 10, 51, 172, 173, 277, 333
properties of matter, 17, 45, 120, 158
protein, 150
proton, 3, 7–9, 44, 45, 49, 51, 59, 213, 116; charge of, 4; acceptor, 302, 303, 304; donor, 300, 303, 306; mass of, 8, 44; numbers of in nucleus, 49, 62, 88, 106, 108, 110, 111, 115
protonation, 304
proton transfer, 298
pure substance, 49
pyramidal geometry, 153–156
quantization of energy, 77
quantum, 77, 79, 137
quantum mechanics, 9, 11, 143
quantum model, 83–90
quantum number, 83–85

quantum state, 83, 85
quantum theory, 77
quartz, 130, 148
Rankine scale, 231
rare-earth element, 105
reactant, 7, 10, 51, 172, 173, 277, 278, 333, 353
reactivity, 42
real gas, 233, 234
red cabbage, 293, 312
redox reaction, 186, 187; 324–357
reducer, 177, 327
reducing agent, 177, 327, 329, 330–332, 334, 350–353, 356, 357
reducing atmosphere, 197
reductant, 177, 327
reduction, 177, 178, 182, 183, 186, 324–357
representative elements, 104, 107
resonance structure, 138–143, 155, 158
rigidity, 208
roasting, 181
rocket fuel, 179, 330
Rolaids, 185
rust, 174
Rutherford, Ernest, 13, 48, 49
Rydberg constant, 82
Rydberg formula, 81, 82
Rydberg, Johannes, 21, 81
sacrificial anode, 356, 357
salt, 185, 293, 294, 299, 335, 348, 349
salt bridge, 344–346, 348, 349, 353
salt flat, 299
saponification, 268
saturated solution, 256, 264
schematic diagram, 345, 346
Schrödinger equation, 83, 85, 87, 122
Schrödinger, Erwin, 83
science, xvi, 10–12
scientific calculator, 36, 37
scientific claims, xvi
scientific fact, 10–12
scientific knowledge, 11, 12
scientific notation, 35–37, 189, 308
Scripture, xiv
self-ionization of water, 306
semiconductor, 160, 200, 214
shell, 4, 84–89, 105, 106, 108, 122, 330
shield effect, 106, 108, 110, 112, 113, 115
significant digits, 24, 29–35, 63, 64, 68, 78, 188, 308, 309
single replacement reaction, 181–184, 326, 327, 342
SI unit system, 18–22, 204, 217, 231, 235
slaked lime, 180
smelting, 177, 324, 344, 357
soap, 268

soda ash, 268
sol, 55
solar radiation, 76
solid, 202, 208, 209; crystal lattice of, 208, 209; physical properties of, 208, 209; with respect to phase transitions, 214, 216, 217
solid aerosol, 55
solid emulsion, 55
solid foam, 55
solid sol, 55
solid solution, 267
solubility, 3, 258, 264–272, 278, 307
solubility equilibrium, 256, 270
solubility guidelines, 265, 276
soluble compound, 340
solute, 256, 258, 266, 274, 278, 279, 281
solution, 56, 129, 210, 249, 256–285, 333, 340, 348; acidic or basic, 135, 136, 296–312; color of, 146, 147; concentration, 272–275, 278; definition of, 54; in equilibrium, 256; precipitate from, 168, 170, 176, 350; reactions in, 182–184, 296, 298, 326, 328
solvation, 259
solvent, 116, 256, 258, 266, 274, 279, 281
spacecraft, 188
spectator ion, 277, 278, 333, 334
spectroscopy, 81, 88
speed distribution, 202, 203, 220
speed of sound, 203, 211
spin, 85
spin projection quantum number, 85
stainless steel, 157, 267, 269
standard hydrogen electrode, 352, 353
Standard Problems List, xv, xviii–xxi
standard reduction potential, 350–353
standard solution, 313, 316–318
standard temperature and pressure, 233, 248
states of matter, 208–223
statistical mechanics, 203
Statue of Liberty, 58
steel, 267, 330
sterling silver, 269
Stock system, 126, 127, 143
stoichiometry, 10, 168–195, 318; of gases, 248, 249
stopcock, 253, 317
STP. *See* standard temperature and pressure
stratosphere,
strigil, 268
subatomic particle, 44, 61
sublimation, 214
submarine, 188

Index

subshell, 2, 84, 85, 87–93, 106, 117, 143, 159; with respect to octet rule, 122, 131
substance, 42, 57, 58, 62, 64, 79, 129; types of, 49
suffix, 134, 136, 143, 295
supercritical fluid, 215
surface tension, 210
suspension, 54, 55
sweat glands, 221
synthesis reaction, 180, 181
temperature, 221, 262, 263, 282; change, 218, 219, 282–285; in gas laws, 228–249; of gases, ; on phase diagrams, 214–217; units, 27, 28; with respect to kinetic energy, 79, 202, 204, 208, 209, 259; with respect to internal energy, 203, 208–210, 216; with respect to solubility, 264, 271, 272; with respect to vapor pressure, 222, 223, 278, 279
temporary dipole, 163
tetrahedral geometry, 153–157, 298
tetrahedron, 153
Texas City disaster, 179
theoretical yield, 194, 195
theory, 11, 12, 142, 203, 207, 230; as modeling, xvi, 10, 83, 203, 234, 295
thermal conductivity, 158, 159
thermal energy, 260, 261
thermal equilibrium, 261, 262
thermite reaction, 359
thermite welding, 359
thermodynamics, 8, 202; second law of, 8, 261
thermoregulation, 221
Thomson, J.J., 46–48
tin experiments, 170, 171
tin pest, 160
titration, 212, 312, 314–318
TNT, 179
torr, 205, 207
Torricelli, Evangelista, 206, 207
transition interval, 313–315
transition metal, 105, 112, 125, 147
Triangulum, 117
triglyceride, 268
trigonal bipyramid, 153
trigonal bipyramidal geometry, 153–155
trigonal planar geometry, 153–155
triple bond, 132, 133, 137, 138, 144, 193
triple point, 214–216, 264
triprotic acid, 296, 304
truth, 11, 12
truth claims, xvi, 11, 234
Tums, 186
Tyndall effect, 56
ultraviolet light, 77, 81

unified atomic mass unit, 60, 61, 65
unit conversion, 22–29, 33, 63, 64, 68
unit conversion factor, 22–28, 63, 68, 188, 190, 218, 247
unit prefix, 22–22
units of measure, 14–37, 64, 69, 234, 235
universal gas constant. *See* ideal gas constant
universal solvent, 258
universe, xiv, 3
uranium enrichment, 250
U.S. Customary Units, 18, 231
vacuum, 206, 207, 233, 241
valence, 102
valence electron, 104, 108–111, 115, 122, 123, 124, 131, 144, 152; in Lewis structures, 138–140; in metallic lattice, 158
valence shell, 108, 112, 113, 123, 179, 302
Valence Shell Electron Pair Repulsion theory, 152–157
Van der Waals force, 160–164, 204, 208, 234, 266, 353
Van der Waals, Johannes Diderik, 234
vapor, 208, 217, 278; with respect to phase transitions, 214, 217, 219, 220
vaporization, 216, 217
vapor pressure, 222, 223, 246, 247, 266, 270, 274, 278; lowering of, 278–281
vapor state, 220, 266, 270
variable, 18
velocity, 77, 204, 211, 232
verdigris, 326
visible spectrum, 76, 77, 81, 89, 159
volatility, 222, 266
Volta, Alessandro, 345
voltage, 350–354
voltaic cell, 345–351, 356
voltmeter, 346, 351
volume, 17, 62, 63, 209; in gas laws, 228–249; of a gas, 10, 65, 203; of solution, 272–275
volumetric flask, 273
volumetric ratio, 248
von Laue, Max, 184
waste heat, 221
water, 281, 317, 353; as acid or base, 300; phase diagram for, 216
water molecule, 4, 8, 66, 68, 115, 122, 210, 259, 279, 280, 340, 341, 347; geometry of, 155, 156; in hydrates, 129; hydrogen bonding in, 5, 161, 162, 265, 266, 269; motion of, 56; polarity of, 5, 6, 115, 129, 146, 258–260, 266
water vapor, 215

water treatment, 175
wavelength, 44, 74, 76–78, 80, 81, 88, 89, 159, 347
West, Texas disaster, 179
wet cell, 345
whole-number ratio, 9, 10, 52, 122, 124
wicking, 162, 210
wind, 206
work, 7, 8, 109
World Trade Center bombing, 179
worldview, xiv
World War I, 49, 332
World War II, 250
yield, 194, 195

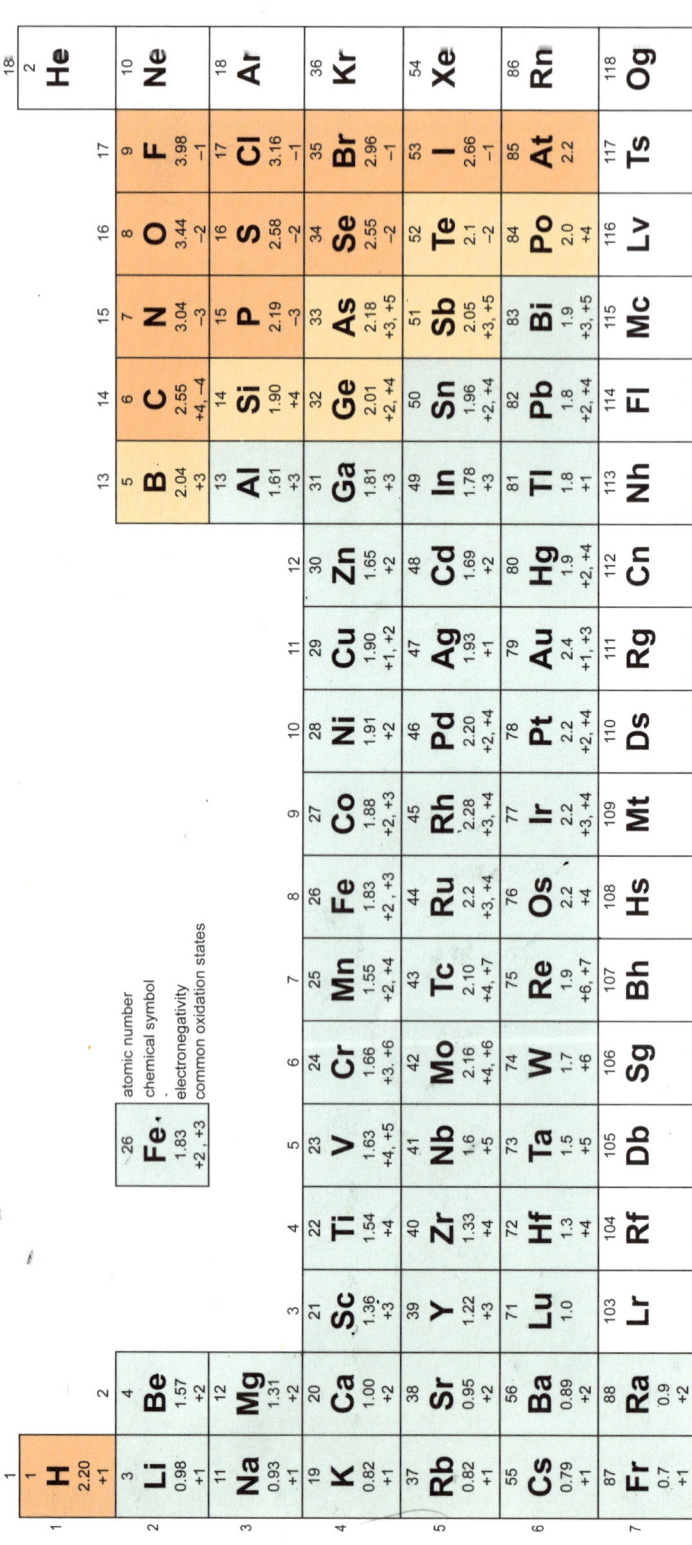